# Investigating Quality of Urban Life

# Social Indicators Research Series

## Volume 45



This new series aims to provide a public forum for single treatises and collections of papers on social indicators research that are long to be published in journal *Social Indicators Research*. Like the journal, the book series deals with statistical assessments of the quality of life from a broad perspective. It welcomes the research on a wide variety of substantive areas, including health, crime, housing, education, family life, leisure activities, transportation, mobility, economcs, work, religion and environmental issues. These areas of research will focus on the impact of key issues such as health on the overall quality of life and vice versa. An international review board, consisting of Ruut Veenhoven, Joachim Vogel, Ed Diener, Torbjorn Moum, Mirjam A.G. Sprangers and Wolfgang Glatzer, will ensure the high quality of the series as a Whole.

For further volumes:
http://www.springer.com/series/6548

Robert W. Marans • Robert J. Stimson
Editors

# Investigating Quality of Urban Life

## Theory, Methods, and Empirical Research

Springer

*Editors*
Robert W. Marans
Institute for Social Research
University of Michigan
Thompson Street 426
Ann Arbor, MI 48106-1248
USA
marans@umich.edu

Robert J. Stimson
Australian Urban Research Infrastructure
Network (AURIN), Faculty of Architecture
Building and Planning
University of Melbourne
VIC 3010, Melbourne
Australia
rstimson@unimelb.edu.au

ISSN 1387-6570
ISBN 978-94-007-1741-1 e-ISBN 978-94-007-1742-8
DOI 10.1007/978-94-007-1742-8
Springer Dordrecht Heidelberg London New York

Library of Congress Control Number: 2011937028



Printed on acid-free paper

Springer is part of Springer Science+Business Media (www.springer.com)

# Preface

All people and all places are concerned with quality of life (QOL). Therefore it is not surprising that the topic has attracted the attention of researchers from many disciplines since the 1960s. While the topic has been studied with regularity for nearly a half century by researchers in psychology, sociology, geography, planning, and other disciplines, the past decade has seen an acceleration of scholarly interest in QOL including a stream of studies investigating individual well-being and happiness. As most of the world's population now lives in urban areas, and with disparate populations in advanced nations being highly concentrated in large cities, it is inevitable that many QOL studies have focused on measuring and modeling aspects of life in urban areas or the quality of *urban* life (QOUL). This book addresses that focus.

In planning for this volume, we established three objectives:

(a) First, we believed it was important to provide a detailed overview of the approaches that have emerged over the past half century in studies of QOL in general and QOUL in particular. This includes, on the one hand, approaches focusing on the objective measurement of QOL and QOUL using secondary analysis of aggregate data and, on the other hand, approaches focusing on the measurement and analysis of subjective evaluations and appraisals of QOL and QOUL. More recently, efforts have been made to integrate the objective and subjective approaches in studies of QOL and QOUL. These efforts have been enhanced with the advent and development of Geographic Information Systems (GIS) technologies. This objective is addressed in Chaps. 1–6.

(b) Second, through our own research on QOUL including empirical studies in Detroit, Michigan (USA), and in Brisbane, Queensland (Australia), we were aware of the emergence of other research employing survey-based approaches to collecting information on aspects of both QOL and QOUL. The research was conducted in a number of urban settings, both large and small, in many parts of the world. Some of those studies used similar sets of questions to those used in the Detroit and Brisbane surveys although the modes of administering the surveys differed. Likewise, the purposes of the various studies, the particular aspects of urban life addressed, and the level or scale of the urban environment varied from place to place. Thus, we wanted to provide a series of case studies conducted by people in our network of research colleagues that illustrated these different situations, approaches, and outcomes. This objective is addressed in Chaps. 7–14.

(c) Third, we wanted to illustrate the application of new methodological approaches to analyzing and modeling QOL in general and QOUL in particular. Furthermore, we wanted to illustrate methodological advances that are being made to integrate the objective and subjective approaches including the increasing use of GIS tools to enhance such investigations. This objective is addressed in Chaps. 15–18.

What we cover in this book is necessarily selective and much of the research in the broader field related to QOL and QOUL has only been touched upon in the chapters that follow. For example, we have not considered the voluminous research that discusses the meaning of happiness and attempts to measure and model the concept. Such limitations are deliberate and may be considered shortcomings by others.

So what are we explicitly presenting in this edited volume?

By way of an *Introduction*, our initial chapter (An Overview of Quality of Urban Life) sets the stage for the investigation of QOUL by reviewing the various approaches that have emerged in research investigating QOL and especially QUOL since the 1960s. The chapter creates a framework for subsequent chapters that are organized into four parts.

*Part I* consists of three chapters that provide detailed reviews of three specific approaches used to investigate QOL and QOUL. These chapters offer detailed reviews of the literature and outline both the theoretical frameworks and methodological approaches that have been used in the research.

In Chap. 2 (Objective Measurement of Quality of Life using Secondary Data Analysis), we (Stimson and Marans) review approaches used in studies of objective QOL and QOUL based on the analysis of secondary aggregate data. Particular emphasis is placed on discussing the social indicators movement, on the use of territorial social indicators, and on the weighting of objective measures in QOUL studies. The chapter also refers to the proliferation in recent years of an industry that sets out to rate cities according to their QOL.

In Chap. 3 (Subjective Measurement of Quality of Life Using Primary Data Collection and the Analysis of Survey Data), Rod McCrea and John Western join us in tracing the evolution of subjective evaluations of QOL in general and the subjective assessment of aspects of QOUL. The subjective approach to QOL studies relies heavily on the use of social surveys to collect information from individuals and thus generate primary data. Much of that work was pioneered by researchers at the University of Michigan's Institute for Social Research. The chapter discusses a range of theoretical frameworks used to measure and model the subjective evaluation of QOL and to appraise aspects of QOUL, including the explicit investigation of urban domains at various levels or scales ranging from dwellings and neighborhoods to city-wide and regional levels. Many of those models are firmly embedded in theories of behavioral psychology. The chapter includes a discussion of the relationship between the subjective appraisal of elements of the urban environment and research into residential location decision process and choice. It concludes with a reference to recently used agent-based modeling.

The evolution of integrative approaches to analyzing QOUL is discussed by McCrea, Stimson, and Marans in Chap. 4. Based on the assumptions that objective urban environments can affect people's assessments of their QOL and that people's satisfaction with urban living may occur at different scales, the chapter outlines the evolution of approaches that have sought to empirically investigate relationships between objective environmental indicators of QOUL and peoples' evaluation of their overall QOUL and their assessments of aspects of their urban environment, both physical and social. Various theories and models are discussed and the relatively meager empirical evidence concerning such relationships is appraised. The chapter also shows how GIS tools are enhancing the capability of researchers to better investigate and, especially when linked with statistical tools of analysis, to model and test hypothesized links between objective and subjective indicators of QOUL. In many ways this represents a cutting-edge of contemporary quantitative modeling approaches in the investigation of QOUL.

*Part II* of the book has two chapters in which empirical information is presented on the objective measurement of QOUL.

Taking a perspective derived from research in regional science, Chap. 5 by Gordon Mulligan and John Carruthers focuses on research that investigates relationships between urban amenities, QOL and regional development. The authors draw on empirical studies conducted mainly in the USA but also in Europe. This research focuses on investigating urban environmental and other amenities using the compensating differentials principle. It also considers the use of hedonic price models to identify the desirable/attractive and the undesirable/unattractive attributes of places that might affect overall urban amenity, and to determine what the effects might be on regional development and employment performance. In that research, natural and other amenity indexes have been constructed and mapped for places across the USA. Finally, the chapter discusses relationships between city size, technology, migration, and urban amenity in the context of QOUL.

In Chap. 6 Subhrajit Guhathakurta and Ying Cao present a case study investigating variations in objective QOUL across Phoenix, Arizona. They discuss the results of their research showing a series of objective indicators of QOUL and highlight the public policy implications of their work.

*Part III* of the book comprises eight chapters covering a series of case studies using survey methods to collect data from individuals and households on subjective evaluations of QOL and the subjective assessments of aspects of QOUL. The studies cover cities and/or regions in different situational settings in the USA, Australia, Europe, and Asia. The design of some of those QOUL studies was in part coordinated so that the survey questionnaires used had common sets of questions. Thus, there is a degree of comparability across some of the case studies. That research effort was initiated through an original collaboration between the research teams led by the editors of this volume who conducted the surveys in metro Detroit and in the

Brisbane-Southeast Queensland region. The chapters presented here detail the research design (e.g., sampling and questionnaire administration) used in each study and review key findings from their surveys. Each chapter provides a brief summary of the situational context for the study reported and the process of gathering and analyzing data. Furthermore, some of the case studies discuss the implications of findings for policy and planning.

In Chap. 7 (The Quality of Life in Metro Detroit at the Beginning of the Millennium), Marans and Byoung-Suk Kweon present results from the Detroit Area Study (DAS2001) that focused on the quality of community life. DAS2001 was significant in that it celebrated the 50th anniversary of the University of Michigan's DAS. The study involves a mixed-mode sample survey design used in collecting information from respondents on their QOL in general and in particular on a comprehensive range of aspects of QOUL across the many and diverse administrative entities comprising the metro Detroit area. There is a discussion of how the findings have been used in a policy context.

In Chap. 8 (The Brisbane-South East Queensland Region, Australia: Subjective Assessment of Quality of Urban Life and Changes over Time), Stimson, McCrea, and Western report on changes that have occurred between 1997 and 2003 in resident perceptions of QOL and QOUL across the Brisbane-Southeast Queensland region using data from surveys conducted in those 2 years. The survey instruments shared a number of questions used in the DAS2001 study. The chapter highlights the spatial variations that exist across 10 sub-regions of SEQ in subjective assessments of QOL domains and on factors that might affect QOUL at various levels or scales.

In Chap. 9, the situational context shifts dramatically to Istanbul, Turkey, a city that straddles Europe and Asia. In conducting the survey of QOUL in Istanbul metropolitan area, Handan Türkoğlu, Fulin Bölen, Perver Korça Baran, and Fatih Terzi, borrow heavily from the survey instrument used in the DAS2001 study, with a focus on investigating the subjective assessment of aspects of community life. In addition, the study offers an objective environmental assessment of the city's residential areas. In particular, the study seeks to investigate how different types of housing might affect the assessment of QOL in general and of aspects of QOUL throughout Istanbul neighborhoods.

In Chap. 10, Derya Oktay and Ahmet Rustemli investigate subjective QOUL and neighborhood satisfaction in Famagusta in Northern Cyprus. Their survey also draws heavily on questions used in the DAS2001 survey. In the Famagusta study particular attention is directed toward looking how subjective assessments of QOUL might effect moving intentions.

In Chap. 11, the situational context moves to Dhaka, Bangladesh, where Abul Mukim Mozammel Haque Mridha and Gary Moore investigate neighborhood quality as a major component of residential satisfaction. The chapter suggests how findings can influence residential design and planning policies.

Chapter 12 returns to a European context where Alexander G. Keul and Thomas Prinz describe a QOUL study in Salzburg, Austria, relying heavily on GIS support. A two-phase research design is used by the authors. In the first phase, a survey investigating people's subjective assessments of several QOUL domains is conducted as a test in one of Salzburg's neighborhoods. In the second phase, six of the city's 24 districts are studied to compare subjective QOUL assessments. The study relies heavily on using GIS to test hypotheses relating to the impact of environmental factors on subjective assessments of QOUL.

In Chap. 13, the situational context changes to a consideration of subjective QOL in Queensland, Australia. McCrea, Mark Western, and Tung-Kai Shyy explicitly focus on investigating differences between three components of the settlement pattern in Queensland, Australia, namely, comparing the metropolitan area, regional cities/towns, and rural areas. The focus is on investigating these differences with respect to four specific attributes of the physical and social urban as well as overall QOL. A series of specific hypotheses are tested.

In Chap. 14, another case study covering the State of Washington in the USA is presented where subjective QOL between urban and rural residents is compared. Benjamin Messer and Don Dillman draw on two statewide studies conducted 37 years apart and focus on subjective community satisfaction across a range of 14 QOL issues/indicators and how those have changed over time. The study makes extensive use of statistical modeling to identify predictors of community satisfaction.

*Part IV* of the book comprises four chapters in which we provide examples of methodological innovations in analyzing and modeling QOUL. These are by no means exhaustive of the innovations that are occurring in QOL/QOUL research in recent years, but they do serve to illustrate the sort of new methodological approaches that are taking place.

In Chap. 16 (Disaggregating the Measurement of Quality of Urban Life Dimensions Across a Complex Metro Region: the Case of Metro Detroit), Byoung-Suk Kweon and Marans propose a new approach for considering geographic scale in QOL research using the data collected in the DAS2001 survey. The concern is to report findings from the analysis of subjective QOUL survey data at different geographic scales to reflect the different types of settlements or places that make up a region, in this case the Detroit metro area.

In Chap. 17 (A Spatial Clustering Approach Analyzing Types of Objective Quality of Urban Life Using Spatial Data for Survey Respondents: South-East Queensland, Australia), McCrea uses the 2003SEQQOL data set to illustrate how an integrated approach using spatial objective data for residents who responded to the survey can be employed to develop a statistical model to analyze types of objective QOUL. The focus is on using spatial clustering of objective indicators to identify different "types" of subjective QOUL relating to the residential locations (neighborhoods) of the survey respondents. The approach uses a number of GIS tools to integrate the survey data with spatial objective information available from a number of sources. Cluster analysis is used to do that and typologies (in this case four "groups") of objective QOUL are derived.

In Chap. 18, Prem Chhetri joins Robert Stimson and John Western in demonstrating how GIS tools may be employed to derive region-wide patterns of QOUL dimensions across a city. The chapter reports on two applications and uses data from the 2003SEQQOL survey. In the first an "ordered weighted average" nonlinear aggregation technique is used to derive generalized patterns of the subjective assessment of QOUL dimensions across sub-regions of the SEQ region. The second identifies and maps generalized spatial patterns of the underlying dimensions (using Principal Components Analysis) of the subjective assessment of "neighborhood attractiveness attributes" that may have affected the choices made by survey respondents in deciding where to live. Those patterns are simulated and mapped using the "neighborhood operation" function in GIS.

The future of QOUL research is discussed in the final chapter (Chap. 19) by the editors. The authors give a recap the book's content and based on their experiences in editing it outline a number of challenges that need to be addressed in future QOUL research

We hope that researchers and students interested in QOL and especially in QOUL will find this volume instructive and that some readers may be inspired to conduct new empirical studies in new situational contexts to help advance this important area of interdisciplinary research linking the social sciences and the environmental design and planning professions. We also hope the book may attract attention among politicians and bureaucrats as the outcomes of well-designed QOL/QOUL research can be used to inform policy and planning in the quest for an improved quality of life in urban areas.

Robert W. Marans and Robert J. Stimson

# Contents

# Contributors (*first author)

**Perver Korça Baran** North Carolina State University, Campus Box 7106, Jordan Hall, Raleigh, NC 27695, USA, perver_baran@ncsu.edu

**Fulin Bölen** Department of Urban and Regional Planning, Faculty of Architecture, Istanbul Technical University, 34347 Istanbul, Turkey, bolen@itu.edu.tr

**Ying Cao** College of Architecture and Environment, Sichuan University, Chengdu, Sichuan Province, China, caoyingscu@hotmail.com

***John I. Carruthers** Sustainable Urban Planning Program College of Professional Studies, The George Washington University, 805 21st St NW, Suite 301, Washington, DC 20052, USA, jic@gwu.edu

***Prem Chhetri** School of Business IT and Logistics, RMIT University, GPO Box 2476V, Melbourne 3001, Australia, prem.chhetri@rmit.edu.au

**Don A. Dillman** Social and Economic Sciences Research Center, Washington State University, P.O. Box 644014, Pullman, WA 99164-4014, USA, dillman@wsu.edu

***Subhrajit Guhathakurta** School of City and Regional Planning, Georgia Institute of Technology, Atlanda, GA 30332-0695, subhro.guha@coa.gatech.edu

***Alexander G. Keul** Psychology Department, Salzburg University, Hellbrunnerstr.34, 5020 Salzburg, Austria, alexander.keul@sbg.ac.at

***Byoung-Suk Kweon** Department of Plant Science & Landscape Architecture, University of Maryland, 2102 Plant Sciences Building, College Park, MD 20742, USA

***Robert W. Marans** Institute for Social Research, University of Michigan, P.O. Box 1248, Ann Arbor, MI 48106, USA, marans@umich.edu

***Rod McCrea** Institute for Social Science Research, The University of Queensland, Level 4, GPN3 Building, Campbell Road, St Lucia, Brisbane, QLD 4072, Australia, r.mccrea@uq.edu.au

*Benjamin L. Messer** Department of Sociology, Washington State University, P.O. Box 644020, Pullman, WA 99164-4020, USA, bmesser@wsu.edu

**Gary T. Moore** Faculty of Architecture, Design and Planning, The University of Sydney, 148 City Road, NSW, Australia, gary.moore@sydney.edu.au

***Abul Mukim Mozammel Haque Mridha** Urban Research Centre, University of Western Sydney, 34 Charles Street, Parramatta, NSW, Australia, M.Mridha@uws.edu.au

**Gordon F. Mulligan** School of Geography and Development, University of Arizona, #408, 1203 Pemberton Avenue, Squamish, BC V8B 0J7, Canada, USA, mulligan@email.arizona.edu

***Derya Oktay** Urban Research & Development Center, Eastern Mediterranean University, Famagusta, North Cyprus, via Mersin 10, Turkey, derya.oktay@emu.edu.tr; de.oktay@gmail.com

**Thomas Prinz** Research Studio iSPACE, Schillerstrasse 25, 5020 Salzburg, Austria, thomas.prinz@researchstudio.at

**Ahmet Rustemli** Faculty of Business and Administration, Eastern Mediterranean University, Erdemkent Funda Sokak No 26, Çankaya, Ankara, Turkey, rustem.ahmet2@gmail.com

**Tung-Kai Shyy** School of Geography, Planning and Environmental Management, The University of Queensland, Brisbane, QLD 4072, Australia, t.shyy@uq.edu.au

**Robert J. Stimson** Australian Urban Research Infrastructure Network (AURIN), Faculty of Architecture, Building and Planning, University of Melbourne, VIC 3010, Melbourne, Australia, rstimson@unimelb.edu.au

**Fatih Terzi** Department of Urban and Regional Planning, Faculty of Architecture, Istanbul Technical University 34347, Istanbul, Turkey, terzifati@itu.edu.tr

***Handan Türkoğlu** Department of Urban and Regional Planning, Faculty of Architecture, Istanbul Technical University 34347, Istanbul, Turkey, turkoglu@itu.edu.tr

**John Western (deceased)**

**Mark Western** Institute for Social Science Research, The University of Queensland, Level 4GPN3 Building, Campbell Road, St Lucia, Brisbane, QLD 4072, Australia, m.western@uq.edu.au

# Chapter 1
# An Overview of Quality of Urban Life

**Robert W. Marans and Robert Stimson**

## Introduction

This book is about quality of life (QOL), particularly as it relates to *place*. By place, we mean the geography or environments of individuals and groups of individuals such as households, neighborhoods and communities. Since most people live in urban environments, and especially in large *urban* environments that we call cities or metropolitan areas, the focus of the book is on the investigation of quality of *urban* life.

In their extensive review of the literature on QOL, Mulligan et al. (2004) broadly interpret QOL as the satisfaction that a person receives from surrounding human and physical conditions, conditions that are scale-dependent and can affect the behavior of individual people, groups such as households and economic units such as firms. For reasons outlined on the following pages, we believe their definition more accurately reflects quality of urban life (hereafter referred to as QOUL) rather than QOL. Accordingly, the book considers the meaning of QOUL as well as how it is measured and assessed.

The measurement and the assessment of QOL, and the investigation of its effects on human behavior are increasingly important topics within the social sciences

R.W. Marans (✉)
Institute for Social Research, University of Michigan, P.O. Box 1248,
Ann Arbor, MI 48106, USA
e-mail: marans@umich.edu

R. Stimson
Australian Urban Research Infrastructure Network (AURIN), Faculty of Architecture,
Building and Planning, University of Melbourne, VIC 3010, Melbourne, Australia
e-mail: rstimson@unimelb.edu.au

R.W. Marans and R. Stimson (eds.), *Investigating Quality of Urban Life: Theory, Methods, and Empirical Research*, Social Indicators Research Series 45,
DOI 10.1007/978-94-007-1742-8_1, © Springer Science+Business Media B.V. 2011

(Dissart and Deller 2000; Diener and Suh (2000); Diener and Biswas-Diener 2008). And as discussed by Lambiri et al. (Lambiri 2007), QOL has increasingly become a concept researches theoretical and empirical in economics.

Investigating QOUL is important not only because it affects how people behave but also their life satisfaction and happiness. And it has broader implications for research and urban policy. For example, QOL in general and QOUL in particular can:

- Underlie the demand for public action (Dahmann 1985; Lu 1999)
- Directly affect the liveability of cities for residents and provide a set of metrices which allow policy makers and planners to assess the effectiveness of their efforts (Marans 2002)
- Motivate residential location decisions and choices (Campbell et al. 1976b; Golledge and Stimson 1987; Zehner 1977)
- Have broad implications for patterns of regional migration, regional economic growth, and environmental sustainability (Kemp et al. 1997)

Indeed it is well established that, at least in part, migration patterns and urban growth arise in response to differences in QOL between places (Keeble 1990; Ley 1996; Glaeser et al. 2000; Liaw et al. 2002), which may reflect the nature of employment opportunities (Brotchie et al. 1985; Grayson and Young 1994; Rogerson 1999), and the competitiveness of a city or metropolitan area (Sirgy et al. 2000). Patterns of intra-urban mobility are also related to differences in both the objective characteristics of neighborhoods and the subjective evaluations people make about aspects of the QOUL, and how that may vary across urban space (Keeble 1990; Ley 1996).

It is, then, not surprising that there is widespread interest in QOL, particularly within the context of the places where people live.

In order to understand the QOL in a particular setting, such as a city, we need to measure conditions in that place using sets of *indicators*. Furthermore, we need to monitor changes in those conditions over time in order to appraise or determine if and how those conditions have changed. And if they have changed, we need to determine if they have improved or deteriorated and by how much. This effort might include evaluating the impact of various public or private interventions which sought to improve conditions.

We know that different people may have different *perceptions* and therefore make different *subjective* judgments about the things which impinge on their QOL including specific attributes of their urban environment. To adequately investigate those aspects of QOL, we need to use model frameworks and collect data to operationalize those frameworks within a particular context.

This book includes sections that provide an overview of the evolution and application of theoretical frameworks and methodologies that have been used to investigate QOL. As discussed by Andelman et al. (1998), investigation has been pursued predominately through two approaches:

(a) The *objective* approach which is most typically confined to the analysis and reporting of *secondary data* – usually *aggregate data* at different geographic or

**Table 1.1** Examples of QOL indicators that can be used to investigate QOUL in cities and neighborhoods

| Objective indicators | Subjective indicators | Behavioral indicators |
|---|---|---|
| Employment rates | Housing and neighborhood satisfaction | Public transit use |
| Educational attainment | Desire to move | Participation in sports |
| Per capita income | Perceptions of crime | Amount of walking and bicycling |
| Crime statistics | Perceptions of school quality | Visits to cultural amenities and events |
| Domestic violence | Perceptions of health care services | Visits to parks |
| Death rates | Feelings about neighbors | Visits to health clinics/doctors |
| Incidence of chronic diseases | Feelings about rubbish collection | Amount of neighboring |
| Air quality | Feelings about congestion and crowding | Participation in voluntary organizations |
| Residential density | Feelings about government | Participation in local decision-making organizations |
| Housing vacancy rates | Satisfaction with health | Residential mobility |
| Amount of parkland | Satisfaction with family, friends, job etc | |
| Number of public transit riders | Life satisfaction, overall happiness (overall well-being) | |
| Distance to transit stop | | |
| Availability of grocery/food stores | | |
| Vehicle kilometers/miles traveled | | |

Source: The authors

spatial scales – that are available mainly from official governmental data collections, including the census. This is an approach that is often associated with *social indicators* research.

(b) The *subjective* approach which is specifically designed to *collect primary data* at the *disaggregate or individual* level using *social survey* methods where the focus is on the peoples' behaviors and *assessments, or evaluations* of aspects of QOL in general and of QOUL in particular.

We might identify a set of *objective indicators* and *subjective indicators* that may be used to evaluate QOL in a city or neighborhoods within a city such as those attributes listed in Table 1.1. As illustrated in the third column in Table 1.1, we might want to also identify explicitly *behavioral indicators* of QOL.

But it is the nature and the strength of the links between broad objective dimensions and subjective evaluations of the urban environment which has represented a challenge for researchers The nature and strength of linkages need to be tested as understanding them may be important in informing how planning and other policy interventions might contribute to improving the QOUL.

Much of this book is devoted to a discussion of QOUL in a number of places or environments throughout the world, reporting the outcomes of recent empirical research that has used survey methods to collect primary data on aspects of QOUL. In most of the studies, a relatively common set of core questions were included to measure perceptions of QOL domains, including those dealing explicitly with place. In many, information relating to the sociophysical environmental context of those places was also collected. Some of the case studies present the results of modeling that explores relationships between subjective and objective aspects of QOUL, including the use of geographic information systems (GIS) technology to integrate survey-based subjective information with spatial objective information.

In the remainder of this introductory chapter, we provide an overview of approaches to the investigation of QOL in general and of QOUL in particular.

## Quality of Life and Living Environments

For many years, scholars in both the social sciences and the environmental design professions have been arguing that "quality" of any entity has a *subjective* dimension that is *perceptual* as well as having an *objective reality*. Central to that assertion is the notion that the environment may be defined as having built, natural, and sociocultural dimensions (Marans 2005: p. 315), and different environmental settings will have specific characteristics with respect to those dimensions. But the places in which people live consist of all three of those dimensions, and research findings have clearly demonstrated that all three form important components of the QOL or subjective *well-being* of people living in a specific place.

In the introduction to their comprehensive book on well-being, Kahneman et al. (1999: p. x) indicated that the *quality of life experience* is embedded in the social and cultural context of the subject and the evaluator. Those researchers also suggest that the objective characteristics of society – such as poverty, crime rates and pollution – contribute predominately to peoples' *judgments* of their lives.

QOL is certainly a multi-faceted concept that seems to defy precise definition. Often it is difficult to differentiate between the notions of *QOL*, *well-being*, *satisfaction*, and *happiness*. Over the years, the study of QOL has attracted the attention of researchers from a wide range of academic disciplines as well as the interest of politicians, policy makers, planners and others in the environmental professionals. It is certainly an interdisciplinary field of study.

Many QOL studies have tended to examine attributes of individuals, such as their employment, age, health, and interpersonal relationships. However, people live their lives in *place*s or series of places, each of which has particular environmental characteristics. Those places might be viewed at various levels or scales – from the dwelling to the local area or neighborhood, to the city, to the broader region or even to a state or a nation – and it may be argued that where people live will influence their lives and, therefore, their QOL. As such, a fundamental assumption underlying many approaches to planning is that urban environments (places) may be designed to increase the level of satisfaction with the lives of residents. Given that most people in advanced economies live in the large urban environments that we call cities or metropolitan areas and such areas are expected to grow over the next few decades, it is important to examine the relationships between the characteristics of urban environments and the perceived QOL of the residents.

While social scientists have had a strong interest over a long period of time in investigating aspects of QOL, that intensity of interest, the approaches used and the focus of those investigations have varied. But in recent times, there does seem to have been an upsurge of interest in QOL studies and related phenomena. An indication of that is the formation of the *International Society for Quality-of-Life Studies* (ISQOLS), which holds an annual conference and which launched in 2006 the journal *Applied Research in Quality of Life*. That journal deals with QOL studies in applied areas of the social and natural sciences, and it has the goal to:

> … help decision-makers apply performance measures and outcome assessment techniques based on concepts such as well-being, human satisfaction, human development, happiness, wellness and quality-of-life.

That statement is indicative of breadth of concerns which might be related to the notion of QOL, and it reinforces the "fuzziness" of its meaning.

## Approaching How to Investigate Quality of Life

As mentioned earlier, two basic approaches have been used by researchers to examine QOUL, particularly in the context of people living in cities and metropolitan areas:

(a) The first has involved monitoring QOL/QOUL through a set of *indicators* – usually over time – derived from *aggregated spatial data* using official sources, such as the census, that are said to be related to perceived QOL (for example, level of household income, crime rates, pollution levels, housing costs, and so on).
(b) The second has involved modeling *relationships* between *characteristics of the urban environment* and measures of peoples' *subjective assessments* of QOL domains, including their *satisfaction* with specific phenomena and with life as a whole. This approach typically involves data collected through survey research methods and analyzed using techniques such as regression analysis or structural equation models.

Monitoring indicators over time can provide information on those aspects of QOUL that people see as improving or declining, while survey data can also provide information on individual and community level perceptions, behaviors, subjective evaluations and levels of satisfaction with various aspects of urban living. However, as pointed out by McCrea et al. (2005), while those indicators are useful, they are also limited. That is because they cannot by themselves indicate the relative importance of the different attributes of urban living and environments that contribute to the level of satisfaction of individuals with urban living.

Even if a sample of residents living in a city were asked to rank in order of importance a list of items relating to QOUL, the information thus gathered does not necessarily allow one to estimate the proportion of the level of satisfaction explained by any one factor nor the unique contribution of any one factor. Therefore, it is important to develop models to analyze the data and to test hypotheses about those relationships using methods to establish the relative and unique importance of various aspects of urban living in contributing to the QOUL of various groups of residents. These methods could range from regression analysis to more sophisticated structural equation modeling techniques.

The complex relationships between the characteristics of urban environments at different scales and the satisfaction of the residents of a city with QOUL domains are certainly difficult to model without a theoretical framework to guide the process. In addition to the complexities just discussed, Schwirian et al. (1995) have identified an "urbanism" construct which consists of four related dimensions, namely:

- Demographic characteristics
- Economic stress
- Social stress
- Environmental stress

The notion is that economic, social, and environmental conditions in an urban setting might create stressful situations or experiences for some of the people living there.

It might appear that it is difficult to incorporate such a complex set of factors into one model. However, in seminal work more than 30 years ago, Marans and Rodgers (1975) proposed a model of satisfaction with residential environments and adapted in Campbell et al. (1976a). The literature in QOL studies seems to most frequently cite the Campbell et al. reference as providing an overarching model framework for the investigation of QOL which can readily incorporate a range of demographic, social, economic and environmental relationships, while taking into account satisfaction with different levels of living or domains of life (see Fig. 1.1).

The model rested on the following four principles:

(a) The experiences of people are derived from their interactions with the surrounding environment.
(b) The subjective experiences of people are different from the objective environment.
(c) People respond to their experiences with the environment.
(d) The level of satisfaction in various life domains contributes to the overall QOL experience.

In essence, the model specified a series of linkages between various objective attributes of each life domain and satisfaction measures of those domains, which in turn could be influenced by a range of individual characteristics and individual standards of comparison.

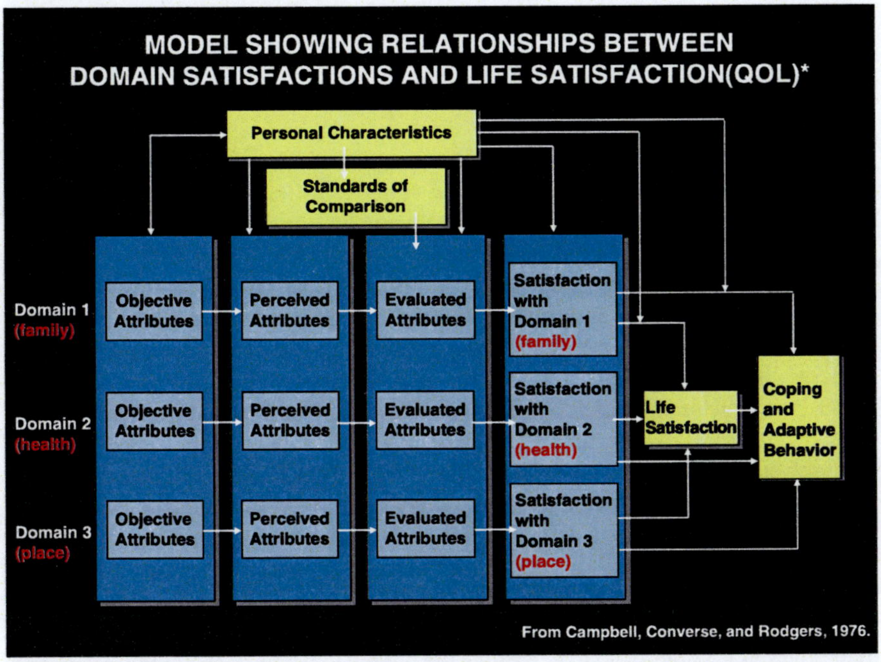


**Fig. 1.1** Model showing the relationship between domain satisfactions and life satisfaction (Source: Campbell et al. 1976a)

The approach proposed by Campbell et al. (1976a) suggested that satisfaction with life could be viewed at multiple levels of analysis (or for different life domains). As suggested by Marans and Rodgers (1975), that might include, for example:

- Satisfaction with housing
- Satisfaction with neighborhood
- Satisfaction with the wider community (or broader region).

This was thus a *bottom-up* model framework in which urban characteristics (such as perceived crime) might contribute to satisfaction in a specific domain (for example, neighborhood satisfaction) which, in turn, might contribute to overall satisfaction with life. Paths could thus be mapped from economic, social and environmental characteristics of urban living to satisfaction with different living domains, and those paths are mostly between variables at the same level of analysis. However, the Campbell et al. (1976a, b) model did provide for relationships between the various QOL domains and geographic levels of urban scale to be analyzed.

## Some Related Concepts: Well-Being, Satisfaction, and Happiness

One of the confusing things in the QOL literature is the proliferation of terms used to relate to the concept QOL. Those terms include *well-being*, *satisfaction*, and *happiness* when talking about investigating aspects of *life experiences* and QOL.

In their seminal study of the quality of American life, Campbell et al. (1976a) conceptualized the "QOL experience" as being about *individual well-being*. They measured peoples' perceptions, evaluations and satisfaction with domains of QOL including urban domains using scales incorporated in questionnaires that were administered to a sample of more than 2,000 US residents. Primarily, the focus was on measuring the *global evaluations of life* rather than on *actual conditions* of life. In doing so, the Campbell et al. approach addressed the concept of *satisfaction* rather than *happiness*, which had been considered in earlier studies of well-being (such as those by Bradburn and Caplowitz 1965; Bradburn 1969). "Satisfaction" was viewed as being more definable and implied judgment or cognitive experience, whereas "happiness" reflected a relative short-term mood of elation or gaiety. And "satisfaction" was also considered by Campbell et al. (1976a) to be a more plausible and realistic objective for policy makers than "happiness" if research outcomes were to be used by policy makers. The intent of Campbell et al. was, then, to measure and compare peoples' assessments of several domains of their lives as well as their "lives as a whole," and to determine the degree to which each domain explained overall well-being or QOL. The seven domains considered were: health, marriage, housing, family, financial situation, leisure, and community or place of residence.

In addition, Campbell et al. (1976a) considered that *context* and evaluator or *person characteristics* were important in understanding QOL, with "context" being the actual conditions of life or "objective attributes." But their attempts to measure those attributes were modest.

With respect to *domain satisfactions*, Campbell et al. (1976a, b) suggested that they were a reflection of peoples' perceptions and assessments of the many attributes of each domain and that these in turn were influenced by the objective attributes themselves. For example, job satisfaction was seen as a function of a person's assessment of the many attributes of a job, such as the degree of autonomy, relationships with co-workers, wages and so on. Furthermore, the assessment of the wage attribute was considered a function of the level of a person's actual salary and his/her expectations and standards of comparison. Similarly, perceptions of crowding in a dwelling were expected to be associated with an objective measure (such as the number of people per room or another measure of housing density) and individual standards relating to crowding. That was similar to the later views of Kahneman et al. (1999) on the role of the objective world in understanding subjective well-being.

In recent years, it has again become fashionable for writers and media commentators to talk about "happiness" For example, on ABC NEWS.COM (May 29, 2008), Bob Cummins, a psychologist in Australia, said:

> … When happiness was considered a mysterious, ephemeral state of mind, it was not worthy of serious consideration. But over the last few decades, science has begun to lift the veil of mystery, revealing happiness as an ordinary state of mind that can be studied and understood.

There has in fact been a proliferation of research and writings on "happiness" as seen in the recent publication of a range of books (such as those by White 2006; Thaler and Sunstein 2008; Eid 2007; van Praag 2004; Lyubomirsky 2008; and Weiner 2008; Diener and Biswas-Diener 2008). Some of that research on happiness (for example, van Praag 2004) reflect what has been perused by economists, and that type of research is considered by some to be at the frontier of that discipline using econometric analysis to deal with variables including income, health, marriage, gender, social comparison norms and the dynamics of satisfaction. However, as has already been noted, most of those factors have long been considered in research by psychologists and sociologists in the study of happiness and satisfaction, often in the context of studying well-being.

The quantitative analysis of happiness by social scientists has resulted in the development of sophisticated scales to measure individual and collective norms that include satisfaction with life as a whole as well as with various domains of life, such as health and income. And there are a number of on-going surveys that attempt to measure "happiness," one being the *Australian Unity Wellbeing Index*, which has been measuring the happiness of Australians since April 2001 (see Cummins et al. 2003). It uses the *Subjective Wellbeing Homeostasis* management system, which suggests that we hold happiness within a relatively narrow range of values. It has been shown to be resilient. It would seem that two key factors relating to peoples' happiness are:

- An internal factor, namely relationships (for which one may read as having an emotionally intimate relationship)
- An external factor, namely resources (for which one may read as "money")

That Australian study suggests that happiness rises only marginally beyond a household income of about A$100,000 a year and that after A$150,000, there is no more rise in happiness. And having more money is not a substitute for not having a good relationship. This is the so-called Easterlin Paradox, which says that once people have met their basic needs, they do not become happier as they become richer.

## Place and Environmental Setting Do Matter

There is considerable evidence to show that "place" matters when it comes to QOL concerns, and studies focusing on QOUL enable us to better understand the meaning of QOL and how it might be measured (Marans 2002: p. 2). For example, Marans and his collaborators (Marans and Rodgers 1975; Lee and Marans 1980; Connerly and Marans 1985, 1988) have built on the seminal work by Campbell et al. (1976a) to explore the *objective–subjective relationships* in investigating QOUL, asserting that the quality of a place or the geographic setting at various levels of *scale* (the region, the city as a whole, the neighborhood, the dwelling) is in fact a subjective phenomenon and that each person occupying that setting might differ in their views about it. Further, it has been suggested that those views would reflect each individual's

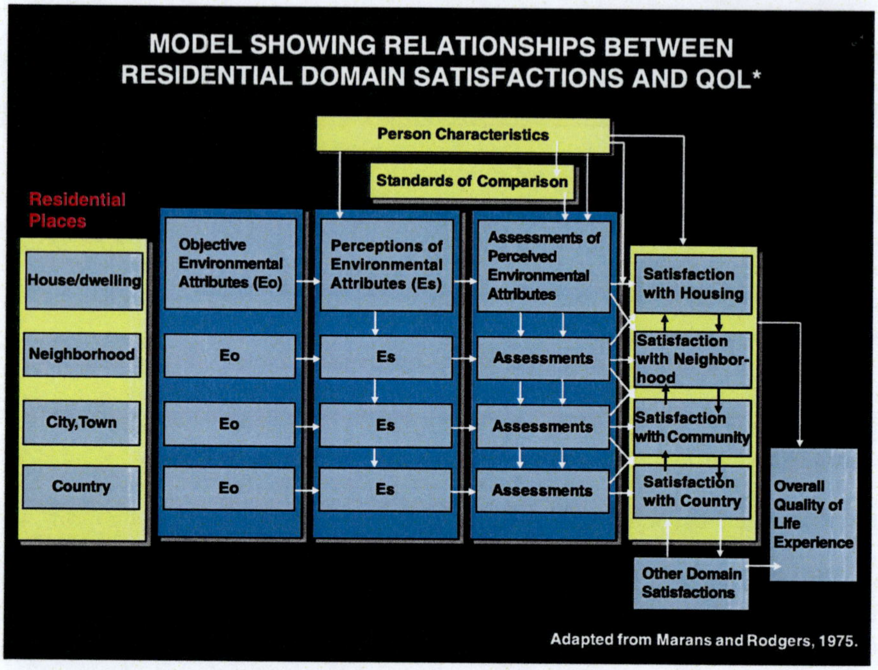


**Fig. 1.2** Model showing the relationships between residential domain satisfactions and quality of life (Source: Marans and Rodgers 1975)

*perceptions* and *assessments* of a number of setting attributes that could in turn be influenced by certain characteristics of the occupant, including their *past experiences*. Those past experiences thus represent a set of standards against which current judgments are being made. Those judgments include other settings experienced by the resident of a place, and they also include their *aspirations*. Finally, it also has been asserted that those assessments and perceptions of setting attributes are associated with the place attributes themselves. Marans (2002) provides this example:

> … the degree to which a person feels crowded at home is expected to be related to some degree to the number of people in his household per room (i.e. housing unit density). At the neighborhood level, assessments of air quality and family health (e.g. the incidence of asthma) are likely to be associated with air quality measures in the neighborhood. (pp. 1–2)

Marans and Rodgers (1975) had proposed a model depicting such relationships for different residential domains of urban environments and how those domains, together with other domains, contribute to QOL (see Fig. 1.2). There are, of course, assumptions underlying the model:

(a) One is that the *quality of the geographical or environmental setting* (the region, the city, the neighborhood or the dwelling) cannot be captured through a single

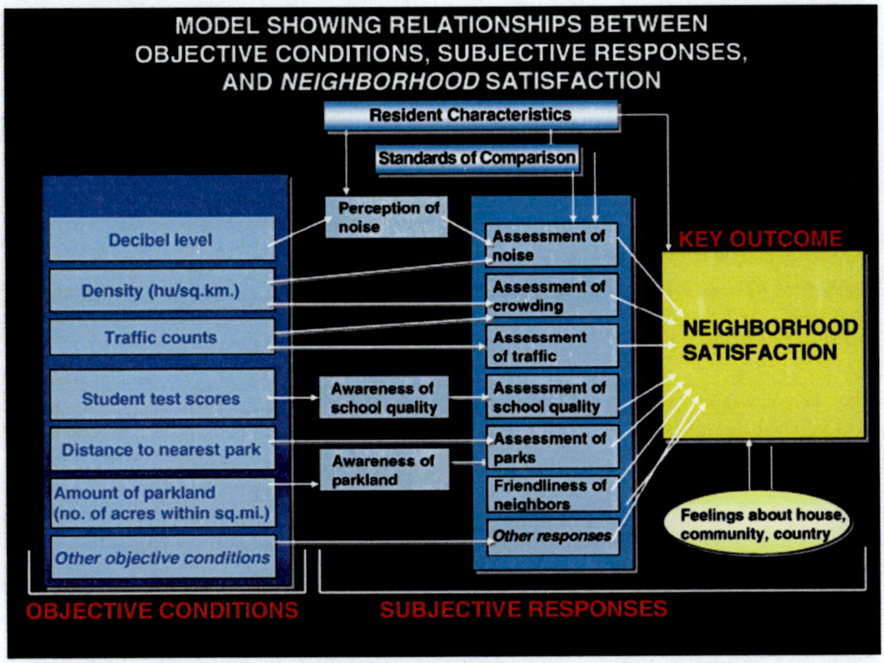


**Fig. 1.3** Model showing the relationships between objective neighborhood conditions, subjective responses and neighborhood satisfaction (Source: Marans 2002)

measure; rather, it requires measures of multiple attributes of the environmental setting in question. In as yet to be specified combination, it reflects the overall quality of the setting.

(b) Another is that quality is a *subjective phenomenon reflecting the life experiences* of the occupants of the setting. The objective conditions of the setting themselves do *not* convey the true quality of the setting; rather, its quality is a reflection of the meaning of those conditions to the occupants.

More recently, Marans (2002) has elaborated on the model by showing the relationships that might account for people's feelings about their neighborhood (that is, "neighborhood satisfaction"), as demonstrated in the example given in Fig. 1.3.

As Marans (2002) has stated:

> … Often, policy-makers want to know the most effective means of enhancing satisfaction. An important part of research therefore is determining the degree to which various objective conditions are associated with satisfaction. There is general agreement that satisfaction as an indicator of individual well-being is an important outcome in quality of life research. Nonetheless, there are other outcomes of importance to well-being that may be examined in quality of life studies. For instance, the physical health of individuals and the amount and type of physical activity they engage in are important to their overall well-being. (p. 3)

**Table 1.2** Additional possible outcomes at the neighborhood and the dwelling level

| Neighborhood | Dwelling |
|---|---|
| Concern for safety | Amount of leisure time spent at home |
| Rating of school quality | Number of accidents |
| Public transit use | Amount of time spent with children |
| Assessment of public transit | Time spent in housekeeping |
| Involvement in governance at a city level | Time spent in home maintenance |
| Amount of neighboring | |
| Number of shopping trips | |
| Where children play | |
| *Park visits* | *Airborne-related illnesses* |
| *Amount of walking* | *Number of meals at home* |
| *Visits to doctors* | |

Note: Items in italics might be used as physical health-related outcomes
Source: The authors

It may be that in investigating QOUL, researchers might want to explicitly focus on outcomes additional to those indicated in Fig. 1.2. By way of example, that might include the outcomes for people at the neighborhood level and at the dwelling level that are listed in Table 1.2. Particular measures might be used to relate to an outcome on a particular domain, such as the items in italics that could relate to physical health outcomes.

A further conceptual model proposed by Marans and Mohai (1991) suggests how health may be linked to a number of objective conditions associated with a set of leisure resources including environmental quality, as illustrated in Fig. 1.4. It showed that environmental and urban amenities are related to community quality and individual activities, satisfactions, and physical health:

(a) Environmental amenities include both:

- Natural recreation resources (for example, rivers, lakes, wetlands, forests)
- The quality of the ambient environment (air, water, noise, solid, and hazardous waste)

(b) Urban amenities include both:

- Man-made recreation resources (swimming pools, bicycle paths, walking trails, golf courses)
- Cultural resources (cinemas, concert halls, orchestras, museums, galleries, sports teams)

The model hypothesized that the perceptions or awareness of these environmental and urban amenities will influence peoples' evaluation and their use of them. And the model also suggested that in the case of the man-made recreational resources and the natural recreational resources, their use or non-use by an individual is associated with physical health.

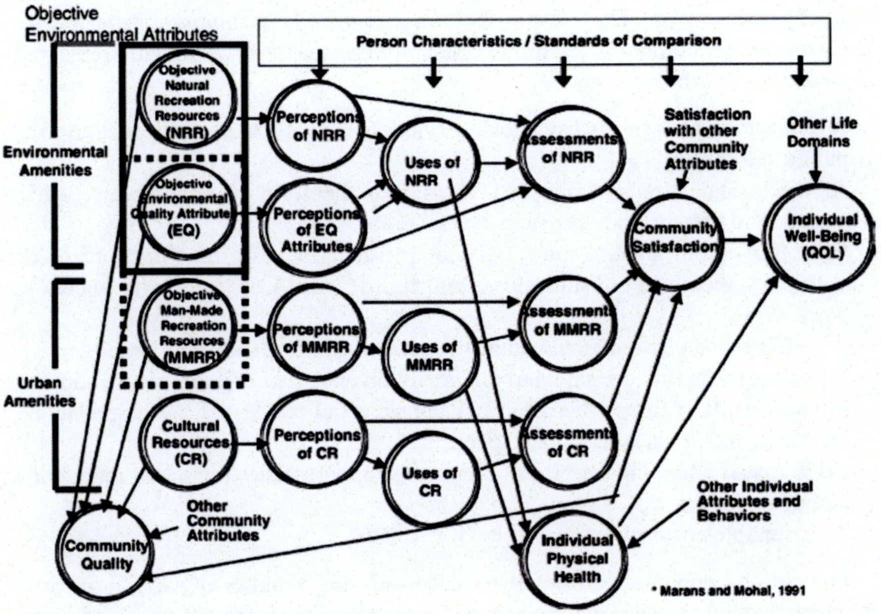


**Fig. 1.4** A model linking recreation resources and activities to individual well-being, health and community quality (Source: Marans & Mohai 1991)

Models such as that depicted in Fig. 1.4 provide opportunities to explore many relationships including the role of recreational, environmental, and cultural resources in understanding QOL and in particular QOUL. Conceptualizing similar models can serve to guide data collection and analyses for other outcomes associated with QOL and with *quality of place*.

But the importance of relationships between urban characteristics and the perceived or subjective assessment of QOUL extends far beyond the satisfaction of individual residents with their living environments. Indeed, that has been the focus of much of the aggregate level analysis and modeling that had been conducted using spatial objective data. For example, as discussed at the outset to this chapter, migration patterns are often attributed in part to QOUL considerations associated with particular cities or regions that may either be places from which people move or places to which people are attracted, and there is a rich literature in geography and regional science investigating migration flows using aggregate data derived from the census. Such models typically use as explanatory variables *place-related attributes*, including measures of local labor market characteristics (such as industry structure and wages) and urban amenities, which might include, for example, climate, the amount of public open space and the number of recreational facilities, museums and art galleries, sports teams, health services and facilities, and public transport characteristics.

By way of an example, in a study for the US Department of Housing and Urban Development, Glaeser et al. (2000) made the claim that urban growth is driven by a variety of QOUL issues that relate to *urban consumption experiences*. Their research identified seven urban consumption areas that are said to predispose an area to rapid urban growth:

- A rich variety of high-quality public services (especially in health, education and public safety)
- Aesthetic and attractive physical settings in the form of architecture, urban design, and natural features, such as a favorable climate.
- Easy movement around the city, with resident location having more to do with easy access to consumption opportunities and less to do with access to work
- A housing stock that is architecturally distinctive, affordable, and varied
- Neighborhoods that are safe and ethnically diverse, that offer transport choices, that have a mix of compatible uses (for example, retail, residential, and commercial), and that contain parks and open spaces
- Civic spaces and civic activities that provide opportunities for social interaction among residents
- A reasonable cost of living

In addition to population flowing to places offering a higher QOUL, so too does economic capital as additional investment is needed to cater for an increasing population, and this in turn is likely to enhance the region's economic growth and possibly its competitiveness (see, for example, Grayson and Young 1994). And Rogerson (1999) has suggested that the consumption experiences afforded in such places of in-migration and urban growth are key factors determining QOUL (as seen, for example, in research by Ley 1996).

A concern for these issues is clearly evident in strategy formulation for regional and local economic development, where a focus on business development and attraction through policy and programs that seek to enhance place amenity, create a business-friendly culture, and build human capital are common-place, with the objective being to improve overall QOL (Blakely 1994; Wong 2006; Roberts and Stimson 1998; Mathur 1999).

The importance of "place" is also demonstrated by Richard Florida (2008) in his book, *Whose Your City?*, in which he had this to say:

> … The place we choose to live affects every aspect of our being. It can determine the income we earn, the people we meet, the friends we make, the partners we choose, and the options available to our children and families. People are not equally happy everywhere, and some do a better job of providing a high quality of life than others. Some places offer us more vibrant labor markets, better career prospects, higher real estate appreciation, and stronger investment earning opportunities. Some places offer more mating markets. Others are better environments for raising children.

In his writings, Florida (2002, 2008) has claimed that the opportunities offered by certain places for achieving a high QOL has become the impetus for attracting and retaining what he calls the "creative class." That class, he says, has been the key to the emergence of the dynamic contemporary of the creative economy which is highly spatially concentrated in particular large cities or urban regions, the top three of which, he says, are the San Francisco Bay Area, Boston and Seattle. The high QOL experiences by the creative class are achieved by the individual consumption of necessities – like health, housing, education – and the consumption of goods and services that help satiate the hedonistic values of the contemporary age, but Florida (2002) claims that most importantly, it is acquired through

> … a passionate quest for experiences. The idea… is to 'live the life' - a creative life packed full of intense, high quality multidimensional experiences. And the *kinds* of experiences they crave reflect and reinforce their identities as creative people … [T]hey favour active, participatory recreation over passive spectator sports. They like indigenous street-level culture – a teeming blend of cafes, sidewalk musicians, and small galleries and bistros, where it is hard to draw the line between participant and observer, or between creativity and its creators. (p 166)

Florida puts the argument that the communities (or "neighborhoods") in which these people live are chosen because they offer opportunities for those experiences. They are close to work in the inner city, where these experiences are most likely to be concentrated. In this way, the post-industrial community of the creative class has some semblance to the occupational community of the earlier industrial age. Both types of community were or are located in the inner city, close to places of work, with residents of what is called "the occupational community" being the key class (the working class) of the industrial era. Florida has called this community an "occupational community" because of the close ties that exist between home/the community and place of employment (for example, a ship building yard). Yet there are also fundamental differences between the communities of the industrial and post-industrial ages. Whereas those residing in industrial age communities had close-knit ties and thus – in Putnam's (1993) terms – held considerable *social capital*, the post-industrial communities of the creative class have weak ties and thus limited social capital. Putnam's lament about the decline of social capital in the contemporary world is not shared by Florida. Indeed Florida (2002, 2008) has shown that places with extensive social capital are not only where the creative economy is largely absent, but those places contain remnants of the earlier and now defunct industrial economy and its working class. Conversely, places with dynamic, contemporary, creative economies – and thus the creative class – have weak ties and low levels of social capital. Moreover, according to Florida the cities and urban regions where the creative economy (and thus the creative class) is concentrated are characterized by diversity and tolerance, with this openness providing a climate for innovation.

Florida (2008) has gone as far as listing those places in the USA which his research shows as offering the "best" places for particular groups of people according to their stage in the life cycle. For example, Table1.3 lists his "best places" for (a) mid-career professionals who are single or married, without children and aged 30 to 44 years, and (b) empty-nesters aged 45 to 64 years.

Florida (2008) says the key is to find a place that fits, that makes one happy, and that enables one to achieve one's life goals.

Table 1.3 Overall best places for people to live in the USA

| *Large regions* | *Medium-size regions* | *Small regions* |
|---|---|---|
| **(a) For mid-career professionals, single or married, without children, aged 30–44** | | |
| San Jose, CA | Stamford, CT | Durham, NC |
| Minneapolis | Portland, ME | Prova, UT |
| Austin | Madison, WI | Reno, NV |
| San Diego | Omaha, NE | Fayetteville, AR |
| Denver | Des Moins, IA | Boulder, CO |
| **(b) For empty-nesters aged 45–64** | | |
| San Francisco | Stamford, CT | Boulder, CO |
| Seattle | Portland, ME | Trenton, NJ |
| Boston | Madison, WI | Fort Collins, CO |
| Minneapolis | Honolulu, HI | Santa Rosa, CA |
| Hartford, CT | Rochester, NY | Norwich, CT |

Source: Florida (2008)

## Benefits of a Modeling Approach

From the discussion so far, it is evident that the relationship between QOL and urban environments is undoubtedly complex, and people's satisfaction with living in urban environments is influenced by their personal characteristics, such as (individual or group) values, expectations, perceptions and evaluations, and their demographic and socioeconomic characteristics. People vary in what they may consider to be important when judging their satisfaction with life in general and their QOUL in particular (Hsieh 2003), and different people might perceive things differently in the same situation or setting. The complexity of the relationships between urban characteristics and those subjective judgments or evaluations might help explain why research finds that a low correlation is often found to exist between individual subjective evaluations and objective measures of QOL (see, for example, Warr 1987, 1999; Schwarz and Strack 1999).

The benefits of using the modeling approaches that have evolved from the original Campbell et al. (1976a) model are numerous, as discussed by McCrea et al. (2005):

(a) The models have been able to accommodate a large number of factors thought to influence levels of satisfaction with urban living as well as personal characteristics of people.
(b) The models also have allowed for the comparison of a number of different geographic levels of urban living. Including different levels of urban living in QOL models has been important because different planning, urban development and service provision policies may target different levels of urban living. When only one level of analysis is incorporated into a model, the results may be confounded because the other levels of analysis are not controlled (Gyourko and Tracy 1991). For example, relationships depicted in a model that are confirmed when

analyzing data covering an entire region may not hold up when analyzing data for separate communities within that region. That is because of what geographers refer to as the "aggregation/disaggregation problem" whereby greater clarity in terms of spatial differentiation in a phenomenon is more likely to be evident at a more disaggregated level of scale than will be the case at a more aggregated level of scale.

(c) In addition to incorporating different levels of urban living, models have allowed for characteristics of one particular satisfaction domain to contribute to satisfaction in another domain. For example, a public transport system may be a characteristic of a city and contribute to its overall quality, but it may also influence neighborhood satisfaction and people's ability to move easily throughout a region.

(d) Finally, it is possible for the level of satisfaction in one domain to influence (or color) satisfaction in other domains. For example, housing and neighborhood satisfaction have been shown to predict community satisfaction. Such links between satisfaction domains are "spillover effects" (Jeffres and Dobos 1995).

## Social Indicators, Urban Amenity and Livability Studies

As indicated earlier in this chapter, one of the common approaches to the investigation of QOL and of QOUL has been concerned with the secondary analysis of aggregate data at different levels of spatial scale typically using information available from official data collections, including the census. Mulligan et al. (2004) have provided a comprehensive review of the research – published predominately in the economics, regional science and geography literature – which has analyzed and modeled QOL in a spatial context using those approaches.

### *Social Indicators*

In an early seminal study, Thorndyke (1939) assessed QOL using a wide variety of single-variable indicators grouped into six categories. Then much later in the 1960s and 1970s, a vast amount of research was conducted using Thorndyke's approach in what became known as the *social indicators* movement, which grew out of a long tradition of social and social policy research. That was reflected in the publication of an array of books written by social scientists from many disciplines, including for example, contributions by geographer David Smith (1973; 1977; 1979), planner Judith Innes de Neufville (1975), and sociologists Otis Dudley Duncan (1969) and Peter Rossi (1972).

The social indicators movement reflected an increasing policy interest in and a concern for an array a range of social issues – such as poverty and crime – which reflected the incidence of social disadvantage in society. The research was conducted at a variety of levels of spatial scale, from the analysis of social indicators at the national level using data for states and counties in the USA, to the analysis of patterns within cities, to a focus on the study of neighborhoods. Often the research involved the benchmarking of performance at the local or regional level against a national performance figure (or standard), and it was common for geographers to do this by mapping the location quotient scores of regions on a particular social indicator.

An example of that approach was a study by Liu (1976) in which more than 240 US metropolitan areas in 1970 were assessed using five general categories of well-being, namely:

- Economic health
- Political performance
- Environmental conditions
- Health and education
- Social concerns.

The raw data were standardized and then added so that once assigned to three size-groups, the cities could be designated as being "sub-standard," "adequate," "good," "excellent," or "outstanding."

In addition, many studies have used multivariate data reduction techniques, such as principal components analysis, to reduce to a small number of significant social dimensions the scores of regions on a battery of social indicator variables (see, for example, Hadden and Borgatta 1965; Berry and Kasarda 1977). Others have used discriminant analysis to classify places according to their social disadvantage problem using measures of inequality and deprivation. An example is the Cheshire et al. (1986) study of more than 100 of the largest metropolitan areas in the European Economic Community countries using functional urban regions and using variables relating to income, unemployment, migration and travel demand. They were able to compare the rank position of cities and calculate a change in health score over the period 1971 to 1988.

It has been relatively common for studies such as those referred to above to claim that they are measuring QOL. A more detailed discussion of social indicators, and in particular of territorial social indicators, is provided in Chap. 2.

## *Urban Amenity*

Social indicator studies fell into disfavor after the 1970s, especially in the USA. However, in recent decades, there has been a resurgence of interest among some social scientists in conducting QOL studies using spatial social data sets with the focus being on the evaluation of regional or city level performance, including the study of *urban amenity* – including public goods – as an attractor for migration in the context of researching labor regional markets and regional economic development, and the role of amenities in urban life, including their impact on housing markets (see, for example, Bartik and Smith 1987; Gyourko et al. 1999). This interest – particularly among regional scientists – is elaborated in more detail in what follows.

The substantial literature that has developed investigating urban amenity has focused on both the inter-urban and the intra-urban scales. In the context of QOUL studies, the role of *urban amenity* has typically been seen from an economic perspective and as being capitalized in real estate and rents – and even wages – which will reflect non-market amenities. That influence has been explained through the *compensating differentials* concept. Over time, there has developed a considerable literature on the capitalization of urban environmental amenity (see, for example, Geoghegan et al. 1997; Orford 1999; Mahan et al. 2000; Esparza and Carruthers 2000; Hardie et al. 2001; Johnston et al. 2001; Bastian et al. 2002; Smith et al. 2002).

The range of amenity variables used in such studies has been considerable and includes surrogate measures of the following:

- Wages and income
- Industrial structure and employment
- Housing prices
- Quality of education facilities and production of human capital
- Cultural facilities
- Wetlands
- Scenic views
- Proximity to farmland
- Climate (especially average temperatures).

It is interesting to note that urban amenity thus incorporates landscape diversity and the built environment in addition to cultural and economic factors. The notion is that all such factors – many of which are intangibles – can operate to enhance QOUL. But spatial variation in the provision of and in access to those place amenity factors is highly variable not only on an inter-regional basis but also on an intra-regional basis (that is, within a city). Similarly, QOUL will be spatially highly variable. This is an indication of the Tiebout principle at work (1956), where residents relocate to maximize their satisfaction with public goods by matching their individual preferences and circumstances with the substantial spatial variations in institutional capacity and capability as is clearly seen in the differential performance of local government across urban space and the way in which property taxes are capitalized (Logan and Molotch 1987; Ladd 1994; Brasington 2002).

Many such studies on urban amenity have used *hedonic analysis* to model the effect of location-specific amenities on wages and housing prices, while controlling for effects such as education and race (see, for example, Rosen 1974, 1979; Roback 1982, 1988). And there has also been focus on investigating revealed preferences (see, for example, Kahn 1995). Much of that modeling has operated at the intra-urban scale.

Thus, there is now a vast literature on urban amenity which incorporated a variety of research approaches some of which are explicitly linked to the study of QOL and QOUL, while in others, the link is more implicit than explicit. A more detailed discussion is provided in Chap. 5, in which models dealing with urban amenity – mainly in the USA – are presented.

*Benchmarking* national, state and regional "performance" has almost become an industry in itself as there is an increasing concern, particularly among local officials, to see how their jurisdictions "rate" against other places, often those places seen as being "competitors." In 1989, the State of Oregon in the USA began a statewide performance measurement initiative called *Benchmarking Oregon*, which attempted to benchmark its performance on measures related to economic, social and environmental "livability" on phenomena as diverse as the educational attainment of Oregonians, income levels and much exercise people were getting every week.

One of the best known of these types of studies is the *Places Rated Almanac* (in the U.S. and Canada) which ranks more than 350 cities across nine general categories and a variety of sub-categories (Savageau and D'Agostino 2000).

It has thus now become common place for states, cities and regions to rate themselves on so-called livability indicators, and it has been estimated by the on-line www.governing.com (the resources for States and Localities) that there are now more than 170 efforts going on at the state, regional and local levels across the USA. In some ways, such efforts might be regarded as exercises in "feelgood," but nonetheless they seem to be taken seriously by public officials, and when the performance of a place is favorable, are widely used in city marketing and promotion campaigns.

Similarly, there has been an increasing interest in rating cities according to their QOL/"liveability"/"amenity." This is reflected in the emergence in recent years of a number of "city rating" ranking studied published by private firms including media outlets (such as Mercer, *Monocle Magazine*, and *The Economist* Intelligence Unit). A discussion of those approached to the investigation of QOUL is provided in Chap. 2.

## Environmental Quality of Life

There has, of course, been a long interest among researchers in issues to do with environmental quality in the context of urbanization and city planning and how to deal with urban problems (see, for example, Schmandt and Bloomberg 1969).

One of the earlier books dealing specifically with the quality of the urban environment and its effect on QOL was a set of papers in a book edited by Harvey Perloff (1969) in which a diverse range of issues were canvassed, including pollution, open space, amenity resources and transport. In addition, the relationships between the environmental factors and the role of location, size and shape of cities were addressed. In the introduction to that book, Perloff (1969) wrote this:

> ... The current interest in the quality of the urban environment is in large part a convergence of two other evolving public concerns. One is a concern with the quality of the natural environment – the quality of air, water, land, wilderness areas, and other resources. The other is a concern with the development of our urban communities – with all the matters coming under the rubric of more traditional city planning, but recently refocused to a special concern for the human beings in the city. The quality of life of all the people who are clustering into urban communities is clearly influenced by what happens to both the natural and the man-made environments in direct interrelationship with each other. (p. 3)

Of particular significance was a concern about the possible arrays of environmental elements and the trade-offs that give rise to the urban-environment system and the implications for policy decision-making.

Much of the focus of research on urban environmental quality has been on pollution (mainly air and water quality), the costs of pollution, pollution as a *disamenity*, and abatement measures (see, for example, Gleaser 1998; Smith and Huang 1995; Kahn 2002). And some research has investigated the relationship between environmental quality and a range of socioeconomic attributes and other attributes, such as political activism (see Millimet and Slottje 2002).

Concerted attempts to address environmental quality also have been evident in the area of land-use planning and zoning through mechanisms to protect, restrict and exclude (see, for example, Knaap 1998; Mayer and Somerville 2000; Pendall 2000; and Malpezzi 2002). However, spillover effects may occur as often poorer people and some firms are displaced to less-restricted localities (Landis 1986; 1992). Through the differential policies and actions of local governments, for example, there can thus be generated considerable differentials in QOUL across large urban regions. At the metropolitan regional level of scale, planning interventions – such as the imposition of an urban growth boundary and other measures aimed at restricting urban sprawl – can also have spillover effects and thus change amenity and perceived QOUL. Knaap (1985) has found that these may influence expectations about when land will be developed, thus distorting price gradients and the like. Metropolitan urban planning policy interventions like these are now widespread and may also lead to supply-restricted increases in property prices detrimentally affecting housing affordability and helping to create stress on the QOL, especially of lower income households.

In more recent times, with the increasing interest in and concern for issues to do with the environment, sustainable development, and the challenges of climate change, it is not surprising that there has emerged a considerable literature on environmental quality of life [Rehdanz and Maddison 2008; Schaffer and Vollmer 2010; Westaway 2009].

This increasing emphasis on environmental QOL is seen in public policy and planning responses as demonstrated, for example, in the assertion by Choonyong (2008) in the Korean Research Institute for Human Settlements publication *KRIHS Gazette* that:

> ... paradigm shift is needed in the policy for road space towards prioritizing human beings, environmental quality of life, as well as public transport and walking.

### *Summing up*

In the conclusion to their comprehensive review of the multidisciplinary literature addressing the complex relationships that exist between the urban environment and QOL, Mulligan et al. (2004) had this to say:

> … Hedonic models have been emphasized but other perspectives have been included. Natural amenities like climate and topography remain important in household migration and are partially responsible for the high housing costs of some cities. However, fiscal prudence, cultural and lifestyle tolerance, and the responsible management of key human-made amenities - especially crime, education, and land use - are increasingly seen as being critical for the continued success of cities. In order to be competitive in a global, high-tech economy, firms must be able to attract high human-capital workers. But these people prefer to live in large cities with broad QOL appeal or smaller places with specific QOL appeal. These same persons avoid areas of high crime, locally if not regionally, and they want their children to be educated in high-quality school districts. Housing costs are bid up accordingly and high taxes ensure the provision of high-quality public goods and services. Especially in large urban areas, these same people tolerate a wide diversity of lifestyles and, increasingly, they demand an orderly and aesthetically pleasing urban landscape. With non-interventionist state- and national-level public policies, and political fragmentation in metropolitan areas, existing resource and life-opportunity gaps between the most advantaged and the most deprived will only widen in our largest cities. (p.787)

## The Introduction of Geographic Information Systems

Increasingly Geographic Information Systems (GIS) technology is being used in social research including QOUL studies, and it is certain that this will become more common in the future.

GIS technology has been employed widely by researchers in re-examining the entire issue of accessibility in urban environments to assess how overall proximity to diverse opportunities such as employment, education, shopping, health and recreation might directly affect something such as personal health (Witten et al. 2003). Studies have examined the relationships between health levels and urban lifestyles, assessing – among other things – how transportation infrastructure affects longevity (Handy et al. 2002; Boarnet et al. 2003).

But it has been the use of GIS technology which now permits the integration of survey-based data on subjective QOUL at the level of the individual with spatial objective information about the urban environment where currently, profoundly dramatic innovations are being made in QOUL studies. By geocoding the residential location of respondents to QOL surveys, it is possible to integrate:

- Survey-based information on individuals' attitudes, preferences, behaviors and expectations with respect to QOL domains and of aspects of QOUL at different spatial scale levels
- Spatial objective information on the demographic and socioeconomic characteristics of populations and of housing of local areas derived from census data

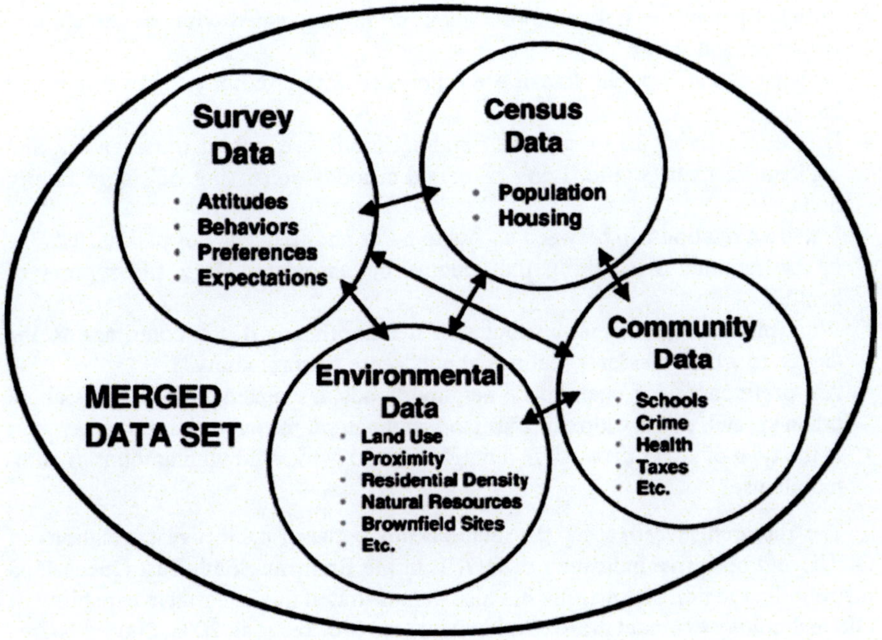


**Fig. 1.5** Using GIS to integrate data sources to generate a merged data set (Source: Marans 2002: p. 7, Figure 5)

- Spatially objective environmental data relating to land use, proximity to urban services and facilities and natural resources like parks, residential density, brownfield sites and noxious industry
- Community data relating to schools, crime, health, taxes, etc.

This integrative capability of GIS is conceptualized in Fig. 1.5 and has been used in a number of QOUL studies recently, including in research that has investigated QOUL in metro Detroit (see Chap. 7) and in the Brisbane-South East Queensland region in Australia (see Chaps. 8 and 18). Data integration gives rise to the possibility of investigating many aspects of the relationships between contextual data and responses to survey questionnaires on QOL using bi-variate analysis and multivariate statistical modeling. For example, an analysis might address a question such as how density (as reflected by multiple density measures) affects peoples' responses to crowding, knowing the names of neighbors, and their interactions with them.

Marans (2002) suggests that this integrative capability potentially is most useful to help researchers address issues that might have policy significance or be useful in helping us to address urban and environmental planning issues. Marans gives the following examples of questions that might be addressed:

- Where do people live who feel negatively about their cities and their neighborhoods?

- To what extent are police reports about crime associated with concerns about neighborhood safety?
- To what extent is public transit use influenced with proximity to bus routes and bus stops?
- Do people living in mixed-use neighborhoods use public transit more and walk more than people living in neighborhoods consisting of single family homes?
- Is there a relationship between access to parks and frequency of park visits?
- Is the amount of walking that people do associated with self-reports of health?
- What physical and social attributes of neighborhoods if any contribute to the degree to which residents feelings about "sense of community"?
- Are preferences for open space neighborhoods associated with feelings about urban sprawl, preservation of farmland, and natural resource conservation?
- Is the type of street network in a neighborhood associated with amount of automobile use?

The research investigating the relationships between subjective evaluations of QOUL and objective indicators of QOUL in the Brisbane-South East Queensland metropolitan region in Australia has also demonstrated this integrative capability of GIS technology to model those relationships (see McCrea et al. 2005, 2006; McCrea 2007). The possibilities for innovation in this regard are also demonstrated in some of the chapters in Part IV of this book.

## Conclusion

This chapter has provided a broad overview of the evolution of approaches to research on QOL and on QOUL. The review has demonstrated that the research is both diverse and complex, incorporating both objective approaches and subjective approaches. Both have evolved over time in terms of theories and methodologies, and in more recent times, the availability of geographic information system (GIS) technology has enhanced the evolution and development of integrated approaches. GIS provides exciting new opportunities for the development of more holistic frameworks for the analysis and modeling of the complexities of QOUL – its nature, characteristics and the complex inter-relationships between the objective and subjective elements that influence overall QOUL and specific QOUL outcomes over time.

There is no doubt that QOL and QOUL studies are experiencing something of a resurgence of interest in contemporary times, driven not only by the research community but also by public policy and the concern in urban governance, planning and management with how to make cities more competitive, achieve sustainable development, and enhance the well-being of residents. Central to those concerns is a whole set of issues that can be informed through QOUL studies.

## References


Andelman, R., Board, R., Carman, L., Cummins, R., Ferriss, A., Friedman, P., et al. (1998). *Quality of life definition and terminology: A discussion document from the International Society of Quality of Life Studies, (Monograph)*. Blacksburg: International Society of Quality of Life Studies.

Bartik, T. J., & Smith, V. K. (1987). Urban amenities and public policy. In E. Mills (Ed.), *Handbook of regional and urban economics: Vol. 2. Urban economics* (pp. 1207–1254). Amsterdam: North-Holland.

Bastian, C., McLeod, D., Germino, M., Reiners, W., & Blasko, B. (2002). Environmental amenities and agricultural land values: A hedonic model using geographic information systems data. *Ecological Economics, 40*, 337–349.

Berry, B. J. L., & Kasarda, J. D. (1977). *Contemporary urban ecology*. New York: Macmillan.

Blakely, E. J. (1994). *Planning local economic development: Theory and practice* (2nd ed.). Sherman Oaks: Sage Publications.

Boarnet, M. G., Greenwald, M. and McMillan, T. E. (2003). *Planning and health promotion: quantifying links between land use, walking, and physical activity*. Unpublished Manuscript, pp. 39.

Bradburn, N. M. (1969). *The structure of psychological well-being*. Chicago: Aldine.

Bradburn, N. M., & Caplowitz, D. (1965). *Reports on happiness*. Chicago: Aldine.

Brasington, D. M. (2002). Edge versus center: Finding common ground in the capitalization debate. *Journal of Urban Economics, 52*, 524–541.

Brotchie, J., Newton, P., Hall, P., & Nijkamp, P. (1985). *The future of urban form*. London: Coom Helm.

Campbell, A., Converse, R., & Rodgers, W. (1976a). *The quality of American life: Perceptions, evaluations and satisfactions*. New York: Russell Sage Foundation.

Campbell, A., Converse, P., Rodgers, W., & Marans, R. W. (1976b). The residential environment. In A. Campbell, P. Converse, & W. Rodgers (Eds.), *The quality of American life: Perceptions, evaluations and satisfactions* (pp. 217–266). New York: Russell Sage Foundation.

Cheshire, P., Carbonaro, G., & Hay, D. (1986). Problems of urban decline and growth in EEC countries. *Urban Studies, 23*, 131–150.

Choonyong, Y. (2008). Space and environment. *KRIHS Gazette Korean Research Institute for Human Settlements, 32*, 12–14.

Connerly, C., & Marans, R. W. (1985). Comparing global measures of neighborhood quality. *Social Indicators Research, 45*, 29–47.

Connerly, C., & Marans, R. W. (1988). Neighborhood quality: A description and analysis of indicators. In E. Huttman & W. Vleit (Eds.), *The US handbook on housing and built environment* (pp. 37–62). Westwood: Greenwood Press.

Coppock, J. T., & Wilson, C. B. (Eds.). (1974). *Environmental quality: With emphasis on urban problems*. New York: A Halsted Press Book, Wiley.

Cummins, R. A., Eckersley, R., van Vugt, J., & Misajon, R. (2003). developing a national index of subjective wellbeing: The Australian wellbeing unity index. *Social Indicators Research, 63*, 159–190.

Dahmann, D. C. (1985). Assessments of neighborhood quality in metropolitan America. *Urban Affairs Quarterly, 20*(4), 511–535.

Diener, E., & Biswas-Diener, R. (2008). *Happiness: Unlocking the mysteries of psychological wealth*. Malden: Blackwell Publishing.

Diener, E., & Suh, E. M. (Eds.). (2000). *Culture and subjective well-being*. Cambridge: The MIT Press.

Dissart, J.-C., & Deller, S. C. (2000). Quality of life in the planning literature. *Journal of Planning Literature, 1*, 36–61.

Duncan, O. D. (1969). *Towards social reporting: Next steps*. New York: Russell Sage Foundation.

Eid, M. (2007). *The science of subjective wellbeing*. New York: Guilford Publications.

Esparza, A., & Carruthers, J. (2000). Land use planning and exurbanization in the rural Mountain West: Evidence from Arizona. *Journal of Planning Education and Research, 20*, 26–39.

Florida, R. (2002). *The rise of the creative class; and how it's transforming work, leisure, and everyday life*. New York: Basic Books.

Florida, R. (2008). *Who's your city? How the creative economy is making where to live the most important decision of your life*. New York: Basic Books.

Geoghegan, J., Wainger, L., & Bockstael, N. (1997). Spatial landscape indices in a hedonic framework: An ecological economics analysis using GIS. *Ecological Economics, 23*, 251–264.

Gilbert, D. (2005). *Stumbling on happiness*. New York: Vintage Books.

Glaeser, E. L. (1998). Are cities dying? *Journal of Economic Perspectives, 12*, 139–160.

Glaeser, E., Kolko, J., & Saiz, A. (2000). *Consumer city* (Working Paper 7790). Cambridge: National Bureau of Economic Research.

Golledge, R. G., & Stimson, R. J. (1987). *Analytical behavioural geography*. London: Croom Helm.

Grayson, L., & Young, K. (1994). *Quality of life in cities*. London: British Library.

Gyourko, J., & Tracy, J. (1991). The structure of local public-finance and the quality-of-life. *Journal of Political Economy, 99*(4), 774–806.

Gyourko, J., Kahn, M., & Tracy, J. (1999). Quality of life and environmental conditions. In P. Cheshire & E. S. Mills (Eds.), *Handbook of regional and urban economics: Vol. 3. Applied urban economics* (pp. 1413–1454). Amsterdam: North-Holland.

Hadden, J. K., & Borgatta, E. F. (1965). *American cities: Their social characteristics*. Chicago: Rand McNally.

Handy, S., Boarnet, M. G., Ewing, R., & Killingsworth, R. E. (2002). How the built environment influences physical activity: Views from urban planning. *American Journal of Preventative Medicine, 23*, 64–73.

Hardie, I., Narayan, T., & Gardner, B. (2001). The joint influence of agricultural and nonfarm factors on real estate values: An application to the Mid-Atlantic region. *American Journal of Agricultural Economics, 83*, 120–132.

Hsieh, C. (2003). Counting importance: The case of life satisfaction and relative domain importance. *Social Indicators Research, 61*, 227–240.

Innes de Neufville, J. (1975). *Social indicators and public policy: Interactive processes of design and application*. Amsterdam: Elsevier.

Jeffres, L. W., & Dobos, J. (1995). Separating peoples satisfaction with life and public perceptions of the quality-of-life in the environment. *Social Indicators Research, 34*(2), 181–211.

Johnston, R., Opaluch, J., Grigalunas, T., & Mazzotta, M. (2001). Estimating amenity benefits of coastal farmland. *Growth and Change, 32*, 305–325.

Kahn, M. (1995). A revealed preference approach to ranking city quality of life. *Journal of Urban Economics, 38*, 221–235.

Kahn, M. (2002). Smog reduction's impact on California county growth. *Journal of Regional Science, 40*, 565–582.

Kahneman, D., Deiner, D., & Schwartz, N. (Eds.). (1999). *Well-being: The foundations of hedonic psychology*. New York: Russell Sage Foundation.

Keeble, D. (1990). Small firms, new firms, and uneven regional development in the United Kingdom. *Area, 22*, 234–245.

Kemp, D., Manicaros, M., Mullins, P., Simpson, R., Stimson, R., & Western, J. (1997). *Urban metabolism: A framework for evaluating the viability, livability and sustainability of South East Queensland*. Brisbane: The Australian Housing and Urban Research Institute.

Knaap, G. (1985). The price effects of urban growth boundaries in metropolitan Portland, Oregon. *Land Economics, 61*, 26–35.

Knaap, G. (1998). The determinants of residential property values: Implications for metropolitan planning. *Journal of Planning Literature, 12*, 267–282.

Ladd, H. F. (1994). Fiscal impacts of local population growth: A conceptual and empirical analysis. *Regional Science and Urban Economics, 24*, 661–686.

Lambiri, D., Biagi, B., & Rpyuela, V. (2007). Quality of life in the economic and urban economic literature. *Social Indicators Research, 84*, 1–25.

Landis, J. (1986). Land regulation and the price of new housing: Lessons from three California cities. *Journal of the American Planning Association, 52*, 9–21.

Landis, J. (1992). Do growth controls work? A new assessment. *Journal of the American Planning Association, 58*, 489–508.

Lee, T., & Marans, R. W. (1980). Subjective and objective indicators; scale discordance and interrelationships. *Social Indicators Research, 6*, 47–64.

Ley, D. (1996). *The new middle class and the remaking of the central city*. Oxford: Oxford University Press.

Liaw, K. L., Frey, W. H., & Lin, J. P. (2002). Location of adult children as an attraction for black and white elderly primary migrants in the United States. *Environment and Planning A, 34*(2), 191–216.

Liu, B. C. (1976). *Quality of life indicators in U.S. metropolitan areas: A statistical analysis*. New York: Praeger.

Logan, J., & Molotch, H. (1987). *Urban fortunes: The political economy of place*. Berkeley: University of California Press.

Lu, M. (1999). Determinants of residential satisfaction: Ordered logit vs. regression models. *Growth and Change, 30*(spring), 264–287.

Lyubomirsky, S. (2008). *The how of happiness: A scientific approach to getting the life you want*. New York: The Penguin Press.

Mahan, B., Polasky, S., & Adams, R. (2000). Valuing urban wetlands: A property price approach. *Land Economics, 76*, 110–113.

Malpezzi, S. (2002). Urban regulation, the "new economy", and housing prices. *Housing Policy Debate, 13*, 323–349.

Mathur, V. (1999). Human capital-based strategy for regional economic development. *Economic Development Quarterly, 13*, 203–216.

Marans, R. W. (2002). Understanding environmental quality through quality of life studies: The 2001 DAS and its use of subjective and objective indicators. *Landscape and Urban Planning, 991*, 1–11.

Marans, R. W. (2005). Modeling residential quality using subjective and objective measures. In D. C. Verstbro, Y. Hurol, & N. Wilkinson (Eds.), *Methods in housing research* (pp. 314–328). Newcastle-Upon-Tyne: Urban International Press.

Marans, R. W., & Mohai, P. (1991). Leisure resources, recreation activity, and the quality of life. In B. L. Driver, P. Brown, & G. L. Peterson (Eds.), *The benefits of leisure* (pp. 351–363). State College, PA.: Venture Publishing.

Marans, R. W., & Rodgers, W. (1975). Towards an understanding of community satisfaction. In A. Hawley & V. Rock (Eds.), *Metropolitan America in contemporary perspective* (pp. 299–352). New York: Halsted Press.

Mayer, C., & Somerville, T. (2000). Land use regulation and new construction. *Regional Science and Urban Economics, 30*, 639–662.

McCrea, R. (2007) *Urban quality of life: Linking objective dimensions and subjective evaluations of the urban environment*. Unpublished PhD thesis, The University of Queensland, Brisbane.

McCrea, R., Stimson, R. J., & Western, J. (2005). Testing a general model of satisfaction with urban living using data for South East Queensland. *Australia, Social Indicators Research, 72*, 121–152.

McCrea, R., Shyy, T.-K., & Stimson, R. (2006). What is the strength of the link between objective and subjective indicator of urban quality of life? *Applied Research in Quality of Life, 1*, 79–96.

Millimet, D. L., & Slottje, D. (2002). Environmental compliance costs and the distribution of emissions in the U.S. *Journal of Regional Science, 42*, 87–105.

Mulligan, G., Carruthers, J., & Cahill, M. (2004). Urban quality of life and public policy: A survey. In R. Capello & P. Nijkamp (Eds.), *Advances in urban economics* (pp. 729–802). Amsterdam: Elsevier Science B.

Orford, S. (1999). *Valuing the built environment: GIS and house price analysis*. London: Ashgate.

Pendall, R. (2000). Local land use regulation and the chain of exclusion. *Journal of the American Planning Association, 66*, 125–142.

Perloff, H. S. (Ed.). (1969). *The quality of the urban environment: Essays on "New Resources" in an urban age*. Baltimore: Resources for the Future, The Johns Hopkins Press.

Putnam, R. (1993). *The prosperous community: Social capital and public life* (pp. 35–42). Spring: The American Prospect.

Rehdanz, K., & Maddison, D. (2008). Local environmental quality and life-satisfaction in Germany. *Ecological Economics, 64*, 787–797.

Roback, J. (1982). Wages, rents, and the quality of life. *Journal of Political Economy, 90*, 1257–1278.

Roback, J. (1988). Wages, rents, and amenities: Differences among workers and regions. *Economic Inquiry, 26*, 23–41.

Roberts, B., & Stimson, R. J. (1998). Multi-sectoral qualitative analysis: a tool for assessing the competitiveness of regions and formulating strategies for economic development. *The Annals of Regional Science, 32*, 469–494.

Rogerson, R. (1999). Quality of life and city competitiveness. *Urban Studies, 36*(5–6), 319–328.

Rosen, S. (1974). Hedonic prices and implicit markets: Product differentiation in pure competition. *Journal of Political Economy, 82*, 34–55.

Rosen, S. (1979). Wage-based indexes of urban quality of life. In P. Mieszkowski & M. Straszheim (Eds.), *Current issues in urban economics* (pp. 74–104). Baltimore: Johns Hopkins Press.

Rossi, P. H. (1972). Community social indicators. In A. Campbell & P. E. Converse (Eds.), *The human meaning of social change*. New York: Russell Sage Foundation.

Savageau, D., & D'Agostino, R. (2000). *Places rated almanac*. New York: Hungry Minds.

Schaffer, D., & Vollmer (Rapporteurs), D. (2010). *Pathways to urban sustainability: Research and development on urban systems*. Washington DC: National Academy Press.

Schmandt, H. J., & Bloomberg, W. M. (Eds.). (1969). *The quality of urban life*. Beverley Hills.: Sage Publications.

Schwarz, N., & Strack, F. (1999). Reports of subjective well-being: Judgmental processes and their methodological implications. In D. Kahneman, E. Diener, & N. Schawrz (Eds.), *Well-being: The foundations of hedonic psychology*. New York: Russell Sage Foundation.

Schwirian, K. P., Nelson, A. L., & Schwirian, P. M. (1995). Modeling urbanism: Economic, social and environmental-stress in cities. *Social Indicators Research, 35*(2), 201–223.

Sirgy, M. J., Rahtz, D. R., Cicic, M., & Underwood, R. (2000). A method for assessing residents' satisfaction with community-based services: A quality-of-life perspective. *Social Indicators Research, 49*(3), 279–316.

Smith, D. M. (1973). *The geography of social well-being in the United States: An introduction to territorial social indicators*. New York: McGraw Hill.

Smith, D. M. (1977). *Human geography: A welfare approach*. London: Edward Arnold.

Smith, D. M. (1979). *Where the grass is greener: Living in an unequal world*. London: Penguin Books.

Smith, D. M. (1994). *Geography and social justice*. Cambridge: Blackwell.

Smith, V. K., & Huang, J. C. (1995). Can hedonic models value air quality? A meta-analysis. *Journal of Political Economy, 103*, 209–227.

Smith, K., Poulos, C., & Kim, H. (2002). Treating open space as an urban amenity. *Resource and Energy Economics, 24*, 109–119.

Thaler, R. H., & Sunstein, C. (2008). *Nudge: Improving decisions about health, wealth, and happiness*. New Haven: Yale University Press.

Thorndike, E. L. (1939). *Your city*. New York: Harcourt, Brace and Company.

Tiebout, C. (1956). A pure theory of local expenditures. *The Journal of Political Economy, 64*(5), 416–424.

van Praag, B. (2004). *Happiness quantified: A satisfaction calculus approach*. Oxford: Oxford University Press.

Warr, P. B. (1987). *Work, unemployment, and mental health*. Oxford: Oxford University Press.
Warr, P. B. (1999). Well-being and the workplace. In D. Kahneman, E. Diener, & N. Schawrz (Eds.), *Well-being: The foundations of hedonic psychology* (pp. 392–412). New York: Russell Sage Foundation.
Weiner, E. (2008). *The geography of bliss*. New York: Twelve (Hatchett Book Group).
Westaway, M. S. (2009). Aspects of environmental quality of life that affect neighbourhood satisfaction in disadvantaged and advantaged Johannesburg communities. *Development Southern Africa, 26*, 447–458.
White, N. P. (Ed.). (2006). *A brief history of happiness*. Malden: Blackwell Publishing.
Witten, K., Exeter, D., & Field, A. (2003). The quality of urban environments: Mapping variation in access to community resources. *Urban Studies, 40*, 161–177.
Wong, C. (2006). *What is local development? A conceptual overview* (*Occasional Paper 49*). Manchester: Department of Planning and Landscape, University of Manchester.
Zehner, R. (1977). *Indicators of quality of life in new communities*. Cambridge: Ballinger Publishing Co.

# Part I
# Approaches to the Study of Quality of Life and Quality of Urban Life

# Chapter 2
# Objective Measurement of Quality of Life Using Secondary Data Analysis

**Robert Stimson and Robert W. Marans**

## Introduction

The literature is replete with attempts to measure and analyze quality of life (QOL), although there is no single model, nor a comprehensive set of measures, that is widely accepted by researchers and policy makers. There are many debates around how best to assess QOL or its reflection in various aspects of daily life.

As discussed in Chap. 1, approaches to measure QOL are typically either:

- *Objective*, based on secondary analysis of data such as that derived from official statistical collections such as the census; or
- *Subjective*, based on primary data collected through sample surveys in which people's perceptions of quality of life "domains" are measured on scaled attributes relating to those QOL domains.

In discussing *objective* QOL, Diener and Suh (1997) have had this to say:

> … [objective QOL] reflects objective circumstances in a given cultural or geographic unit… (and) are based on objective, quantitative statistics. (p. 192)

For example, a city might use various health measures, crime statistics, levels of educational attainment, work force participation and the proportion of welfare recipients in a given area as indicators of its objective quality of urban life (QOUL).

---

R. Stimson (✉)
Australian Urban Research Infrastructure Network (AURIN), Faculty of Architecture, Building and Planning, University of Melbourne, VIC 3010, Melbourne, Australia
e-mail: rstimson@unimelb.edu.au

R.W. Marans
Institute for Social Research, University of Michigan, POB 1248, Ann Arbor, MI 48106, USA
e-mail: marans@umich.edu

R.W. Marans and R. Stimson (eds.), *Investigating Quality of Urban Life: Theory, Methods, and Empirical Research*, Social Indicators Research Series 45,
DOI 10.1007/978-94-007-1742-8_2, 

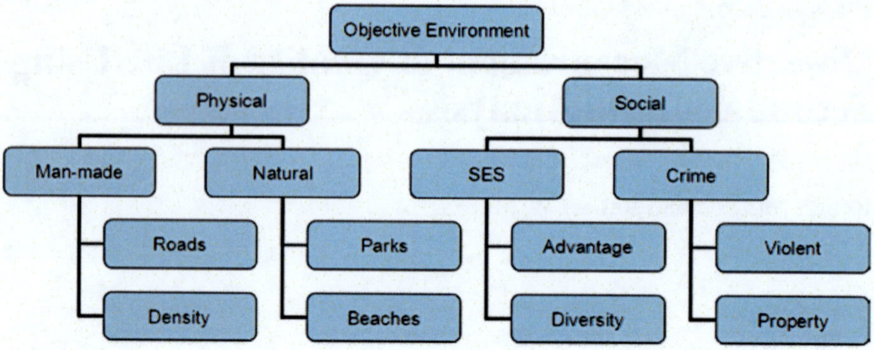


**Fig. 2.1** Types of objective indicators of the urban environment (Source: Provided by Rod McCrea)

In the context of urban environments, it may be useful to think of both *physical* and *social* dimensions, which may affect or be related to the QOL of individuals, or as aggregate level measures that may be indicators of QOUL. Examples of those physical and social dimensions that have been included in the investigation of objective QOUL are represented in Fig. 2.1.

Objective QOUL research incorporates a number of approaches:

(a) Probably the simplest approach has been the *social indicators* approach in which mostly objective indicators of aspects of life in general and of the urban environment in particular are measured and monitored to identify the nature of those phenomena at a point in time and to identify trends over time, particularly with respect to achieving a desired standard of achievement or performance. Occasionally, subjective measures have been included as social indicators.
(b) A common approach for deriving estimates of objective QOL and QOUL has been the *weight* objective indicators of QOL and of the urban environment to rank places by those objective measures. That has sometimes included the use of hedonic models using implicit amenity prices as theoretical weights for amenities based on assumptions of behavior of individuals and households.

Those approaches are discussed in this chapter.

## The Social Indicators Movement

An important approach to the investigation of objective QOL was the *social indicators movement* which became popular from the 1960s and throughout the 1970s. It represented a shift in the concern of public attention away from a consideration of mostly economic phenomena to a consideration of the social state of society as well. Some referred to this as the move to develop a system of "social accounts." An early leader of the social indicators movement (Gross 1966) had this to say:

> … Economic statistics, as a whole, emphasize the monetary value of goods and services. By doing so they tend to discriminate against nonmonetary values and against public services for which costs invariably serve as surrogates for output value. Because figures on health and life expectancy are not in national accounts, progress in those areas may be seriously ignored, either in formulating goals or in evaluating performance.

That shift coincided with debate in the US on policies for the "Great Society." And, it reflected an increasing tendency for governments to produce regular "social reports" at the national, state and city levels of scale.

## *Evolution and Definition*

The social indicators movement may well have had its origins in the 1929 Hoover Presidency's Committee on Social Trends in the USA and its report *Recent Trends in the United States* (1933). But as discussed by David Smith (1973) in his book *The Geography of Social Well-Being in the United States: An Introduction to Territorial Social Indicators*, the movement really took off between 1959 and 1966 with the publications *HEW Indicators* and *HEW Trends* by the US Department of Health, Education and Welfare.

That was followed by the publication of four important books:

- A study sponsored by NASA in the USA by Bauer (1967)
- A study by Gross (1966) for the Tavistock Institute in London
- A Russel Sage Foundation study by Sheldon and Moore (1968)
- A synopsis of social goals and indicators published in the *Annals of the American Academy of Political and Social Science* (Gross 1967).

During the presidency of Lyndon B. Johnson in the USA, there was also the publication by the HEW of the landmark report *Toward a Social Report* from the government, which was an attempt to produce a social equivalent of the annual Economic Reports. The Nixon administration that followed then committed to an annual reporting on social goals and indicators.

In the U.K., there was also a parallel succession of government reports on social indicators, including the *Social Trends* report (General Statistical Office 1970) and books by Moser (1970) and Scheonfield and Shaw (1972). And from the mid-1960s, there was a succession of scholarly papers on social indicators by sociologists, political scientists and economists (for example, the work by Agoe 1970; McVeigh 1971).

An important question is *What is a social indicator?* A simple definition was that provided in the US Department of Health, Education and Welfare (1969) in its *Toward a Social Report*, namely:

> … A social indicator, as the term is used here, may be defined to be a statistic of direct normative interest which facilitated concise, comprehensive and balanced judgement about the condition of major aspects of society. It is in all cases a direct measure of welfare and is

subject to the interpretation that, if it changes in the 'right' direction, while other things remain equal things have gotten better, or people are 'better off'. (p. 97)

Another definition offered by Perloff (1969) suggested that a social indicator was:

... normally used to describe the condition of a single element, factor, or the like, which is part of a complex interrelated system. (p. 20)

Further, Land (1970) said that a social indicator is:

a component (i.e., a parameter or variable) in a social system or some segment thereof. (p.35)

And Culyer et al. (1972) made the distinction between indicators of the present condition or state, indicators of the gap between the existing and the desired state (what might be describes as "need") and indicators of the effectiveness of programs designed to close the gap.

Smith (1973) suggested that ideally social indicators should achieve the following:

.. (1) measure the state of and changes over time in (2) major aspects or dimensions of (3) social conditions that can be judges normatively, as (4) part of a comprehensive and inter-related set of measures embedded in a social model, and (5) their compilation and use should be related to public policy goals. (p. 54)

Embedded in that approach was the notion that through appropriate interventions it would be possible to improve the state of things for the benefit of the people and hence the link between social indicators movement and the incorporation of objectives related to equity, equality and social justice into public policy. As pointed out by Hoffenberg (1970), that required a broad consensus on what those goals and objectives meant; and that was a difficult task. Added to that was the problem of how to address the task of *measurement*, which involves the assignment of numbers to the variables that operationally define the concepts which social indicators are meant to shed light on. As Smith pointed out (1973: pp. 59–61), in operationalizing social indicators, it was common to run into problems of data availability, data inaccuracies and coverage deficiencies, data validity and reliability and the incompatibility of different sources of data.

## Territorial Social Indicators

As discussed by David Smith (1973), the social indicators movement also incorporated an explicitly spatial perspective through:

... the geographic notion of social well-being as a condition with areal variations. (p. 63)

Hence, the term territorial social indicator has been used to refer to approaches that explicitly subsume what Smith had referred to as:

... the concepts of 'local', 'regional', 'metropolitan', and 'urban' indicators. (p. 63)

Indeed, the idea of *territorial social indicators* involving spatially disaggregated analysis was part of the social indicators movement from the beginning. For example, in the USA, Gross (1965a, b) had suggested that an annual social reporting by the President could have incorporated reporting on the performance of states and cities. The Department of Health, Education and Welfare initiated inquiry into metropolitan regions, and The Urban Institute (1971) began to report on urban indicators. A study in Michigan by Pearle (1970) was one of the first examples of using territorial social indicators to describe patterns of social well-being at a more spatially disaggregated level. As discussed by Wilson (1969), the development of interest in territorial social indicators in the USA was encouraged because of the more dominant roles local and state governments play and the immediate impact they can have on peoples' QOL and especially on the QOUL.

Similarly, in the U.K., there was an increasing interest in territorial social indicators, and from the early 1970s in its *Social Trends* report, the Central Statistical Office (1970) began to map regional disparities in aspects of social conditions across the U.K. highlighting areal concentrations of special social significance.

Not surprisingly, the task of deciding the level of spatial disaggregation (what Smith 1973: p. 65 referred to as "territorial division") to use in addressing territorial social indicators was not straight forward and inevitably involved compromise. On the one hand, in the USA, there was the macro-level approach focusing on conditions and trends in broad regions such as States, State Economic Regions and Standard Metropolitan Statistical Areas (SMSAs). On the other hand, the micro-level approach would focus on investigating socially significant territorial units by combining census enumeration districts, blocks or tracts. As Smith (1973) pointed out:

> … ultimately there should be some correspondence between territories defined for social reporting and those used for public policy implementation. (p. 66)

But whatever the scale used, Smith (1973: p. 66) indicated that difficulties would arise because of the problem of the "ecological fallacy." That arises because of the difficulty in attributing aggregate characteristics of spatial units to individuals or groups of people living in them. There also were difficulties to do with autocorrelation which will complicate statistical analysis of spatial data. And if the data being used were compiled from sample surveys, there could be significant inaccuracies in using that data for sub-national, smaller geographic areas because of sample coverage deficiencies across the disaggregated spatial units.

## *Summary*

In some ways, the social indicators approach was perhaps the simplest approach to the study of objective QOL and QOUL. It used mainly objective indicators based on secondary data relating to aspects of the urban environment, including such things as employment, housing, health and life expectancy, pollution, traffic flows, and so forth. Those indicators were monitored either separately or sometimes collectively through the use of multivariate statistical analysis, the main purpose being to provide both a cross-sectional snapshot and/or analyzing trends over time (see, for example, Cicerchia 1996; D'Andrea 1998; Perz 2000). While there have been calls to include subjective indictors on an equal basis as objective indicators (see, for example, Cutter 1985; Diener and Suh 1997; Santos and Martins 2007), the social indicators approach has been based primarily on objective measures.

## Operationalizing the Measurement of Social Well-Being Through Territorial Social Indicators

It has not been easy to get a consensus on what to include in social indicators and in territorial social indicator studies which aimed at measuring aspects of social well-being. There has been no accepted or uniformly adopted model from which a "correct" set of variables could be derived and then be used to measure the nature of and the changes in social conditions that needed to be improved.

As discussed by Smith (1973):

> … the concept of social well-being is sometimes thought of as synonymous with the quality of life. But it may be preferable to regard it as being at the more concrete or specific end of the continuum of abstraction that descends from human happiness through the concept of the quality of life to social well-being. "Quality of life" implies a rather personalized concept, whereas reference to aggregates of people defined by area of residence more appropriately addresses the welfare of some social group. (p. 66)

Smith went on to suggest that:

> … any specific operational definition of the concept of social well-being ought eventually relate to human happiness or the capacity of individuals to realize their perception of the good life, for this is the ultimate criterion for determining whether a society is well or sick. (p. 67)

Indeed, Stagner (1970) had proposed a set of "psychological urban indicators" focusing on the frequency or level of perceived satisfaction/dissatisfaction people had with aspects of urban file, which of course involved the collection of information through survey approach. However, studies associated with the social indicators movement almost exclusively used officially collected aggregate level data and, in rare instances, used individual level behavioral and perceptual data collected through surveys.

### *Consensus on Criteria*

There did seem to be a degree of consensus apparent among the proliferation of studies that occurred as the social indicators movement gathered momentum from the 1960s and into the 1970s. In his book, Smith (1973: pp. 66–70) reports the results of a survey he made of the degree to which ten key official studies published in the US and in the U.K. between 1960 and 1970 covered 20 topics as major categories of social well-being or social indicators.

On getting a consensus on what to include, Smith (1973) had the following to say:

(a) There was almost complete agreement on the inclusion of four conditions, namely income and wealth, employment, health, and education.
(b) Then, there was a high degree of agreement on four additional conditions: social status and mobility, public order and safety, the state of the family, and the living environment.
(c) There followed seven conditions that frequently occurred as major items in more than half of the studies, namely science and technology, participation and alienation, leisure and recreation, social disorganization (or social pathologies), the natural environment, access to services, and culture and the arts.
(d) Other topics included in no more than three of the ten studies were: the production of goods and services, demographic characteristics, the political process, the mass media, and religion.

Smith also surveyed text books and journals discussing social indicators and compiled similar lists of topics being investigated. From this "meta-analysis" of the literature, Smith (1973) reached this conclusion:

> In a well society, people will have incomes adequate for their basic needs of food, clothing and shelter, and a 'reasonable' standard of living; people will not live in poverty. The status and dignity of the individual will be respected, and he will be socially and economically mobile. Good quality education and health services will be available to all, and their use will be reflected in a high level of physical and mental health and an informed populace able to perform their societal roles in a satisfactory manner. People will live in decent houses, in decent neighborhoods, and will enjoy a good quality of physical environment. They will have access to recreational facilities, including culture and the arts, and adequate leisure time in which to enjoy these things. Society will show a low degree of disorganization, with few personal social pathologies, little deviant behavior, low crime, and high public order and safety. The family will be a stable institution, with few broken homes. Individuals will be able to participate in social, economic and political life and will not be alienated on the basis of race, religion, ethnic origin, or any other cause. (p. 69)

Smith (1973) acknowledged that this "begs far more questions than it answers" (p. 69) and that almost every word "requires definition, clarification or reservation" (p. 69). But out of that, Smith was able to propose some general criteria for a list of social well-being (see Table 2.1) for which territorial social indicators could be derived at various levels of spatial scale.

## *Measures*

The issue of deciding on appropriate numerical measures of social well-being involved addressing:

- The limitations in sources of official data that are available at particular levels of spatial scale necessary for the construction of territorial social indicators
- The decisions on how those indicators would be measured

**Table 2.1** General criteria of social well-being

| | |
|---|---|
| I. Income, wealth, and employment | V. Social order (and disorganization) |
| i. Income and wealth | i. Personal pathologies |
| ii. Employment status | ii. Family breakdown |
| iii. Income supplements | VI. Social belonging (alienation and participation) |
| II. The living environment | i. Democratic participation |
| i. Housing | ii. Criminal justice |
| ii. The neighborhood | iii. Segregation |
| III. Health | VII. Recreation and leisure |
| i. Physical health | i. Recreation facilities |
| ii. Mental health | ii. Culture and the arts |
| IV. Education | iii. Leisure available |
| i. Achievement | |
| ii. Duration and quality | |

Source: Smith (1973: p. 70)

Typically that involved the choice of an *absolute* or a *relative* indicator measure:

(a) An *absolute indicator* is a "scientifically" established maximum or minimum level for a specified condition, as in a poverty line.
(b) A *relative indicator* has no absolute limit or optima but is simply a measure of the relative position of a territorial unit with respect to the specified condition, such as the percentage of households living below a poverty line. A relative territorial indicator might be benchmarked against the national incidence of the condition through the use of the Location Quotient (LQ) to indicate the extent to which the incidence of the condition in a territorial unit is above or below the national incidence of that condition where the national benchmark is LQ = 1.

## *Modeling*

An important methodological development by Smith (1973: pp. 73–77) was the proposal of a framework by which researchers could undertake simple mathematical modeling of territorial social indicators.

Smith (1973) started with an individual member of society and assumed that the person had a set of expectations about the way life ought to be, or what were a set of needs. The personal set of well-being is determined by the difference between that and a set of perceptions as to the way things ought to be. If $O(1, 2,\ldots., i\ldots. m)$ is the observed state on the various conditions of life, and $E(1, 2,\ldots., i,\ldots. m)$ is the expectations, and if unit linear pay-offs between the individual conditions are assumed, then the overall well-being (*B*) of any individual (j) could be expressed as:

$$B_j = (O_{1j} - E_{1j}) + (O_{2j} - E_{2j}) + \ldots + (O_{mj} - E_{m,j})$$

or

$$B_j = {}_{j=1}\Sigma^m (O_{ij} - E_{ij})$$

For all the individuals (*n*), living in a given territorial unit that aggregated social well-being would be:

$$S = {}_{j=1}\Sigma^n{}_{i=1}\Sigma^m (O_{ij} - E_{ij})$$

to which a term to express the effect of human interaction and the fact that group social well-being may be more or less than the sum of the well-being of its individual members considered in isolation.

Smith (1973) presented this model of the determination of individual well-being in the diagram reproduced in Fig. 2.2 in which some elements of the system within which the levels of well-being are determined. The individual's conditions or observed states are shown to derive from the general state of the social system, and the person's needs or expected state come from personal knowledge of life and whatever concepts are suggested by national ideals of justice, equity or the like. Smith showed that the interactions of individuals and the aggregation of their personal experiences produce a group social well-being. He said that:

> … extending the model into the policy field, the overall level of social well-being, when related to national ideals, may simulate social policy and remedial programs which affect the state of the system and level of the individual. (p. 74)

Of course, to properly operationalize this type of model, it was necessary to collect data through a large scale sample survey to generate the measures of social well-being. The development of the territorial social indicators approach tended to rely heavily on some aggregate measures of the individual or group condition – often using a surrogate variable measure – with respect to the various criteria of well-being that were being included in a study. Thus, as Smith (1973: pp. 74–75) pointed out, the focus has thus been on the *O* in Fig. 2.1 as some territorial measurements on this do exist in national data bases, including census data, whereas without the existence of detailed sample survey data for the territorial unit(s) being studied, not much will be known about individual expectation as a spatial variable. For example, if *O* relates to conditions such as income, education and health, then the levels of *O* may be thought of as outputs of some sub-system within the wider social system.

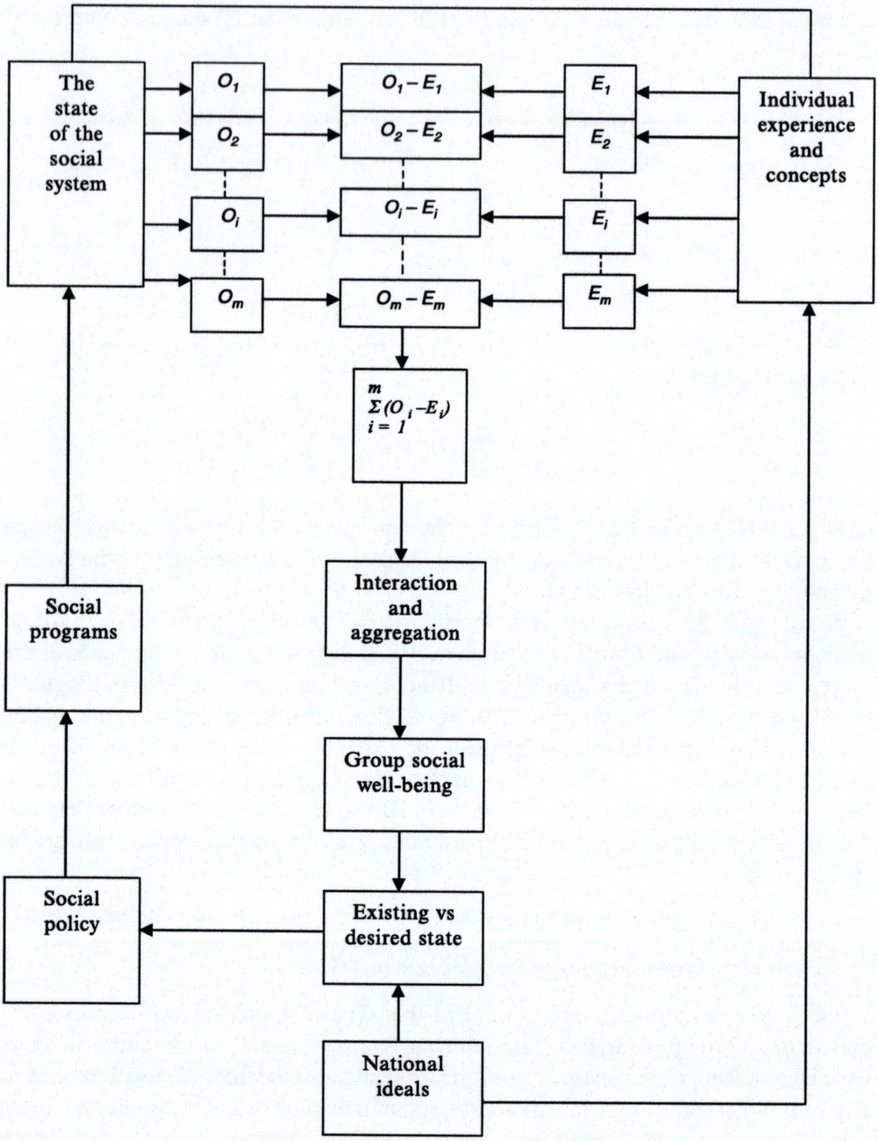


**Fig. 2.2** A model of the determination of individual well-being (Source: Smith 1973: p. 75)

Smith (1973: pp. 75–77) gave the example of education to illustrate such an approach. The outcome of an educational system might be measures by student scores on various tests which, when aggregated for people living in a particular area, might represent a territorial social indicator of social well-being on the assumption that educational attainment is something that matters as providing a means of access to a good life. The actual level of educational attainment ($O_E$) is likely to be some function of the inputs to the system, such as, for example, expenditures on physical plant (P), curriculum development (C) and teacher training (T),

$$O_E = f\left(I_P, I_C, I_T\right).$$

To accommodate the effect of other variables known to influence educational attainment – like family socioeconomic status (*F*) – other terms are added, such as

$$O_E = f\left(I_P, I_C, I_T\right), f(F).$$

This expression is a simplified shorthand expression of the system determining the magnitude of the education indicator of a group of people defined by where they live. The model may then be generalized for a social system (or for sub-systems) in this way:

$$O_i = f\left(I_{i1}, I_{i2}, \ldots I_{ik}\right),$$

where

$O_i$ is the output of the system *i*;

$I_i$ (1, 2,…., *k*) are the inputs to the system, including those that are outputs of other systems and

*f* is "some function of."

The wider system is made up of many of these sub-systems – all of which has inputs in the form of investments and interactions with other sub-systems – which may combine in some functional mechanism to produce a set of outputs. Because the inputs and the functional relationships vary between territories, then so too will the level of outputs. Smith says these social outputs may be thought of as transformed, weighted, and combined to produce the overall territorial level of well-being (*S*). Identifying the magnitude of *S* as it varies spatially then becomes the focus of the research effort. The level of social well-being in any territory may be thought of as some function of the output of the various social sub-systems in that territory, that is

$$S_j = f\left(O_{1j}, O_{2j}, \ldots, O_{mj}\right),$$

where the output measures are reasonable surrogates for the collective individual experiences of people.

That type of heuristic modeling of society was used by the US Office of Education (1969) in a report on master social indicators.

Nowadays, with the evolution of sophisticated micro-simulation techniques, it is possible to merge national sample survey data with small area aggregate data from the census to generate synthetic small-area spatial variables that reflect the behavioral and attitudinal variables that are in the national survey dataset, and this is now a common place approach.

## Examples of Empirical Applications

There are many examples of territorial social indicator studies, particularly in the USA, and the last three chapters of David Smith's (1973) book, *The Geography of Social Well-Being in the United States: An Introduction to Territorial Social Indicators*, provides a detailed overview of studies conducted in the 1960s and early 1970s. In that book, Smith provides details of studies he and other researchers conducted investigating territorial social indicators for US States (see Smith 1972), inter-city analyses for metropolitan cities (see, for example, Jones and Flax 1970; Coughlin 1970) and an intra-city analysis of Tampa, Florida, by Smith and Gray (1972).

Those studies of territorial social indicators were characterized by the inclusion of sets of variables that were measures of the sorts of criteria of social well-being, set out earlier in Table 2.1, with the variables being derived from official sources. Typically the data variables were transformed into some type of standard scores by conversion into rankings or by using *Z* scores. That enabled the territorial unit scores on different variables to be combined by simple addition to develop a Standard Score Additive Model by which a composite social indicator was calculated for each of the major criteria of social well-being. The patterns of territorial unit scores on those composite social indicators could then be mapped to identify places with positive or negative performance or above or below average performance on a particular composite social indicator. Correlation analysis was also used to investigate the ecological associations between territorial scores on pairs of the composite territorial social indicators.

Those studies also featured methodological innovation including the use of the data reduction tools Principal Components Analysis (PCA) or Factor Analysis to identify the underlying dimensions of variance in the spatial data matrices by extracting those generalized dimensions that account for the decreasing amounts of the cumulative variance in the data.

For example, in his study of territorial social indicators across the States of the USA, Smith (1972) used PCA to extract six principal components that cumulatively explained 77.49% of the variance in the territorial social indicators at the State level in the USA. Those components were:

- Component #1: General socioeconomic well-being (38.56% of the variance)
- Component #2: Social pathology (13.74%)
- Component #3: Mental health (11.89%)
- Component #4: Racial discrimination (5.94%)
- Component #5: Public assistance/unionization (3.72%)
- Component #6: Social disruption (3.55%)

Smith then used the State scores on those leading components to map the pattern of spatial variation in those generalized territorial social indicators.

In his study of 31 territorial social indicators across 109 US SMSAs with 250,000 or greater population in 1960, Coughlin (1970) used PCA to identify two leading components of social well-being:

- Component #1 which identified an "affluence" dimension accounting for 27% of the variance
- Component #2 which identified a "crime" dimension accounting for 16.5% of the variance.

Another methodological innovation used in some of the studies was to use cluster analysis to group territorial unit scores on the territorial social indicator components derived from PCA. And, it also became common-place to use correlation analysis and multiple regression analysis to explore the links between territorial unit scores on generalized social well-being dimensions and other characteristics of territorial units, including, for example, population size and growth rates, net migration and the concentration of employment in various industry sectors. For example, in the analysis of 109 SMSAs across the USA, Coughlin (1970) used a step-wise multiple regression analysis to find that:

- On the "affluence" dimension component score, the correlation with population size, non-white population and migration produced an $R^2=0.514$, and the addition of manufacturing growth and population density variable raised it to $R^2=0.555$
- On the "crime" dimension component score, the non-white population, manufacturing employment and population density produces an $R^2=0.604$, and the addition of population size and migration raised it to $R^2=0.685$.

In many cities, the above approaches were employed to develop territorial social indicators. In the case of Tampa, Florida (summarized in Smith 1973), the focus was on intra-urban dimension\s and patterns of social well-being. That city had been selected as one of 20 to participate in a new federal program called "Planned Variation," which was basically a model cities project. In 1967, data had been collected on 21 variables relating to 6 criteria, namely:

- Housing conditions
- Physical conditions
- Health
- Crime and delinquency
- Unemployment
- Welfare services.

Scores on those 21 variables were used to rank census tracts to identify "primary target areas" as with social well-being problems.

Then in 1971, a territorial social indicators project was began assembling a comprehensive set of spatial data variables measuring aspects of six criteria of social well-being:

- Economic status
- Environment
- Health

- Education
- Social disorganization
- Participation and equality.

Standard scores on the variables were calculated for census tracts and summed for each of the six criteria with the mapped patterns of those scores revealing where there were marked spatial concentrations of poor performance on the criteria.

The PCA that followed identified four leading components of social well-being, namely:

- Component #1 identifying "social problems" (explaining 17.8% of the variance)
- Component #2 identifying "socio-economic status" (11.3%)
- Component #3 identifying "racial segregation" (9.5%)
- Component #4 identifying "social deprivation" (8.3%).

Census tract scores on those of leading dimensions of social well-being were mapped, and those patterns were then used to help identify those tracts which were potential Planned Variation program target areas. This approach drew heavily on the methodology that was popular at the time in factorial ecology studies of large cities.

Smith (1973) demonstrated how the Tampa study, which focused on intra-city social conditions, was illustrative of how a spatially disaggregated approach to territorial social indicators might be used in the formulation and implementation of public urban planning policy. That represents what Smith saw as being a particularly important objective of the social indicators movement, namely that

> … territorial social indicators should be developed within an action framework…. [which]… should arise from specific practical planning needs…. and their use should be guided by a philosophy dedicated to change and not to the preservation of an unsatisfactory status quo. (p. 135)

## Investigating Quality of Urban Life by Weighting Objective Indicators

Many studies inquiring into objective QOUL typically have incorporated numerous objective measurements of characteristics of the urban environment, often combining or weighting objective indicators to generate an objective QOUL ranking for places (see, for example, Liu 1975; Boyer and Savageau 1981, 1985, 1989; Cutter 1985; Blomquist et al. 1988; Stover and Leven 1992; Cicerchia 1996; Savageau 2007).

However, the use of weighting systems has often been criticized because of their seemingly *ad hoc* nature and because the place rankings can change markedly by using alternative set of weights (Cutter 1985; Landis and Sawicki, 1988; Rogerson et al. 1989), and that raised questions about the objectivity of "objective" QOUL estimates.

But, there have been significant methodological developments to derive more objective weights for estimating QOUL and to rank it for cities, including using hedonic price equations (see, for example, Blomquist et al. 1988; Stover and Leven 1992).

Some empirical studies of objective QOUL have emphasized the trade-off between positive and negative aspects of urban living. Examples include the following:

(a) Blomquist et al. (1988) modeled the trade-off between housing costs, wages and amenity to develop QOUL indexes for 253 Standard Metropolitan Statistical Areas in the US Hedonic wage and rent equations were used to derive implicit prices for various urban amenities, which in turn were used as weights in compiling an objective QOUL index. The underlying idea was that the "amenity value" of an area is implied from areal variation in housing costs and wages. Such models have used implicit amenity prices as theoretical weights for amenities which, it is argued, is a more objective weighting system. However, those weights have also been criticized because they rely on a range of assumptions that can be challenged. For example, households might maximize their well-being, and markets may accurately reflect a trade-off between land costs, wages and amenity values. Moreover, there is the question about whether the weights can be truly "objective" and, if so, how they are to be derived.

(b) Rogerson et al. (1989) derived a subjective set of weights by taking the average subjective importance of various attributes of the urban environment obtained from a national opinion survey which were then used to weight objective attributes of places to produce a ranked list of QOUL in British cities. In part, that approach recognized the subjective nature of QOUL. However, in averaging importance measures across residents, the derived estimates of QOUL also smooth over individual variations in what residents consider important in the urban environment. Thus any resident may disagree with the rankings of QOUL for those British cities. That limitation is true of all the weighting systems for ranking places regardless of how the weights are derived, and it highlights the ultimately subjective nature of QOUL. Thus, while estimating objective QOUL for places may have particular uses for ranking places and monitoring change in objective QOUL over time, it is important to recognize the ultimately subjective nature of QOUL, and so to try and understand the links between objective indicators and subjective evaluations of the urban environment.

(c) Cicerchia (1996) theorized about the trade-off between city effect and urban load. In that study, city effect related to "access to superior urban functions, opportunities and services" available by virtue of the size of a city. Urban load related to a number of negative consequences of urban growth (for example, congestion and environmental degradation). The underlying idea was to ascertain the "optimum centrality" for a city, which was the size of a city where city effect is less as urban load is maximized. If the city size becomes too large, then an escalating urban load might exceed the city effect and create urban overload.

(d) In another study, Schwirian et al. (1995) were to show how population size and density can contribute to economic, social and environmental load or stress in cities.

## Contemporary Rating of Cities and Their QOL

In recent years, there seems to have been an upsurge of public and corporate interest in the QOL cities around the world, and commercial rating agencies have published lists of cities according to their QOL or "livability." This is usually done by rating or ranking cities according to how they perform on a set of objective measures of QOL usually comprising statistics relating to things such as:

- Average salaries
- Housing costs
- The cost of health care
- School performance
- Crime rates
- Public transport systems
- Planned infrastructure improvements
- Neighborhood diversity
- Access to public parksand the like.

Sometimes such measures might be augmented by estimates of factors such as the following:

- Physical and technological connectivity
- Tolerance
- The strength of local media and culture
- The quality, range and independence of restaurants and retail shopping
- Late night entertainment
- Business investment climate
- Planned infrastructureand so on.

Assessing city performance on such issues is sometimes done by the rating agency soliciting so-called "expert opinion," which might, for example, comprise business and professional expatriates who have worked in cities, as is the case with the Mercer quality of living survey for 215 cities across the world (www.mercer.com/qualityofliving). The survey released in April 2009 ranked Vienna on top, followed by Zurich, Geneva, Vancouver and Auckland. The top Asian city was Singapore at rank #256, and the top US city was Honolulu at rank #29. In the U.K., London was the lead city at rank #38.

Another well-known city rating agency is *The Economist* Intelligence Agency whose June 2009 list ranks Vancouver on top place out of 140 cities, followed by Vienna (www.citymayors.com/environment/eiu_bestcities.html).

Amore recent entry to the city QOL rating game is *Monocole* magazine (www.monocle.com/sections/edits/Web-Articles/Top-25-Cities/) which, in 2007, first published a list of the world's "most 'livable' cities." The ranking of the top 25 most livable cities is reproduced in Table 2.2. About 40 cities are considered. The list in 2009 is topped by Zurich, followed by Copenhagen, Tokyo, Munich and Helsinki. The list in 2008 was topped by Copenhagen, followed by Munich, which had won first rank in 2007. However, there is a tendency for the city rating criteria to change from year-to-year and for the list of cities also to be variable.

**Table 2.2** The Monocle list of the world's 25 "most 'livable' cities"

| Rank 2009 | City | Rank 2008 |
|---|---|---|
| 1 | Zurich | 4 |
| 2 | Copenhagen | 1 |
| 3 | Tokyo | – |
| 4 | Munich | 2 |
| 5 | Helsinki | – |
| 6 | Stockholm | 7 |
| 7 | Vienna | 6 |
| 8 | Paris | 10 |
| 9 | Melbourne | – |
| 10 | Berlin | 14 |
| 11 | Honolulu | 12 |
| 12 | Madrid | 13 |
| 13 | Sydney | 11 |
| 14 | Vancouver | 8 |
| 15 | Barcelona | – |
| 16 | Fukuoka | 17 |
| 17 | Oslo | – |
| 18 | Singapore | 22 |
| 19 | Montreal | 16 |
| 20 | Auckland | – |
| 21 | Amsterdam | 18 |
| 22 | Kyoto | 20 |
| 23 | Hamburg | 21 |
| 24 | Geneva | 23 |
| 25 | Lisbon | 25 |

Source: *Monocle Magazine* as reported by Brule (2009: p. 1)
First time on list: Oslo, Auckland
Dropped off: Minneapolis, Portland

The *Monocle* editor-in-chief, Tyler Brule (2009), wrote that in undertaking the city ratings, in addition to considering criteria such as those just referred to:

> … more broadly, we consider the way in which locals and visitors are able to navigate and use everything from public parks to the local property market. In our view, places with the best quality of life are those with the fewest daily obstructions, allowing residents to be both productive and free of unnecessary stress. (p. 1)

On the performance of Zurich, Brule had this to say:

> … Zurich did move into top spot, thanks to outstanding and still improving public transport, including an expanding tram system and main rail station, ample leisure activities, including 50 museums and excellent restaurants; environmental activism in setting new emissions targets; good business culture, with local authorities offering both advise and low-cost office space; and its airport, which serves 170 destinations and is now in line for a revamp. (p. 1)

The *Monocle* top 25 list is dominated by European cities, but London is not included. Outside Tokyo and Singapore, Asian cities fare poorly, and cities in

North America do not appear except for Vancouver. From Australia and New Zealand, Melbourne ranks #9, Sydney #13 and Auckland #20 in 2009.

Undertaking such city QOL/livability ratings or rankings is, of course, a difficult business and it is not uncommon for such rating agencies to come up with different results; not surprising since they tend to use different indicators, and all too often there is something of a lack of transparency in exactly what measures are being used and in how the rating scores for cities are derived.

There does seem to be some consistency in the results from the various city QOL/livability rating agencies and that is that the top rating cities tend to be medium-sized cities in developed countries, and they offer culture and recreation, have low crime, and have fewer infrastructure problems than do larger cities.

## Conclusion

It will become apparent to the reader that from this review and from Chap. 3 there are fundamental differences in objective and subjective measures of QOUL. That is why it is important (as will be discussed in Chap. 4) to include both types of indicators or measures when examining QOUL for places, particularly when using a social indicators approach discussed earlier in this chapter. Objective indicators measure objective changes in QOUL and indicate whether objective standards have been met, whereas subjective indicators of satisfaction measure the extent to which subjective standards or expectations have been met. Often the problem for researchers is that it is costly to collect subjective measures of QOUL through survey research. In contrast, it is relatively easy and relatively inexpensive to use existing aggregate spatial data sources such as the census to generate measures of objective QOUL.

However, much of the QOL literature recommends the inclusion of subjective evaluations of QOUL alongside the objective indicators of QOUL (see, for example, Cutter 1985; Diener and Suh 1997; Marans 2003; Santos and Martins 2007) since these measures often serve different purposes. McCrea (2007) makes this point:

> … measuring the extent to which expectations have been met is as important to urban planners and policy makers as measuring objective changes over time. For example, subjective evaluations are related to residential relocation (Clark and Ledwith 2006; Lu 1998) and participating in community action (Dahmann 1985). (p. 183)

From the perspective of the integrated explanation of subjective QOUL (to be discussed in Chap. 4), it would also be worthwhile to separately analyze subjective evaluations of those residents recently arriving and those intending to move to assist in understanding the dynamics of subjective QOUL for a region.

It may well be that attempts to rate or rank places (be it cities or particularly neighborhoods within a city) according to their objective QOUL are somewhat futile if different types of residents are attracted to a place or a local area according to what is important to them. For example, Cummins (2000) relates the anecdote of media interviews with residents in Lawrence, Massachusetts, which was at the time rated as having the worst QOL of any place in the USA (Boyer and Savageau 1981). However, residents living there generally reported positive evaluations of their community and an attachment to their town. Cummins explained those positive evaluations of Lawrence at the time in terms of adaptive psychological processes based on a need to maintain self-esteem (Cummins 2000; Cummins and Nistico 2002). In contrast, the integrated explanation of subjective QOUL suggested that the residents of Lawrence chose to live in Lawrence because the town had attributes that were important to them and which met their standards.

Thus, the notion of using an "objective" set of weights to calculate objective QOUL for a place may not be valid. In their study of British cities, Rogerson et al. (1989) partially addressed this problem of the "arbitrariness" of objective weights by recommending that objective attributes of the urban environment be multiplied by the average subjective importance of various attributes for residents. However, that approach still assumed that residents had similar sets of subjective weights across the different British cities, and it was still not valid to assume that residents chose different places to live based on the attributes of places and their subjective importance and standards of comparison.

## References


Agoe, C. (1970). Social indicators: selected readings. *Annals of the Association of American Academy of Political Science, 388*, 127–132.

Bauer, R. A. (Ed.). (1967). *Social indicators*. Cambridge: MIT Press.

Blomquist, G. C., Berger, M. C., & Hoehn, J. P. (1988). New estimates of quality of life in urban areas. *The American Economic Review, 78*(1), 89–107.

Boyer, R., & Savageau, D. (1981). *Places rated almanac*. Chicago: Rand McNally.

Boyer, R., & Savageau, D. (1985). *Places rated almanac*. Chicago: Rand McNally.

Boyer, R., & Savageau, D. (1989). *Places rated almanac*. New York: Prentice-Hall.

Brule, T. (2009). The city of your dreams, *Financial Times*, 'House and home', June 13/14, p. 1.

Cicerchia, A. (1996). Measures of optimal centrality: Indicators of city effect and urban overloading. *Social Indicators Research, 46*, 276–299.

Clark, W. A. V., & Ledwith, V. (2006). Mobility, housing stress, and neighborhood contexts: Evidence from Los Angeles. *Environment and Planning A, 38*(6), 1077–1093.

Coughlin, R. E. (1970). *Goal attainment levels in 101 metropolitan areas* (Discussion Paper No. 41). Philadelphia: Regional Science Research Institute, .

Culyer, A. J., Lavers, R. J., & Williams, A. (1972). Health indicators. In S. Shaw & A. Scheonfield (Eds.), *Social indicators and social policy* (pp. 94–118). London: Heinemann.

Cummins, R. (2000). Objective and subjective quality of life: An interactive model. *Social Indicators Research, 52*(1), 55–72.

Cummins, R., & Nistico, H. (2002). Maintaining life satisfaction: The role of positive cognitive bias. *Journal of Happiness Studies, 3*, 37–69.

Cutter, S. (1985). *Rating places: A geographer's view on quality of life*. Washington, DC: Association of American Geographers.

D'Andrea, S. S. (1998). Italian quality of life. *Social Indicators Research, 44*(1), 5–39.

Dahmann, D. C. (1985). Assessments of neighborhood quality in metropolitan America. *Urban Affairs Quarterly, 20*(4), 511–535.

Diener, E., & Suh, E. (1997). Measuring quality of life: Economic, social and subjective indicators. *Social Indicators Research, 40*(1), 189–216.

General Statistical Office. (1970). *Social Trends*, No. 1. London: Her Majesty's Stationery Office.

Gross, B. M. (1966). *The state of the nation: Social systems accounting*. London: Tavistock Publications.

Gross, B. M. (1967). *Social goals and indicators for American Society*. Philadelphia: American Academy for Political and Social Science.

Hoffenberg, M. (1970). Comments on 'Measuring progress towards social goals: Some possibilities and at national and local levels' [Terelekyj, 1970]. *Management Science, 16*, B779–B783.

Jones, M. V., & Flax, M. J. (1970). *The quality of life in metropolitan Washington, D.C: Some statistical benchmarks*. Washington, DC: The Urban Institute.

Land, K. C. (1970). Social indicators. In R. B. Smith (Ed.), *Social science methods* (pp. 216–232). New York: The Free Press.

Landis, J. D., & Sawicki, D. S. (1988). A planners guide to the places rated almanac. *Journal of the American Planning Association, 54*(3), 336–346.

Liu, B. (1975). *Quality of life indicators in U.S. metropolitan areas, 1970*. Washington: Washington Environmental Research Center.

Lu, M. (1998). Analyzing migration decision making: Relationships between residential satisfaction, mobility intentions, and moving behavior. *Environment and Planning A, 30*(8), 1473–1495.

Marans, R. W. (2003). Understanding environmental quality through quality of life studies: The 2001 DAS and its use of subjective and objective indicators. *Landscape and Urban Planning, 65*(1–2), 73–83.

McCrea, R. (2007). *Urban quality of life: Linking objective dimensions and subjective evaluations of the urban environment*. Unpublished PhD thesis, The University of Queensland, Brisbane.

McVeigh, T. (1971). *Social indicators: A bibliography*. Monticello: Council of Planning Librarians Exchange Bibliography No. 52.

Moser, C. (1970). Measuring quality of life. *New Society, 428*, 1042–1043. Dec.10.

Perle, E. D. (1970). *Social reporting in Michigan: Problems and issues, Office of Planning Coordination, Bureau of Policies and Programs* (Technical Report A 37). State of Michigan.

Perloff, H. (1969). A framework for dealing with the urban environment: Introductory statement. In H. Perloff (Ed.), *The quality of the urban environment* (pp. 3–25). Washington, DC: Resources for the Future.

Perz, S. G. (2000). The quality of urban environments in the Brazilian Amazon. *Social Indicators Research, 49*(2), 181–212.

Rogerson, R. J., Findlay, A. M., Morris, A. S., & Coombes, M. G. (1989). Indicators of Quality of Life – Some Methodological Issues. *Environment and Planning A, 21*(12), 1655–1666.

Santos, L. D., & Martins, I. (2007). Monitoring urban quality of life: The Porto experience. *Social Indicators Research, 80*(2), 411–425.

Savageau, D. (2007). *Places Rated Almanac: The classic guide to finding the best places to live*. Washington, DC: Places Rated Books LLC.

Scheonfield, A., & Shaw, S. (Eds.). (1972). *Social indicators and social policy*. London: Heinemann.

Schwirian, K. P., Nelson, A. L., & Schwirian, P. M. (1995). Modeling urbanism – Economic, social and environmental-stress in cities. *Social Indicators Research, 35*(2), 201–223.

Sheldon, E. B. & Moore, W. E. (Eds.). (1968) *Indicators of social change: Concepts & measurements*. New York: Russell Sage Foundation.

Smith, D. M. (1972). Towards a geography of social well-being: inter-state variations in the United States. In R. Peet (Ed.), *Geographical perspectives on American poverty*, Antipode monographs in Social Geography, No. 1 (pp. 17–46).

Smith, D. M. (1973). *The geography of social well-being in the United States: An introduction to territorial social indicators*. New York: McGraw Hill.

Smith, D. M., & Gray, R. J. (1972). *Social indicators for Tampa, Florida* (Mimeo Report). Urban Studies Bureau, University of Florida.
Stagner, R. (1970). Perceptions, aspirations, frustrations, and satisfactions: An approach to urban indicators. *The Annals of the American Academy of Political and Social Science, 388*, 59–68.
Stover, M. E., & Leven, C. L. (1992). Methodological issues in the determination of the quality-of-life in urban areas. *Urban Studies, 29*(5), 737–754.
The Urban Institute. (1971, May–June). Developing urban indicators: Some first steps, *Search, 1*(3). Washington, DC.
U.S. Department of Health, Education and Welfare. (1969). *Toward a social report*. Washington, DC: USGPO.
U.S. Office of Education. (1969). *Toward master social indicators, Research Memorandum 67*. Washington, DC: Educational Policy Research Center.
Wilson, J. (1969). *Quality of life in the United States*. Kansas City: Midwest Research Institute.

# Chapter 3
# Subjective Measurement of Quality of Life Using Primary Data Collection and the Analysis of Survey Data

**Rod McCrea, Robert W. Marans, Robert Stimson, and John Western**[†]

## Introduction

As noted in Chap. 2, dealing with objective QOL, the literature is replete with attempts to both measure and analyze quality of life (QOL). But there is no single agreed model nor is there comprehensive set of measures universally accepted by researchers and policy makers. Subjective QOL is also measured in a variety of ways, and as Diener and Suh (1997) have said, these are assumed to be defined by

> … people's conscious experiences - in terms of hedonistic feelings or cognitive satisfactions (p. 191).

Typically, the *subjective* assessment of QOL is based on primary data collected through sample surveys in which people's evaluations of QOL "domains" are

R. McCrea (✉)
Institute for Social Science Research, The University of Queensland,
Level 4, GPN3 Building, Campbell Road, St Lucia, Brisbane, QLD 4072, Australia
e-mail: r.mccrea@uq.edu.au

R.W. Marans
Institute for Social Research, University of Michigan,
POB 1248, Ann Arbor, MI 48106, USA
e-mail: marans@umich.edu

R. Stimson
Australian Urban Research Infrastructure Network (AURIN), Faculty of Architecture,
Building and Planning, University of Melbourne, VIC 3010, Melbourne, Australia
e-mail: r.stimson@unimelb.edu.au

[†]Deceased

R.W. Marans and R. Stimson (eds.), *Investigating Quality of Urban Life: Theory, Methods, and Empirical Research*, Social Indicators Research Series 45,
DOI 10.1007/978-94-007-1742-8_3, 

measured on scaled attributes relating to those QOL domains. As stated by Carley (1981), subjective QOL is

> … based on reports from individuals on the "meaning" of aspects of their reality, and as such represent psychological variables (p. 31).

Those subjective approaches to examining QOL in general and QOUL in particular are the focus of this chapter.

## Subjective QOL as an Empirical Positivist Investigation

It was pointed out in Chap. 1 that there is no standard definition or measure of QOL, nor is there a standard model for investigating subjective QOL. And it was also pointed out that various terms such as *happiness*, *life satisfaction* and *well-being* have been used in the context of investigating QOL.

Historically, the terms "happiness" and "quality of life" have tended to be used interchangeably. However, in early studies, the focus seemed to be on happiness which could be viewed from either a *eudaemonistic* or a *hedonistic* philosophical perspective. In fact, the eudaemonistic view of happiness may be traced back to Aristotle who recommended living a "good and virtuous" life which led to a happy and successful life (Aristotle, [circa 350 BC] 1998). In contrast, the hedonistic view of happiness recommended maximizing pleasure or satisfaction which can also be traced back a long way to Jeremy Bentham ([1789] 1998) and John Stuart Mill ([1863] 1998).

As discussed by Crotty (1998), the philosophical perspective adopted in research has implications for methodology. The eudaemonistic view of happiness was certainly a "normative" view prescribing what "should be done" to be happy and lends itself to moralistic approaches. However, the hedonistic view focusing on *satisfaction* is a "positive" view asking "what is it" that makes one satisfied. That view lends itself to empirical approaches. Consistent with the latter philosophical view, most of the QOL literature investigating subjective QOL have tended to adopt an empirical positivist methodology.

Diener (1984) has reminded us that it was in 1973 when "happiness" was first listed as an index term in *Psychological Abstracts International*, and then in 1974, the first journal dedicated to QOL research, *Social Indicators Research*, was started. Empirical positivist research into subjective QOL really began in earnest around that time. For example, in the mid-1970s, researchers in the Institute for Social Research (ISR) at the University of Michigan published seminal works empirically investigating subjective QOL (Andrews and Withey 1976; Campbell et al. 1976) and quality of the residential environment (Marans and Rodgers 1975).

A related concept to QOL which is more well-defined and which has had a generally accepted meaning is *subjective well-being* (Diener 1984; Diener et al. 1999). It is seen as having three dimensions:

- Pleasant affect (for example, joy, elation, contentment or indeed happiness as a feeling)
- Unpleasant affect (for example, shame, sadness, anxiety, etc.)

- Life satisfaction (either overall life satisfaction or satisfaction in particular life domains)

The pleasant and unpleasant affective dimensions can be thought of as positive and negative *feelings*, whereas life satisfaction can be thought of as a *subjective evaluation* or cognitive judgment.

In QOL research, an important distinction has been made between *feelings* and *judgments*. Judgments refer to particular targets or objects, and another is that judgments are influenced by standards of comparison (Abele and Gendolla 1999; Campbell et al. 1976; Kahneman 1999; Michalos 1985; Schwarz and Strack 1999). In contrast, feelings are often generalized and may not be easily related to specific targets (Forgas 1995); for example, a depressed or low mood may not relate to any specific target or may be generalized across all targets. While feelings are an important component of subjective well-being, research into specific QOL domains like QOUL is more focused on judgments or subjective evaluations of various targets in those domains (for example, judgment targets in the urban environment, such as housing or transportation).

Empirical research into QOL was certainly to grow exponentially, and at the turn of the twenty-first century, more than 35,000 publications on QOL were identified in five main electronic databases (Evans and Huxley 2002). Not surprisingly, in 2000, a second international journal dedicated to QOL research – *The Journal of Happiness Studies* – had been established, and in 2006, a third international journal – *Applied Research in Quality of Life* – was established.

But despite that flurry of empirical research into QOL, there still remains no generally accepted definition of QOL nor agreement about how best to measure it even after considerable debate within the *International Society of Quality of Life Studies* (Andelman et al. 1998). However, in the broadest sense, QOL seems to mean some evaluation of human circumstances.

## The Subjective Evaluation of QOL

### *Measuring Subjective QOL Through Surveys*

It has been most common to collect information on people's subjective assessments and evaluations of QOL and of QOUL using social survey methods. In such survey research, QOL is often measured by asking respondents to evaluate or assess various aspects of their lives, which often are presented as QOL *domains* (for example, health, work, etc.). Usually, responses such as levels of satisfaction are captured using a standard response format, such as a Likert scale, which yields a numerical rating (see, for example, Cummins 1996; Headey and Wearing 1992; Salvatore and Mu uz Sastre 2001; Trauer and Mackinnon 2001). However, aspects of quality of life might not hold the same importance for everybody, and as a result, the evaluation of the importance of each aspect has sometimes been built into questionnaires used to collect data in a survey (Gill and Feinstein 1994).

## *The "Domains of Life" Approach*

As mentioned, it has been common in the subjective analysis of QOL for a "domains of life" approach to be used. Within the "domain of life" approach, there exists considerable debate surrounding the number of independent domains; nevertheless, partitions based on parsimony, meaningfulness and usefulness underpin their success (Rojas 2004). For example, Cummins et al. (1996) have identified seven main domains, namely:

- Material well-being
- Health
- Productivity
- Intimacy
- Safety
- Community
- Emotional well-being

Satisfactions in these domains are often used to predict a more global evaluation of life.

A global evaluation of life satisfaction is an aggregate concept which may also be represented as the sum or average of the QOL domain components. Sums or averages effectively give each domain the same weight in the overall score (called "unit weighting"). Alternatively, satisfaction ratings for various life QOL domains may be multiplied by their importance ratings and then summed, thus generating a weighted sum or weighted average indicating overall QOL. This reflects the notion that individuals will assign unique weightings to various life satisfaction domains, with lower weights being allocated to those domains that contribute little to the person's life satisfaction (Oliver et al. 1995). However, as discussed below, most research shows that both unit weightings and weighting by subjective importance ratings produce similar results. Consequently, most subjective QOL research does not weight subjective evaluations by importance ratings.

Subjective evaluations of life domains can be both internal and external to the individual (Veenhoven 2000). For example, evaluations of neighborhood crime would be an evaluation of an external neighborhood quality, while how safe people feel in their neighborhood would be an evaluation of an internal quality within the individual. Veenhoven (2000) identifies four interdependent categories of QOL by distinguishing between these "inner and outer" qualities and between "life chances and life outcomes":

- The livability of the environment
- The life-ability of the individual
- The external utility of life
- The inner appreciation of life.

It is thus apparent that aspects of QOL are indeed imbedded within a situational setting where a person lives and will thus have an environmental context.

## *Explicitly "Urban" Domains of QOL*

In conceptualizing QOUL, it is useful to make a distinction between QOL that is "derived from" the urban environment (that is, satisfaction derived in urban domains, such as housing, neighborhood, community and region) and QOL as "experienced *in*" the urban environments (which would include satisfaction across all life domains; for example, work, social relationships, health, neighborhood, and so on). The former – QOUL derived from the urban environment – is related to links between the objective dimensions of the urban environment and people's subjective evaluations of the urban environment, which is attracting attention in some of the more recent research into QOUL. This is an issue that is specifically addressed in Chap. 4.

In a review of QOL research in urban geography in which there has been an explicitly spatial orientation, Pacione (2003) identified various dimensions of QOUL, which have been investigated in the QOL literature. Reference is made to what may be regarded as objective and subjective planes which relate to objective and subjective measures of QOL. Those objective and subjective QOL measures may also vary by social group. On each of these planes, there are also two other dimensions, namely domain specificity (for example, life domains and sub-domains) and levels of geographic scale:

(a) *Levels of domain specificity*. Levels of satisfaction with overall life consist of satisfaction across a range of important *life domains* of the type referred to in the previous section, and there have been many studies that have investigated levels of satisfaction in each of the main life domains. For example, studies have investigated satisfaction with:

- Work (Hart 1999; Heller et al. 2002)
- Relationships (Acock and Hurlbert 1993; Evans et al. 1993; Foroughi et al. 2001)
- Health (John 2004; Michalos et al. 2001).

Pacione's (2003) framework suggests that QOL may be conceptualized as satisfaction with overall life or satisfaction across the various life domains, including *urban* domains. Similarly, the QOUL may be conceptualized as satisfaction across various urban sub-domains.

(b) *Geographic scale*. Levels of satisfaction at explicit levels of spatial scale relate to the environment in which people live. These include people's satisfaction with:

- Housing (the dwelling people occupy)
- The neighborhood where people live
- The community to which people relate
- The wider urban region in which people live.

The three most commonly studied urban domains are housing satisfaction, neighborhood satisfaction and community satisfaction (see Bruin and Cook 1997; Campbell et al. 1976; Marans and Rodgers 1975; Lu 1999; Parkes et al. 2002; Sirgy and Cornwell 2002). However, regional satisfaction is much less studied (for example, McCrea et al. 2005; Turksever and Atalik 2001). Regions may be viewed as areas consisting of a number of communities linked together by a shared geography (for example, a shared water catchment area), by shared organizations (for example, regional development organizations) and by shared major service centers.

Of course, rather than being distinct, these four urban domains have been found to be interrelated. Research has found the following:

(a) Housing satisfaction is predicted by features of the home; for example, dwelling age, size, structure and tenure (Campbell et al. 1976; Lu 1999). That is not a surprising finding. Housing satisfaction is also predicted by surrounding features, such as neighbors, housing in the local area and community size (Campbell et al. 1976; Lu 1999; Parkes et al. 2002). And housing satisfaction is also predicted by community satisfaction (Campbell et al. 1976) and even by regional characteristics, such as geographic location within the metropolitan region (Lu 1999). Thus, housing satisfaction is linked not only with attributes of the house where people live but also with the surrounding urban environment.

(b) Similarly, while neighborhood satisfaction is predicted by a wide range of physical, economic and social features of neighborhoods (see Sirgy and Cornwell 2002), neighborhood satisfaction is also linked to satisfaction in other urban domains, such as housing and community. Sirgy and Cornwell (2002) found that satisfaction with economic features of the neighborhood was also a good predictor of housing satisfaction and that satisfaction with a neighborhood's social features was a good predictor of community satisfaction. And Campbell et al. (1976) also found that neighborhood and community satisfaction were strongly related.

(c) Community satisfaction, neighborhood satisfaction and housing satisfaction all seem to be interrelated (Sirgy and Cornwell 2002). For example, relationships with neighbors predict each of these three urban domains (Campbell et al. 1976; Lu 1999; Turksever and Atalik 2001). However, community satisfaction is related more to neighborhood satisfaction than housing satisfaction (Campbell et al. 1976).

(d) Regional satisfaction is not a commonly studied domain in subjective QOUL. However, a study by Turksever and Atalik (2001) found that a range of factors predicted both regional and community satisfaction (health, climate, crowding, sporting facilities, housing conditions and environmental pollution). But only overcrowding and travel to work were uniquely related to regional satisfaction. Thus, regional satisfaction also seems to be associated with satisfaction in other urban domains.

## Model Approaches

Some of the general models used in empirical investigations of subjective QOL and QOUL have been summarized by McCrea (2007) and are discussed below.

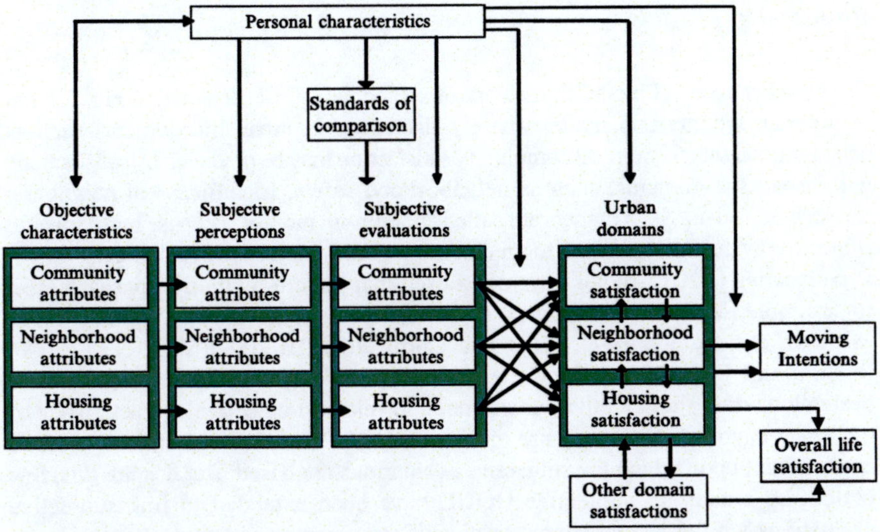


**Fig. 3.1** A broad model framework for investigating subjective assessment of determinants of satisfaction with the residential environment (Source: Derived from Marans and Rodgers 1975; Campbell et al. 1976)

## *A Broad Conceptual Model Framework*

Even though subjective QOL is often researched using different urban domains, it is clear that satisfaction in those urban domains is interrelated, and that is reflected in the conceptual model frameworks proposed by Marans and Rodgers (1975) and Campbell et al. (1976) that was discussed in Chap. 1 and shown again in Fig. 3.1. In that figure, two-way arrows are shown between the different urban domains making up the urban environment. Because of those interrelations, subjective QOUL might also be viewed as a composite of housing, neighborhood and community satisfaction together with regional satisfaction.

This broad conceptual model framework has been applied to the study of subjective QOUL in the context of investigating the determinants of satisfaction with the urban environment. This includes predicting satisfaction in urban domains from subjective evaluations of various attributes in the urban environment. It also provides an underlying rationale for linking objective characteristics of the urban environment and people's subjective evaluations of the urban environment – as is discussed in detail in Chap. 4 – with QOUL ultimately viewed as subjective.

This broad conceptual model can also incorporate other models that relate to various parts of the broad conceptual model. For example, satisfaction judgments of people can be influenced by both bottom-up and top-down processes (as demonstrated by Lance et al. 1989; Lance et al. 1995; Michalos and Zumbo 1999). Specific models relating to parts of this broad conceptual model are discussed below.

## Bottom-Up Models

The broad conceptual model framework of subjective QOUL depicted in Fig. 3.1 can incorporate a bottom-up model where satisfactions in urban life domains such as neighborhood satisfaction and community satisfaction may be predicted by satisfactions with urban sub-domains, such as neighborhood safety, friendliness of neighbors, services, and so on. Such models are called "bottom-up models" because more specific subjective evaluations are used to predict more global subjective evaluations.

Subjective QOUL studies have often used that bottom-up model approach (see, for example, Marans and Rodgers 1975; Campbell et al. 1976; Cummins 1996; Michalos and Zumbo 1999; Sirgy and Cornwell 2001, 2002; Sirgy et al. 2000; Ibrahim and Chung 2003; McCrea et al. 2005). Ultimately however, subjective evaluations of the urban environment need to relate to objective characteristics of the urban environment if they are to be meaningful since it is the objective urban environment upon which the subjective evaluations are based. But despite this, few bottom-up models of subjective QOUL have been extended to link subjective evaluations with objective characteristics of the urban environment (for example, Marans and Rodgers 1975; Campbell et al. 1976; Galster and Hesser 1981; McCrea et al. 2006), and that research gap is discussed in detail in Chap. 4.

## Top-Down Models

Top-down models may be incorporated in Fig. 3.1 by the arrow from personal characteristics to satisfaction in urban and other life domains. Those models would seek to predict satisfaction with overall life and satisfaction in the life domains from personality characteristics such as extroversion, neuroticism and self-esteem (see, for example, Diener 1984; Diener et al. 1999; Hart 1999; Vitterso 2001; Hayes and Joseph 2003; Vitterso and Nilsen 2002). Such top-down models reflect stable individual differences which may influence subjective evaluations (Headey and Wearing 1989).

## Mood Bias Models

Several psychologists have found that mood biases a wide variety of subjective judgments. That includes:

- Persuasion (Petty et al. 1993)
- Stereotyping (Roesch 1999)
- Self-conceptions (Sedikides 1995)
- Life satisfaction (Abele and Gendolla 1999; Schwarz and Clore 1983; Schwarz and Strack 1999; Schwarz et al. 1987).

In models, mood bias may be controlled using measures for positive and negative *affect*, which also assists in controlling for personality since positive and negative affect are highly correlated with the personality traits of extroversion and introversion, respectively (see Diener et al. 1999).

There has been some disagreement about the main underlying mechanisms causing mood bias, which may differ at different levels of domain specificity. With the "affect-as-information" mechanism, positive and negative moods may be used as information if we consult our feelings when making a subjective judgment (Schwarz and Clore 1988; Schwarz and Strack 1999; Clore and Tamir 2002). Furthermore, affect may be more likely to be consulted as information in more global subjective judgments like overall life satisfaction because these judgments are more complex and consulting affect can be used as a simplifying heuristic to reduce the cognitive burden (Schwarz and Strack 1999). Conversely, affect may be less likely to be consulted in more specific and less complex evaluations involving specific life domains. With the "affect priming" mechanism, positive moods facilitate retrieval of positive memories and negative moods facilitate retrieval of negative memories when making subjective judgments, irrespective of domain specificity (Bower 1981; Clark and Williamson 1989; Forgas 1995). However, McCrea (2007) found little mood bias when predicting more specific subjective evaluations of the urban environment, while finding approximately 3% bias when predicting more global urban domains (that is, subjective QOUL). This supports the "affect-as-information" mechanism in QOUL research.

## Subjective Judgment Models

Subjective judgment models may also be incorporated into the broad conceptual model framework shown in Fig. 3.1 via standards of comparison. Michalos (1985) has reviewed a wide range of theories that incorporate "standards of comparison" into subjective judgment models; for example, aspirations theory, equity theory, cognitive dissonance theory, reference group theory and social comparison theory. Moreover, subjective judgment theories are commonly used in QOL research (for example, Brickman and Campbell 1971; Marans and Rodgers 1975; Brickman et al. 1978; Wright 1985; Michalos 1985, 1986; Schwarz et al. 1987; Meadow et al. 1992; Abele and Gendolla 1999).

In subjective judgment models, it is hypothesized that the subjective evaluations of people depend on the difference between the attributes of a judgment target and the standards of comparison rather than simply on the attributes of a judgment target alone. Thus, individual variations in standards of comparison have the potential to significantly influence subjective evaluations. For example, two people might evaluate the same urban environment differently – one positively, the other one negatively – by virtue of having different standards from different prior experiences with urban environments. Since "standards of comparison" have the potential to significantly influence subjective evaluations in a way that weakens relationships between objective dimensions and subjective evaluations of the urban environment, subjective judgment models may be used to investigate the effect of psychological processes on people's subjective evaluations of objective characteristics.

## *Adaptation Models*

Although not reflected in the broad conceptual model framework in Fig. 3.1, adaptation is another psychological process that may potentially weaken the relationships between objective dimensions and people's subjective evaluations of the urban environment. With adaptation, a person's perceptions and standards of comparisons merge over time such that initially strong positive or negative subjective evaluations become more moderate over time. Thus, if the perceptions of a person equal their "standards of comparisons" over time, then they will be satisfied since their expectations have been met.

Kahnemann (1999) has proposed that perceptions and standards of comparison can merge over time through two different processes:

(a) First, by *adjusting sensory perceptions*, where, for example, initially striking perceptions of an urban environment may become less noticeable over time with increasing familiarity with the urban environment (the "hedonic treadmill").
(b) Second, by *adjusting standards of comparison* where, for example, standards become increasingly influenced by every-day expectations associated with living in a particular urban environment (the "satisfaction treadmill").

Another adaptation model is Cummins's (2000) *theory of homeostasis*. In the homeostatic model, cognitive biases serve as functional adaptive devices to promote positive subjective evaluations of a respondent's life circumstances. Failure of positive cognitive biases in undesirable objective circumstances (for example, undeniably bad objective circumstances which cannot be cognitively reframed) can lead to depression and withdrawal from life, which is seen as dysfunctional. Positive cognitive biases occur when constructing targets and standards underlying subjective evaluations (Cummins and Nistico 2002). Targets may be selectively perceived so as to provide a positive bias. Standards of comparison may also be selective so as to give positive subjective evaluations by making downward comparisons. Thus, Cummins' homeostasis model also involves adjustments of targets and standards which weaken the relationship between targets and subjective evaluations.

Empirical evidence has been found for adaptation in various life domains (for a review, see Diener et al. 2006). However, it is not clear how important adaptation is in urban domains, and that has implications for urban planning and whether changes in the urban environment are associated with long-lasting changes in subjective QOUL (Brickman and Campbell 1971).

## *Subjective Importance Models*

Aside from psychological processes, individual and social group differences in the subjective importance of various attributes of the urban environment may influence the links between objective dimensions and people's subjective evaluations of the urban environment. That may be investigated by weighting objective dimensions by subjective importance when predicting subjective evaluations of the urban environment. That is a novel approach and has been attempted only recently by McCrea (2007).

In contrast, there has been a considerable amount of research on weighting satisfaction in life domains or sub-domains by the subjective importance people ascribe to them when predicting more global subjective evaluations of satisfaction. That subjective weighting of subjective evaluations is encapsulated in the broad conceptual model framework depicted in Fig. 3.1 in the arrow from personal characteristics to relationships *between* subjective evaluations and satisfaction in urban domains. However, there is a continuing debate as to whether it is really necessary to weight subjective evaluations by subjective importance when predicting more global subjective evaluations (Trauer and Mackinnon 2001; Hsieh 2003, 2004; Wu and Yao 2006b).

Most research has shown that weighting subjective evaluations by importance does not significantly improve prediction of satisfaction in more global domains (for example, Andrews and Withey 1976; Campbell et al. 1976; Mastekaasa 1984; Cummins et al. 1994; Russell et al. 2006). Some authors have suggested that this is because subjective evaluation measures already include a component of subjective importance since very positive or very negative subjective evaluations imply a level of subjective importance (Trauer and Mackinnon 2001; Wu and Yao 2006a). However, that argument would seem less applicable to weighting objective dimensions of the urban environment since objective indicators do not imply a component of subjective importance in their measurement. Thus, individual and social group differences in subjective importance of various attributes of the urban environment could hypothetically assist in explaining relationships between objective dimensions and subjective evaluations of the urban environment. However, this has now been investigated by McCrea (2007) who found that weighting by subjective importance ratings was not important in this context either.

## *Residential Relocation Models*

In the broad conceptual model framework depicted in Fig. 3.1, the residential relocation decision and choice process is reflected in the intentions of people to move, which could be further demonstrated by drawing another arrow from the "intentions to move" box back to the objective characteristics of the urban environment in a feedback loop. Residential relocation is of interest because differences in the subjective importance of various attributes of the urban environment may also influence the links between objective dimensions and subjective evaluations of the urban environment by influencing where residents choose to live.

As discussed by Golledge and Stimson (1997), there have been two main ways of classifying approaches taken in examining residential relocation: a *macro* versus *micro* approach and a *functionalist* versus *behavioral* approach:

(a) *Macro approaches* use aggregated secondary data to examine population flows between places, usually by examining asymmetric patterns in flows between places of origin and destination (Quigley and Weinberg 1977). However, macro approaches focus on aggregated objective data for places, and so, are of limited use in examining links between objective dimensions and subjective evaluations of the urban environment.
(b) In contrast, *micro approaches* focus more explicitly on the processes underlying residential relocation, and can be classified as either functionalist or behaviorist (Golledge and Stimson 1997).
(c) *Functionalist approaches* make assumptions about the underlying residential relocation decision processes in order to model residential relocation outcomes. Those assumptions are usually based on maximizing utility by rational residents and on market based principles, with theories falling into three main types: minimizing travel cost; trading off travel cost and housing cost; and maximizing house expenditure (Balchin et al. 1995), or more recently, trading off housing quality and location status (Phe and Wakely 2000). However, functional approaches have limited usefulness in explaining links between objective dimensions and subjective evaluations of the urban environment because they *assume* the underlying processes of rational residents.
(d) In contrast, *behavioral approaches* describe or examine the underlying processes of residents during the residential relocation process. This may be in relation to either longer distance relocation (for example, interstate) or shorter distance relocation (for example, intra-urban). Push-pull models generally focus on longer distance relocation (for example, Walmsley et al. 1998; Longino et al. 2002; Stimson and McCrea 2004).

In his investigation of subjective QOUL, McCrea (2007) has used a behavioral model which focuses on describing and examining processes underlying intra-urban relocation. That research made use of an early and detailed behavioral model of intra-urban residential relocation proposed by Brown and Moore (1970) (see Fig. 3.2) in which the intra-urban residential relocation process is classified into two main phases: deciding whether to move and deciding where to move:

(a) In deciding whether to move (Phase 1), residents compare their objective residential environment with their subjective needs and aspirations. If their cognitive appraisal is unfavorable, a resident becomes dissatisfied with their place utility and may decide to move.
(b) In deciding where to move (Phase 2), the criteria for evaluating and choosing a new residential location are based on the subjective importance of various attributes of the urban environment, though a resident's needs and aspirations may adjust as part of the residential search process.

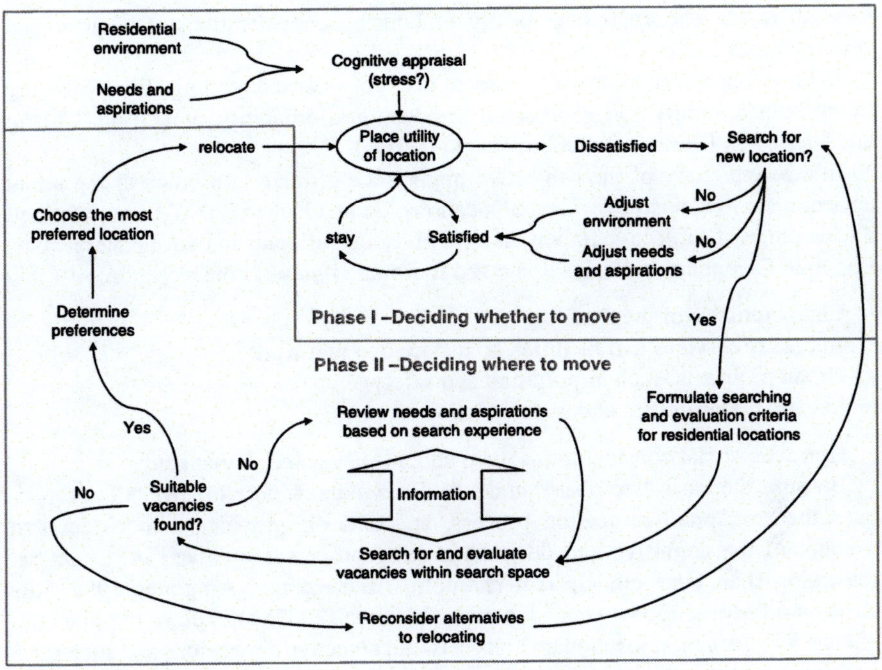


**Fig. 3.2** Behavioral model of the residential relocation processes (Source: McCrea 2007, adapted from Brown and Moore 1970; Golledge and Stimson 1997)

This basic model can be extended in a number of ways, including:

- Incorporating the costs of moving into the decision to move or stay (see, for example, Fredland 1974; Speare et al. 1975)
- Taking into consideration the effect of social norms and institutional constraints on intentions to move (see, for example, Desbarats 1983)
- Extending the notion of residential preferences into residential lifestyles (see, for example, Ge and Hokao 2006).

However, the residential relocation process is still encapsulated in this basic model (see, for example, Pacione 1990; Golledge and Stimson 1997) where residents choose where they live based on what is important to them, subject to any constraints which may result in adjusting needs and expectations.

As mentioned already, Brown and Moore's model divides the residential relocation process into two stages: deciding whether to move and deciding where to move. Factors underlying decisions about whether to move include changing household space requirements relating to life course changes. This was the main finding in Peter Rossi's seminal work on *Why Families Move* (Rossi 1955), and since then, a body of research has confirmed the role of changing housing space requirements in the decision to move (see, for example, Clark and Huang 2003, 2004; Clark and Ledwith 2006, Michelson 1977). However, once deciding to move, neighborhood attributes play an important role in deciding where to move, though a residential location is not chosen independently of housing considerations (Dieleman and Mulder 2002).

In choosing *where* to move, a resident may consider many things like proximity to workplace, family and good schools; housing affordability, and so on (Chiang and Hsu 2005; Clark et al. 2000; Kim et al. 2005).

In a recent study of the subjective importance of many attributes of the urban environment in choosing residential locations, Ge and Hokao (2006) have examined a wide range of subjective importance variables for residents in two Japanese cities. Principle Components Analyses showed four main types of consideration:

- Characteristics of the dwelling
- Access to services and facilities, transportation and work
- Urban problems, such as pollution and safety
- Leisure opportunities and social relationships.

However, social considerations were under-represented in that study.

Because the basic two-phase model by Brown and Moore (1970) still encapsulates the residential relocation process, and since it provides a more detailed account of the cognitive and decision making processes involved in residential relocation than later models (for example, Amerigo and Aragones 1997), the Brown and Moore model was adopted by McCrea (2007) to explore the potential effects of residential relocation on links between objective dimensions and subjective evaluations of the urban environment. McCrea makes the point that many of the general processes discussed in the above sections have the potential to weaken relationships between objective dimensions and subjective evaluations of the urban environment.

## Using Agent-Based Modeling to Analyze Subjective QOUL Data

One of the recent methodological developments in the analysis of survey-based subjective QOUL data has been the use of agent-based modeling employing powerful computational tools which:

> … facilitates the examination of system-level outcomes of the heterogeneous outcomes of a set of heterogeneous agents. (Fernandez et al. 2005: p. 799)

That approach has the potential to richly enhance the analysis of QOUL data, particularly in providing insights into how agents act in the urban development process and how that interfaces with subjective – and indeed objective – QOUL.

Agent-based models have been used in studies investigating a wide range of behaviors in a spatial context, and they range from econometric and hedonic models based on individual household and price data (for example, Bell and Irwin 2002) and discrete choice models (for example, Landis 1994), to cellular automata models which have a degree of calibration of rules to data (for example Batty and Xie

1997), to physical analogue models which may be based on diffusion-limited aggregation and correlated percolation (for example, Batty and Longley 1994).

Recently, developments in agent-based simulation of land use patterns has been creating the opportunity to simulate micro-level individual and household behaviors and evaluate their implications at the aggregate level which has required new forms of data (Parker et al. 2003). Generally, one of two approaches is taken by either:

- Using observations from aggregate outcomes to evaluate or calibrate the model (as seen, for example, with models of city formation) or
- Using data on individual or agent decisions to generate agent behaviors, which might involve role playing games to collect information on the agents (as demonstrated by Barreteau 2003) or the use of deductive reasoning (as demonstrated by Brown et al. 2003).

An example of agent-based modeling using subjective QOUL survey data was reported by Fernandez et al. (2005). The data were collected as part of a study in 2001 of QOUL in Detriot (see Chap. 7). The agent-based modeling was developed in the context of a broader study investigating land use and land cover change in the Detroit metropolitan area as a means of better understanding human–environment interactions at the urban–rural fringe. The objective was to model land development patterns as a function of agent preferences for nearness to services and jobs, landscape aesthetic quality and different levels of urban density, and to look at the influence of agent characteristics, such as age, marital status, race and income. The data collected in the Detroit QOUL survey covered people's residential preferences and their demographic and socio-economic characteristics. The analysis involved the comparison of two alternative approaches to characterizing the heterogeneous preferences of agents, both based on factor analysis of resident responses to questions about their reasons for moving to their current residential location:

(a) Cluster analysis was used to determine the number of different types of residents according to their preferences.
(b) The relationships between residents' demographic and socio-economic characteristics and their location preferences were evaluated using regression analysis. That involved evaluating the "fit" of the relationship to ascertain the degree to which the demographic and socio-economic characteristics of the residents predicted their preferences.

That enabled the researchers to then identify the degree to which preferences are independent of agents' demographic and socio-economic characteristics. When that question was answered then it was possible to decide how to represent residents as agents; or, in other words, to determine whether to use preference groups or whether to assign preferences to agents based on their demographic and socio-economic characteristics.

The Fernandez et al. (2005: p. 818) study found that:

(a) The preferences of exurban single-family home residents were heterogeneous and could be grouped into seven distinct clusters representing unique preference groups in resident choice.

(b) Preferences themselves could be grouped into four general factors (related to social comfort, openness/naturalness, residential aesthetics, and schools and work) representing major topics considered by exurban residents in choosing a residence.
(c) Further issues related to cost and value, access to services and community size were observed across all preference groups and did not distinguish among agent groups.
(d) By analyzing the relationships between the empirically derived preference clusters and six socioeconomic and demographic variables, some significant univariate relationships between the two were found, but those variables were incomplete predictors of residential preference.

Fernandez et al. (2005) concluded that

> … the delineation of agents might be most profitably done on the basis of agent types with separate distributions of preferences. (p. 818)

## Conclusion

Empirical research into subjective QOL has grown exponentially over the last 40 years. While subjective QOL involves both feelings and subjective judgments, research into specific life domains like QOUL focuses more on subjective judgments. These judgments may relate to how important various aspects of life are to individuals or individual preferences; however, typically, they relate to evaluations of satisfaction with various aspects of life. As such, measures of subjective QOL involve psychological processes.

Many of the general processes discussed in this chapter have the potential to weaken the relationships between objective dimensions of QOUL and subjective evaluations of the urban environment. Notwithstanding this, measures of subjective QOUL gain meaning because people interact with their objective urban environments, and so, integrated approaches to examining QOUL are evolving which incorporate both objective and subjective indicators.

## References

Abele, A. E., & Gendolla, G. H. E. (1999). Satisfaction judgments in positive and negative moods: Effects of concurrent assimilation and contrast producing processes. *Personality and Social Psychology Bulletin, 25*(7), 883–895.

Acock, A. C., & Hurlbert, J. S. (1993). Social networks, marital-status, and well-being. *Social Networks, 15*(3), 309–334.

Amerigo, M., & Aragones, J. I. (1997). A theoretical and methodological approach to the study of residential satisfaction. *Journal of Environmental Psychology, 17*(1), 47–57.

Andelman, R., Board, R., Carman, L., Cummins, R., Ferriss, A., Friedman, P., et al. (1998). *Quality of life definition and terminology: A discussion document from the international society of quality of life studies (Monograph)*. Blacksburg: International Society of Quality of Life Studies.

Andrews, F., & Withey, S. B. (1976). *Social indicators of well-being: Americans perceptions of quality of life*. New York: Plenum Press.

Aristotle. ([circa 350 BC] 1998). Nicmachean ethics. In L. P. Pojman (Ed.), *Ethical theory: Classical and contemporary readings* (3rd ed., pp. 365–378). Melbourne: Wadsworth Publishing Company.

Balchin, P., Kieve, J., & Bull, G. (1995). *Urban land economics*. London: Macmillan.

Barreteau, O. (2003). The joint use of role-playing games and models regarding negotiation processes: Characterization of associations. *Journal of Artificial Societies and Social Simulation*, 6, http://jasss.soc.surrey.ac.uk/6/2/3.html

Batty, M., & Longley, P. (1994). *Fractal cities*. San Diego: Academic.

Batty, M., & Xie, Y. (1997). Possible urban automata. *Environment and Planning B: Planning and Design, 24*, 275–292.

Bell, K. P., & Irwin, E. G. (2002). Spatially explicit micro-level modeling of land use change at the rural-urban interface. *Agricultural Economics, 27*, 217–232.

Bentham, J. ([1789]1998). An introduction to the principles of morals and legislation. In L. P. Pojman (Ed.), *Ethical theory: Classical and contemporary readings* (3rd ed., pp. 149–151). Melbourne: Wadsworth Publishing Company.

Bower, G. H. (1981). Mood and memory. *American Psychologist, 36*, 129–148.

Brickman, P., & Campbell, D. T. (1971). Hedonic relativism and planning the good society. In M. H. Appley (Ed.), *Adaptation-level theory: A symposium* (pp. 287–304). New York: Academic.

Brickman, P., Coates, D., & Janoff-Bulman, R. (1978). Lottery winners and accident victims: Is happiness relative? *Journal of Personality and Social Psychology, 36*, 917–927.

Brown, L. A., & Moore, E. G. (1970). The intra-urban migration process: A perspective. *Geografiska Annaler, 52*, 368–381.

Brown, D. G., Page, S. E., Riolo, R., & Rand, W. (2004). Agent-based and analytical modeling to evaluate the effectiveness of greenbelts. *Environmental Modelling and Software, 19*, 1097–1109.

Bruin, M. J., & Cook, C. C. (1997). Understanding constraints and residential satisfaction among low-income single-parent families. *Environment and Behavior, 29*(4), 532–553.

Campbell, A., Converse, P. E., & Rogers, W. L. (1976). *The quality of American life: Perceptions, evaluations and satisfactions*. New York: Russell Sage.

Carley, M. (1981). *Social measurement and social indicators: Issues of policy and theory*. London: George Allen and Unwin.

Chiang, L. N., & Hsu, J. R. (2005). Locational decisions and residential preferences of Taiwanese immigrants in Australia. *GeoJournal, 64*(1), 75–89.

Clark, W. A. V., & Huang, Y. Q. (2003). The life course and residential mobility in British housing markets. *Environment and Planning A, 35*(2), 323–339.

Clark, W. A. V., & Huang, Y. Q. (2004). Linking migration and mobility: Individual and contextual effects in housing markets in the UK. *Regional Studies, 38*(6), 617–628.

Clark, W. A. V., & Ledwith, V. (2006). Mobility, housing stress, and neighborhood contexts: Evidence from Los Angeles. *Environment and Planning A, 38*(6), 1077–1093.

Clark, M. S., & Williamson, G. M. (1989). Mood and social judgments. In H. L. Wagner & A. S. R. Manstead (Eds.), *Handbook of social psychophysiology: Emotion and social behavior* (pp. 347–370). Chichester: Wiley.

Clark, W. A. V., Deurloo, M. C., & Dieleman, F. M. (2000). Housing consumption and residential crowding in US housing markets. *Journal of Urban Affairs, 22*(1), 49–63.

Clore, G. L., & Tamir, M. (2002). Affect as embodied information. *Psychological Inquiry, 13*(1), 37–45.

Crotty, M. (1998). *The foundations of social research*. Sydney: Allen and Unwin.

Cummins, R. A. (1996). The domain of life satisfaction: An attempt to order chaos. *Social Indicators Research, 38*, 303–332.

Cummins, R. (2000). Objective and subjective quality of life: An interactive model. *Social Indicators Research, 52*(1), 55–72.

Cummins, R., & Nistico, H. (2002). Maintaining life satisfaction: The role of positive cognitive bias. *Journal of Happiness Studies, 3*, 37–69.

Cummins, R. A., McCabe, M. P., Romeo, Y., & Gullone, E. (1994). Comprehensive quality of life scale (ComQol): Instrument development and psychometric evaluation on college staff and students. *Educational and Psychological Measurement, 54*, 372–382.

Desbarats, J. (1983). Spatial choice and constraints on behavior. *Annals of the Association of American Geographers, 73*(3), 340–357.

Dieleman, F. M., & Mulder, C. H. (2002). The geography of residential choice. In J. I. Aragones, G. Francescato, & T. Garling (Eds.), *Residential environments: Choice, satisfaction, and behavior* (pp. 35–54). London: Bergin and Garvey.

Diener, E. (1984). Subjective well-being. *Psychological Bulletin, 95*(3), 542–575.

Diener, E., & Suh, E. (1997). Measuring quality of life: Economic, social and subjective indicators. *Social Indicators Research, 40*(1), 189–216.

Diener, E., Suh, E. M., Lucas, R. E., & Smith, H. L. (1999). Subjective well-being: Three decades of progress. *Psychological Bulletin, 125*(2), 276–302.

Diener, E., Lucas, R. E., & Scollon, C. N. (2006). Beyond the hedonic treadmill: Revising the adaptation theory of well-being. *American Psychologist, 61*(4), 305–314.

Evans, S., & Huxley, P. (2002). Studies of quality of life in the general population. *International Review of Psychiatry, 14*(3), 203–211.

Evans, D., Pellizzari, J. R., Culbert, B. J., & Metzen, M. E. (1993). Personality, marital, and occupational factors associated with quality-of-life. *Journal of Clinical Psychology, 49*(4), 477–485.

Fernandez, L. E., Brown, D. G., Marans, R. W., & Nassauer, J. I. (2005). Characterizing location preference in an exurban population: Implications for agent-based modeling. *Environment and Planning B: Planning and Design., 32*, 799–820.

Forgas, J. P. (1995). Mood and judgment: The affect infusion model (Aim). *Psychological Bulletin, 117*(1), 39–66.

Foroughi, E., Misajon, R., & Cummins, R. A. (2001). The relationships between migration, social support, and social integration on quality of life. *Behaviour Change, 18*(3), 156–167.

Fredland, D. R. (1974). *Residential mobility and home purchase: A longitudinal perspective on the family life cycle and the housing market*. Lexington: Lexington Books.

Galster, G. C., & Hesser, G. W. (1981). Residential satisfaction: Compositional and contextual correlates. *Environment and Behavior, 13*(6), 735–758.

Ge, J., & Hokao, K. (2006). Research on residential lifestyles in Japanese cities from the viewpoints of residential preference, residential choice and residential satisfaction. *Landscape and Urban Planning, 78*(3), 165–178.

Gill, T. M., & Feinstein, A. R. (1994). A critical appraisal of the quality-of-life measurements. *The Journal of the American Medical Association, 272*(8), 92–101.

Golledge, R. G., & Stimson, R. J. (1997). *Spatial behavior: A geographic perspective*. New York: Guilford Press.

Hart, P. M. (1999). Predicting employee life satisfaction: A coherent model of personality, work and nonwork experiences, and domain satisfactions. *Journal of Applied Psychology, 84*(4), 564–584.

Hayes, N., & Joseph, S. (2003). Big 5 correlates of three measures of subjective well-being. *Personality and Individual Differences, 34*(4), 723–727.

Headey, B. W., & Wearing, A. (1989). Personality, life events, and subjective well-being: Toward a dynamic equilibrium model. *Journal of Personality and Social Psychology, 57*(4), 731–739.

Headey, B. W., & Wearing, A. (1992). *Understanding happiness: A theory of subjective well-being*. Melbourne: Longman Cheshire.

Heller, D., Judge, T. A., & Watson, D. (2002). The confounding role of personality and trait affectivity in the relationship between job and life satisfaction. *Journal of Organizational Behavior, 23*(7), 815–835.

Hsieh, C. M. (2003). Counting importance: The case of life satisfaction and relative domain importance. *Social Indicators Research, 61*(2), 227–240.

Hsieh, C. M. (2004). To weight or not to weight: The role of domain importance in quality of life measurement. *Social Indicators Research, 68*(2), 163–174.

Ibrahim, M. F., & Chung, S. W. (2003). Quality of life of residents living near industrial estates in Singapore. *Social Indicators Research, 61*(2), 203–225.

John, L. H. (2004). Subjective well-being in a multicultural urban population: Structural, and multivariate analyses of the Ontario health survey well-being scale. *Social Indicators Research, 68*(1), 107–126.

Kahneman, D. (1999). Objective happiness. In D. Kahneman, E. Diener, & N. Scharwz (Eds.), *Well-being: The foundations of hedonic psychology* (pp. 3–26). New York: Russell Sage.

Kim, T. K., Horner, M. W., & Marans, R. W. (2005). Life cycle and environmental factors in selecting residential and job locations. *Housing Studies, 20*(3), 457–473.

Lance, C. E., Lautenschlager, G. J., Sloan, C. E., & Varca, P. E. (1989). A comparison between bottom-up, top-down, and bidirectional models of relationships between global and life facet satisfaction. *Journal of Personality, 57*(3), 601–624.

Lance, C. E., Mallard, A. G., & Michalos, A. C. (1995). Tests of the causal directions of global life facet satisfaction relationships. *Social Indicators Research, 34*(1), 69–92.

Landis, J. D. (1994). The California urban futures model: A new generation of metropolitan simulation models. *Environment and Planning B: Planning and Design, 21*, 399–420.

Longino, C. F., Perzynski, A. T., & Stoller, E. P. (2002). Pandora's briefcase: Unpacking the retirement migration decision. *Research on Aging, 24*(1), 29–49.

Lu, M. (1999). Determinants of residential satisfaction: Ordered logit vs. regression models. *Growth and Change, 30*(2), 264–287.

Marans, R. W., & Rodgers, W. (1975). Toward an understanding of community satisfaction. In V. Rock & A. Hawley (Eds.), *Metropolitan America in contemporary perspective* (pp. 299–352). New York: Halsted.

Mastekaasa, A. (1984). Multiplicative and additive-models of job and life satisfaction. *Social Indicators Research, 14*(2), 141–163.

McCrea, R. (2007). *Urban quality of life: Linking objective dimensions and subjective evaluations of the urban environment*. Unpublished PhD thesis, The University of Queensland, Brisbane.

McCrea, R., Stimson, R., & Western, J. (2005). Testing a moderated model of satisfaction with urban living using data for Brisbane-South East Queensland, Australia. *Social Indicators Research, 72*, 121–151.

McCrea, R., Shyy, T. K., & Stimson, R. (2006). What is the strength of the link between objective and subjective indicators of urban quality of life? *Applied Research in Quality of Life, 1*, 79–96.

Meadow, H. L., Mentzer, J. T., Rahtz, D. R., & Sirgy, M. J. (1992). Life satisfaction measure based on judgment theory. *Social Indicators Research, 26*(1), 23–59.

Michalos, A. C. (1985). Multiple discrepancies theory (MDT). *Social Indicators Research, 16*(4), 347–413.

Michalos, A. C. (1986). An application of multiple discrepancies theory (MDT) to seniors. *Social Indicators Research, 18*(4), 349–373.

Michalos, A. C., & Zumbo, B. D. (1999). Public services and the quality of life. *Social Indicators Research, 48*(2), 125–156.

Michalos, A. C., Hubley, A. M., Zumbo, B. D., & Hemingway, D. (2001). Health and other aspects of the quality of life of older people. *Social Indicators Research, 54*(3), 239–274.

Michelson, W. (1977). *Environmental choice, human behavior, and residential satisfaction*. New York: Oxford University Press.

Mill, J. S. ([1863]1998). Utilitarianism. In L. P. Pojman (Ed.), *Ethical theory: Classical and contemporary readings* (3rd ed., pp. 189–210). Melbourne: Wadsworth Publishing Company.

Oliver, N., Holloway, F., & Carson, J. (1995). Deconstructing quality of life. *Journal of Mental Health, 4*, 1–4.

Pacione, M. (1990). Urban livability: A review. *Urban Geography, 11*(1), 1–30.

Pacione, M. (2003). Quality-of-life research in urban geography. *Urban Geography, 24*(4), 314–339.

Parker, D. C., Manson, S. M., Janssen, M. A., Hoffman, M. J., & Deadman, P. (2003). Multi-agent systems for the simulation of land-use and land-cover change: A review. *Annals of the Association of American Geographers, 93*, 314–337.

Parkes, A., Kearns, A., & Atkinson, R. (2002). What makes people dissatisfied with their neighbourhoods? *Urban Studies, 39*(13), 2413–2438.

Petty, R. E., Schumann, D. W., Richman, S. A., & Strathman, A. J. (1993). Positive mood and persuasion: Different roles for affect under high- and low-elaboration conditions. *Journal of Personality and Social Psychology, 64*(1), 5–20.

Phe, H. H., & Wakely, P. (2000). Status, quality and the other trade-off: Towards a new theory of urban residential location. *Urban Studies, 37*(1), 7–35.

Quigley, J., & Weinberg, D. (1977). Intra-urban residential mobility: A review and synthesis. *International Regional Science Review, 2*, 41–66.

Roesch, S. C. (1999). Modelling the direct and indirect effects of positive emotional and cognitive traits and states on social judgements. *Cognition and Emotion, 13*(4), 387–418.

Rojas, M. N. (2004) *Well-being and the complexity of poverty: A subjective well-being approach* (WIDER Research Paper). pp. 2004–2029.

Rossi, P. H. (1955). *Why families move: A study in the social psychology of urban residential mobility*. Clencoe: The Free Press.

Russell, L. B., Hubley, A. M., Palepu, A., & Zumbo, B. D. (2006). Does weighting capture what's important? Revisiting subjective importance weighting with a quality of life measure. *Social Indicators Research, 75*(1), 141–167.

Salvatore, N., & Muňuz Sastre, M. T. (2001). Appraisal of life: "Area" versus "dimension" conceptualisations. *Social Indicators Research, 53*, 229–255.

Schwarz, N., & Clore, G. L. (1983). Mood, misattribution, and judgments of well-being: Informative and directive functions of affective states. *Journal of Personality and Social Psychology, 45*(3), 513–523.

Schwarz, N., & Clore, G. L. (1988). How do I feel about it? The informative function of affective states. In K. Fiedler & J. P. Forgas (Eds.), *Affect, cognition, and social behavior* (pp. 44–62). Gottingen, Germany: Hogrefe.

Schwarz, N., & Strack, F. (1999). Reports of subjective well-being: Judgmental processes and their methodological implications. In D. Kahneman & E. Diener (Eds.), *Well-being: The foundations of hedonic psychology* (pp. 61–84). New York: Russell Sage.

Schwarz, N., Strack, F., Kommer, D., & Wagner, D. (1987). Soccer, rooms, and the quality of your life: Mood effects on judgments of satisfaction with life in general and with specific domains. *European Journal of Social Psychology, 17*(1), 69–79.

Sedikides, C. (1995). Central and peripheral self-conceptions are differentially influenced by mood: Tests of the differential sensitivity hypothesis. *Journal of Personality and Social Psychology, 69*(4), 759–777.

Sirgy, M. J., & Cornwell, T. (2001). Further validation of the Sirgy et al.'s measure of community quality of life. *Social Indicators Research, 56*(2), 125–143.

Sirgy, M. J., & Cornwell, T. (2002). How neighborhood features affect quality of life. *Social Indicators Research, 59*(1), 79–114.

Sirgy, M. J., Rahtz, D. R., Cicic, M. A., & Underwood, R. (2000). A method for assessing residents' satisfaction with community-based services: A quality-of-life perspective. *Social Indicators Research, 49*(3), 279–316.

Speare, A., Goldstein, S., & Frey, W. H. (1975). *Residential mobility, migration, and metropolitan change*. Cambridge: Ballinger.

Stimson, R. J., & McCrea, R. (2004). A push-pull framework for modelling the relocation of retirees to a retirement village: The Australian experience. *Environment and Planning A, 36*, 1451–1470.

Trauer, T., & Mackinnon, A. (2001). The role of importance ratings in quality of life measurement. *Quality of Life Research, 10*, 579–585.

Turksever, A. N. E., & Atalik, G. (2001). Possibilities and limitations for the measurement of the quality of life in urban areas. *Social Indicators Research, 53*(2), 163–187.

Veenhoven, R. (2000). The four qualities of life: Ordering concepts and measures of the good quality. *Journal of Happiness Studies, 1*, 1–39.

Vitterso, J. (2001). Personality traits and subjective well-being: emotional stability, not extraversion, is probably the important predictor. *Personality and Individual Differences, 31*(6), 903–914.
Vitterso, J., & Nilsen, F. (2002). The conceptual and relational structure of subjective well-being, neuroticism, and extraversion: Once again, neuroticism is the important predictor of happiness. *Social Indicators Research, 57*(1), 89–118.
Walmsley, D. J., Epps, W. R., & Duncan, C. J. (1998). Migration to the New South Wales North coast 1986–1991: Lifestyle motivated counterurbanisation. *Geoforum, 29*(1), 105–118.
Wright, S. J. (1985). Health satisfaction: A detailed test of the multiple discrepancies theory model. *Social Indicators Research, 17*(3), 299–313.
Wu, C. H., & Yao, G. (2006a). Do we need to weight item satisfaction by item importance? A perspective from Locke's range-of-affect hypothesis. *Social Indicators Research, 9*(3), 485–502.
Wu, C. H., & Yao, G. (2006b). Do we need to weight satisfaction scores with importance ratings in measuring quality of life? *Social Indicators Research, 78*(2), 305–326.

# Chapter 4
# The Evolution of Integrative Approaches to the Analysis of Quality of Urban Life

**Rod McCrea, Robert Stimson, and Robert W. Marans**

## Introduction

In the discussion in the preceding chapters, we have already seen that quality of life (QOL) is a broad term which encompasses notions of a good life, a valued life, a satisfying life, and a happy life. Often the notions of a satisfying life and happy life are combined into the notion of subjective well-being, which includes life satisfaction, positive affect, and negative affect (Andrews and Withey 1976; Diener et al. 1999). The term *subjective* quality of urban life (QOUL) focuses on the notion of satisfaction with place or where one lives in the urban environment. The notion of an explicitly "urban" context for QOL, as is designated by the title of this book, is deliberate and realistic as the vast majority of people in developed nations live in urban settings, and most of them live in large urban environments including big cities and large metropolitan areas. Even in the developing world, a rapidly increasing proportion of people lives in urban settings.

Thus, it is reasonable to suggest that the nature of objective urban environments affects people's evaluations or assessments of their QOL. For instance, urban

R. McCrea (✉)
Institute for Social Science Research, The University of Queensland,
Level 4, GPN3 Building, Campbell Road, St Lucia, Brisbane, QLD 4072, Australia
e-mail: r.mccrea@uq.edu.au

R. Stimson
Australian Urban Research, Infrastructure Network (AURIN), Faculty of Architecture,
Building and Planning, University of Melbourne, VIC 3010, Melbourne, Australia
e-mail: r.stimson@unimelb.edu.au

R.W. Marans
Institute for Social Research, University of Michigan,
POB 1248, Ann Arbor, MI 48106, USA
e-mail: marans@umich.edu

R.W. Marans and R. Stimson (eds.), *Investigating Quality of Urban Life: Theory, Methods, and Empirical Research*, Social Indicators Research Series 45,
DOI 10.1007/978-94-007-1742-8_4, 

planners are charged with planning and managing the objective urban environment (for example, designating and regulating land uses, planning parks and open spaces, transport systems and urban infrastructure, and approving buildings), and they are interested in how their decisions affect resident's satisfaction with more generally, the quality of urban living. So in strategic urban planning documents, it is now commonplace for there to be explicit statements that an objective of planning and of strategic plans is to enhance the QOL for urban residents. That is seen, for example, in the context of a rapidly growing region in Australia, where the South East Queensland Regional Plan (Office of Urban Management 2005) stated that regional planning:

> … presents the opportunity to manage this growth and associated change to protect and enhance our *quality of life* and achieve sustainability. [italics added] (p. 4)

As we saw in the discussion in Chap. 3, *satisfaction with urban living* may occur at different levels of *geographical scale*, for example, the local area or "neighborhood," wider local "community" which might be a council area or a town, and the larger "region" which might be a large city or metropolitan area. Research has shown that satisfaction at different levels of urban living influences overall life satisfaction of people (see, for example, Andrews and Withey 1976; Campbell et al. 1976; Michalos and Zumbo 1999; Sirgy and Cornwell 2002; Sirgy et al. 2000; McCrea et al. 2005; see also Chaps. 7, 8 and 15).

However, often assumptions are made that improvements in objective indicators of QOUL are associated with improvements in the subjective experience of QOUL when little is actually known about the strength of these associations. The strength of those associations needs to be empirically established in order to validate inferences from objective indicators of QOUL to subjective QOUL and to better understand the relationships between the two. But they should not be considered two measures of the same thing (for example, measuring objective access to services and facilities is different from subjectively evaluating whether that access is good or bad). The distinction between objective and subjective QOL indicators is also useful because it relates the public experience of QOUL with the private subjective experiences (Cummins 2000).

Somewhat surprisingly, relatively little work has been done to empirically test the links between satisfaction with urban living and objective characteristics of the urban environment (Cutter 1985; Rogerson et al. 1989). Linking requires research designs in QOUL studies which explicitly seek to investigate the links between characteristics of the objective urban environment and people's subjective evaluations of the urban environment because there needs to be a methodology for linking them operationally. However, it is somewhat surprising that there have been relatively few attempts to integrate data obtained at these two levels.

In this chapter we explore the nature of the links between the *objective* urban environment and the *subjective* evaluation or assessment of the urban environment, and in particular how that integration may enhance our understandings of relationships between them. With the exception of studies discussed in this volume, links between objective and subjective indicators are not often investigated in QOUL studies. More commonly, either objective indicators (see Chap. 2) or subjective indicators (see Chap. 3) are examined independently, resulting in two main paradigms of QOL

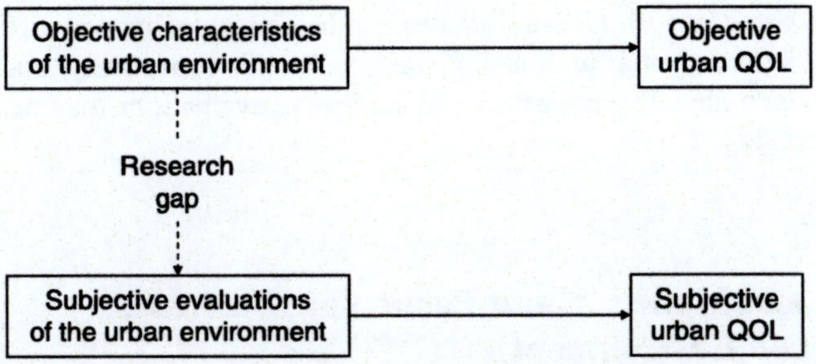


**Fig. 4.1** A research gap in the QOUL literature (Source: McCrea 2007)

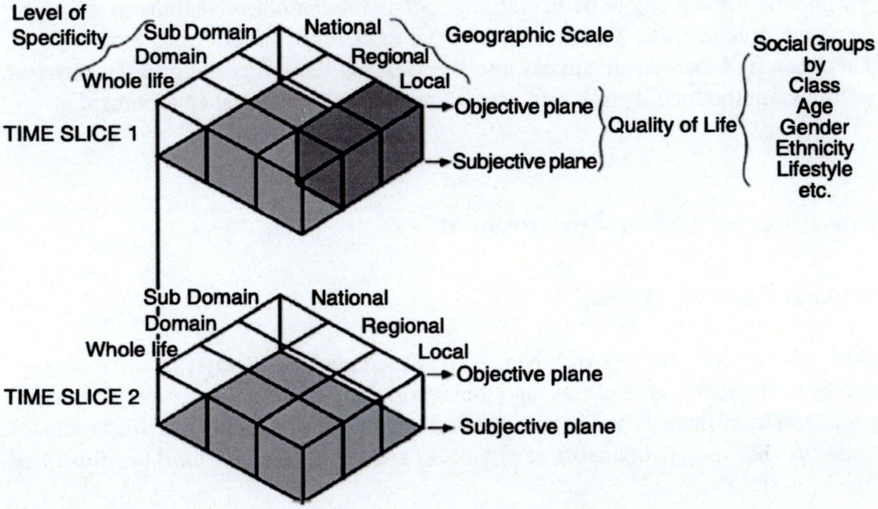


**Fig. 4.2** A five-dimensional framework for QOUL research (Source: Pacione 2003)

research (Andelman et al. 1998). Linking the two approaches as illustrated in Fig. 4.1 has been identified as a significant research gap (McCrea 2007).

However, objective and subjective indicators are only two dimensions of QOL. Pacione (2003) proposes a five-dimensional framework that integrates various dimensions of QOL research: objective, subjective, time, domain specificity, geographic scale, and social group dimensions. This is reproduced in Fig. 4.2. In each time slice, there are objective and subjective planes which relate to objective and subjective measures of QOUL. Objective and subjective QOUL may also vary by social group. In addition, each of those objective and subjective planes has two other dimensions: geographic scale and levels of specificity. Looking at the levels of specificity on the subjective plane, *whole life* may be conceptualized as satisfaction with overall life, which consists of satisfaction across a range of important *life domains* (such as satisfaction with work, social relationships, community, and so on), as well as satisfaction with explicitly *urban domains*. Urban domains can be conceptualized at different levels of geographic scale, such as housing, neighborhood, community, and the wider region for which objective measures and subjective evaluations may be derived, as discussed in Chap. 3.

## Theories and Models About Particular Dimensions of the Urban Environment

In Chap. 3 we discussed some general theories and models linking objective and subjective indicators in QOL domains. In this section we examine in greater detail theories and models relating to particular *physical* and *social* objective dimensions of the urban environment and associated subjective evaluations of the urban environment. These can be broadly categorized into theories and models relating to the physical environment (both built and natural environments) and the social environment.

### *The Physical Urban Environment*

#### Optimal Centrality Theory

Optimal centrality theory (Archibugi 2001; Cicerchia 1999) relates to urban density, access to services and facilities, and the overloading of urban structure. The theory postulates that there is an optimum urban scale or urban size which maximizes trade-offs between the benefits of *city effect* and costs of *urban load* (as illustrated in Fig. 4.3):

(a) "City effect" relates to access to opportunities, services, and facilities available by virtue of a city's size.
(b) "Urban load" relates to the negative consequences of urban growth (for example, congestion, overcrowding, cost of housing, and environmental degradation).

The theory postulates that there will be net benefits to QOUL as small urban centers grow, and additional services and facilities are provided to a growing critical mass of population, while at the same time, relatively low costs are incurred in terms of increased urban load. However, as urban growth continues past the hypothesized optimum, the rate of increase in city effect slows, and the rate of increase in urban load quickens, eventually leading to "overload" where additional growth is hypothesized to decrease net benefits in QOUL.

This theory may be extended from considering the influence of urban scale on QOUL to considering the influence of *urban density* on QOUL. As with urban scale, urban density is associated with increasing access to services and facilities, and it can also be associated with increasing urban load-associated problems such as pollution, traffic congestion and cost of housing. Using that extended theoretical

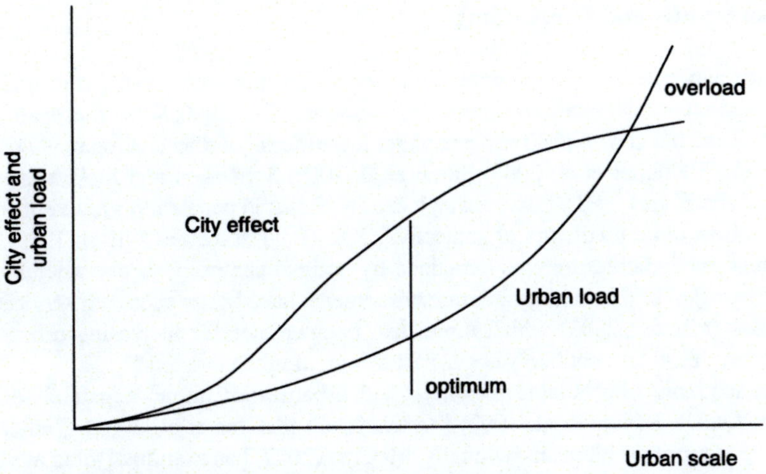


**Fig. 4.3** Model of optimal urban scale (Source: Cicerchia 1999)

framework, objective density should be positively related to both subjective access and subjective overloading.

## Access to Services and Facilities

It has been shown that access to services and facilities is an important component of subjective QOUL (see, for example, Glaeser et al. 2000; Rogerson et al. 1989; 1996). For example, community satisfaction has been predicted by the provision of services and facilities such as education services, emergency services, public transport, parks, shopping, and leisure opportunities (for example, Campbell et al. 1976; Sirgy and Cornwell 2001; Sirgy et al. 2000; Turksever and Atalik 2001). Satisfaction with access to services and facilities is also important in making residential location decisions (for example, Chiang and Hsu 2005; Dokmeci and Berkoz 2000; Ge and Hokao 2006; Mitrany 2005). However, despite the well-established connection between subjective access to services and facilities and subjective QOUL, little research has been conducted examining the strength of relationships between broad objective access and subjective evaluations of access in urban environments.

Recent research by McCrea (2007) examined the strength of relationships between broad objective access and subjective access in terms of overall proximity to services and facilities and overall satisfaction with access to services and facilities. Even though subjective access involves more than simply proximity to services and facilities, proximity is a main component of access, and so, a moderately strong relationship could be expected between objective and subjective access. However, the relationship may be weakened by any of the processes discussed in the previous chapter, such as individual variations found in the importance of access to various services and facilities between residents (see, for example, Dokmeci and Berkoz 2000; Kim et al. 2005).

### Urban Density and Overloading

High-density and rapidly growing urban environments have been associated with increased economic, social and environmental stress (Perz 2000; Schwirian et al. 1995), and research has shown that residents prefer lower-density urban environments (Brown et al. 1997; Cramer et al. 2004; Filion et al. 2006; Schwanen and Mokhtarian 2004; Senecal and Hamel 2001). However, high density and rapid population growth have also been shown to be predictors of subjective QOL (Baldassare and Wilson 1995). That apparent contradiction may be explained by optimal centrality theory whereby residents living in higher-density urban environments have better access to services and facilities (Mitrany 2005) which more than compensates for increasing urban load. However, this would not likely occur past a point of urban overload.

A wide range of problems associated with urbanization impact negatively on subjective QOUL (Kemp et al. 1997; Marans 2002; McCrea et al. 2005). Rather than looking at those problems individually, McCrea (2007) has examined them at a broad level by investigating the relationships between objective density and subjective overloading where the latter was a composite measure of a wide range of urban problems such as pollution, loss of natural areas, traffic congestion, and cost of housing.

### Natural Environments

Close proximity to natural environments (for example, rural and coastal environments) has been found to facilitate recovery from stress (Berto 2005; Kaplan 1995; Ulrich et al. 1991). That is in contrast to higher levels of stress frequently found in more dense and crowded urban environments (see Walmsley 1988 for a review). Preferences for suburban and low-density living might then be explained, in part, by attraction to natural environments due to their restorative effects on stress associated with urban living (van den Berg et al. 2007). Thus, close proximity to rural and coastal (or other water) environments should be associated with favorable subjective evaluations of the natural environment which in turn should be positively associated with subjective QOUL.

Notwithstanding these general expectations, preferences for the natural environment do appear to vary between people (Vogt and Marans 2004). For example, families with children are more likely to prefer neighborhoods with green space and recreational opportunities in choosing where to live (Kim et al. 2005). Differences between people's preferences and their residential location choices may weaken the relationships found between proximity to natural environments and subjective evaluations of the urban environment.

## *The Social Urban Environment*

Subjective evaluations of the social environment are related to subjective QOUL via the satisfaction of the social needs of residents, such as favorable neighborly relations and a sense of community (Davidson and Cotter 1991; Farrell et al. 2004;

Sirgy and Cornwell 2002). Favorable neighborly relations might include social capital, which incorporates trust and reciprocity (Coleman 1988; Putnam 1995), as well as general friendliness between neighbors, while sense of community might include a faith that needs would be meet through a shared commitment and a sense of belonging (McMillan and Chavis 1986). It has been shown that sense of community is closely related to positive relations among neighbors (Farrell et al. 2004; Prezza et al. 2001). That supports examining the subjective social environment as a broad construct.

Objective dimensions of social environments often relate to household structure, socioeconomic status and ethnicity. In their factorial ecology of the social and spatial structure of the Brisbane-South East Queensland metropolis in Australia, Western and Larnach (1998) found these dimensions, as well as a disadvantage dimension (that is, unemployment, single parenthood, and public housing occupancy). These objective social dimensions of the urban environment were examined by McCrea (2007) in relation to respondents' subjective evaluations of their QOUL.

#### Social Disorganization Theory

Social Disorganization Theory (SDT) predicts that neighborhood social ties would be stronger (that is, more organized) in neighborhoods that are:

- More stable (that is, with less residential mobility)
- More affluent (for example, more community facilities and resources)
- Less disadvantaged (for example, fewer social problems)
- More ethnically homogeneous (for example, fewer ethnic minorities)as shown in pioneering research by Shaw and McKay (1942) and in later research by Sampson and Groves (1989)

Even though SDT is normally associated with studying the effects of social organization on juvenile crime via the impact of informal social control over youths and their development (for example, Cullen and Agnew 2003; Kubrin and Weitzer 2003), SDT can also be used in studying how objective social dimensions may impact on subjective evaluations of the social environment since "socially organized" neighborhoods also have more favorable neighborly interactions and sense of community.

In testing STD, evidence has been found that supports relationships between objective social dimensions and subjective evaluations of the social environment. For example:

- Less social capital and sense of community have been found in more disadvantaged neighborhoods (Cantillon et al. 2003; Kawachi et al. 1999)
- Less social cohesion among neighbors has been found in disadvantaged and less residentially stable neighborhoods (Sampson et al. 1997)
- Higher neighborhood attachment and involvement have been found in higher class and more residentially stable neighborhoods (Taylor 1996)

However, the direct effects of objective social dimensions on subjective evaluations of the social environment have not been found to be strong.

### Subculture Theory

Subculture theory postulates that in urban environments, the population becomes large enough for the formation of subcultures to manifest spatially by allowing people of similar social backgrounds and with similar lifestyles to live in close proximity, and once areas become associated with particular subcultures, they attract others of similar backgrounds through selective residential location decisions (Savage et al. 2003). The underlying process is driven by a preference of many residents to live in neighborhoods with similar others, which is a form of "homophily" (see Lazarsfeld and Merton 1954; McPherson et al. 2001), and facilitates the generation of intra-urban spatial variation in subcultures (Fischer 1984) as opposed to a more general urban way of life (see Simmel 1950; Wirth 1938). Homophily can be literally translated as "love of the same" and encapsulated in the phrase "birds of a feather flock together."

Urban subcultures and homophily might also assist in explaining weak relationships between objective social dimensions and subjective evaluations of the social environment. As with other dimensions of the urban environment, the strength of relationships may depend on the extent to which people consider an attribute of the urban environment to be important. With regard to social dimensions, the similarity with others also depends on the social characteristics of particular residents. For example, people considering that living near similar others is important may well evaluate the social environment more favorably if they live in a neighborhood which has social dimensions similar to their own social characteristics. This type of homophily is what Lazarsfeld and Merton (1954) call "status homophily."

Lazarsfeld and Merton (1954) have also distinguished between "status homophily" and "value homophily" (McPherson et al. 2001), and intra-urban spatial variation in subcultures may not only develop around social characteristics of residents such as socioeconomic status or ethnicity. Subcultures may also be based around different values or lifestyles (Curry et al. 2001; Ge and Hokao 2006; Walmsley et al. 1998). However, those two types of homophily are not mutually exclusive, and research by McCrea (2007, 2009) has examined the role of status homophily based on social characteristics of people and their neighborhood in investigating the relationships between objective social dimensions and subjective evaluations of the social environment.

## *Other General Theories Linking Objective and Subjective QOL*

In Chap. 3, a number of more general theories linking objective and subjective QOL indicators were discussed. These were bottom-up models, top-down models, mood bias models, subjective judgment models, adaptation models, subjective importance models, residential relocation models, and agent-based models. A common theme across many of these theories was that psychological processes, individual differences, and residential relocation processes serve to weaken direct links between objective and subjective indicators of QOL and QOUL. Nonetheless, we may still expect significant relationships between subjective evaluations and objective characteristics of the urban environment.

## Empirical Evidence Linking Objective and Subjective QOUL

There have been some attempts to be overcome the apparent dichotomy between the *subjective* and the *objective* approaches to the study of QOUL by including both types of indicators, but relatively few investigations have related them together by developing research designs to explicitly empirically examine the links between objective characteristics of the urban environment and people's subjective evaluations of the urban environment. Studies that have attempted to bridge this research gap include those by Campbell et al. (1976), Marans and Rodgers (1975); Marans (2002); McCrea et al. (2006); and McCrea (2007). With few exceptions, direct links between objective and subjective indicators have been found to be weak in these studies.

Weak relationships between objective and subjective measures are common in QOL research covering other life domains. In reviews of QOL studies by Evans and Huxley (2002) and McCrea (2007), it is concluded that objective circumstances do not greatly influence subjective QOL (see, for example, Bowling and Windsor 2001; Headey et al. 1984; Schwarz and Strack 1999) and that life domain satisfactions are better predictors of overall life satisfaction than objective circumstances (see, for example, Andrews and Withey 1976). However, it is also the case that that objective circumstances are more related to satisfaction in related specific life domains than to overall life satisfaction (refer back to the discussion on bottom-up models in Chap. 3). This does suggest that relationships between objective circumstances and overall life satisfaction are mediated via satisfaction in various life domains (Evans and Huxley 2002).

In the context of enhancing both objective and subjective QOUL, it is important to continue to develop research designs that explicitly seek to investigate the links between the objective characteristics of the urban environment and the subjective evaluations of the urban environment, as has been undertaken in recent research by McCrea and colleagues (McCrea 2007; McCrea et al 2006). Unfortunately, most research in QOUL focuses on either subjective evaluations or objective characteristics of the urban environment. While the objective and subjective approaches have been discussed in Chaps. 2 and 3, respectively, in the sections that follow, we summarize some of the findings from research that has specifically focused on the analysis of the urban environment.

## *A Focus on Subjective Evaluations of the Urban Environment*

Studies focusing primarily on the subjective evaluation of QOUL have found that people's subjective evaluations of many aspects of the urban environment can contribute to satisfaction with urban living and overall life satisfaction. Some examples are given below:

(a) A study by Michalos and Zumbo (1999) predicted life satisfaction from 14 life domains for seven different time periods between 1979 and 1997. Of those explicitly urban domains relating to QOL, it was found that housing was significant in six time periods, recreational activity in five, transportation in four, government services in three, and residential area was significant in two time periods (although it was not included in one time period). Thus, satisfactions in various urban domains were shown to be predictors of overall life satisfaction.
(b) Research by McCrea et al. (2005) has examined different geographic levels of subjective QOUL in the Brisbane-South East Queensland metropolis in Australia. It was shown that regional satisfaction was best predicted by people's evaluation of regional services (such as health and education) and the cost of living; neighborhood satisfaction was best predicted by evaluations of social interactions, neighborhood crime and public facilities (parks, libraries, etc); while housing satisfaction was predicted best by age of home and home ownership.
(c) A study by Turksever and Atalik (2001) has predicted life satisfaction in seven districts of Istanbul as well as the Istanbul region as a whole using satisfaction with 18 different aspects of living in the region. The significance of different predictors varied across the districts; however, for the Istanbul region as a whole, the significant predictors were health, climate, crowding, sporting, housing conditions, travel to work, and environmental pollution.
(d) Research by Sirgy and colleagues (Sirgy and Cornwell 2001; Sirgy et al. 2000) has shown that a wide range of specific services provided by government, business and nonprofit institutions contributed to community satisfaction and ultimately to overall life satisfaction in combination with satisfaction in other life domains.

Additional studies could be cited, but those referred to above show that subjective evaluations of many aspects of urban living influence subjective QOUL and overall QOL, and that the importance of those factors can vary with geographic scale within urban regions.

## *A Focus on Objective Characteristics of the Urban Environment*

Studies where the focus is on objective QOUL typically include many objective characteristics of the urban environment, often combining or weighting objective indicators to generate an objective QOUL index for ranking places (see, for example, Blomquist et al. 1988; Boyer and Savageau 1981; Cicerchia 1999; Stover and Leven

1992). Interestingly, some studies also emphasize the trade-off between positive and negative aspects of urban living. For example, consider the following:

(a) Blomquist et al. (1988) have modeled the trade-off between housing costs, wages and amenity to develop QOUL indexes for 253 Standard Metropolitan Statistical Areas in the United States. Hedonic wage and rent equations were used to derive implicit prices for various urban amenities, which in turn were used as weights in compiling an objective QOUL index. The underlying idea was that the amenity value of an area is implied from areal variation in housing costs and wages.
(b) Cicerchia (1999) has theorized about the trade-off between city effect and urban load. City effect relates to "access to superior urban functions, opportunities and services" available by virtue of a city's size. Urban load relates to a number of negative consequences of urban growth (for example, congestion and environmental degradation). The underlying idea in this study was to ascertain the "optimum centrality" for a city, which is the size of a city which maximizes the difference between city effect over urban load. However, if the city size becomes too large, then escalating urban load may exceed city effect and create urban overload.

## An Example of Modeling Links Between Objective and Subjective Indicators of QOUL

As discussed above, attempts to investigate the links between objective characteristics of urban environments and individuals' subjective evaluations of QOUL have been relatively few, and the development and implementation of operational models are rare. However, recently, McCrea et al. (2006) used structural equation modeling to do so.

That first involved setting up a model framework. The general framework used by McCrea et al. (2006) was adapted from Campbell et al. (1976) (see Fig. 4.4 which is reproduced from Fig. 1.1 in Chap. 1). However, this general framework was too complex to model in one study. So a simplified bottom-up operational model was developed to investigate paths between broad objective dimensions of the urban environment, subjective evaluations relating to those dimensions, and overall subjective QOUL.

The bottom-up part of the general framework in Fig. 4.4 is reflected in the relationships between the shaded boxes. Objective characteristics (for example, traffic noise) predict subjective perceptions (for example, hearing the noise) which predict subjective evaluations (for example, the traffic noise is too loud), which predict satisfaction in various urban domains. In the operational model outlined later (see Fig. 4.5), this bottom-up model is simplified by skipping the process of sensory or subjective perception (that is, by predicting subjective evaluations directly from objective characteristics), as well as combining the various geographic levels of subjective QOUL (housing, neighborhood, community, etc.) into an overall measure of subjective QOUL.

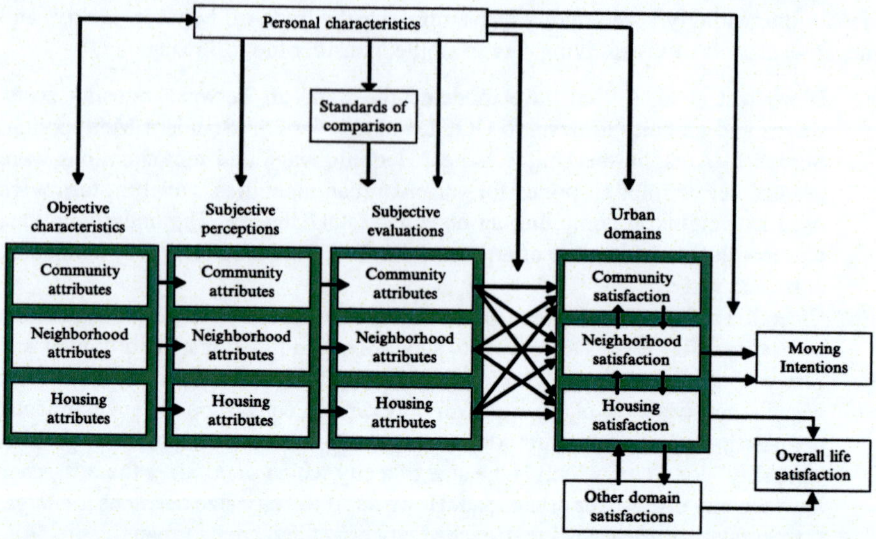


**Fig. 4.4** Model of determinants of satisfaction with the residential environment (Source: McCrea 2007: p. 1/6. Adapted from Campbell et al. 1976)

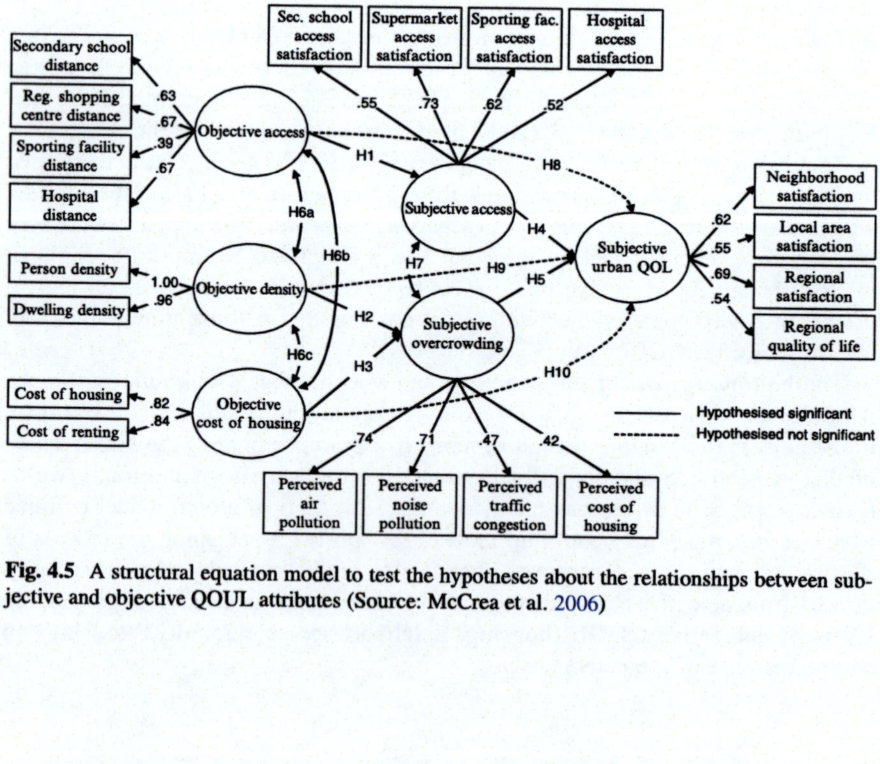


**Fig. 4.5** A structural equation model to test the hypotheses about the relationships between subjective and objective QOUL attributes (Source: McCrea et al. 2006)

## *Using GIS Technology to Link Objective and Subjective Indicators*

In QOL studies in other life domains, the objective measures have often been demographic and socioeconomic characteristics of the *individual* rather than objective characteristics of the *environment* or *situational setting* in which a person lives. This is suitable for domains such as satisfaction with income or health where both the objective and subjective indicators relate to individuals. However, when examining links in QOUL, characteristics of urban *environments* need to be linked to *individuals*. This second method was taken by McCrea et al. (2006) in a study of the Brisbane-South East Queensland metro-region (SEQ) in Australia where they link objective and subjective measures of the urban environment using geographic information systems (GIS). This involved obtaining residential addresses of respondents, geocoding these addresses, and relating their subjective responses with spatially defined objective data sets using GIS.

GIS technology is now providing a sophisticated set of tools for storing, displaying and analyzing geographic information and is being increasingly used in the social sciences (Goodchild 2000). It is a particularly powerful tool in integrating different types of data – such as point-located data and area or polygon data – at different levels of spatial scale. After determining the residential location of survey respondents, GIS can relate other objective spatial data to those locations, for example, by calculating distances to other urban features or by linking a resident to other geographically based data sets such as population census information. Therefore, GIS methodology can be used to link objective information about urban environments with subjective indicators of the urban environment efficiently and cost-effectively by taking advantage of readily available geographic information and linking it to residential locations of survey respondents. Given the utility of this tool, the wider use of GIS has been urged for the social sciences to gain new insights and to develop a more "spatially integrated social science" (Goodchild et al. 2000).

## *The Modeling Approach and Procedures*

In a study by McCrea et al. (2006), a spatially stratified probability sample design was used to generate a random sample of households across the SEQ region with a total of 1,471 respondents living in urban areas. The survey collected information on the subjective assessments of survey respondents to a wide range of QOUL issues. Data from the survey was then linked to objective information about the urban environment using GIS. Every survey respondent's street address was associated with geometric information (that is, digital street data for SEQ) using MapInfo Professional GIS and the MapInfo StreetPro national roads database. That process of attaching geometric location information to the QOL database is referred to as *geocoding*.

GIS software packages such as MapInfo Professional GIS and ESRI Arcview GIS can also be used, as in the case of the study by McCrea et al. (2006), to locate survey respondents and relate them to objective spatial information, such as the location of urban facilities like shopping centers, schools and parks. In addition, the location of survey respondents' residences may be readily interfaced with spatially defined social data sets, such as aggregate information on local areas from the census data, thus enabling each survey respondent to be matched to the demographic and socioeconomic characteristics of, say, the neighborhood in which they live. Once the location of the survey respondents is geocoded, it is possible to use GIS technology to link the individual-level survey data to virtually any spatial objective information that is available from a wide range of spatial data sets.

In the study of the SEQ region, GIS was used to generate objective variables associated with a resident's urban environment which were then tested in a bottom-up structural equation model. The modeling approach proposed that objective measures of access and overcrowding predicted subjective evaluations of access and overcrowding, which in turn predicted subjective urban QOL (or subjective QOUL). The strength of those associations between the objective and subjective measures was explored as well as the extent to which the relationship between the objective measures and subjective QOUL was mediated by subjective measures of access and overcrowding.

McCrea et al. (2006) proposed the following hypotheses for testing (see also Fig. 4.5):

H1: Higher objective access significantly predicts higher subjective access.

H2: Higher objective density significantly predicts higher subjective overcrowding.

H3: Higher objective cost of housing significantly predicts higher subjective overcrowding.

H4: Higher subjective access significantly predicts higher subjective QOUL.

H5: Lower subjective overcrowding significantly predicts higher subjective QOUL.

H6: Objective access, objective density and objective cost of housing are significantly correlated with each other.

H7: Subjective access and subjective overcrowding are significantly correlated.

H8: The relationship between subjective QOUL and objective access is mediated by subjective access (that is, there is no direct relationship between subjective QOUL and objective access).

H9: The relationship between subjective QOUL and objective density is mediated by subjective overcrowding (that is, there is no direct relationship between subjective QOUL and objective density).

H10: The relationship between subjective QOUL and objective cost of housing is mediated by subjective overcrowding (that is, there is no direct relationship between subjective QOUL and objective cost of housing).

These hypotheses were tested simultaneously through structural equation modeling.

## *Measures*

The structural equation model developed by McCrea et al. (2006) used 12 *manifest* subjective variables derived from a survey to measure 3 *latent* subjective variables (subjective access, subjective overcrowding and subjective QOUL), and 8 *manifest* objective variables, to measure 3 *latent* objective variables (objective access, objective residential density, and objective cost of housing). These measures are described in what follows.

### The Subjective Variables

Subjective QOUL (or subjective urban QOL) was a latent variable measured using four manifest or measured variables:

(a) Neighborhood satisfaction was based on a question asking survey respondents how much they agreed or disagreed with the statement "I feel satisfied living in this neighborhood," and answered on a 5-point scale from 1 = "strongly disagree" to 5 = "strongly agree."
(b) Local area satisfaction was based on a question of survey respondents "How satisfied are you with living in your local council area (that is, city or shire council area)?" and answered on a 5-point scale from 1 = "very dissatisfied" to 5 = "very satisfied."
(c) Regional satisfaction was based on how satisfied the resident was with "living in the Brisbane-South East Queensland Region," and answered on a 5-point scale from 1 = "very dissatisfied" to 5 = "very satisfied."
(d) Regional QOL was based on the question "In general, how would you rate the overall quality of life in the Brisbane-South East Queensland Region?" and answered on a 5-point scale from 1 = "very poor" to 5 "very good."

A subjective access latent variable was measured using the four manifest variables:

- Secondary school access satisfaction
- Supermarket access satisfaction
- Sporting facility access satisfaction
- Hospital access satisfaction

Those manifest variables were based on a question asking respondents how satisfied they were with their access to these facilities on a 5-point scale from 1 = "very dissatisfied" to 5 = "very satisfied."

The subjective overcrowding latent variable was measured with four manifest variables: air pollution, noise pollution, traffic congestion, and cost of housing. These manifest variables were based on the degree to which residents thought these things were a problem in the Brisbane and South East Queensland region, responding on a 5-point scale from 1 = "not a problem" to 5 = "a very great problem."

### The Objective Variables

The latent variable objective access was measured with four manifest variables:

- Secondary school distance
- Regional shopping center distance (there were 12 in the study region)
- Sporting facility distance (including parks, swimming centers, bowling centers, golf courses, rifle ranges, soccer fields, and tennis courts)
- Hospital distance

Each manifest variable was measured as the straight line distance between the respondent's residence and the closest facility of that type. The source of information on the facility locations was MapInfo StreetPro, and distances were calculated using GIS. Note that lower scores indicate better access (that is, shorter distances to facilities).

The objective density latent variable was measured with two manifest variables: population density per square kilometer and dwelling density per square kilometer. The density measures for each resident were based on the area of the Census Collection District in which each respondent resided (as determined by GIS) and the associated population and dwelling counts from the 2001 Australian Census of Population and Housing (Australian Bureau of Statistics 2001a). A Census Collection District is the smallest geographic area used in the census, containing an average of about 225 dwellings in the 2001 census (Australian Bureau of Statistics 2001b).

The cost of housing was a latent variable measured with two manifest variables relating to the cost of renting and the cost of purchasing dwellings in the resident's Census Collection District. Cost of renting was the percentage of rented dwellings costing $200 per week or more, and cost of purchasing was the percentage of owner-occupied dwellings with housing repayments of over $1,000 per month. Both of these benchmarks – $200 and $1,000 – are related to the national medians for renting and purchasing dwellings, respectively.

## *Structural Equation Modeling*

Structural equation modeling was used by McCrea et al. (2006) to test the hypotheses linking objective dimensions and subjective evaluations of the urban environment for two main reasons: First, structural equation models can easily test a number of mediated bottom-up paths simultaneously, and second, structural equation models include measurement models for the latent variables which separate out measurement error from the main structural model of latent variables.

The full structural equation model is shown diagrammatically in Fig. 4.5. In that figure:

- The boxes represented *manifest variables*
- The ovals represented *latent variables*

- In the *measurement model*, the straight arrows from the latent variables to the manifest variables indicated factor loadings
- In the *structural model*, the arrows from one latent variable to another reflected regression coefficients, and a dashed line indicates that a nonsignificant path was hypothesized
- The paths between the latent objective measures of the urban environment and subjective urban QOL (or subjective QOUL) were hypothesized to be mediated by the latent subjective measures of the urban environment, and so the direct or unmediated paths were hypothesized to be nonsignificant
- The double-headed curved arrows in the structural model indicated relationships with no directionality hypothesized (that is, simple correlations)
- Where there were no paths (that is, arrows), no relationships were hypothesized

McCrea et al. (2006) undertook the analysis in three stages using the correlation matrix of manifest variables as data:

(a) In Stage 1, a mediated path structural model was hypothesized where the relationships between objective latent variables and subjective QOUL were mediated by subjective measures of the urban environment, and the fit of the model was evaluated (that is, paths for H8, H9 and H10 were not included).
(b) In Stage 2, direct paths between the objective latent variables of the urban environment and subjective QOUL were added to see whether these paths were significant (that is, H8, H9 and H10 were added to test whether subjective access and subjective overloading fully mediated these paths).
(c) In Stage 3, other paths which significantly improved the fit of the model were identified by examining the modification indices and the standardized residuals, and then added to the model.

The measurement part of the structural equation model was satisfactory. By way of example, for Stage 1 of the analyses, the measurement model factor loadings are shown in Fig. 4.5, represented by the arrows from latent variables (in circles) to manifest variables (in rectangles). All manifest variables loaded well onto their latent variables, with the lowest factor loading being 39.

## *Summary of the Findings*

The findings from the modeling undertaken by McCrea et al. (2006) may be summarized as follows:

(a) It was actually found that there was *not* a strong link between the objective and subjective latent measures of the urban environment as there was not a strong relationship between objective access and subjective access, and the relationships were weak between objective density and subjective overcrowding, as well as between objective cost of housing and subjective overcrowding. As a consequence, those objective latent variables explained very little variance in subjective access and subjective overcrowding. That finding was in accord with results from studies in other life domains where objective indicators were generally found to be weak predictors of satisfaction in related life domains (see, for example, Cummins 2000; Evans and Huxley 2002). Consequently the McCrea et al. (2006) study suggests that we should take care when making inferences from improvements in objective indicators of urban QOUL to improvements in subjective QOUL.

(b) As expected, the objective latent variables (objective access, objective density, and objective cost of housing) were intercorrelated, with objective access and objective density being highly correlated. That finding aligned with a review of studies by Cummins (2000) who found that objective variables were more closely associated with each other than with subjective variables. Cummins also found subjective variables were more closely associated with each other than with objective variables. However in the McCrea et al. (2006) study, no significant correlation was found between subjective access and subjective overcrowding despite a strong correlation between objective access and objective density. That finding indicates that relationships in the objective world may not always hold in the subjective world. Within the subjective world, however, subjective access and subjective overcrowding both predicted subjective QOUL, with subjective access being the better predictor. In terms of optimal centrality theory (Cicerchia 1999), this suggests that subjective access is more important than subjective overcrowding in the trade-off between the two when maximizing subjective QOUL, and that helps explain the trend toward increasing urbanization in the study area.

(c) The relationship between objective access and subjective QOUL was fully mediated by subjective access, as hypothesized. However, the relationship from objective density to subjective QOUL was only partially mediated by subjective overcrowding, suggesting that other factors associated with higher urban density negatively impact on subjective QOUL.

(d) Similarly, the relationship from objective cost of housing to subjective QOUL was only partially mediated by subjective overcrowding, suggesting other factors. However the direct relationship between cost of housing and subjective QOUL was positive, indicating any other factors associated with cost of housing would positively impact on subjective QOUL (for example, neighborhood quality factors).

The suggestion of other mediating factors highlights one of the limitations of the McCrea et al. (2006) study in that only a small number of objective and subjective latent variables were included in the model. When linking objective and subjective indicators of QOUL, other factors also need to be investigated. Perhaps, the most obvious are social factors, such as social interaction and crime. Another limitation with the study was the bottom-up model tested does not take into account psychological processes and the systemic nature of QOUL reflected in residential relocation decisions. These two limitations were addressed in doctoral research by McCrea (2007) which resulted in proposing an integrated framework which modifies the general framework first proposed by Campbell et al. (1976) and Marans and Rodgers (1975).

Even though the relationships found in McCrea et al. (2006) between objective indicators of QOUL and subjective QOUL were weak, this is an insufficient justification for complacency in continuing to improve the objective urban environment for residents (Kahneman 1999). Moreover, it is important to understand why the links are weak because different explanations have different urban policy implications. In a systemic model, this means understanding the impact of subjective judgment processes, individual differences in what is important for QOUL, and the residential relocation process on weakening links between objective and subjective indicators. Such integrated model is discussed in the next section.

## An Integrated Approach to Investigating QOUL

In the above example, a simplified bottom-up model of QOUL was tested, and the links between objective and subjective indicators of QOUL were found to be weak. However, the general framework of QOUL proposed by Campbell et al (1976) and Marans and Rodgers (1975) is much more complex and therefore needs a number of studies to test different elements of this general framework (refer back to Fig. 4.4). Recent doctoral research by McCrea (2007) has empirically tested in great detail the different elements of this framework as part of an integrated approach to the investigation of QOUL. McCrea (2007) set out to answer two research questions, namely:

(a) What are the strength of the links between the objective dimensions of the urban environment and associated subjective evaluations?
(b) What are the effects of psychological processes, individual and social group differences, and residential relocation processes on these links?

Using GIS technology, McCrea (2007) first linked objective and subjective data sets:

- Survey data collected from a 2003 QOL survey in the SEQ region in Australia which included people's subjective evaluations of QOL and QOUL
- Objective QOUL factors from GIS and population census data sets

Then, measures for ten objective dimensions of the physical and social environments and associated subjective evaluations of the urban environment were derived (see Table 4.1).

The impact of various processes on links between these objective dimensions and associated subjective evaluations was then explicitly modeled. Finally, the findings were drawn together into an integrated explanation of weak relationships between objective dimensions and subjective evaluations of the urban environment. This integrated explanation involved aligning individual expectations with residential choices as part of the residential relocation process, subjective judgment models, and differences in the subjective importance of various attributes of the urban environment in choosing where to live. From that empirical testing, McCrea then suggested a modified version of the broad conceptual model, as shown in Fig. 4.6.

**Table 4.1** Objective dimensions and subjective evaluations of the urban environment examined for South East Queensland

| Objective dimensions of the urban environment | Associated subjective evaluations of the urban environment |
|---|---|
| *Physical dimensions of the urban environment* | |
| Objective access | Subjective access |
| Objective density | Subjective overloading |
| Objective rural environment | Subjective natural environment |
| Objective coastal environment | Subjective natural environment |
| *Social dimensions of the urban environment* | |
| Objective younger nonnuclear households | Subjective social environment |
| Objective nuclear family households | Subjective social environment |
| Objective older nonnuclear households | Subjective social environment |
| Objective socioeconomic environment | Subjective social environment |
| Objective disadvantaged environment | Subjective social environment |
| Objective ethnic environment | Subjective social environment |

Source: McCrea (2007)

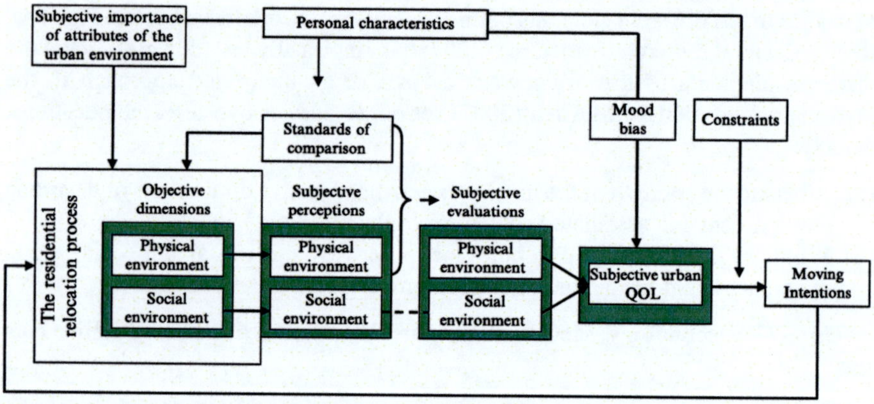


**Fig. 4.6** An integrated model of QOUL (Source: McCrea 2007: p. 176)

As is commonly found, subjective evaluations of the physical and social environments contribute to subjective QOUL (for example, Michalos and Zumbo 1999; Sirgy et al. 2000), which in turn contributes to moving intentions (for example, Clark and Ledwith 2006; Lu 1998), even though moving intentions are also influenced by constraints facing individuals (for example, Desbarats 1983; Li 2004). This part of the integrated explanation is well established in previous research.

The right brace (bracket) in Fig. 4.6 indicates that relationships between objective dimensions and subjective evaluations of the physical environment are not direct relationships, but indirect relationships where subjective evaluations arise from differences between objective dimensions and individuals' standards of comparison. McCrea (2007) found that differences between objective dimensions and standards of comparison were usually small. This aligning of standards and their judgment targets was an essential part of the explanation for weak relationships between objective dimensions and subjective evaluations of the urban environment.

Such aligning of standards and targets could not be explained by adaptation to residential environments over time *after* residential relocation. In contrast, there was support for the aligning of standards and targets *during* the residential relocation process in relation to the *physical* urban environment. That suggests that this aligning is primarily driven by residential relocation processes and individual differences in the subjective importance of various aspects of the physical urban environment in choosing where to live.

However, there was no support for an aligning of standards and targets in relation to the *social* urban environment. Objective dimensions in the social environment primarily arose because residents with similar social characteristics found similar attributes of the physical environment important rather than wishing to be near other residents with similar social characteristics (also see McCrea 2009). Consequently, a dashed line was used in Fig. 4.6 to show the inherently weak links between objective dimensions and subjective evaluations of the social environment.

McCrea's integrated model of QOUL modified the original general framework of Campbell et al (1976) and Marans and Rodgers (1975) in a number of ways:

(a) McCrea has added the residential relocation process to the left of the model in Fig. 4.6 to indicate that a resident's location on broad objective dimensions of the urban environment commonly arises out of the residential relocation decision and choice process and is frequently influenced by their subjective assessment of the importance of various attributes of the urban environment and their standards of comparison, both of which were found to be correlated with broad objective dimensions of the urban environment.

(b) McCrea does not include moderating influences of personal characteristics on subjective perceptions or sensory perceptions in this modified model because the links between objective characteristics and subjective perceptions of the urban environment were relatively direct as proposed by Campbell et al. (1976) and Marans and Rodgers (1975). That assists in simplifying the modified version.

(c) In the modified model in Fig. 4.6, it is suggested that subjective evaluations of the urban environment are predicted by the *difference* between subjective perceptions of objective dimensions of the physical environment and standards of comparison. That notion of the difference between perceptions and standards was also in the original model, although in the modified version, McCrea makes this explicit by using a right brace symbol rather than direct arrows from standards of comparison and subjective perceptions to subjective evaluations. This also serves to emphasize the indirect nature of links between objective dimensions and subjective evaluations of the physical environment since the direct arrow between the two has been removed.

(d) In his research, McCrea (2007) found that mood predicted subjective urban QOL (or subjective QOUL), but not subjective evaluations of the urban environment. In the modified model in Fig. 4.6, mood bias predicts subjective QOUL while the arrow showing direct effects of personal characteristics on subjective evaluations of the urban environment has been removed. That finding supported Schwarz and Strack's (1999) view that mood bias is only important in more global judgments relating to life satisfaction judgments rather than specific life domain judgments. While mood was the only direct personal characteristic on subjective evaluations examined by McCrea, in other research, personality traits of extroversion and neuroticism have been found to be highly correlated with positive affect and with negative affect, and the impact of personality may be mediated by mood (for a review, see Diener et al. 1999). So, these personality traits would also seem not important in directly influencing subjective evaluations of the urban environment.

(e) In the modified model in Fig. 4.6, the moderating arrow from personal characteristics to the relationships between subjective evaluations of the urban environment and subjective urban QOL (subjective QOUL) was removed by McCrea because most studies had found that subjective importance does not moderate subjective evaluations when predicting satisfaction in more global domains (see, for example, Mastekaasa 1984; Andrews and Withey 1976; Campbell et al. 1976; Cummins et al. 1994; Russell et al. 2006). Some authors have suggested this because very favorable and very unfavorable subjective evaluations inherently include subjective importance in their measurement (Trauer and Mackinnon 2001; Wu and Yao 2006).

(f) McCrea also tested whether there were moderating effects of the subjective importance of various attributes of the urban environment on objective dimensions when predicting subjective evaluations of the urban environment. However, no such moderating effects were found. Hence, there are no arrows in the modified version of the model in Fig. 4.6 reflecting moderating effects of subjective importance. Rather, the effects of subjective importance of various attributes of the urban environment impact on objective dimensions of the urban environment via the residential relocation process.

(g) In the modified version of the model in Fig. 4.6, moving intentions are linked back to the residential relocation process to emphasize the dynamic and systemic nature of relationships between objective dimensions and subjective evaluations of the urban environment. An intention to move is part of Phase I of the residential relocation model by Brown and Moore (1970) – refer back to Fig. 3.2 in Chap. 3 – encompassing the decision *whether* to move while Phase II encompasses deciding *where* to move. This linking of intentions to move with the residential relocation process makes explicit that the objective dimensions are not exogenous factors, but are part of a dynamic model of residential satisfaction (Amerigo and Aragones 1997).

## Conclusion

Researchers have taken an empirical positivistic approach to examining the relationships between objective dimensions of urban environments and people's subjective evaluations of the urban environment. Even though QOUL can be measured either objectively or subjectively, it may well be that QOUL is ultimately more subjective, which is consistent with the conceptual model first proposed by Campbell et al. (1976) and Marans and Rodgers (1975) as set out in Fig. 1.1 in Chap. 1, a model which is still being used as the basis for examining links between objective dimensions of urban environments with people's subjective evaluations of urban environments. Also, subjective QOUL has generally been conceptualized as satisfaction in various urban domains – housing, neighborhood, community, and regional satisfaction – rather than considering overall life satisfaction which takes into account all life domains (such as work, health, partner, standard of living, and so on) because the former relates better to examining links between objective dimensions and subjective evaluations of the urban environment. The model framework proposed by Pacione (2003) as reproduced earlier in Fig. 4.2 has also provided a useful framework for investigating such links.

In this chapter, we have reviewed a range of theories and empirical research findings relating to particular objective dimensions and subjective evaluations of the urban environment, as well as highlighting a range of theories and findings relating to more general processes that may weaken relationships between objective dimensions and subjective evaluations of the urban environment. The potential impacts of these more general processes on those relationships raise the question as to what are the strength of direct links between objective dimensions and subjective evaluations of the urban environment.

McCrea (2007: pp. 1–2) has suggested that a range of possible explanations might be used to account the strength of these relationships, each with different implications for QOUL and urban planning. For example:

(a) If moderate to strong direct relationships were found, then McCrea says that it would imply an environmentally deterministic model with changes in broad objective dimensions of the urban environment directly impacting on subjective QOUL.
(b) If weak relationships were found, the implications may depend on the explanation found for weakness. For example, if the weakness was best explained by psychological adaptation whereby residents simply adjust psychologically to changes in the objective urban environment, then that would imply that changes in broad objective dimensions of the urban environment have relatively little impact on subjective QOUL after a period of time. However, if the weakness was explained by adjustment via residential relocation whereby dissatisfied residents tended to move to other locations while satisfied residents tended to stay, then that would imply a significant impact on subjective QOUL. McCrea (2007) found most support for this last explanation of weak links.

Thus:

> … it is important not only to examine the strength of links between objective dimensions and subjective evaluations of the urban environment, but also to examine a range of explanations that may account for the strength of these links. (McCrea 2007: p. 1)

After examining in detail the links between the objective dimensions and subjective evaluations of the urban environment in his study of the SEQ region in Australia, McCrea (2007) has proposed a modification of the original Campbell et al. (1976) model, as shown in Fig. 4.6. The detailed modeling conducted by McCrea (2007) found that there were generally weak *direct* relationships between objective dimensions and subjective evaluations of QOUL, which also seems to be the case in much of the relatively limited literature in the field.

However, modeling direct relationships was found to be too simplistic. Objective and subjective indicators of QOUL influence each other *indirectly* via an integrated system involving subjective judgments, residential relocation processes and individual differences in what is considered important in choosing where to live.

McCrea (2007) concludes as follows:

> … By emphasizing that individual residents find satisfaction in different types of urban environment or local areas, the integrated explanation of subjective QOUL suggests that future research in QOUL should pay attention to the distinctiveness of objective and subjective QOUL in different local areas, including examining different subcultures and lifestyles. Enhancing subjective QOUL in different areas may be pursued by enhancing the main attributes of different areas that attracted residents to them (for example, building upon the unique character of areas); and when other more generic QOUL initiatives are taken, at least not detracting from the main attributes of areas. Any changes significantly detracting from these main attributes are likely to create dissatisfaction among existing residents and a disequilibrium which would then stimulate the residential relocation process, reflecting the systemic nature of relationships between objective characteristics and subjective evaluations of the urban environment. (p. 187)

Of course many challenges remain in developing operational frameworks to analyze QOUL through integrated approaches incorporating GIS-enabled modeling despite the substantial recent advances that have been made and that have been discussed in this chapter. That will involve more carefully conceptualized but more complex research designs that will have resourcing implications.

In particular, specific attention will need to be paid to spatial stratification in sampling designs for the survey research. This is essential to:

- obtain the necessary data on the subjective assessment of QOUL domains, and to
- spatially generalized patterns of those subjective assessments

which can then be integrated with spatial objective data relating to the complexities of urban environments, including well-established objective QOUL indicators at a spatially disaggregated level of urban scale, such as neighborhoods.

## References


Amerigo, M., & Aragones, J. I. (1997). A theoretical and methodological approach to the study of residential satisfaction. *Journal of Environmental Psychology, 17*(1), 47–57.

Andelman, R., Board, R., Carman, L., Cummins, R., Ferriss, A., & Friedman, P. (1998). *Quality of life definition and terminology: A discussion document from the international society of quality of life studies (Monograph).* Blacksburg: International Society of Quality of Life Studies.

Andrews, F., & Withey, S. B. (1976). *Social indicators of well-being: Americans perceptions of quality of life*. New York: Plenum Press.

Archibugi, F. (2001). City effect and urban overload as program indicators of the regional policy. *Social Indicators Research, 54*(2), 209–230.

Australian Bureau of Statistics (2001a) *Census of population and housing: Basic community profiles*, Retrieved September 2004.

Australian Bureau of Statistics. (2001b). *2001 Census dictionary*. Canberra: Australian Bureau of Statistics.

Baldassare, M., & Wilson, G. (1995). More trouble in paradise: Urbanization and the decline in suburban quality-of-life ratings. *Urban Affairs Review, 30*(5), 690–708.

Berto, R. (2005). Exposure to restorative environments helps restore attentional capacity. *Journal of Environmental Psychology, 25*(3), 249–259.

Blomquist, G. C., Berger, M. C., & Hoehn, J. P. (1988). New estimates of quality of life in urban areas. *The American Economic Review, 78*(1), 89–107.

Bowling, A., & Windsor, J. (2001). Towards the good life: A population survey of dimensions of quality of life. *Journal of Happiness Studies, 2*, 55–81.

Boyer, R., & Savageau, D. (1981). *Places rated almanac*. Chicago: Rand McNally.

Brown, L. A., & Moore, E. G. (1970). The intra-urban migration process: A perspective. *Geografiska Annaler, 52*, 368–381.

Brown, D. L., Fuguitt, G. V., Heaton, T. B., & Waseem, S. (1997). Continuities in size of place preferences in the United States, 1972–1992. *Rural Sociology, 62*(4), 408–428.

Campbell, A., Converse, P., & Rodgers, W. (1976). *The quality of American life: Perceptions, evaluations and satisfactions*. New York: Russell Sage.

Cantillon, D., Davidson, W. S., & Schweitzer, J. H. (2003). Measuring community social organization: Sense of community as a mediator in social disorganization theory. *Journal of Criminal Justice, 31*(4), 321–339.

Chiang, L. N., & Hsu, J. R. (2005). Locational decisions and residential preferences of Taiwanese immigrants in Australia. *GeoJournal, 64*(1), 75–89.

Cicerchia, A. (1999). Measures of optimal centrality: Indicators of city effect and urban overloading. *Social Indicators Research, 46*(3), 276–299.

Coleman, J. (1988). Social capital in the creation of human capital. *The American Journal of Sociology, 94*, S95–S120.

Clark, W. A. V., & Ledwith, V. (2006). Mobility, housing stress, and neighborhood contexts: evidence from Los Angeles. *Environment and Planning A, 38*(6), 1077–1093.

Cramer, V., Torgersen, S., & Kringlen, E. (2004). Quality of life in a city: The effect of population density. *Social Indicators Research, 69*(1), 103–116.

Cullen, F. T., & Agnew, R. (2003). The Chicago School: The city, social disorganization, and crime. In F. T. Cullen & R. Agnew (Eds.), *Criminological theory: Past and present* (pp. 95–103). Los Angeles: Roxbury Publishing Company.

Cummins, R. A. (2000). Objective and subjective quality of life: An interactive model. *Social Indicators Research, 52*(1), 55–72.

Cummins, R., McCabe, M., Romeo, Y., & Gullone, E. (1994). The comprehensive quality of life scale (ComQol): Instrument development and psychometric evaluation on college staff and students. *Educational and Psychological Measurement, 54*, 372–382.

Curry, G. N., Koczberski, G., & Selwood, J. (2001). Cashing out, cashing in: Rural change on the south coast of Western Australia. *Australian Geographer, 32*(1), 109–124.

Cutter, S. L. (1985). *Rating places: A geographer's view on quality of life*. Washington, DC: Association of American Geographers, Resource Publication in Geography.

Davidson, W. B., & Cotter, P. R. (1991). The relationship between sense of community and subjective well-being: A first look. *Journal of Community Psychology, 19*(3), 246–253.

Desbarats, J. (1983). Spatial choice and constraints on behaviour. *Annals of the Association of American Geographers, 73*(3), 340–357.

Diener, E., Suh, E. M., Lucas, R. E., & Smith, H. L. (1999). Subjective well-being: Three decades of progress. *Psychological Bulletin, 125*(2), 276–302.

Dokmeci, V., & Berkoz, L. (2000). Residential-location preferences according to demographic characteristics in Istanbul. *Landscape and Urban Planning, 48*(1–2), 45–55.

Evans, S., & Huxley, P. (2002). Studies of quality of life in the general population. *International Review of Psychiatry, 14*(3), 203–211.

Farrell, S. J., Aubry, T., & Coulombe, D. (2004). Neighborhoods and neighbors: Do they contribute to personal well-being? *Journal of Community Psychology, 32*(1), 9–25.

Filion, P., McSpurren, K., & Appleby, B. (2006). Wasted density? The impact of Toronto's residential-density-distribution policies on public-transit use and walking. *Environment and Planning A, 38*(7), 1367–1392.

Fischer, C. S. (1984). *The urban experience*. New York: Harcourt, Brace Jovanovich.

Ge, J., & Hokao, K. (2006). Research on residential lifestyles in Japanese cities from the viewpoints of residential preference, residential choice and residential satisfaction. *Landscape and Urban Planning, 78*(3), 165–178.

Glaeser, E., Kolko, J., & Saiz, A. (2000). *Consumer city* (Working Paper 7790), Cambridge: National Bureau of Economic Research.

Goodchild, M. F. (2000). New horizons for the social sciences: Geographic information systems. In *Social sciences for a digital world: Building infrastructure and databases for the future (2004)* (pp. 163–172). Paris: Organisation for Economic Cooperation and Development.

Goodchild, M. F., Anselin, L., Appelbaum, R. P., & Harthorn, B. H. (2000). Toward spatially integrated social science. *International Regional Science Review, 23*(2), 139–159.

Headey, B., Holmstrom, E., & Wearing, A. (1984). The impact of life events and changes in domain satisfactions on well-being. *Social Indicators Research, 15*(3), 203–227.

Kahneman, D. (1999). Objective happiness. In D. Kahneman, E. Diener, & N. Schawrz (Eds.), *Well-being: The foundations of hedonic psychology*. New York: Russell Sage.

Kaplan, S. (1995). The restorative benefits of nature: Toward an integrated framework. *Journal of Environmental Psychology, 15*, 169–182.

Kawachi, I., Kennedy, B. P., & Wilkinson, R. G. (1999). Crime: Social disorganization and relative deprivation. *Social Science & Medicine, 48*(6), 719–731.

Kemp, D., Manicaros, M., Mullins, P., Simpson, R., Stimson, R., & Western, J. (1997). *Urban metabolism: A framework for evaluating the viability, livability and sustainability of South East Queensland*. Brisbane: The Australian Housing and Urban Research Institute.

Kim, T. K., Horner, M. W., & Marans, R. W. (2005). Life cycle and environmental factors in selecting residential and job locations. *Housing Studies, 20*(3), 457–473.

Kubrin, C. E., & Weitzer, R. (2003). New directions in social disorganization theory. *Journal of Research in Crime and Delinquency, 40*(4), 374–402.

Lazarsfeld, P. F., & Merton, R. K. (1954). Friendship as a social process: A substantive and methodological analysis. In M. Berger (Ed.), *Freedom and control in modern society* (pp. 18–66). New York: Van Nostrand.

Li, S. M. (2004). Life course and residential mobility in Beijing, China. *Environment and Planning A, 36*(1), 27–43.

Lu, M. (1998). Analyzing migration decisionmaking: Relationships between residential satisfaction, mobility intentions, and moving behavior. *Environment and Planning A, 30*(8), 1473–1495.

Marans, R. W. (2002). Understanding environmental quality through quality of life studies: The 2001 DAS and its use of subjective and objective indicators. *Landscape and Urban Planning, 991*, 1–11.

Marans, R. W., & Rodgers, W. (1975). Toward an understanding of community satisfaction. In A. Hawley & V. Rock (Eds.), *Metropolitan America in contemporary perspective* (pp. 299–352). New York: Halsted.

Mastekaasa, A. (1984). Multiplicative and additive-models of job and life satisfaction. *Social Indicators Research, 14*(2), 141–163.

McCrea, R. (2007). *Urban quality of life: Linking objective dimensions and subjective evaluations of the urban environment*. Unpublished PhD thesis, The University of Queensland, Brisbane.

McCrea, R. (2009). Explaining regional socio-spatial patterns in South East Queensland, Australia: Social homophily versus other attributes of neighborhoods. *Environment and Planning A, 41*(9), 2201–2214.

McCrea, R., Stimson, R., & Western, J. (2005). Testing a moderated model of satisfaction with urban living using data for Brisbane-South East Queensland, Australia. *Social Indicators Research, 72*, 121–152.

McCrea, R., Shyy, T.-K., & Stimson, R. (2006). What is the strength of the link between objective and subjective indicator of urban quality of life? *Applied Research in Quality of Life, 1*, 79–96.

McMillan, D. W., & Chavis, D. M. (1986). Sense of community: A definition and theory. *Journal of Community Psychology and Health, 14*(1), 6–23.

McPherson, M., Smith-Lovin, L., & Cook, J. M. (2001). Birds of a feather: Homophily in social networks. *Annual Review of Sociology, 27*, 415–444.

Michalos, A. C., & Zumbo, B. D. (1999). Public services and the quality of life. *Social Indicators Research, 48*(2), 125–156.

Mitrany, M. (2005). High density neighborhoods: Who enjoys them? *GeoJournal, 34*, 131–140.

Office of Urban Management. (2005). *South East Queensland regional plan 2005–2026*. Brisbane: Department of Local Government, Planning, Sport and Recreation, Queensland Government.

Pacione, M. (2003). Quality-of-life research in urban geography. *Urban Geography, 24*(4), 314–339.

Perz, S. G. (2000). The quality of urban environments in the Brazilian Amazon. *Social Indicators Research, 49*(2), 181–212.

Prezza, M., Amici, M., Roberti, T., & Tedeschi, G. (2001). Sense of community referred to the whole town: Its relations with neighboring, loneliness, life satisfaction, and area of residence. *Journal of Community Psychology, 29*(1), 29–52.

Putnam, R. (1995). Tuning in, tuning out: The strange disappearance of social capital in America. *PS: Political Science and Politics, 28*(4), 664–683.

Rogerson, R. J., Findlay, A. M., & Morris, A. S. (1989). Indicators of quality of life: Some methodological issues. *Environment and Planning A, 21*(12), 1655–1666.

Rogerson, R. J., Findlay, A. M., Paddison, R., & Morris, A. S. (1996). Class, consumption and quality of life. *Progress in Planning, 45*(1), 1–66.

Russell, L. B., Hubley, A. M., Palepu, A., & Zumbo, B. D. (2006). Does weighting capture what's important? Revisiting subjective importance weighting with a quality of life measure. *Social Indicators Research, 75*(1), 141–167.

Sampson, R. J., & Groves, W. B. (1989). Community structure and crime: Testing social-disorganization theory. *The American Journal of Sociology, 94*(4), 774–802.

Sampson, R. J., Raudenbush, S. W., & Earls, F. (1997). Neighborhoods and violent crime: A multilevel study of collective efficacy. *Science, 277*(5328), 918–924.

Savage, M., Warde, A., & Ward, K. (2003). *Urban sociology, capitalism and modernity* (2nd ed.). New York: Palgrave MacMillan.

Schwanen, T., & Mokhtarian, P. L. (2004). The extent and determinants of dissonance between actual and preferred residential neighborhood type. *Environment and Planning B: Planning and Design, 31*(5), 759–784.

Schwarz, N., & Strack, F. (1999). Reports of subjective well-being: Judgmental processes and their methodological implications. In D. Kahneman & E. Diener (Eds.), *Well being: The foundations of hedonic psychology* (pp. 61–84). New York: Russell Sage.

Schwirian, K. P., Nelson, A. L., & Schwirian, P. M. (1995). Modeling urbanism: Economic, social and environmental-stress in cities. *Social Indicators Research, 35*(2), 201–223.

Senecal, G., & Hamel, P. J. (2001). Compact city and quality of life: Discussions of the Canadian approach to sustainability indicators. *Canadian Geographer-Geographie Canadien, 45*(2), 306–318.

Shaw, C. R., & McKay, H. D. (1942). *Juvenile delinquency and urban areas*. Chicago: University of Chicago Press.

Simmel, G. (1950). *The sociology of Georg Simmel*. New York: Free Press.

Sirgy, M. J., & Cornwell, T. (2001). Further validation of the Sirgy et al.'s measure of community quality of life. *Social Indicators Research, 56*(2), 125–143.

Sirgy, M. J., & Cornwell, T. (2002). How neighborhood features affect quality of life. *Social Indicators Research, 59*(1), 79–114.

Sirgy, M. J., Rahtz, D. R., Cicic, M., & Underwood, R. (2000). A method for assessing residents' satisfaction with community-based services: A quality-of-life perspective. *Social Indicators Research, 49*(3), 279–316.

Stover, M. E., & Leven, C. L. (1992). Methodological issues in the determination of the quality-of-life in urban Areas. *Urban Studies, 29*(5), 737–754.

Taylor, R. B. (1996). Neighborhood responses to disorder and local attachments: The systemic model of attachment, social disorganization, and neighborhood use value. *Sociological Forum, 11*(1), 41–74.

Trauer, T., & Mackinnon, A. (2001). Why are we weighting? The role of importance ratings in quality of life measurement. *Quality of Life Research, 10*(7), 579–585.

Turksever, A. N. E., & Atalik, G. (2001). Possibilities and limitations for the measurement of the quality of life in urban areas. *Social Indicators Research, 53*(2), 163–187.

Ulrich, R. S., Simons, R. F., Losito, B. D., Fiorito, E., et al. (1991). Stress recovery during exposure to natural and urban environments. *Journal of Environmental Psychology, 11*(3), 201–230.

van den Berg, A. E., Hartig, T. M., & Staats, H. (2007). Preference for nature in urbanized societies: Stress, restoration, and the pursuit of sustainability. *Journal of Social Issues, 63*(1), 79–96.

Vogt, C. A., & Marans, R. W. (2004). Natural resources and open space in the residential decision process: A study of recent movers to fringe counties in southeast Michigan. *Landscape and Urban Planning, 69*(2–3), 255–269.

Walmsley, D. J. (1988). *Urban living: The individual in the city*. New York: Wiley.

Walmsley, D. J., Epps, W. R., & Duncan, C. J. (1998). Migration to the New South Wales north coast 1986–1991: Lifestyle motivated counterurbanisation. *Geoforum, 29*(1), 105–118.

Western, J., & Larnach, A. (1998). The social and spatial structure of South-East Queensland. *Australasian Journal of Regional Studies, 4*(2), 215–237.

Wirth, L. (1938). Urbanism as a way of life. *The American Journal of Sociology, 44*(1), 1–24.

Wu, C. H., & Yao, G. (2006). Do we need to weight item satisfaction by item importance? A perspective from Locke's range-of-affect hypothesis. *Social Indicators Research, 79*(3), 485–502.

# Part II
# Empirical Studies of the Objective Measurement of Quality of Urban Life

# Chapter 5
# Amenities, Quality of Life, and Regional Development

**Gordon F. Mulligan and John I. Carruthers**

## Introduction

The purpose of this chapter is to review recent advances in research on the role of amenities – broadly defined, to include both the natural and human varieties – in urban growth and regional development. The focus is on approaches that use secondary data at the aggregate level of scale to generate objective measures of quality of urban life (QOUL). The chapter extends and complements a survey by Mulligan et al. (2004) on QOUL and public policy, which reviewed hundreds of studies from a very broad pool of evidence. The main goal of this survey is to synthesize the literature in a way that informs an interdisciplinary audience of researchers and practitioners in the social sciences and public policy fields.

*Amenities* are key to understanding quality of life (QOL) because they are precisely what make some places attractive for living and working, especially relative to other places that do not have them and/or are burdened with their opposites, *disamenities*. Because they influence where households and firms choose to locate within and among regions – plus, in part, determine the costs incurred in doing so – amenities and disamenities exert an exceptionally strong organizing force within advanced economies. And, in response to the strength and reach of this force, a corresponding nexus between amenities and public policy has also emerged: Indeed, it is now common for urban and regional planning efforts to actively address quality

G.F. Mulligan
School of Geography and Development, University of Arizona,
#408, 1203 Pemberton Avenue, Squamish, BC V8B 0J7, Canada, USA
e-mail: mulligan@email.arizona.edu

J.I. Carruthers (✉)
Sustainable Urban Planning Program College of Professional Studies,
The George Washington University,
805 21st St NW, Suite 301, Washington, DC 20052, USA
e-mail: jic@gwu.edu

R.W. Marans and R. Stimson (eds.), *Investigating Quality of Urban Life: Theory, Methods, and Empirical Research*, Social Indicators Research Series 45,
DOI 10.1007/978-94-007-1742-8_5, © Springer Science+Business Media B.V. 2011

of life. For example, in 2008, the Puget Sound region of Washington State in the USA, via the *Puget Sound Regional Council*, adopted a long-range planning document entitled *Vision 2040*[1]:

> …Vision 2040 is an integrated, long-range vision for the future that lays out a strategy for maintaining a healthy region — one that promotes the wellbeing of people and communities, economic vitality, and a healthy environment.

The plan explicitly reflects the fact that environmental quality and social well-being are fundamental to economic prosperity.[2]

What follows is an overview of recent research that has potential for guiding such activities. One empirical example shows how amenities have recently affected population and per capita income growth – two widely accepted measures of regional development – across the numerous and diverse counties of the USA. A second empirical example shows how housing values in the USA have come to depend upon the geographic incidence of amenities. Both studies are designed so that the behavioral and spatial econometric treatment of natural and human amenities is highlighted. Because of space limitations, most, but not all, of the discussion addresses interregional, as opposed to intraregional (local), issues. As an initial step, the chapter sets out some preliminary concepts – it is intended to clarify the nature of amenities and the behavioral mechanisms through which they influence both the process and the outcome of regional development. It then builds on this by providing a brief history of research on the consideration of environmental amenities as "compensating differentials" and further clarifies how amenity valuation is related to regional development. The remaining sections of the chapter deal more narrowly with:

- An empirical example of how natural amenities influence regional development
- The dual roles of production versus consumption in the development process
- Migration

Finally, the chapter closes with a summary and brief discussion of the material that is most relevant to planning and other forms of public policy.

## Some Preliminaries

### *Amenities and Disamenities*

Amenities are site- or region-specific goods and services that make some locations particularly attractive for living and working. Their opposites, disamenities, make places unattractive. Natural amenities are those, like climate, which are (for the most part) not influenced or produced by people, while human amenities are those, like culture, which are. Both types exist and are experienced at various geographic

---

[1]Excerpted from: http://www.psrc.org/projects/vision/index.htm

[2]See: http://www.psrc.org/projects/vision/pubs/V2040execsumm.pdf

scales and, in large part, determine relative QOL or social well-being (Smith 1977). Moreover, amenities influence the consumption decisions of households, the production decisions of firms, and the location decisions of both economic agents. Consequently, their implications for both local and regional development outcomes – and public policies aimed at shaping those outcomes – are enormous. Amenities were once thought to be mainly natural, as in the case of sunshine and/or landscapes, but the human-created variety is increasingly of interest to researchers and policy makers (Wong 2002; Welch et al. 2007).

Urban societies create and maintain many different kinds of human amenities, which often generate spillovers or so-called external economies (Harvey 1973; Tolley 1974; Diamond and Tolley 1982; Brueckner et al. 1999). At the most general level, these amenities include:

- Public goods and services (like education)
- Private consumption goods (like restaurants)
- Transportation and communication (transit)
- Cultural institutions (like museums)

Social capital, in various forms, seems to qualify as yet another type (Putnam 2000). In the USA, city-based information on such amenities is regularly updated in the *Places Rated Almanac* (Savageau 2007). Cities increasingly compete with one another in providing these amenities – consider the more visible examples of green spaces and public transportation systems (Henderson 1974; Fujita et al. 1999). In fact, some analysts (Glaeser and Mare2001) have argued that perhaps too much urban research in the USA has focused on the production of goods and services instead of on their consumption. Amenities are also of great interest to planners and policy makers working in nonmetropolitan regions, including micropolitan (emerging metropolitan) and low-density rural areas (Elliott and Perry 1996). Like metropolitan areas, nonmetropolitan regions compete for firms and households – sometimes on a seasonal basis – and these areas often market their comparative advantages in terms of recreation, landscapes, and waterscapes accordingly (Power 1996; Power and Barrett 2001). In fact, the most popular measure of natural amenities in the United States was specifically designed to assist planners and policy makers in addressing the problems that are endemic to nonmetropolitan counties (McGranahan 1993). Deller et al. (2001) discuss many issues of concern to regional development practitioners in these more peripheral economies.

Amenities (disamenities) influence urban growth and regional development by increasing (decreasing) the level of competition between different places. Other things being equal, people are particularly drawn to (away from) attractive (unattractive) settings – and because of this, expect to pay a premium (discount) via increased (decreased) housing prices and/or forgone (extra) wages. In Seattle – the so-called Emerald City – people colloquially refer to the combination of higher than expected rents and lower than expected wages as the "Mt. Rainier effect," meaning that it is owed to the region's world-renown natural beauty. In sum, by determining relative QOL, amenities and disamenities directly influence where people choose to live and at what cost.

## *The Economic Concept of Value*

Because most, though certainly not all, environmental amenities are nonmarket goods – meaning that they are not bought and sold in conventional markets: there is no store where one can go to purchase a few additional "days of sunshine" – their value can only be estimated, not measured directly. The economic concept of value (for an in-depth discussion, see Bockstael and Freeman 2005) is derived from neoclassical welfare economics, which holds that the purpose of economic activity is to promote individual and, by extension, societal, well-being. In this context, people's well-being comes from consuming goods and services delivered via private markets, nonmarket goods and services delivered via governments, and nonmarket goods and services derived from the natural environment and the kind of external economies mentioned above. The neoclassical framework further assumes that people have well-defined and well-known, or understood, preferences and that these preferences have the property of substitutability. Substitutability is key because it establishes how people make trade-offs between alternative bundles of goods and services. In the context of QOUL, people clearly make trade-offs between what various cities have to offer in terms of both natural amenities and human amenities.

There are two distinct ways of observing these trade-offs and measuring the values associated with them:

- Stated preference methods, such as contingent valuation
- Revealed preference methods, such as hedonic price analysis (see Freeman 2003).

Economists generally prefer the latter approach because it is based on what people actually do, not on what they say they would do. Hedonic price analysis, discussed in detail later, involves estimating the transacted price of housing (or people's wages, the transacted price of their labor), taking into account the most important attributes of that housing (those workers). Thus:

- Housing prices are regressed on characteristics of the home, its site, its location vis-à-vis various points of attraction (downtown, parks) and repulsion (undesirable land uses), and so on.
- Wages are regressed on characteristics of the worker (experience, educational attainment) and the job (hardship, level of danger), plus characteristics of where the job is located (number of sunny days, annual temperature extremes).

In this way, the implicit values of various attributes, which are rolled up in the overall value of homes and wages, are revealed.

So, looking out across the country, other things being equal, people are expected to pay a premium for housing and/or work for lower wages in places having nice weather – that is, if good weather has economic value. Again, the reason this happens is that there is increased competition to live in places offering a high QOL, so the demand is high relative to supply in the housing market, and the supply is high relative to demand in the labor market.

## *Spatial Equilibrium*

The outcome of these kinds of value-related trade-offs is a (theoretical) state of spatial equilibrium, wherein households are indifferent among locations. This situation – which is achieved via differentials in population, employment, and wage growth (see Carruthers and Mulligan 2008) – implies that, roughly speaking, the value of wages plus the value of quality of life minus the value of housing is more or less equivalent across the country (Glaeser 2007).

To understand how the equilibrating process works, it is useful to characterize the regional development process as happening in two interconnected ways:

- Via "demand-induced growth," which occurs when firms require additional labor, causing an increase in the demand for workers
- Via "supply-induced growth," which occurs when households move from one place to another for reasons that do not have to do with employment, causing an increase in the supply of labor

The classic example of demand-induced growth is when an export-oriented employer, like Boeing's commercial airliner operation, increases production and people move from elsewhere to fill newly created jobs. An example of supply-induced growth is when people relocate – for reasons having to do with personal preference – to the Puget Sound region because they value its temperate climate and abundant opportunities for outdoor recreation. In practice, both mechanisms matter, and neither happens in isolation (Borts and Stein 1964; Muth 1969; Carlino and Mills 1987; Clark and Murphy 1996; Mulligan et al. 1999).

Carruthers and Mulligan (2008) have recently extended the methodology for empirically modeling this process from one that focuses on only the two demographic outcomes (population and employment growth) to one that encompasses all three outcomes at the core of the equilibrating process. In particular, only demand-induced growth is precipitated by gains in the export market, but both mechanisms place pressure on the real estate market, raising rents and at the same time population and employment densities due to more intense competition over urban space. Expressing population and employment in terms of the density of land use ties the modeling framework directly to land rent and gives rise to the third equation for wages. Land use density measures the spatial intensity of activity, which is influenced by the average annual wage because of its relationship to land consumption: For people, land is a normal good, so the more they earn in wages, the more space they are able to consume, leading to a lower population density; for profit-maximizing firms, land is a factor of production, so the more they pay in wages, the less space they are able consume, leading to a higher employment density. Meanwhile, population density, which measures how concentrated the supply of labor is, and employment density, which measures how concentrated the demand for labor is, simultaneously drive the average annual wage. Working from Roback's (1982) model of compensating differentials, Mueser and Graves (1995) show how labor demand, labor supply, and wages combine to form a kind of "moving equilibrium" that calls for more or less continuous migration as the space economy searches for an optimal organization of activity (for detailed expositions, see Mulligan et al. 1999; Carruthers and Mulligan 2007, 2008).

## Compensating Differentials

For those new to environmental amenities research, this section provides an historical perspective on its role in QOL-oriented research. British observers commented on the economic growth of "fashionable watering places" like Bath and Brighton in the UK as early as 1811 (see Kendall and Pigozzi 1994). While largely forgotten today, Goodrich et al. (1936) identified a data based "plane of living" for US counties during the Great Depression. But the comprehensive assessment of urban amenities really began with Thorndike (1939), who rated the "goodness of life" in 300 cities.[3] While he did not actually use the term amenity, Thorndike included many variables that are commonly adopted today as QOL indicators, including per capita expenditures on schools, homicide rates, and infant mortality rates. Rogerson et al. (1988) provide a contemporary application of such indicator analysis for cities in the UK.

Ullman (1954) was the first to popularize the notion that amenities, or "pleasant living conditions," were in part responsible for differential growth rates in US regions. Careful to avoid being accused of environmental determinism, he believed that factors like climate and landscape had increasingly important roles to play in societies where affluence and mobility were both high. Perloff et al. (1960) reinforced this perspective when drawing out differences in the trajectories of economic development experienced by two Sunbelt states, California and Florida. Berry and Horton (1970) then argued that amenities were playing a very important role in the ongoing urban transformation of the entire American Sunbelt. Likewise, a key role for amenities is implicit in:

- Borts and Stein's (1964) theory of economic growth
- Galbraith's (1967) new industrial state
- Rostow's (1968) age of high mass consumption
- Bell's (1973) post-industrial society

By the mid-1960s, most social scientists recognized that affluent households were increasingly gauging their QOL by the level and variety of services they received instead of just the quantity of goods they consumed.

Some of the best known QOL research originates from the so-called compensating differential framework and hedonic price methodology developed by Rosen (1974, 1979) and extended by Roback (1982, 1988). In these studies, desirable (undesirable) living conditions negatively (positively) influence wages because, everything else

[3]The earliest example of all is apparently Ravenstein 1885 – see Greenwood and Hunt's 2003 review of early migration research.

being equal, people living in attractive (unattractive) places demand less (more) pay for their work; conversely, desirable (undesirable) living conditions positively (negatively) influence housing values because people living in attractive (unattractive) places are willing to pay more (less) for their homes. Both Rosen (1979) and Roback (1982, 1988) used their estimates to develop place-to-place QOL rankings, which rated regions having nice weather and other natural amenities, like San Francisco, at the top. Key studies that have extended this approach to developing quality-of-life rankings include Berger et al. (1987), Hoehn et al. (1987), Blomquist et al. (1988), and Gyourko and Tracy (1989, 1991). These are reviewed in Mulligan et al. (2004) and elsewhere.

In order to correct for "counter-intuitive" rankings, Albouy (2008) has recently recommended making three adjustments to the traditional approach:

- Incorporate other household cost items besides housing.
- Account for differential federal taxes in the payment of wages.
- Include non-earnings income in the resources available to households.

These adjustments serve to simultaneously narrow the disposable income differences and widen the cost-of-living differences across American cities. Or to look at things differently, more weight is placed on housing-cost differences, and less weight is placed on wage differences. Both exogenous (precipitation, sunshine, coastal location, and so on) and endogenous (for example, air quality, violent crimes, incidence of bars and restaurants) amenities are accounted for in the new imputed QOL estimates. Honolulu and places like Santa Barbara and San Francisco in California are given the highest scores. Moreover, the overall rankings that come out of the analysis resemble those listed in the *Places Rated Almanac* (Savageau 2007). In fact Albouy (2008) argues that popular publications like the *Almanac* should place even greater weight on climate factors and on location. It is important to note that the size of a city does not appear to have an effect on its (adjusted) QOL. The various amenities and disamenities of urban life appear largely to cancel one another out. So, any calls for restricting the sizes of large cities because of estimated welfare losses seem premature. If anything, in the interest of national economic efficiency, areas with favorable natural amenities should make greater attempts to accommodate larger populations.

The hedonic approach is demanding both in terms of analytics and the amount of data required. So, it is not surprising that a somewhat simpler approach to estimating the impact of amenities has been devised. Following Harris et al. (1968), Glaeser et al. (2001) have recommended using the residuals from a simple, bivariate regression model wherein median house value is the dependent variable and median household income is the independent variable. Geographic variation in this relationship should depend in part on the interplay existing between natural advantages and any scale effects in consumption (Krupka 2008). Observations above (below) the trend line – that is, positive (negative) residuals – indicate places where households spend more (less) than an average amount of their income on housing. The sizes of the positive (negative) residuals are a measure of the value placed on place-specific amenities (disamenities). In a study of almost all counties in the USA, Carruthers and Mulligan (2006) have extended this logic to include natural amenities as a second independent variable. So, one possible interpretation is that positive (negative) residuals point to places having a surplus (deficit) of human-created amenities.

Regional scientists have had somewhat different perspectives on the uneven access to amenities found in capitalist societies. Economists have stressed that different people command different resources, and therefore, households must place different demands on amenities. Demographers have recognized that factors like age, race, and gender often mediate these demands, and geographers have often pointed out that amenities are inequitably distributed across space (Massam 1975, 1993; Dicken and Lloyd 1981; Diamond and Tolley 1982). But these disparate views have slowly converged, and now, social scientists generally recognize that accessibility *per se* impacts a wide range of social and economic issues at different spatial scales (Glaeser and Kohlhase 2004; Des Rosiers et al. 2005; Partridge et al. 2008a). Comprehensive literature reviews on amenities – some being more technical than others – have already been compiled by Bartik and Smith (1987), Gyourko et al. (1999), Dissart and Deller (2000), Mulligan et al. (2004), and Lambiri et al. (2007). Moreover, discussion in the social sciences and public policy fields continues with regard to the conceptualization and measurement of QOL, where it is recognized that biases have sometimes occurred in the selection of amenity indices. Serious discussion is also needed to clarify how amenities relate to various facets of social justice, especially geographic and intergenerational equity (Nussbaum and Sen 1993; Smith 1994; Sen 1999; van Praag and Frijters 1999; Lee 2006; Rothschild 2009).

## Natural Amenities and Regional Development: Empirical Examples

There has emerged a considerable literature measuring natural amenity and its relationship to regional development. That has included attempts to derive natural amenities indices and studies which attempt to model the relationship between amenities and population density and growth.

### *The McGranahan Natural Amenity Index*

McGranahan's (1993) well-known county-level study was in some ways a turning point for the study of natural amenities in the American space economy. From a somewhat wider pool of potential candidates, he selected six different measures of natural amenities, which, as a set, exhibit surprisingly little intercorrelation. Along with the signed preferences by households, these were:

- Average January temperature (+)
- Average January sunshine (+)

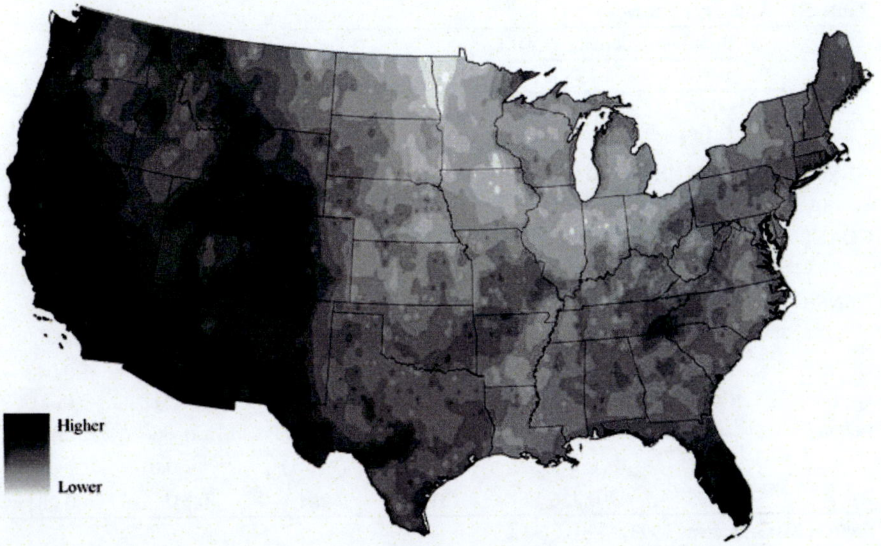


**Fig. 5.1** Natural amenity index: U.S. (Source: Carruthers and Mulligan 2009)

- Average July temperature, computed as a residual (–)
- Average July humidity (–)
- Topography (+)
- Percentage of land in water (+)

Standardized $z$-scores were computed for each subindex and then for an overall index – mapped in Fig. 5.1 – based on the summation of the signed subindices.

These measures have allowed various studies to determine the role of natural amenities in various facets of regional development, particularly in the so-called jobs versus amenities debate, as the prime engine for local and regional growth. Prior to these indices being available, much analysis was confined to metropolitan areas, where the information resources of the *Places Rated Almanacs* could be conveniently drawn upon.

## *Modeling Relationships Between Amenities and Population Density Growth*

An analysis of county-level growth in the US during the closing decades of the twentieth century highlights various issues that are of central interest to researchers in economic geography and regional economics.

Two widely adopted measures of local or regional change are the focus of the analysis:

- Growth in population density
- Growth in per capita income

**Table 5.1** Amenity elasticities

| | Population density growth | | | Per capita income growth | | |
|---|---|---|---|---|---|---|
| | 1970s | 1980s | 1990s | 1970s | 1980s | 1990s |
| Intercept | –0.031 | –0.022 | –0.413 | 3.464 | 3.276 | 1.615 |
| | (–1.0) | (–0.2) | (–3.2) | (36.6) | (30.2) | (16.2) |
| NAMEN | 0.174 | 0.528 | 0.570 | –0.045 | –0.059 | –0.057 |
| | (20.9) | (18.7) | (22.2) | (–1.8) | (–2.4) | (–2.9) |
| COAST | 0.021 | 0.096 | –0.033 | 0.004 | 0.091 | -0.015 |
| | (5.7) | (7.8) | (–2.9) | (0.4) | (8.4) | (–1.7) |
| DENSE | 0.009 | 0.016 | 0.017 | 0.000 | 0.026 | 0.012 |
| | (19.4) | (9.9) | (11.0) | (0.1) | (18.8) | (9.8) |
| DEGRE | 0.014 | 0.051 | 0.004 | 0.078 | 0.111 | 0.048 |
| | (5.8) | (7.0) | (0.5) | (10.9) | (17.5) | (8.8) |
| PCINC | –0.004 | –0.030 | 0.027 | –0.352 | –0.358 | –0.165 |
| | (–1.2) | (–2.3) | (1.87) | (–33.1) | (–30.6) | (–15.1) |
| Adj $R^2$ | 0.285 | 0.226 | 0.181 | 0.284 | 0.315 | 0.081 |

Source: McGranahan (1993)

Note: *t*-scores are shown in parentheses; all county estimates ($n=3{,}107$) are controlled for prior population density, percentage college educated, and per capita income (in $2000). Sources: U.S. Bureau of the Census; U.S. Bureau of Economic Analysis

Density is often preferred to size because counties vary so much in areal extent. Practitioners of local and regional development often use sustained population (density) growth as an indicator of economic development, whereas regional economists favor sustained per capita (personal) income growth as an indicator of local or regional economic development. In any case, the two are used in combination here to examine how economic growth across the US space economy changed during the last three decades of the twentieth century.

Table 5.1 presents estimates from an ordinary least squares (OLS) regression model where McGranahan's composite index of natural amenities is supplemented by a second natural amenity, coastal location (Rappaport and Sachs 2003). The various regressions also include a number of so-called initial conditions, including per capita income, population density, and percentage of the population aged 25 plus with a college degree. Initial conditions provide some required context for the estimates of decadal growth and also address, at least in part, thorny issues related to circularity or endogeneity (see below). Moreover, two of these measures are widely known to represent human-created amenities. Population density captures opportunity or variety in the availability of both private and public goods, while the incidence of college education among residents reflects the availability of human capital in the region. So, the regressions, although somewhat underspecified compared to some in the literature, shed useful, if simplified, light on several amenity-related growth issues, including the "jobs versus amenities" debate in the USA

The left-hand panel of Table 5.1 lists estimates of (instantaneous) population density growth over each decade, and the right-hand panel lists estimates of (instantaneous) per capita income growth over each decade. In both sets of models, a single composite index for natural amenities has been used instead of the six separate amenity subindices. All of the variables are transformed into natural logarithms – so, all parameter estimates are elasticities.

Although there are some interdecadal changes, the overall story is one of remarkable stability, echoing the remarks of Rappaport (2004) about persisting patterns in US population movements. On the one hand, local population (density) growth has depended on a mixture of natural and human amenities; on the other hand, local per capita income growth has largely depended on human amenities. For population growth, the positive effect of natural amenities only grew over time. During the 1970s, a 1% shift upward in the composite index brought forth a 0.17% increase in the population growth rate, while in the 1990s, this natural-amenity effect was a remarkable 0.57%, some three times as great. Coastal locations had higher population growth rates through the first two decades, and dense, well-educated regions – or major metropolitan areas – enjoyed higher population growth through all three decades. In contrast, bountiful natural amenities appear to have diminished any regional growth that was experienced in per capita (personal) income during the 30-year study period, a result entirely consistent with Rosen's (1974, 1979)/Roback's (1982, 1988) compensating differentials theory outlined above. Improvements here were largely confined to dense, highly educated areas, though the negative signs on prior per capita income levels suggest a sustained pattern of convergence.

During the middle decade – the 1980s – the effects of both prior density and prior education benefitted from their concentrations of human capital, and urban regions found in desirable environments were especially well-off. Rural regions with prized natural amenities experienced population growth but not per capita income growth; rural regions in undesirable locations simply lagged behind along both dimensions of local and regional growth.

Although these are both very simple models, together they establish that the role of QOL in regional economic development will always be complicated by the fact that amenities impact the two main components of economic development – growth in people and growth in income – in somewhat different ways.

The findings of the research outcomes presented in Table 5.1 probably raise as many issues as they answer. For example, on the one hand, population growth depends upon relative change in various components of change, including natural increases, interregional in-migration and out-migration, and immigration. Moreover, none of these components are homogeneous; for example, migration rates intimately depend upon personal characteristics like age, race, and income. On the other hand, per capita income growth is not homogenous either. Income levels depend upon the balance between earnings and non-earnings streams, where the former in turn depends upon things like wage levels and unemployment rates and the latter depends upon the local mixture of both (private) factor returns and (public) transfer payments. Of course, bringing more demographic or economic detail to the growth models only complicates our interpretation of the role of amenities in regional development. Nevertheless, many of these effects are highlighted in the literature reviewed in subsequent sections.

For now, the discussion returns to county-level population growth and then address a couple of other issues of great interest to regional scientists.

**Table 5.2** Natural amenity elasticities: population density growth

| | OLS | | | 2SLS | | |
|---|---|---|---|---|---|---|
| | 1970s | 1980s | 1990s | 1970s | 1980s | 1990s |
| Intercept | –0.115 | –0.188 | –0.388 | –0.573 | 1.041 | –0.265 |
| | (–3.1) | (–1.3) | (–2.7) | (–13.9) | (6.08) | (–1.5) |
| JATEM | 0.178 | 0.539 | 0.507 | 0.161 | 0.524 | 0.508 |
| | (18.8) | (16.7) | (16.9) | (18.0) | (16.6) | (16.9) |
| JASUN | 0.043 | 0.190 | 0.147 | 0.052 | 0.135 | 0.145 |
| | (3.1) | (4.0) | (3.3) | (4.0) | (2.9) | (3.2) |
| JUTEM | –0.225 | –0.733 | –0.754 | –0.172 | –0.725 | 0.757 |
| | (–7.9) | (–7.6) | (–8.4) | (–6.5) | (–7.7) | (–8.5) |
| JUHUM | –0.015 | –0.068 | –0.134 | –0.006 | –0.065 | –0.133 |
| | (–2.2) | (–3.0) | (–6.7) | (–0.9) | (–2.9) | (–6.6) |
| TOPOG | 0.012 | 0.021 | 0.053 | 0.004 | 0.024 | 0.053 |
| | (2.9) | (1.5) | (4.1) | (1.0) | (1.8) | (4.2) |
| WATER | 0.023 | 0.077 | 0.044 | 0.019 | 0.076 | 0.045 |
| | (3.2) | (3.1) | (1.9) | (2.9) | (3.1) | (1.9) |
| COAST | 0.012 | 0.068 | –0.027 | 0.007 | 0.067 | –0.026 |
| | (2.7) | (4.6) | (–2.0) | (1.7) | (4.7) | (–1.9) |
| DENSE | 0.006 | 0.009 | 0.013 | 0.099 | 0.226 | -0.007 |
| | (9.9) | (4.3) | (6.4) | (22.0) | (–11.6) | (–0.4) |
| DEGRE | 0.023 | 0.074 | 0.019 | 0.045 | 0.004 | 0.014 |
| | (9.0) | (9.4) | (2.5) | (17.4) | (0.5) | (1.6) |
| PCINC | 0.006 | 0.008 | 0.036 | 0.040 | –0.097 | 0.026 |
| | (1.5) | (–0.5) | (2.4) | (10.2) | (–6.1) | (1.5) |
| EMPLY | –0.094 | 0.235 | 0.021 | (–20.8) | (12.1) | (1.2) |
| Adj Rsq | 0.292 | 0.232 | 0.174 | 0.379 | 0.266 | 0.175 |

Source: McGranahan (1993)

Note: *t*-scores are shown in parentheses; all county estimates ($n$=3,107) are controlled for prior population density, percentage college educated, and per capita income (in $2000). Sources: U.S. Bureau of the Census; Bureau of Economic Analysis

## *Disaggregating the Amenity Index into Its Component Parts*

The first step is to disaggregate McGranahan's single composite index into its various components. This is done in order to determine if American households responded differently to the array of natural amenities during each of the three decades of the study period – that is, the purpose here is to examine if human *behavior* changed in the face of a constant stock of natural amenities. The left-hand panel of Table 5.2 shows the results for the various subindices, with the human amenities still included. As noted before, however, population growth is embedded in a variety of other growth processes, a fact that means there is a degree of endogeneity in the growth model. In order to address this problem, the models employ an adjustment mechanism that includes employment as another exogenous variable (Carlino and Mills 1987). The right-hand panel of Table 5.2 shows these results, which are the second stage of a two-stage least squares (2SLS) estimation.

The results in the left-hand panel of Table 5.2 indicate that a remarkable stability existed in the preferences of American households during the 1970–2000 timeframe.

With only two exceptions – for income in the second decade and coastal location in the third decade – there is complete consistency in the signs; moreover, all of the natural amenities are signed according to McGranahan's a priori expectations (see above). Evidently, among these amenities, the most important driving forces of growth were temperature in January (+) and residual temperature in July (–). Here again, if anything, the role of the *individual* natural amenities strengthened over time as the positive and negative amenities played off against one another. This, of course, is the underlying reason for the significant rise in the estimate for NAMEN noted earlier in Table 5.1.

Disaggregation of McGranahan's index also leads to some recalibration of the elasticities for human capital, with the estimates for density moving lower and those for college education moving higher at each point in time. However, the results in the right-hand side of Table 5.2 are even more interesting. Note that employment is included as an endogenous variable, and in the adjustment model, this is the "targeted" level of employment reached at the end of each decade. The estimates entirely endorse the findings for natural amenities, as just discussed. But as might be expected, the introduction of employment into the population growth estimation shifts the estimates – and generally diminishes the importance – of the human-created amenities, at least in the later decades. Moreover, the prior level of per capita income now appears to be an important factor in local demographic change. All in all, the results of Table 5.2 suggest that natural amenities played a steady and consistent role in population growth throughout the US counties during the late twentieth century.

## A Recent US Study

Perhaps the best recent study of the role of natural amenities in US regional growth is by Partridge et al. (2008b). Here, county-level employment growth was examined between 1990 and 2004 using the well-known REIS data generated by the Bureau of Economic Analysis.

A series of data vectors was first created, accounting in all for nearly 40 variables, and these were arrayed along four main dimensions:

- Amenity (with 5 of McGranahan's 6 indices),
- Demography (recent immigration, education levels, age shares, race and ethnic composition),
- Economic (industry mix, initial unemployment), and
- Distance (five distances to size-tiered nearest urban centers).

Next, in addition to the standard linear regression (OLS) approach, the authors estimated both a spatial error model (SEM) and a geographically weighted regression (GWR) model. This was done to address *spatial heterogeneity*, where both the mean and the variance of job growth could vary geographically (Anselin 1988; Fotheringham et al. 2002). The thinking – which is entirely correct – was that many marginal impacts could vary across space, depending upon the location-specific preferences expressed by firms and households. In fact, as GWR recognizes, these preferences may not even be global, for amenities like topography or water cover and factors like unemployment or racial composition could have larger effects on job growth at some locales than at others.

In the Partridge et al. (2008b) research, two tables of estimates and a series of maps developed separately for nonmetropolitan and metropolitan counties proved to be very instructive. In nonmetropolitan counties, natural amenities like January sun and July humidity turn out to be significant locally but not significant globally as indicated by either the OLS or the SEM approach. And in both types of counties, a college degree proves to be significant locally but not significant globally again by either method. Moreover, here, the greatest positive impacts are distributed across the western part of the country, suggesting a possible brain drain from the east. All methods indicate that access to water has much more important implications for job growth in rural than in urban areas. Partridge et al. (2008b) conclude that public policy might well be incorrect unless local conditions are fully accounted for. But these local conditions can in fact be very difficult to assess because they involve not only the levels of different factors but also the interactions among those different factors.

## *European Studies*

Similar analyses using a different mix of variables have been recently carried out in Europe.

Building on earlier work, Cheshire and Magrini (2006) have examined annualized population growth rates across some 100 of the largest cities (Functional Urban Regions (FURs)) in the European Union (EU-12). Variables were first introduced into a family of basic models to address the attributes of prior economic bases, port locations, ongoing national population trends, and the integration gains – measured as a form of economic potential – that resulted from lower tariffs and lower transportation costs across much of the continent (Bruinsma and Rietveld 1993). Then another family of models was developed having several measures of climate – including rainfall, frost, cloudiness, and temperature – all expressed in ratio form. When these climate variables were expressed using EU-wide means, the various ratios were insignificant; however, when those variables were expressed using nationwide means, the ratios were significant. Even though intercity mobility is much lower in Europe than in the USA, it is very clear that cold, cloudy weather impeded urban population growth and warm, dry weather promoted population growth. Moreover, the climate variables proved to be superior to any location proxies in modeling recent urban population growth.

Cheshire and Magrini (2008) then returned to their topic in order to test the comparative drivers of population growth and per capita income (GDP) growth. Introducing a new variable to allow interaction between FURs and adjusting for spatial lags improved the econometrics of their original population growth model.

A few new variables, including closeness to R&D facilities and densities of university students, were used along with the basic variables to estimate income growth across the cities during the same period, 1980–2000. Human capital clearly promoted income growth, but the role of natural amenities proved to be insignificant at best. In fact, these results are very similar to those illustrated in the United States example above. The authors concluded by saying that productivity differences would persist among large European cities even though the population might be equilibrating in space.

## Jobs Versus Amenities

Various research efforts focusing on matters like urban–rural wage gaps, the importance of skills and education, and the nature of agglomeration economies have come together to inform the so-called debate of jobs versus amenities. Here, the interest is in identifying the relative roles of production (firms) and consumption (households) in driving regional development. Depending on the purposes of the study, a variety of indicators – including population growth, job growth, wage growth, and GDP per capita growth – have been used to measure differential levels of urban or regional performance.

Following the path-breaking ideas of Rosen and Roback (referred to earlier), the early literature was developed by Beeson (1991), Rauch (1993), and others who analyzed the so-called urban wage gap (for a review, see Mulligan et al. 2004). An important distinction was eventually drawn between individual-level and city-level effects in how factors like education and work experience affect worker productivity. Glaeser and Maré (2001) have reported that workers in American metropolitan areas earned 33% more than their nonmetropolitan counterparts, but only one third of this wage premium was attributable to any differences in ability or skills. Clearly then, large cities must have substantial production advantages based on agglomeration economies that raise the wages of (most, if not all) urban workers. Moreover, these high wages might also include payment for a variety of large-city disamenities, including violent crime and congestion.

### *The Relationship Between Education and City Size*

Adamson et al. (2004) have examined the relationship between education and city size in some detail. A wage equation was specified, which addressed more than 4,300 workers aged 23–36 in the 1988–1993 National Longitudinal Survey of Youth. Educational attainment was captured by five indicators running from high school graduates to professional degrees. Dropouts with little human capital were simply omitted. Included amenities existed along various dimensions:

- Population in quadratic form to capture urban scale in the variety of private (for example, entertainment) and public goods

- Six climate variables (including annual heating and cooling degree days) and indices of water coverage and topography
- Serious crimes per capita
- Local government fiscal conditions.

Other variables in the wage equation captured demographic differences among the workers and labor market differences among the cities, including employment density.

Adamson et al. (2004) interpreted their results by focusing on the various regression interactions between education and population. If the demand for labor dominated – where skill biases and agglomeration drive wage change – then this interaction term was positive. On the other hand, if households (and, to a lesser extent, firms) enjoyed many amenities, then the supply of labor was affected, and the education–population interactions were negative.

The results suggest that the latter effect dominated US cities in the early 1990s. While urban workers clearly have a nominal wage advantage, this gap is largely due to the presence of highly educated workers. However, having said that, it seems that returns to investment in education fall steadily with city size. Even in the nation's very largest cities, urban amenities, largely reflecting scale effects, dominate skill-based advantages in driving metropolitan productivity. Adamson et al. (2004) reach the conclusion that policy makers should pay more attention to the role of urban amenities in driving or maintaining city employment and population growth. Unfortunately, many amenities are used only as control variables and are not included in the discussion of the regression estimates. Consequently, their impact on wage growth cannot be discerned. In fact, the all various relationships discussed in detail in the paper have only marginally significant estimates even though the overall model manages to account for much of the wage variance. That finding suggests that other variables, those not highlighted in the paper, may be responsible for most of the wage spreads among US counties. In the end, the authors endorse the findings of Glaeser et al. (2001), arguing that, to be more successful, cities should focus as much on QOL issues as on training and skill enhancement programs.

## *Amenities and High-Tech Manufacturing*

Other aspects of the jobs versus amenities debate are evident in the recent research by Dorfman et al. (2008) on high-tech manufacturing in the USA. Here, a nonparametric smoothing method is used to assist in discerning how amenities of various types might affect county-level job growth. The thinking was that, while rural areas often offer an array of natural amenities, they usually lack human-created amenities that are associated with either localized (clusters) or urbanization economies. In particular, rural areas often lack knowledge spillovers and pooling in labor markets. Job growth between 2000 and 2006 across 14 4-digit NAICS industries is examined over nearly 3,000 urban and rural counties; those counties with fewer than 10 employees in high-tech industries were simply eliminated. Numerous variables were introduced as explanatory or control variables:

- The incidence of college education in the adult population
- McGranahan's composite index expressed in ordinal form
- Housing prices to control for the cost advantages of low-density areas
- Initial high-tech employment in those 14 industries
- The various distances to three different tiers of metropolitan places and also distance to the nearest major university.

The main finding of the analysis was that local job creation was strongly enhanced when there was a prior pool of college graduates. In fact, the estimates suggest that a 1% increase in such workers induced more than 50 new high-tech jobs over the 6-year study period. The anticipated effects for natural amenities and for accessibility to larger places were largely unrealized, however. Moreover, high-tech firms did not always seem to benefit from locations adjacent to large metropolitan centers once the initial level of high-tech employment was taken into account. So, starting or enhancing a university research park would likely not be a successful strategy in most rural areas. But widening and deepening the local pool of human talent likely would be a good strategy even though natural amenities (usually a normal good) did not seem to be responsible for attracting high human-capital pools to rural areas.

Ferguson et al. (2007) repeated this basic methodology in an analysis of population change in some 2,400 census consolidation subdivisions during 1991–2001. These units – typically much smaller than US counties – were classified as urban (21% of the total), where they were parts of census agglomerations or census metropolitan areas or as rural (79%), where they comprised small towns or rural areas.

The model incorporates the following vectors:

(a) A comprehensive amenity vector contains eight natural and 10 human-created (modern) varieties:
   - Five of the natural amenity variables represent climate, and the others relate to coastline, forest cover, and topography.
   - The modern amenities include violent and property crime and accessibility to various private and public (for example, distance to the nearest hospital, college) goods.

(b) The economic vector has 11 variables (per capita income, overall industry mix, industry shares).
(c) The demography vector has four variables.
(d) The geography vector has four regional dummies.
(e) The human vector, three variables (measures of human capital).
(f) The social vector, four variables (percent home ownership, social capital).
(g) The agglomeration vector, three variables (population size, proximity to large cities).

The analysis, which uses variance decomposition, identifies those bundles of local factors that are chosen by different age cohorts. Economic factors tend to dominate in both urban and rural areas. Moreover, amenities count more in urban areas, where they are perhaps responsible for as much as 22% of the overall variation in population growth. But this figure falls to 9% in rural areas, many of which are particularly vulnerable during economic downturns (Frenette 2008). Amenities are valued especially high by the two youngest age cohorts, those aged 5–19 years and from 20 to 34 years. Surprisingly, seniors (aged 60+) do not appear to demand locations with mild winters as much as might be expected, although weak effects indicate that urbanites, as a whole, seek warm winters and to avoid humid summers. Violent crime diminishes overall population growth (for all cohorts except the elderly) and this effect is strongest in cities. Young adults show preferences for particular natural amenities, like the presence of mountainous terrain, and seniors dislike the penalties of having to travel great distances to hospitals and physicians.

Ferguson et al. (2007) conjecture that differences with the USA likely arise in part because of cultural factors (there is less mobility across a language barrier) and because incomes are lower in Canada, making the demand for income-elastic amenities somewhat lower.

Alasia et al. (2008) provide a somewhat different perspective on these Canadian communities. Paralleling the work done earlier in Australia (Stimson et al. 2001), the authors examine "vulnerability,", which they claim is a more forward-looking perspective on socioeconomic disadvantage than is economic deprivation. Specifically, they address the likelihood of worsening conditions in these various places, as measured by either population or employment decline. Their conceptual framework is a "stressor–asset–outcome" triad, wherein 29 different stressors (for example, low incomes, weak labor markets) and asset (for example, educational attainment) indicators are used as dependent variables in a probit model. The post-2001 viability of each local economy is predicted based on the earlier 1981–2001 trend. Global exposure and certain conditions of distress, such as high unemployment rates and low participation rates, are shown to increase community vulnerability. Strong community assets, including high human capital, diversification, and proximity to larger places, work to reduce vulnerability. Nearly 20% of the communities are targeted as being vulnerable to long-run population growth, but only 5% are thought to be vulnerable to long-run employment decline. Unfortunately, indicators for natural amenities and social capital (although mentioned) were not developed and brought into the two models. Of course, long-run vulnerability can also vary substantially with the policies of federal and state governments who make decisions regarding the tradeoff between national efficiency and interregional equity (Canaleta et al. 2004).

## Migration

In the past few decades, it has become increasingly appreciated that QOL factors do not uniformly affect migrating populations. In fact, this perspective led in part to the emergence of a "spatial-equilibrium" (described previously) school whose tenets complement, but sometimes counter, those of an earlier "disequilibrium" school in migration research. Households recognize that they can become compensated for lower wages by either better amenities or improved public goods, and they vote with their feet, as Tiebout suggested half a century ago. In some cases, this movement has been sufficient to dramatically lower the local demand for housing (Allinson 2005). The two theoretical perspectives have been resolved, however, in research by Greenwood and Hunt (1989) and Mueser and Graves (1995).

In the USA, as indicated by the earlier numerical example, natural and human-created amenities appear to have had a fairly constant effect on levels of interregional migration but, in contrast, the effect of employment-based opportunities seems to have been much more cyclical (for details, see the review by Mulligan et al. 2004). In recent years, analysts have continually stressed that national or regional populations should be considered heterogeneous, as opposed to homogeneous entities in terms of their composition. Following Clark and Hunter (1992), Cushing (1993), and others, this segmentation of households or families has been developed along various lines, including education, income, age (life cycle), or even gender. Several recent studies in this vein of research are worthy of special attention.

## *Investigating the Drivers of Migration by the College-Educated*

Gottlieb and Joseph (2006) join a growing list of scholars (Herzog et al. 1986; Yousefi and Rives 1987; Hansen et al. 2003) interested in the ever changing geography of human capital in the USA, by analyzing the metropolitan out-migration of the college-educated. The authors use a restricted database supplied by the National Science Foundation that allows them to examine intercity moves made between 1993 and early 1995. The main purpose is to see whether economic opportunities, the traditional driver of migration behavior, or amenities and lifestyle considerations are largely responsible for the out-migration patterns of recent graduates. The results have obvious implications for region- or state-based policies aimed at stemming a "brain drain."

A distinction is made between all college graduates and doctorate-earning graduates, recognizing that migration behavior could vary markedly across these two groups. Doctorate holders tend to participate in narrower labor markets and, as a result, may be less responsive to natural or human-made amenities; on the other hand, they certainly enjoy more bargaining power and might vote with their feet for a better QOL. Fifty large metropolitan destinations are included in the study, and special dummies are introduced for San Francisco, New York, Boston, and Atlanta. Personal characteristics, occupation data, geographic separation, and various properties of the destinations are addressed. Amenities include crime, climate, and recreation, where data are taken from the 1993 *Places Rated Almanac*. Estimates are provided using both the binary logit and the mixed logit (RPL) models, where the former is inferior in addressing substitution issues (because of the IIA property).

## *The Interplay Between Life-Course Attributes and Location-Specific Attributes*

Next, in an update of a well-known paper by Herzog and Schlottmann (1986), Whisler et al. (2008) provide a different perspective in demonstrating that household migration depends upon the interplay between personal characteristics – life-course attributes – and location-specific attributes, including both natural and human-made amenities. Their appreciation of the importance of demographic segmentation was deepened by traditional migration research, including that of Plane and Heins (2003) as well as the identification of such entities as power couples (Costa and Kahn 2000) and the creative class (Florida 2002) in contemporary society. Whisler et al. use a binary logit model to examine metropolitan out-migration ("stayers" versus "leavers") during the time period 1995–2000; the 2000 Census Bureau, 5% of the PUMS controls the estimates for household characteristics. Performance data for cities were drawn from the 1997 *Places Rated Almanac*, where city scores are provided for seven different amenities:

- Climate
- Recreation
- Education
- Crime
- Health care
- Transportation
- The arts

along with city-wide cost-of-living and job outlook variables. The authors include other contextual variables, including population size, density, and recent change in the city's stock of college graduates.

## *Amenities and Migration by Seniors*

A third perspective on migration is given by Jensen and Deller (2007) who, in building on earlier work by Deller et al. (2001), focus specifically on how US seniors evaluate different types of amenities in their migration decisions. The authors are particularly interested in understanding why some older households move to micropolitan and rural areas across the nation. They recognize, of course, that when non-earnings income is introduced into many of these nonmetropolitan places, considerable numbers of jobs are created through local expenditures and subsequent multiplier effects.

The research looks at county-level in-migration and out-migration during 1995–2000 across four adjacent age cohorts: 55–64; 65–74, 75–84; and 85+ years. The prior characteristics of each county include a variety of demographic, economic, and land-use variables, and Beale codes are used to control for the urban-rural continuum. A very wide array of recreational data, capturing built amenities, is reduced through principal component analysis to a handful of indices. Temperature, water, and snowfall data address natural amenities, tax and expenditure data address fiscal policy, and crime rates address social conditions. Two health-care variables and one human-capital variable are also included.

## *Migration and the Stock of Knowledge and Prior Experience*

Krupka (2007) has provided fundamentally new insights into the destination choices of households based on their stock of prior knowledge and experience. He argues that amenities work through household production and not through consumption. Individuals cannot control their locations during preadult years, and they have added incentive to invest in the appreciation of all the amenities that are present in their origin region. These investments are very location-specific and are not similarly valued by households when living in very different areas. Once these location-specific investments are made, the opportunity costs of moving to dissimilar locations are increased. The overall result is that migrants prefer moving to areas that are more similar than not to their childhood residences. So, national population distributions result from people sorting into their most preferred locations as measured by bundled amenities, a point not adequately stressed by the current QOL literature.

The research uses geocoded US data from the National Longitudinal Survey on Youth 1979 cohort to test the prediction that exposure to certain types of areas during youth increases the likelihood of finally settling in such areas, even for long-distance migrants. A very wide array of amenities is used, including natural, cultural, social, and retail-variety types. Even when eliminating nonreturn migrants from the household sample, origin-region exposure appears to trump other factors like human capital in driving interregional migration. The author makes a genuinely new contribution to understanding the behavioral decision-making of households.

## *Push Factors and Migration: Reacting to Big-City Diseconomies and Congestion*

Lastly, another interesting perspective on recent US migration has been provided by Davies et al. (2008). As the nation's population continues to react to big-city diseconomies and congestion, many households have migrated down through the national urban hierarchy to more peripheral communities (Plane et al. 2005). Push factors have also included the presence of foreign-born populations, especially in gateway cities, and the lack of affordable housing in many large metropolitan areas. The authors provide solid evidence that much intercounty migration, especially between states, is being driven by housing-cost adjustments. An index is constructed by forming a ratio between median housing values and median household income. Movement is then studied between the years 1995 and 2000 and compared to the change in that index over the same 5-year period. Special attention is given to the 100 largest intercounty moves, where suburban Washington, Las Vegas, and Phoenix were among the most popular destinations.

Davies et al. (2008) found that while most migration streams were associated with an increase in median income, some 60% of the moves were associated with moves to places having more affordable housing. Nuances on this broad theme are provided: For example, Hispanics are shown to have been more likely to adjust their housing costs than other ethnic groups. Four-quadrant typologies of county-to-county migration are given with change in housing cost indicated on one axis and change in income indicated on the other. The authors argue that different types of households react differently to issues like social capital and labor-market segmentation, hence, the somewhat different hierarchical streams.

## Summary and Conclusion

This chapter has reviewed recent advances in research on the role of amenities in urban growth and regional development. A main goal was to synthesize the literature in a way that informs an interdisciplinary audience of researchers and practitioners in the social sciences and public policy fields. Having met that objective, the few remaining comments are observations for planning activities aimed at shaping the outcome of urban and regional development.

Foremost, amenities – and QOUL more broadly – need to be the central focus of urban and regional planning. As illustrated in this chapter, they direct interregional migration flows, influence intraregional settlement patterns, and generate compensating differentials in labor and housing markets. These forces are powerful and only continue to grow more, so it seems logical that public policy should attempt to leverage them to every extent possible. For example, Power (1996) and Florida (2002) argue persuasively for natural and human amenities, respectively, to be made central components of economic development policy. In many places throughout the country, the natural environment itself is literally the engine of economic growth: Both people and jobs are drawn to scenic landscapes and a favorable climate. While public policy cannot influence the climate in the short run, the growing focus on greenhouse gas emissions is an explicit acknowledgement that it can in the intermediate and long run. Land-use planning does directly influence the character of the built environment and the territory that accommodates it. What's more, culture, human capital, and other human amenities are readily influenced and, indeed, created via public policy. To cite one example, the USA has entered a period of widespread divestment from public education,[4] and

[4]For example, according to the *Census of Governments* between 1992 and 2007, the number of instructional employees at state institutions of higher education fell by 50,497 from 435,789 to 385,292 – a cutback of more than 11%.

this, however forced (by economic circumstance) and/or undesired, may have lasting consequences for urban and regional QOL.

Overall, natural and human environmental amenities are perhaps best thought of – in the public policy context – as both fulcrums that can be used to help leverage desired outcomes and desirable outcomes in and of themselves. Places that are desirable to live and work in develop inertias of their own, but great care has to be taken to ensure that growth and change does not somehow erode the very quality of life that makes them successful. As the Puget Sound region's *Vision 2040* is proof that (QOL) can be at the heart of broad-based urban and regional planning strategies. The question, for that and similar planning efforts, is whether or not there is enough local-level implementation to turn visions into reality.

## References

Adamson, D. W., Clark, D. E., & Partridge, M. D. (2004). Do urban agglomeration effects and household amenities have a skill bias? *Journal of Regional Science, 44*, 201–223.

Alasia, A., Bollman, R, Parkins, J., & Reimer B. (2008). *An index of community vulnerability: Conceptual framework and application to population and employment changes 1981 to 2001* (Agriculture and rural working paper series). Ottawa: Statistics Canada.

Albouy, D. Y. (2008). *Are big cities really bad places to live? Improving quality-of-life estimates across cities* (Working paper 14472). Cambridge, MA: National Bureau of Economic Research.

Allinson, J. (2005). Exodus or renaissance? Metropolitan migration in the late 1990s. *Town Planning Review, 76*, 167–189.

Anselin, L. (1988). *Spatial econometrics: Methods and models*. Boston: Kluwer Academic Publishers.

Bartik, T. J., & Smith, V. K. (1987). Urban amenities and public policy. In E. S. Mills (Ed.), *Handbook of regional and urban economics* (Urban economics, Vol. 2, pp. 1207–1254). Amsterdam: North-Holland.

Beeson, E. (1991). Amenities and regional differences in returns to worker characteristics. *Journal of Urban Economics, 30*, 224–241.

Bell, D. (1973). *The coming of post-industrial society*. New York: Basic Books.

Berger, M. C., Blomquist, G. C., & Waldner, W. (1987). A Revealed-Preference Ranking of Quality-of-life for Metropolitan Areas. *Social Science Quarterly, 68*, 761–778.

Berry, B. J. L., & Horton, F. E. (1970). *Geographic perspectives on urban systems*. Englewood Cliffs: Prentice Hall.

Blomquist, G. C., Berger, M. C., & Hoehn, J. P. (1988). New estimates of quality of life in urban areas. *The American Economic Review, 78*, 89–107.

Bockstael, N. E., & Freeman III, A. M. (2005). Welfare theory and valuation. Handbook of Environmental Economics. Vol. 2, Valuing Environmental Changes, pp. 517–570. Elsevier.

Borts, G. H., & Stein, J. L. (1964). *Economic Growth in a Free Market*. New York: Columbia University Press.

Brueckner, J. K., Thisse, J. F., & Zenou, Y. (1999). Why is central Paris rich and downtown Detroit poor? An amenity-based theory. *European Economic Review, 43*, 91–107.

Bruinsma, F., & Rietveld, P. (1993). European agglomerations in European infrastructure networks. *Urban Studies, 30*, 919–934.

Canaleta, C. G., Arzoz, P. P., & Gárate, M. R. (2004). Regional economic disparities and decentralization. *Urban Studies, 41*, 71–94.

Carlino, G. A., & Mills, E. S. (1987). The determinants of county growth. *Journal of Regional Science, 27*, 39–54.

Carruthers, J. I., & Mulligan, G. F. (2006). Environmental valuation: Connecting theory, evidence, and public policy. In J. I. Carruthers & B. Mundy (Eds.), *Environmental valuation* (pp. 3–25). Aldershot: Ashgate.

Carruthers, J. I., & Mulligan, G. F. (2008). A locational analysis of growth and change in American Metropolitan Areas. *Papers in Regional Science, 87*, 155–171.

Carruthers, J. I, & Mulligan, G. F. (2009). *The "plane of living" and the recent evolution of housing values in the United States* (U.S. Department of housing and urban development working paper, # REP 09–01).

Cheshire, P. C., & Magrini, S. (2006). Population growth in European cities: Weather matters but only nationally. *Regional Studies, 40*, 23–37.

Cheshire, P. C., & Magrini, S. (2008). *Urban growth drivers and spatial inequalities: Europe – a case with geographically sticky people*. Venice: Department of Economics, Ca' Foscari University of Venice.

Clark, D. E., & Hunter, W. (1992). The impact of economic opportunity, amenities, and fiscal factors on age-specific migration rates. *Journal of Regional Science, 32*, 349–365.

Clark, D. E., & Murphy, C. A. (1996). Countywide employment and population growth: An Analysis of the 1980s. *Journal of Regional Science, 36*, 235–256.

Costa, D. L., & Kahn, M. E. (2000). Power couples: Changes in the locational choice of the college educated, 1940–1990. *Quarterly Journal of Economics, 115*, 1287–1315.

Cushing, B. (1993). The effect of the social welfare system on metropolitan migration in the United States, by income group, gender, and family structure. *Urban Studies, 30*, 325–338.

Davies, S. W., Clark, W. V. A., & Ruiz, T. (2008). Demographic variation in housing cost adjustments with United States family migration. *Population, Space and Place, 14*, 305–325.

Deller, S. C., Tsai, T. H., Marcouiller, D. W., & English, D. B. K. (2001). The role of amenities and quality of life in rural economic growth. *American Journal of Agricultural Economics, 83*(2), 352–365.

Des Rosiers, F., Thériaul, M., Florent, J., & Villeneuve, P. (2005). *The heterogeneous nature of accessibility to urban amenities and its impact on property values*. Paper presented at the Annual North American Meeting of the Regional Science Association International, Las Vegas.

Diamond, D. B., & Tolley, G. S. (Eds.). (1982). *The economics of urban amenities*. New York: Academic.

Dicken, P., & Lloyd, P. E. (1981). *Modern western society*. New York: Harper and Row.

Dissart, J. C., & Deller, S. C. (2000). Quality of life in the planning literature. *Journal of Planning Literature, 1*, 36–61.

Dorfman, J. H., Partridge, M. D., & Galloway, H. (2008). *Are high-tech employment and natural amenities linked? Answers from a smoothed Bayesian spatial model*. Paper prepared for presentation at the Annual Meeting of the American Agricultural Association, Orlando.

Elliott, J. R., & Perry, M. J. (1996). Metropolitanizing nonmetro space: Population redistribution and emergent metropolitan areas, 1965–1990. *Rural Sociology, 61*, 497–512.

Ferguson, M., Ali, K., Olfert, M. R., & Partridge, M. D. (2007). Voting with their feet: Jobs versus amenities. *Growth and Change, 38*, 77–110.

Florida, R. (2002). *The rise of the creative class*. New York: Basic Books.

Fotheringham, A. S., Brunsdon, C., & Charlton, M. (2002). *Geographically weighted regression: The analysis of spatially varying relationships*. Chichester: Wiley.

Freeman, A. M. (2003). *The Measurement of Environmental and Resource Values*. Washington, DC: Resources for the Future.

Frenette, J. F. (2008). *The index of community vulnerability (ICV) revisited: Exploring the determinants of rural working age population out-migration in Canada, 2001–2006*. Paper presented at the Annual North American Meeting of the Regional Science Association International, Brooklyn.

Fujita, M., Krugman, P., & Venables, A. J. (1999). *The spatial economy: Cities, regions, and international trade*. Boston: MIT Press.

Galbraith, J. K. (1967). *The new industrial state*. New York: Signet Books.
Glaeser, E. L. (2007). The economics approach to cities. *National Bureau of Economic Research* working paper # 13696.
Glaeser, E. L., & Kohlhase, J. E. (2004). Cities, regions, and the decline of transport costs. *Papers in Regional Science, 83*, 197–228.
Glaeser, E. L., & Maré, D. C. (2001). Cities and skills. *Journal of Labor Economics, 19*, 316–342.
Glaeser, E. L., Kolko, J., & Saiz, A. (2001). Consumer city. *Journal of Economic Geography, 1*, 27–50.
Goodrich, C., Allin, B. W., Thornthwaite, C. W., Brunck, H. K., Tyron, F. G., Creamer, D. B., Vance, R., & Hayes, M. (1936). *Migration and economic opportunity*. Philadelphia: University of Pennsylvania.
Gottlieb, P. D., & Joseph, G. (2006). College-to-work migration of technology graduates and holders of doctorates within the United States. *Journal of Regional Science, 46*, 627–659.
Greenwood, M. J., & Hunt, G. L. (1989). Jobs versus amenities in the analysis of metropolitan migration. *Journal of Urban Economics, 25*, 1–16.
Greenwood, M. J., & Hunt, G. L. (2003). The early history of migration research. *International Regional Science Review, 26*, 3–37.
Gyourko, J., & Tracy, J. (1989). The importance of local fiscal conditions in analyzing local labor markets, *Journal of Political Economy, 97*, 1208–1231.
Gyourko, J., & Tracy, J. (1991). The structure of local public finance and the quality of life. *Journal of Political Economy, 99*, 774–806.
Gyourko, J., Kahn, M. E., & Tracy, J. (1999). Quality of life and environmental conditions. In P. Cheshire & E. S. Mills (Eds.), *Handbook of regional and urban economics* (Applied urban economics, Vol. 3, pp. 1413–1454). Amsterdam: North Holland.
Hansen, S. B., Ban, C., & Huggins, L. (2003). Explaining the 'brain drain' from older industrial cities. *Economic Development Quarterly, 17*, 132–147.
Harris, R. N. S., Tolley, G. S., & Harrell, C. (1968). The residence site choice. *The Review of Economics and Statistics, 50*, 241–247.
Harvey, D. (1973). *Social justice and the city*. Baltimore: Johns Hopkins.
Henderson, J. V. (1974). The sizes and types of cities. *The American Economic Review, 70*, 894–910.
Herzog, H., & Schlottmann, A. (1986). The metro rating game: What can be learned from recent migrants? *Growth and Change, 17*, 37–50.
Herzog, H., Schlottmann, A., & Johnson, D. (1986). High-technology jobs and worker mobility. *Journal of Regional Science, 26*, 445–459.
Hoehn, J. P., Berger, M. C., & Blomquist, G. C. (1987). A hedonuc model of interregional washes, rents, and amenity values. *Journal of Regional Science, 27*, 341–349.
Jensen, T., & Deller, S. C. (2007). Spatial modeling of the migration of older people with a focus on Amenities. *The Review of Regional Studies, 37*, 303–343.
Kendall, J., & Pigozzi, B. W. (1994). Nonemployment income and the economic base of Michigan Counties: 1959–1986. *Growth and Change, 25*, 51–74.
Krupka, D. J. (2007). *Location-specific human capital, location choice and amenity demand* (Discussion Paper No. 2987). Bonn: Institute for the Study of Labor (IZA).
Krupka, D. J. (2008). *On amenities, natural advantage and agglomeration* (Discussion Paper No. 3598). Bonn: Institute for the Study of Labor (IZA).
Lambiri, D., Biagi, B., & Royuela, V. (2007). Quality of life in the economic and urban economic literature. *Social Indicators Research, 84*, 1–25.
Lee, N. (2006). *Ideopolis: Knowledge cities – A review of quality of life measures*. London: The Work Foundation.
Massam, B. H. (1975). *Location and space in social administration*. London: Edward Arnold.
Massam, B. H. (1993). *The right place*. New York: John Wiley and Sons.
McGranahan, D. A. (1993). *Natural amenities drive rural population change* (Agricultural Economic Report No. 781). Economic Research Service, United States Department of Agriculture, Washington.

Mueser, P. R., & Graves, P. E. (1995). Estimating the role of economic opportunity and amenities in explaining population redistribution. *Journal of Urban Economics, 37*, 176–200.

Mulligan, G. F., Vias, A. C., & Glavac, S. M. (1999). Initial diagnostics of a regional adjustment model. *Environment and Planning A, 31*, 855–876.

Mulligan, G. F., Carruthers, J. I., & Cahill, M. (2004). Urban quality of life and public policy: A review. In R. Capello & P. Nijkamp (Eds.), *Urban dynamics and growth: Advances in urban economics* (pp. 729–802). Amsterdam: Elsevier.

Muth, R. F. (1969). *Cities and Housing*. Chicago, IL: The University of Chicago Press.

Nussbaum, M. C., & Sen, A. (Eds.). (1993). *The quality of life*. Oxford: Clarendon.

Partridge, M. D., Rickman, D. S., Ali, K., & Olfert, M. R. (2008a). Lost in space: Population growth in the American hinterlands and small cities. *Journal of Economic Geography, 8*, 727–757.

Partridge, M. D., Rickman, D. S., Ali, K., & Olfert, M. R. (2008b). The geographic diversity of United States nonmetropolitan growth dynamics: A geographically weighted regression approach. *Land Economics, 84*, 243–268.

Perloff, H. S., Dunn, E. S., Lampard, E. E., & Muth, R. F. (1960). *Regions, resources, and economic growth*. Washington, DC: Resources for the Future.

Plane, D. A., & Heins, F. (2003). Age articulation of United States inter-metropolitan migration flows. *The Annals of Regional Science, 37*, 10–130.

Plane, D. A., Henrie, C. J., & Perry, M. J. (2005). Spatial demography special feature: Migration up and down the urban hierarchy and across the life course. *Proceedings of the National Academy of Sciences of the United States of America, 102*, 15313–15318.

Power, T. M. (1996). *Lost landscapes and failed economies*. Washington: Island Press.

Power, T. M., & Barrett, R. N. (2001). *Post-Cowboy Economics: Pay and Prosperity in the New American West*. Washington D.C.: Island Press.

Putnam, R. D. (2000). *Bowling alone*. New York: Touchstone.

Rappaport, J. (2004). Why are population flows so persistent? *Journal of Urban Economics, 56*, 554–580.

Rappaport, J., & Sachs, J. D. (2003). The United States as a coastal nation. *Journal of Economic Growth, 8*, 5–46.

Rauch, J. E. (1993). Productivity gains from geographic concentration of human capital: Evidence from the cities. *Journal of Urban Economics, 34*, 380–400.

Ravenstein, E. G. (1885). The laws of migration. *Journal of the Royal Statistical Society, 48*, 167–235.

Roback, J. (1982). Wages, rents, and the quality of life. *Journal of Political Economy, 90*, 1257–1278.

Roback, J. (1988). Wages, rents, and amenities: Differences among workers and regions. *Economic Inquiry, 26*, 23–41.

Rogerson, R. J., Findlay, A. M., & Morris, A. S. (1988). The best cities to live in. *Town and Country Planning, 57*, 270–273.

Rosen, S. (1974). Hedonic prices and implicit markets: Product differentiation in pure competition. *Journal of Political Economy, 82*, 34–55.

Rosen, S. (1979). Wage-based indexes of urban quality of life. In P. Mieszkowski & M. Straszheim (Eds.), *Current issues in economics* (pp. 74–104). Baltimore: Johns Hopkins.

Rostow, W. W. (1968). *The stages of economic growth*. Cambridge: Cambridge University Press.

Rothschild, E. (2009). The decline of the auto-industrial society. *New York Review of Books, 61*(3), 8–9.

Savageau, D. (2007). *Places rated almanac* (7th ed.). Washington: Places Rated Books.

Sen, A. (1999). *Development as freedom*. New York: Anchor Books.

Smith, D. M. (1977). *Human geography: A welfare approach*. London: Edward Arnold.

Smith, D. M. (1994). *Geography and social justice*. Oxford: Blackwell.

Stimson, R. J., Baum, S. W., Mullins, P. F., & O'Connor, K. (2001). A typology of community opportunity and vulnerability in metropolitan Australia. *Papers in Regional Science, 80*, 45–66.

Thorndike, E. L. (1939). *Your city*. New York: Harcourt, Brace, and Company.

Tolley, G. S. (1974). The welfare economics of city bigness. *Journal of Urban Economics, 1*, 324–345.

Ullman, E. L. (1954). Amenities as a factor in regional growth. *Geographical Review, 44*, 119–132.

Van Praag, B. M. S., & Frijters, P. (1999). The measurement of welfare and well-being: The Leyden approach. In D. Kahneman, E. Diener, & N. Schwarz (Eds.), *Well-being: The foundations of hedonic psychology* (pp. 413–433). New York: Russell Sage.

Welch, R. K., Carruthers, J. I., & Waldorf, B. S. (2007). Public service expenditures as compensating differentials in U.S. Metropolitan Areas: Housing values and rents. *Cityscape, 9*, 131–156.

Whisler, R. L., Waldorf, B. S., Mulligan, G. F., & Plane, D. A. (2008). Quality of life and the migration of the college educated: A life-course approach. *Growth and Change, 19*, 58–94.

Wong, C. (2002). Developing indicators to inform local economic development. *Urban Studies, 39*, 1833–1863.

Yousefi, M., & Rives, J. (1987). Migration behavior of college graduates: An empirical analysis. *Journal of Behavioral Economics, 16*, 35–49.

# Chapter 6
# Variations in Objective Quality of Urban Life Across a City Region: The Case of Phoenix

**Subhrajit Guhathakurta and Ying Cao**

## The View from the Sky

The first experience of flying into the Phoenix Sky Harbor airport is usually an unforgettable one. Well before the pilot begins the descent, the stark undulating and gray landscape below abruptly transforms into a series of tract-home communities punctuated by rocky outcroppings or open land. A few more minutes later, the identifiable contours of these communities blend into a vast sea of cloned homes crisscrossed by a rectilinear network of roads. As the plane starts to descend, the road network shows signs of life. Most likely, it is a clear day with a distant haze of pollution from the automobile fumes. Soon, the blue misshapen dots of backyard pools grab our attention. The repetitive built forms lead the eyes to the distant horizon in all directions. The vastness of the unbroken, undifferentiated expanse of habitation offers a unique perspective that is both awesome and forbidding.

The vastness of this metropolitan region together with its characteristic homogeneity in appearance raises several interesting questions about intra-metropolitan variation in residents' quality of urban life (QOUL). How does it vary within this

S. Guhathakurta (✉)
School of Geographical Sciences and Urban Planning, GeoDa Center for Geospatial Analysis and Computation, Arizona State University, Tempe, AZ 85287-5302, USA

School of City and Regional Planning, Georgia Institute of Technology Atlanda, GA 30332-0695
e-mail: subhro.guha@coa.gatech.edu

Y. Cao
College of Architecture and Environment, Sichuan University, Chengdu, Sichuan Province, China
e-mail: caoyingscu@hotmail.com

R.W. Marans and R. Stimson (eds.), *Investigating Quality of Urban Life: Theory, Methods, and Empirical Research*, Social Indicators Research Series 45,
DOI 10.1007/978-94-007-1742-8_6, © Springer Science+Business Media B.V. 2011

apparently uniform urban landscape? Where are the highs and lows, and what specific attributes of quality of life (QOL) stands out as distinctly uneven or relatively unvarying? What aspects of QOUL most poignantly characterize regional identity? These questions take on a special meaning in the Phoenix metropolitan region given its unique history and evolutionary trajectory as a post–World War II and post air-conditioning Western boomtown.

## Conceptual and Methodological Challenges

A significant challenge in examining variations in intra-metropolitan QOUL indicators is the problem of defining the spatial unit of analysis within the region. This is especially critical in the case of objective data that are aggregated at some level of geography. Secondary data, which form the basis for objective QOUL information, are reported by standard accounting or administrative units, such as those available in the decennial censuses. Given the standardized reporting norms, we are faced with questions such as:

- What is the appropriate spatial unit of analysis?
- How do we conceptualize the space in social terms?
- Is it a neighborhood, a community, an economic market area?
- Is it defined in terms of administrative and accounting regions, such as municipalities, census tracts, and school districts?

In the case of subjective indicators, this is not a serious problem, since individual data are usually collected from primary surveys and can be aggregated by any chosen geographic unit according to the study's objectives.

Previous research has provided very little guidance on the appropriate spatial unit for conducting intra-metropolitan QOUL analysis. The seminal work by Marans and Rodgers (1975) offered a conceptual framework for measuring QOUL that includes both objective and subjective data at three spatial scales. These spatial scales, which they termed "urban domains," include:

- The community
- The neighborhood
- The individual dwelling

An individual's level of satisfaction with each of the three domains is captured in a composite QOL measure. The definition of the spatial unit and its areal extent is important to delineate carefully given that many authors have documented a "Modifiable Areal Unit Problem" (MAUP). The problem appears when different areal definitions lead to different results suggesting that each level of geography is unique and cannot be generalizable to other areal definitions (Openshaw 1984; Haynes et al. 2007).

Of the three domains mentioned above, the neighborhood boundaries have been the most difficult to delineate objectively. Although the term neighborhood evokes immediate recognition and imaging, the edges of this spatial entity are often blurry or open to subjective interpretation. The problem of defining a neighborhood has been acknowledged by many scholars but most ultimately resort to aggregating smaller levels of census enumeration districts into larger blocks according to some chosen criteria to form neighborhoods (Haynes, et al. 2007; Kawachi and Berkman 2003; Flowerdew et al. 2008).

The classical definition of a neighborhood offered by Perry, Heydecker and Adams (1929) and Stein (1966) is based upon a 5-min walking radius (approximately 160 acres). The radius is measured from the center, which is anchored by important public uses, such as a school. More recent empirical research has extended the size of a neighborhood to about two thirds of a mile walking radius, which translates to an area of about 500 acres (Moudon et al. 2006). Regardless of how a neighborhood is defined in the literature, its residents may have varying impressions of the boundaries of their neighborhood (Ellaway et al. 2001; Simons et al. 2004).[1]

The Phoenix metropolitan (statistical) area includes Maricopa and Pinal counties within an areal extent of 14,598 square miles and is home to about 4.2 million people in 2008. Despite the vast urban distention the region is administratively divided into 33 cities and 35 unincorporated communities (see Fig. 6.1). Beneath the surface, it is indeed like other urban areas of its size that exhibit a textured surface of human demographics with different geographic concentrations of young and old families with more and less affluence, and neighborhood clusters of ethnic communities. The region's home county, Maricopa, was the fastest growing county in the USA from 1990 to 2000. The population build-up has sparked the growth of smaller cities in the county as well. Gilbert was ranked 2nd in population increase across all US incorporated places, with Chandler (9th) and Scottsdale (15th) following close behind. According to unofficial projections, Phoenix is currently the fifth largest metropolitan region in the USA having surpassed Philadelphia sometime in 2005 (Wikipedia). Several of these cities, such as Paradise Valley and Sun City, are relatively small enclaves; others such as Phoenix, Mesa, and Scottsdale are large conurbations. These unique features, as well as the pace and character of their growth, make the Phoenix metropolitan area an ideal candidate for the study of intra-metropolitan variations in QOUL.

For this study, it was determined that the optimum strategy would be to select spatial and temporal scales based on the most reliable and accessible publicly

[1] Data from the 2001 Detroit Area Study indicate that residents throughout the metro area varied greatly in their definition of neighbourhood. About a third of the respondents described their neighborhood as their block or the clustering of houses around them, another third defined their neighborhood as anywhere from 2 to 120 blocks around them, and the remainder described their neighborhood as a square mile or more. For a description of the Detroit Area Study, see Chap. 7.

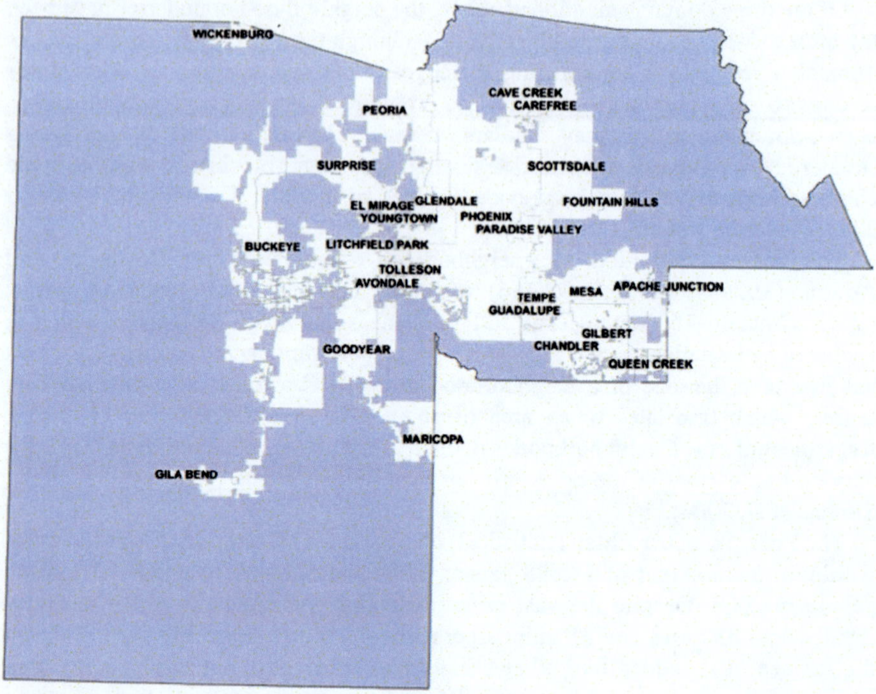


**Fig. 6.1** Map showing municipalities and unincorporated areas in Maricopa County (Source: The authors)

available data. The American Community Survey (ACS) provides such a data set that is readily available, and its frequent updates make it more current in most instances than the decennial censuses. The spatial resolution of the ACS is not as fine grained as in the census. However, it offers place-based information for communities that form the metropolitan region. The ACS also provides data for school districts, which, in the case of Phoenix, are not complete. That is, not all districts are included. Further, school districts are drawn variously as elementary, secondary, or unified (covering K through 12) in different parts of the county, making comparability among different districts problematic. Thus, when all three factors:

- Data access
- Frequency of updates
- Spatial resolution were taken into consideration, the ACS data for places were the clear choice as opposed to using other data sources or another level of spatial resolution.

## Attributes of Quality of Urban Life

Defining the theoretical constructs and identifying a set of practical attributes of QOUL is a significant challenge. The choice of QOUL attributes is intrinsically subject to cultural specificity and individual interpretations. Hence, QOUL researchers need to consider a diversity of possible referents together with a wide array of measurement approaches. Despite the challenges, monitoring of QOUL is often an essential component of evaluation research that seeks to understand the human impact of changes in policy regarding the physical and social environments.

If indeed the goals of QOUL studies involve comparisons across time and space, a common biophysical approach must be adopted that captures the common elements of human nature and human societies (Sadalla et al. 2005). Assessments of QOUL would be less useful if the chosen indicators vary by culture, place, and over time. There has been an increasing interest in developing generally applicable QOUL indicators in the last 20 years. The assessment of QOUL has received special attention from international organizations, such as the United Nations, the World Bank, the World Health Organization, and the International Labor Office. Table 6.1 displays an illustrative set of QOUL indicators used by various groups in different parts of the world.

Several scholars have attempted to extract a number of "value domains" that have universal applicability in a wide range of cultures and societies. For example, (Diener 1995) proposed seven categories of "universal values" based on prior studies conducted by Mukherjee (1989) and Schwartz (1994) (see Table 6.2). Similarly, the Physical Quality of Life (PQLI) index (Morris 1996) and the Human Development Index (HDI) attracted a lot of interest as alternatives to the pure economic measure, like the Gross Domestic Product (GDP) that was often quoted as a reflection of a countries socioeconomic development. The objective of developing PQLI and HDI was to measure the distributional effects of income growth across the cultural and structural differences among countries.

In part, we based the study reported in this chapter on objective QOUL indicators in Greater Phoenix on prior assessments of both objective and subjective QOUL in this region conducted by the Morrison Institute for Public Policy at Arizona State University. These were published in a series of reports generally titled *What Matters in Greater Phoenix*. The first such report was published in 1997, which was followed by similar publications in 1998, 1999, and 2004. This series of reports relied on preference surveys that were conducted to assess the subjective well being of individuals and their impressions about the attractiveness of the region. The nine indicators that were chosen for inclusion in the Morrison Institute's surveys were derived from a lengthy process of focused discussions in which many citizens, as well as leaders in business, government, and civic organizations participated. This process resulted in 318 indicators ranked by priority in nine categories (Morrison Institute for Public Policy 1997: p. 6). The nine indicators are:

- Education
- Public safety and crime
- Health and healthcare

**Table 6.1** Survey of indicators: examples from across the world

| United Nations | OECD | Philippines | Japan |
|---|---|---|---|
| Health | Nutrition | Health, nutrition | Health |
| Food | Clothing | Learning | Education, learning |
| Education | Shelter | Income | Employment |
| Employment | Health | Employment | Quality of work |
| Housing | Education | Environmental resources | Leisure |
| Social security | Leisure | Housing | Income |
| Clothing | Social security | Utilities | Spending |
| Recreation | Social environment | Public safety | Material environment |
| Human freedoms | Physical environment | Justice | Crime |
| Population | Social status | Political values | Law enforcement |
| Income and expenditure | Education | Social mobility | Community life |
| Communication and transport | | | Class, social mobility |
| **Finland** | **Sweden** | **India** | **USA (Calvert-Henderson)** |
| Health | Working conditions | Population | Education |
| Education | Economic resources | Health, nutrition | Employment |
| Physical environment | Political resources | Housing | Energy |
| Housing, habitat | Schooling | Education | Environmental health |
| Working conditions | Health, medical care | Labor, employment | Human health |
| | Family origin, family relations | Income, expenditure, wealth | Human rights |
| | Housing | | Income |
| | Nutrition | | Physical infrastructure |
| | Leisure time and pursuits | | National Security |
| | | | Public safety |
| | | | Recreation |
| | | | Shelter |

Source: Compiled by the authors from various reports

**Table 6.2** Quality of life value regions and their constituent elements as classified by Schwartz

| Value regions | QOO elements |
|---|---|
| Masterly (success, capable, ambitions) | Length of life<br>Infant mortality rate<br>Families at risk |
| Affective autonomy (Enjoying life, pleasure, exciting life) | Suicide rate<br>Self-rated health |
| Intellectual autonomy (Curious, broad-minded, creative) | College education |
| Egalitarian commitment (Equality, social justice, freedom) | Unemployment rate<br>Poverty rate |
| Harmony (Protective of environment, aesthetic appreciation, unity with nature) | Residential density<br>Environmental toxins index |
| Conservation (Social order, self-discipline, family security) | Violent crime rate |
| Hierarchy (Wealth, social power, authority) | Per capita personal income |

Source: Schwartz (1994, 1996) and Diener (1995)

- Economy
- Environment
- Families and youth
- Transportation and mobility
- Community
- Arts, culture, and recreation

The Morrison Institute's study in Greater Phoenix was also the basis for another longitudinal study of QOUL in the USA–Mexico border region. This project, also known as the Border Observatory Project (http://bop.caed.asu.edu), collected information on both perceptual and objective indicators as did the Morrison study. However, the indicators were adjusted to improve the specificity of the content. For example the indicator labeled "families and children" in the Morrison's report was not used for the Border Observatory project but the specific measures under the indicator (for example, median home sale prices, apartment rent, and poverty rates) where incorporated under "housing" or "economy" indicators. In addition, the Morrison Institute indicator called "community indicator" included slightly different measures in the Border Observatory project, reflecting municipal service quality such as condition of roads, street lighting, and responsiveness of emergency services. Finally, an important subjective indicator was added in the border area study that measured the "emotional well-being" of the respondents.

Our study is molded from the two projects mentioned above but with a particular focus on the objective component of the indicators used in those earlier studies. We asked the same questions to screen the applicability of the indicators used in the other studies:

(a) Is the indicator measurable from data that are available at regular intervals from reliable and accessible sources?
(b) Is the indicator relevant to the QOL of the vast majority of the Greater Phoenix residents?
(c) Is the indicator affected by policies, regulation, and social behavior?
(d) Is the indicator relevant for the subregional (place-based) spatial level as part of the Greater Phoenix area?

The last question is important to consider in an intra-metropolitan context where many of the amenities are shared across the region. For example, the professional football team plays in an arena in Glendale but serves fans from the entire Greater Phoenix region. Similarly, the art museums, the symphonies, opera houses, and other cultural centers may be located in one or more large cities within Greater Phoenix but cater to metropolitan area-wide residents. Also, significantly, we decided that an attribute measuring environmental aspects would be unsuitable for this study, since environmental parameters such as air quality, soil and water contamination, have significant spillover effects. Indeed, the concentration of Ozone appears to be highest in the Western parts of the metropolitan region although the sources of that pollution are in the central city areas of Phoenix (see http://sustainability.asu.edu/eatlas/maps/map20.htm).

**Table 6.3** Chosen indicators and their specific measures

| | Indicators of QOUL in Greater Phoenix | Measures contributing to each indicator |
|---|---|---|
| 1 | Education | Graduation rate |
| | | AIIMS Math score |
| | | AIIMS Reading score |
| | | AIIMS Writing score |
| | | AIIMS Science score |
| | | NCE Math (median percentile) |
| | | NCE Reading |
| | | NCE Language |
| 2 | Economy, income, and jobs | Percent unemployed |
| | | Median household income |
| | | Percent of people under poverty |
| 3 | Public safety | Crime index |
| | | Percent violent crimes |
| | | Percent aggravated assault |
| | | Percent burglaries |
| 4 | Housing | Homeowner vacancy rate |
| | | Renter vacancy rate |
| | | Percent owner occupied |
| | | Median value |
| | | Median rent |
| 5 | Transportation and mobility | Mean travel time to work |
| | | Mode of travel |
| 6 | Public health | Percent fatalities from cardiovascular diseases |
| | | Percent fatalities from lung cancer |
| | | Percent fatalities from breast cancer |
| | | Percent fatalities drug induced |
| | | Percent fatalities alcohol induced |

Source: Compiled by authors

After applying the filters for appropriate intra-metropolitan indicators mentioned above, we arrived at six indicators that could characterize sub-regional QOL within municipalities in Greater Phoenix. These indicators, together with the specific measures that contribute to the indicators, are provided in Table 6.3.

## Selecting and Preparing the Data

As explained earlier, this analysis of objective QOL indicators is based on easily accessible data from public sources having regularly scheduled updates. The objective is to provide a template that can be updated over time to see how the indicators change for the cities within the Greater Phoenix metro region. One obvious data source that we looked past and did not include is the decennial census.

The reason for not choosing to use the census is simply to make the analytical template more amenable to more frequent updates than the 10-year census intervals. The downside of this choice is the lower geographic resolution of the selected spatial entities. That is, the trade-off we accepted is to limit our spatial resolution to the level of cities, municipalities, and unincorporated "places" for which more frequent data updates can be obtained. If we had instead opted for a higher spatial resolution, such as the census tract, we would be limited to 10-year cycles for updating the QOL attributes in this study.

Besides the exigencies of data availability, the choice of a municipality or a place is logically consistent with the most significant attachment that people form with their physical environment. Neighborhood boundaries are often amorphous and subjected to varying perceptions of the residents. School districts, do have specified boundaries that have been shown to impact housing prices in many instances, but the influence of these districts has somewhat waned due to more flexible approaches to public education. In Phoenix, due to the rapid growth of the metropolitan population, large unified school districts have been formed in the recently settled areas, and the existing district have allowed open enrollments, thereby making the boundaries more porous. In addition, the recent growth of publicly supported Charter Schools has absorbed many students who would have otherwise opted for a public education. In contrast, cities and municipalities are a strong determinant of the QOL of their residents through their role as a provider of basic services, their ability to extract taxes and fees, and their function as a regulatory authority for zoning, land uses, and building codes.

The potential drawback of using cities as the geographic entity of concern when comparing quality of life in a metropolitan region is the wide variation in city sizes and, consequently, in services and amenities. Larger cities are less homogenous and may include areas that exhibit significantly different QOL attributes than the average for the city. Therefore it is easy to fall into the trap of "ecological fallacy" where we infer smaller area (such as a neighborhood) QOL indicators purely based on average indices for the city. As long as we are aware of such pitfalls and limit our interpretation to the geographic level of cities and not to their constituent spatial units, the indicators can offer useful, comparative, and longitudinal information about sub-regional QOUL.

The data for this study were derived from both federal and local sources. The most important federal source of data used in developing the QOUL indices in this study is the American Community Survey (ACS). Since 1994, the US Census Bureau began to test the feasibility of using the census "long form" on an annual basis to provide more current snapshots of socioeconomic conditions in larger communities. This was the beginning of the ACS, which is now poised to become another form of census that is administered yearly to a sample of approximately three million households throughout the USA and Puerto Rico. The pilot was rolled out in 1996 for four counties and expanded to 36 counties in 1999. As a result of the success of ACS, the census long form will be discontinued from 2010 census onwards. Full implementation began in 2005 when the ACS sample included all of the USA and its territories. The ACS has also been expanded to include group quarters (nursing homes, barracks, college/university housing, among others) in 2006. The frequency of updates of the ACS data is dependent on the size of the community. That is, communities with population of 65,000 and over will find 1 year estimates of economic, social, housing, and demographic characteristics in ACS; communities under 65,000 but above 20,000 persons will be able to rely on 3 year ACS estimates; while 5 year estimates will provide information on all communities. The first 5 year estimate of ACS was due for release in 2010.

In this study we have used the 2006–2008 three year estimate of ACS that includes most, but not all, places and municipalities in the Greater Phoenix region. The analysis can be easily updated to all other communities (places) in Phoenix that are currently excluded once the 2005–2009 five year estimates are released in 2010.

Of the six indicators for which QOUL data are analyzed, we have relied exclusively on ACS data for three: economy, income, and jobs; housing; and transportation. The indicators related to education, public safety, and health are based upon data obtained from state and local sources. For example, the information on test scores for 8th grade students is derived from the Arizona Department of Education's web site. Similarly, the data for public safety and health are obtained from the Arizona Department of Public Safety and Arizona Department of Health Services, respectively. All the documents and files from which the information for this study was obtained are in the public domain and part of the reporting requirements in the state and county. Therefore, the analysis undertaken in the next section can be easily updated over time as long as the format and reporting guidelines remain in place.

## Computing QOUL Indicators

The six indicators selected for the study provide a snapshot of the variation by place or community in the QOL of Greater Phoenix's residents. Given that the specific measures contributing to the indicators need to reflect conditions within the municipalities, they do not include region-wide amenities that can be accessed by all residents. Also, the snapshot of QOUL presented in this study is compiled from the most recent data available, but does not suggest a specific date. Instead, we have adopted a temporal window of about 3 years (2005–2008) as the "time stamp" for this assessment.

The cities in Greater Phoenix have distinct histories and patterns of settlement. Cities like Phoenix, Mesa, Tempe, and Scottsdale include some of the oldest urban developments, while Queen Creek, Buckeye, Gilbert, and Avondale are mostly newer settlements. Figure 6.2 shows that population is mostly concentrated in the central and eastern parts of the metropolitan region. The newer cities are, however, among the fastest growing cities in the county as shown in Table 6.4. Queen Creek grew in population by over 400% between 2000 and 2007 followed closely by Buckeye and El Mirage. Buckeye has become the second largest city in area within Maricopa County, through its aggressive annexation policies.

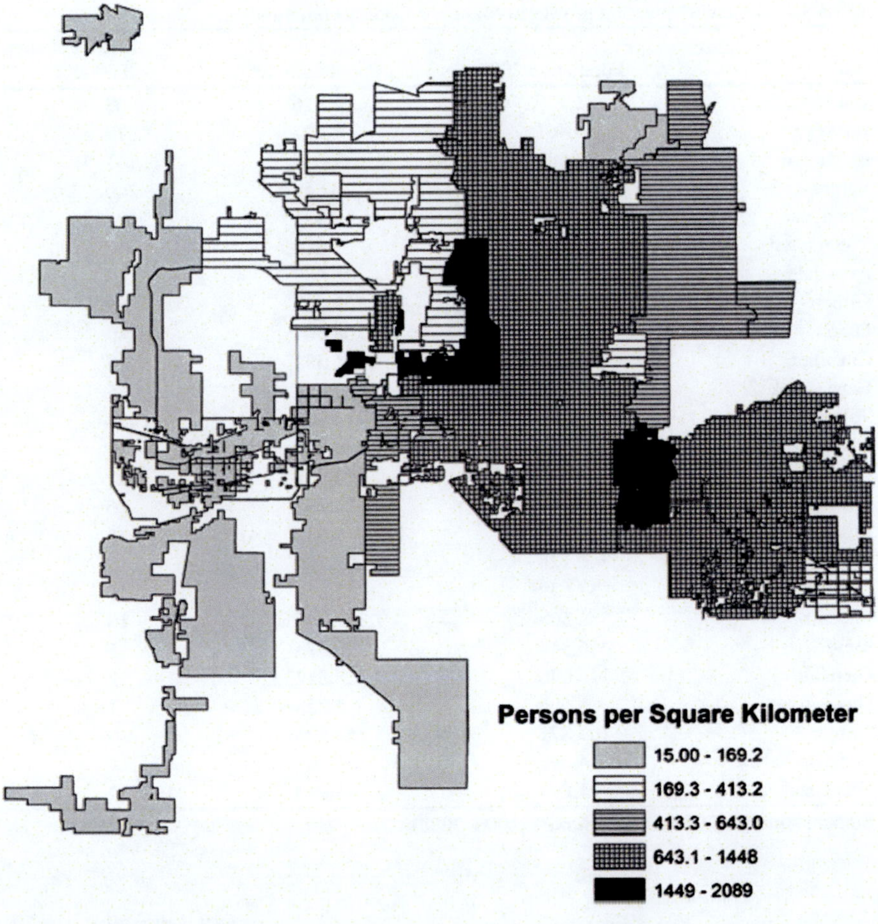


**Fig. 6.2** Population density across the cities in Phoenix: For communities (Source: American Community Survey 2006–2008)

## *Education*

Education has been consistently rated as the most or the second most important QOL attribute for the residents in Greater Phoenix according to the Morrison Institute surveys conducted between 1997 and 2004. Quality of schools is also an important predictor of neighborhood demographic characteristics and housing values. Most QOUL studies include some measure of educational attributes. These measures include outcome measures such as mean or median scores on national assessment tests, graduation rates, and college acceptance. Education quality has also been measured through input measures such as expenditure per student, teacher-to-student ratios, and availability of special programs.

**Table 6.4** Source: Population growth in Maricopa County communities

| Name | Population 2007 | Population 2000 | Percent change 2000–2007 |
|---|---|---|---|
| Queen Creek | 21,729 | 4,316 | 403 |
| Buckeye | 40,467 | 8,497 | 376 |
| El Mirage | 33,583 | 7,609 | 341 |
| Surprise | 104,895 | 30,848 | 240 |
| Goodyear | 55,954 | 18,911 | 196 |
| Youngtown | 6,332 | 3,010 | 110 |
| Avondale | 75,256 | 35,883 | 110 |
| Gilbert | 203,656 | 109,697 | 86 |
| Peoria | 151,544 | 108,364 | 40 |
| Chandler | 241,205 | 176,581 | 37 |
| Cave Creek | 5,028 | 3,728 | 35 |
| Tolleson | 6,680 | 4,974 | 34 |
| Litchfield Park | 5,055 | 3,810 | 33 |
| Carefree | 3,871 | 2,927 | 32 |
| Fountain Hills | 25,540 | 20,235 | 26 |
| Wickenburg | 6,380 | 5,082 | 26 |
| Scottsdale | 240,126 | 202,705 | 18 |
| Apache Junction | 37,539 | 31,814 | 18 |
| Phoenix | 1,538,568 | 1,321,045 | 16 |
| Mesa | 456,344 | 396,375 | 15 |
| Glendale | 246,076 | 218,812 | 12 |
| Guadalupe | 5,606 | 5,228 | 7 |
| Tempe | 167,871 | 158,625 | 6 |
| Paradise Valley | 14,215 | 13,664 | 4 |
| Gila Bend | 1,891 | 1,980 | –4 |

Source: American Community Survey (2006–2008)

In this study, we focus mostly on outcome measures given that variation in input measures such as expenditures per student would be low within a metropolitan region. We measure the performance of eight graders graduating from elementary schools since elementary districts are smaller than high school districts, thus offering greater metropolitan variation. In addition, we derive weighted scores at the level of cities or towns in Greater Phoenix as our analysis is conducted at this level of geography.

Elementary school boundaries within Greater Phoenix often straddle two or more cities. Also, each city can contain several elementary school districts according to the size of its population and its areal extent. The weighted scores by city/place are derived by first intersecting the elementary school boundaries with the place boundaries and recording the area of overlap. The percentage of area an elementary school district occupies within a city is the weight used in attributing the scores for that district. These weighted scores for all districts that overlap the city are then aggregated to determine the city-level score.

**Table 6.5** Academic achievement scores for reading and math for grade 8 students by City

| City/town | Grade 8 AIIMS Math (% meets or exceeds) | Grade 8 AIIMS Reading (% meets or exceeds) | Grade 8 NCE Median Percentile Rank Math | Grade 8 NCE Median Percentile Rank Reading |
|---|---|---|---|---|
| Cave Creek | 76 | 82 | 71 | 65 |
| Carefree | 75 | 82 | 73 | 66 |
| Gilbert | 75 | 78 | 63 | 59 |
| Paradise Valley | 74 | 79 | 58 | 56 |
| Chandler | 74 | 75 | 57 | 52 |
| Fountain Hills | 71 | 77 | 69 | 62 |
| Scottsdale | 70 | 76 | 60 | 56 |
| Peoria | 65 | 71 | 57 | 56 |
| Glendale | 64 | 70 | 50 | 50 |
| Guadalupe | 63 | 71 | 58 | 56 |
| Phoenix | 62 | 69 | 50 | 48 |
| Youngtown | 62 | 68 | 50 | 50 |
| Surprise | 60 | 66 | 53 | 52 |
| El Mirage | 60 | 66 | 49 | 49 |
| Mesa | 59 | 65 | 51 | 48 |
| Avondale | 58 | 65 | 46 | 44 |
| Litchfield Park | 58 | 64 | 44 | 44 |
| Tempe | 58 | 69 | 49 | 48 |
| Buckeye | 55 | 64 | 43 | 45 |
| Wickenburg | 54 | 76 | 54 | 65 |
| Apache Junction | 52 | 66 | 51 | 56 |
| Queen Creek | 49 | 55 | 70 | 66 |
| Tolleson | 46 | 58 | 33 | 35 |
| Goodyear | 42 | 48 | 34 | 32 |
| Gila Bend | 20 | 43 | 28 | 40 |

Source: Arizona Department of Education (hhp://www10.ade.az.gov/ReportCard/Research.aspx) (2009)

Arizona's Instrument to Measure Standards (AIMS) is a state mandated assessment that all Arizona students take at grades 3, 5, and 8, and in high school. It is a Criteria Referenced Test (CRT) that is designed to assess how well the students are performing at grade with respect to the Arizona Academic Standards. The AIMS scores are reported for math, reading, writing, and science, in four performance categories: Falls Far Below the Standard; Approaches the Standard; Meets the Standard; and Exceeds the Standard. The percentages of students in the top two categories, Meets the Standard and Exceeds the Standard, by city and subject, are provided in Table 6.5.

In contrast to AIMS, TerraNova is a national norm-referenced test that measures how well Arizona students compare with their US peers. The scores are usually reported as percentile ranks or as The Normal Curve Equivalent (NCE). NCE are standard scores with a mean of 50 and a standard deviation of 21.06. An NCE score that is one standard deviation above the mean or 71 (50+21) would be higher than approximately 84% of the population with only 16% of the population scoring higher. Similarly, since the normal distribution is symmetrical the reverse is also true. The NCE scale is easily interpreted with reference to all students taking the test as scores above 50 would indicate better than average and vice-versa.

The eight grade achievement scores by city/community generally suggest that affluent communities of Cave Creek, Carefree, Paradise Valley, Fountain Hills, Scottsdale, and Chandler have high outcome measures for education. More than 70% of eighth graders in these communities have met or exceeded the AIMS requirements. A typical eighth grade student from these communities has also outperformed his/her average US peer in the norm-referenced TerraNova tests.

## *Public Safety*

Public safety is usually a critical part of a community's QOL and Maricopa County's residents are especially sensitive to issues of crime and safety. Morrison Institute's perception surveys confirm that public safety and crime has been among the top three factors that determine residents' QOL each time the survey was administered (in 1997, 1998, 1999, and 2002). In fact, the concern about public safety and crime continued to be high even when the index of crime and incidence of crime declined (Morrison Institute for Public Policy 1999: p. 9).

Table 6.6 provides the most recent data on crime currently available from the Arizona Department of Public Safety. The Crime Index reported is the aggregate number of eight offenses used to measure the extent, fluctuation and distribution of crime in a given municipality. The Crime Index includes: murder, rape, robbery, aggravated assault, burglary, larceny-theft, motor vehicle theft, and arson.

Phoenix, the county's iconic city, is clearly leading all other cities in overall crime. Its crime index is over four times that of Mesa, the city with the second highest crime index. When normalized against the population, the crime in Phoenix is well in excess of five times that of Mesa. Glendale, Chandler, and Scottsdale round off the top-five cities in terms of crime in Maricopa County but fall in well behind Phoenix and Mesa.

Although the relative number of overall crimes in Phoenix heavily skews the crime index, the picture is quite different when we consider individual offenses. Violent crimes as a percentage of population are highest in the small town of Youngtown with just over 6,000 inhabitants. Other smaller places like Avondale and Apache Junction also report low crime rates but register high proportion of violent crimes. Similarly, Mesa leads all other Maricopa County cities in the proportion of aggravated assaults and Glendale is ahead of Mesa with the second highest rate of burglaries after Phoenix.

It should also be noted that the incidence of crime is a localized phenomenon with large variation of crime rates within a municipality. A majority of the crime in

**Table 6.6** Incidence of crime in Maricopa county cities

| Cities in Maricopa County | Crime index | Crimes per 100,000 | % Violent crimes | % Burglaries |
|---|---|---|---|---|
| Phoenix | 93,626 | 15.4 | 11.2 | 5.3 |
| Mesa | 19,878 | 4.6 | 11.5 | 7.3 |
| Glendale | 14,845 | 2.5 | 9.0 | 4.4 |
| Tempe | 11,152 | 1.7 | 7.7 | 4.4 |
| Chandler | 8,877 | 2.4 | 9.0 | 5.6 |
| Scottsdale | 8,362 | 2.4 | 5.1 | 3.1 |
| Peoria | 5,672 | 1.5 | 5.6 | 2.7 |
| Avondale | 5,173 | 0.8 | 8.8 | 5.6 |
| Gilbert | 5,150 | 2.0 | 4.6 | 2.9 |
| Surprise | 2,705 | 1.1 | 4.3 | 2.3 |
| Goodyear | 2,539 | 0.6 | 5.7 | 3.4 |
| Buckeye | 1,594 | 0.4 | 3.2 | 2.1 |
| Apache Junction | 1,317 | 0.4 | 7.9 | 5.5 |
| El Mirage | 1,140 | 0.3 | 6.3 | 2.4 |
| Tolleson | 1,031 | 0.1 | 6.4 | 3.3 |
| Paradise Valley | 360 | 0.1 | 2.5 | 2.2 |
| Wickenburg | 208 | 0.1 | 6.3 | 6.3 |
| Youngtown | 109 | 0.1 | 15.6 | 3.7 |

Source: Arizona Department of Public Safety, Crime in Arizona (2008)

Phoenix is limited to its central and south central areas. Most other parts of the city have crime rates that are comparable to, or less than, the average within the Greater Phoenix region. The city- or place-level analysis glosses over these variations apparent at finer spatial scales.

## *Public Health*

Health is critical component of an individual's QOL. It therefore follows that any measure of community well-being would involve residents' health status. All QOL indices noted in Table 6.1 included a measure of health status. Health also features prominently in the Morrison Institute surveys on QOL, and the concerns have grown stronger over the time frame of those surveys (in 1997, 1998, 1999, and 2004). In 2004, the last survey in that series, over 60% of the population claimed that health is an important component of their QOL (Morrison Institute for Public Policy 2004: p. 23).

Public health status is measured in various ways in different surveys. These measures include outcome measures that reflect the state of health and mortality statistics in the population. Indicators of access to healthcare also feature prominently in measuring public health status. Access indicators such as percentage of population

**Table 6.7** Mortality rates from serious health and behavioral health conditions

| Community | Cardio-vascular diseases (%) | Lung and breast cancer (%) | Drug-induced deaths(%) | Alcohol-induced deaths(%) | Median age |
|---|---|---|---|---|---|
| Litchfield Park | 0.85 | 0.26 | 0.00 | 0.04 | 44.7 |
| Youngtown | 0.65 | 0.16 | 0.02 | 0.00 | 34.2 |
| Wickenburg | 0.44 | 0.19 | 0.02 | 0.02 | 48.4 |
| Cave creek | 0.42 | 0.24 | 0.00 | 0.04 | 44.7 |
| Gila Bend | 0.32 | 0.05 | 0.00 | 0.11 | 29.3 |
| Scottsdale | 0.27 | 0.07 | 0.01 | 0.01 | 41.0 |
| Peoria | 0.25 | 0.06 | 0.01 | 0.00 | 35.6 |
| Mesa | 0.25 | 0.05 | 0.01 | 0.01 | 32.0 |
| Tolleson | 0.24 | 0.06 | 0.01 | 0.03 | 29.4 |
| Carefree | 0.21 | 0.03 | 0.00 | 0.05 | 55.2 |
| Glendale | 0.21 | 0.07 | 0.01 | 0.01 | 30.8 |
| Fountain Hills | 0.20 | 0.05 | 0.00 | 0.00 | 46.4 |
| Paradise Valley | 0.20 | 0.07 | 0.01 | 0.00 | 46.3 |
| Surprise | 0.19 | 0.06 | 0.01 | 0.00 | 46.1 |
| Buckeye | 0.15 | 0.05 | 0.01 | 0.01 | 30.0 |
| Tempe | 0.14 | 0.02 | 0.01 | 0.01 | 28.8 |
| Phoenix | 0.13 | 0.04 | 0.01 | 0.01 | 30.7 |
| Goodyear | 0.12 | 0.04 | 0.01 | 0.01 | 36.5 |
| Chandler | 0.12 | 0.03 | 0.00 | 0.01 | 31.2 |
| Queen Creek | 0.12 | 0.03 | 0.00 | 0.00 | 30.9 |
| Gilbert | 0.09 | 0.03 | 0.00 | 0.00 | 30.1 |
| Avondale | 0.07 | 0.03 | 0.01 | 0.00 | 29.0 |
| Guadalupe | 0.07 | 0.00 | 0.02 | 0.09 | 25.2 |
| Apache Junction | 0.06 | 0.02 | 0.01 | 0.00 | 44.1 |
| El Mirage | 0.06 | 0.03 | 0.00 | 0.00 | 24.6 |

Source: Arizona Department of Health Services (2008)

with health insurance, number and quality of hospitals, and the number of physicians have been used in the Morrison Institute's QOL surveys. The access to healthcare indicators is less relevant in an intra-metropolitan study. Regionally, municipality boundaries do not make a difference to individual's ability to access healthcare facilities in another municipality. Therefore, this study will only include a limited set of outcome measures such as mortality rates from serious health conditions (cancer, heart disease) and fatalities from adverse behavioral health habits (alcohol induced or drug induced deaths).

As shown in Table 6.7, the smaller cities with above average median age of residents record higher mortality rates from cardio-vascular diseases and cancer. These cities include Litchfield Park (population 5,055), Youngtown (population 6,332), Wickenburg (population 6,380), and Cave Creek (population 5,028). Median age of residents is, however, not the most critical factor determining mortality from cardio-vascular diseases and cancer. Affluence mediates the effect of age as can be seen in communities like Paradise Valley and Fountain Hills whose typical residents are older but have lower mortality rates from the two disease categories. Apache Junction is the anomaly in this apparent relationship between age, affluence, and disease rates since it has very low disease rates and below average median incomes, yet a higher median age.

While drug induced deaths are generally low in all communities in Greater Phoenix, alcohol induced deaths seem to be a serious problem in small communities with concentrated population of minorities. Guadalupe and Gila Bend are notable in this category. Hispanics constitute over half the population in both these cities. In contrast, suicide rates are significantly higher in two small and overwhelmingly Non-Hispanic White communities of Cave Creek and Wickenburg.

## *Housing*

The quality and availability of housing are critical components of a household's relationship with the community. The location of housing determines the neighbors, the amenities that the households can access, and, importantly, the schools that the children can attend. Housing is, therefore, a composite good, the choice of which is related to other important lifestyle choices. The desirability of a neighborhood (and the amenities they offer) is reflected in the home values and rental rates. Therefore most indicators of housing quality include these measures. In addition, a high level of owner occupancy is considered to be a reflection of households' long-term investment in the neighborhood, thereby ensuring that the housing stock is maintained adequately.

Paradise Valley is a unique residential community that is almost entirely zoned for large lot (one acre) single-family dwellings (R-43). Its exclusive credentials are reflected in the high median home values (US$1.5 million) and high median income of resident households (US$187,128) as reported in Table 6.8. Carefree and Cave Creek are small mostly residential communities in the northeast corner of the urbanized region. Both these neighboring communities have median home values that are above US$0.5 million.

Immediately below the three small communities mentioned above in median home value rankings are the larger cities with a more diverse population and land use characteristics such as Scottsdale, Fountain Hills, and Queen Creek. Tempe, Peoria, Surprise, and Phoenix are among the oldest communities in the region with home values close to the regional median of US$250,000. Home ownership rates are generally high throughout the region with 11 communities registering homeownership rates at higher than 80%. The most inexpensive homes are located in the communities with high minority or senior populations such as Gila Bend, Guadalupe, or Youngtown.

## *Income, Economy, and Jobs*

Indicators of household income, jobless rates, and levels of human capital such as education can provide a good measure of the economic health of residents in Greater Phoenix.

**Table 6.8** Housing values and home ownership rates in the communities in Greater Phoenix

| Communities in Metro Phoenix | Home ownership (%) | Median rent | Median home value |
|---|---|---|---|
| Paradise Valley | 97 | $1,172 | $1,493,288 |
| Carefree | 88 | $934 | $844,703 |
| Cave Creek | 81 | $880 | $541,070 |
| Scottsdale | 70 | $999 | $462,900 |
| Fountain Hills | 84 | $1,112 | $423,276 |
| Queen Creek | 92 | $748 | $369,827 |
| Litchfield Park | 87 | $959 | $349,887 |
| Goodyear | 85 | $971 | $313,500 |
| Gilbert | 85 | $1,088 | $304,400 |
| Chandler | 74 | $909 | $277,600 |
| Tempe | 51 | $806 | $260,700 |
| Peoria | 84 | $1,023 | $250,400 |
| Surprise | 88 | $968 | $249,700 |
| Phoenix | 61 | $747 | $234,700 |
| Wickenburg | 64 | $599 | $232,093 |
| Glendale | 65 | $725 | $228,600 |
| Avondale | 78 | $1,003 | $226,000 |
| Mesa | 67 | $738 | $218,400 |
| Apache Junction | 83 | $685 | $163,421 |
| Tolleson | 66 | $553 | $159,936 |
| El Mirage | 72 | $507 | $158,497 |
| Buckeye | 68 | $496 | $150,686 |
| Youngtown | 63 | $661 | $133,829 |
| Guadalupe | 69 | $440 | $118,822 |
| Gila Bend | 61 | $391 | $13,310 |

Source: www.city-data.com. 2008

Other indicators of economic health of a region that have been used in the past, specifically by the Morrison Institute studies, included cost of living index and wages. These two indicators do not allow appropriate intra-metropolitan comparisons of QOL since they are mostly invariant across the metropolitan region. In addition, even if some variations did exist, they would have a minor impact given that residents can easily cross municipal borders to access better bargains depending upon the cost of travel.

Cities in Greater Phoenix that include significant numbers of residents with high economic status seem to also be those that have few jobs within their borders. Paradise valley and Carefree are both small residential communities that intend to remain mostly high-end residential (see Table 6.9). Among the large cities in the metropolitan region with a diverse land use and employment mix, Scottsdale and Chandler stand out as the economic high-flyers. The median incomes of residents of the other large cities like Phoenix, Tempe, and Mesa are well below the median for the metropolitan region (US$62,337). Once again, Guadalupe, Gila Bend, and Youngtown comprise the tail of the median income ranking of cities in metro Phoenix.

**Table 6.9** Indicators of economic health in Greater Phoenix

| Communities in Metro Phoenix | Percent unemployed (%) | Percent of residents with income below poverty level (%) | Percent with bachelor's degree or higher (%) | Median household income |
|---|---|---|---|---|
| Apache Junction | 7 | 9 | 9 | $45,820 |
| Avondale | 4 | 12 | 16 | $63,533 |
| Buckeye | 6 | 21 | 10 | $44,074 |
| Carefree | 1 | 4 | 50 | $110,489 |
| Cave Creek | 2 | 9 | 41 | $74,660 |
| Chandler | 3 | 5 | 33 | $69,278 |
| El Mirage | 7 | 18 | 6 | $42,118 |
| Fountain Hills | 2 | 5 | 38 | $76,754 |
| Gila Bend | 8 | 28 | 8 | $33,501 |
| Gilbert | 3 | 5 | 36 | $82,726 |
| Glendale | 5 | 17 | 21 | $54,063 |
| Goodyear | 4 | 7 | 23 | $71,613 |
| Guadalupe | 8 | 30 | 5 | $37,480 |
| Litchfield Park | 4 | 5 | 47 | $89,529 |
| Mesa | 4 | 10 | 22 | $51,433 |
| Paradise Valley | 2 | 3 | 69 | $187,128 |
| Peoria | 4 | 7 | 22 | $63,662 |
| Phoenix | 6 | 18 | 23 | $50,140 |
| Queen Creek | 5 | 10 | 17 | $79,349 |
| Scottsdale | 4 | 5 | 44 | $69,172 |
| Surprise | 6 | 10 | 21 | $62,337 |
| Tempe | 4 | 20 | 40 | $52,157 |
| Tolleson | 4 | 15 | 6 | $48,297 |
| Wickenburg | 2 | 13 | 20 | $39,506 |
| Youngtown | 8 | 15 | 11 | $28,854 |

Source: American Community Survey (2005–2008)

## *Transportation and Mobility*

Transportation and mobility impact people's QOL in very significant ways. The ability to quickly and conveniently reach the place of work, shopping, and play allows a person to have enough time left over to attend to family and friends or to pursue leisure activities. In addition, if indeed the destinations are close by, the use of non-motorized modes of travel such as walking or biking, become more viable. It is also known that an active lifestyle that includes walking and biking leads to better health outcomes (Lathey et al. 2009; Frank et al. 2004).

In this study we provide two forms of transportation indicators; mode choice and mean travel time to work. These two measures have been consistently used as valid measures for determining the quality of transportation access and have been used in the earlier QOL reports published by the Morrison Institute. As can be noted from Table 6.10, most commuters prefer to drive alone. The percentage of commuters

**Table 6.10** Mode choice and travel time to work by communities in Greater Phoenix

| Communities in Metro-Phoenix | Drove a car alone (%) | Bus or trolley bus (%) | Worked at home (%) | Mean travel time to work |
|---|---|---|---|---|
| Carefree | 68 | N/A | 17 | 34.2 |
| Cave Creek | 77 | N/A | 8 | 32.5 |
| Queen Creek | 78 | N/A | 4 | 32.3 |
| Apache Junction | 75 | 0 | 2 | 31.9 |
| Surprise | 76 | 0 | 3 | 30.2 |
| Goodyear | 78 | 0 | 3 | 29.8 |
| El Mirage | 69 | 0 | 1 | 29.3 |
| Peoria | 80 | 0 | 3 | 28.9 |
| Fountain Hills | 80 | 0 | 8 | 28.8 |
| Gilbert | 82 | 0 | 4 | 28.6 |
| Glendale | 75 | 2 | 3 | 28.2 |
| Buckeye | 71 | N/A | 1 | 27.3 |
| Avondale | 72 | 0 | 2 | 26.9 |
| Youngtown | 77 | 1 | 2 | 26.6 |
| Phoenix | 72 | 3 | 3 | 26.1 |
| Litchfield Park | 75 | 0 | 5 | 26.0 |
| Mesa | 75 | 1 | 3 | 25.9 |
| Chandler | 80 | 1 | 3 | 25.1 |
| Scottsdale | 80 | 1 | 7 | 24.3 |
| Paradise Valley | 82 | N/A | 11 | 22.3 |
| Guadalupe | 48 | 3 | 3 | 22.1 |
| Tolleson | 72 | 0 | 3 | 21.7 |
| Tempe | 73 | 3 | 3 | 20.4 |
| Gila Bend | 61 | 1 | 5 | 18.6 |
| Wickenburg | 72 | N/A | 5 | 17.6 |

Source: American Community Survey (2005–2008) and www.City-data.com

who drive alone to work varies from a low of 61% in Gila Bend to a high of 82% in both Gilbert and Paradise Valley. It is also interesting to note that the comparatively high-income communities of Paradise Valley and Carefree also record the highest percentage of people who work from home. The use of public transportation, buses in this case, is highest in the contiguous region of Phoenix, Tempe, and Guadalupe, an area that has a relatively high level of service compared to other cities in metro-Phoenix. El Mirage and Guadalupe, two cities with high concentrations of minorities record the highest rates of walking to work.

Not surprisingly, the residents in communities that are furthest from the central areas of Phoenix are spending more time commuting. Residents of five communities that are in the fringes of this metropolitan region record more than 30 minutes in commute time to work. These communities include Carefree, Cave Creek, Queen Creek, Apache Junction, and Surprise. Besides the small and somewhat detached communities from the urban agglomeration of Gila Bend and Wickenburg, the lowest commute times are recorded in Tempe, Tolleson, Guadalupe, and Paradise Valley.

## *Developing an Overall QOUL Index*

There are serious challenges in developing an *aggregate index of QOUL* from the individual components of QOUL described above. Given that the component attributes have different measures, combining the measures would be akin to comparing "apples and oranges." In addition, finding the appropriate weights for the components included is problematic. Not all attributes are equally important in determining residents' QOL. Regardless, an overarching index of QOUL such as the Mercer survey (Quality of Living global city rankings), the Kiplinger's "city rankings," and *The Economist* "Livability rankings" (referred to in Chap. 2) has some popular appeal.

Economists have for long wrestled with the problem of weighting amenities by their utility to city residents. Rosen (1979) and Roback (1982), among several others who followed, used implicit amenity prices as weights for a QOL index. These weights were estimated through hedonic regression analysis. The studies noted above were the precursors to later work on the theory of compensating differentials, which suggests that assuming low mobility costs, local public goods will be capitalized into wages and rentals (Kahn 1995). Thus, the variation in wages and rentals by cities would provide an indicator of comparative preference for particular places.

None of the economic theories mentioned above is appropriate for integrating the different QOL measures into a single QOUL index for a metropolitan region. The *theory of compensating differentials* breaks down when wages and rentals are separated by adjacent municipalities that function as one integrated region. Therefore, in this study we take a more intuitive and straightforward approach to aggregating the different indicators of QOUL:

- First, instead of the actual measures in the various component indices, we apply a ranking system by cities with the first ranked cities having the best QOL measure in that attribute and lower ranked cities are comparatively worse-off.
- Second, we weight the key six components according to the ranking assigned by the survey respondents to the 1997–2004 surveys of QOL conducted by the Morrison Institute.

Rather than use very fine distinctions (for which there would be far less support in empirical observations), our approach was to address three broad categories of weighting based on the overall rankings of the attributes in all four surveys. These four surveys provided mostly consistent rankings with small variations, if any, from one year to another. Education and public safety have consistently received the top rankings, and therefore, form our highest weighted category. Similarly, economy and public health have come soon after the top two, thereby receiving the next category of weights. The remaining two attributes, housing and transportation, form the final weighted category. The final assigned weights are as follows:

Education: 25%
Public safety: 25%
Economy and jobs: 15%

**Table 6.11** Developing an overall index of QOUL for communities in greater Phoenix

| City/town | Education rank | Public safety rank | Public health rank | Housing rank | Economy and jobs rank | Travel and mobility rank |
|---|---|---|---|---|---|---|
| Apache Junction | 13 | 11 | 3 | 14 | 18 | 19 |
| Avondale | 20 | 14 | 15 | 13 | 14 | 11 |
| Buckeye | 22 | 2 | 21 | 22 | 21 | 12 |
| Carefree | 1 | N/A | 1 | 2 | 2 | 22 |
| Cave Creek | 2 | N/A | 16 | 6 | 5 | 21 |
| Chandler | 7 | 15 | 10 | 12 | 8 | 7 |
| El Mirage | 18 | 10 | 23 | 15 | 23 | 25 |
| Fountain Hills | 3 | N/A | 2 | 5 | 4 | 15 |
| Gila Bend | 25 | N/A | 17 | 25 | 24 | 2 |
| Gilbert | 4 | 4 | 11 | 8 | 6 | 14 |
| Glendale | 11 | 16 | 20 | 18 | 16 | 13 |
| Goodyear | 24 | 6 | 7 | 7 | 9 | 17 |
| Guadalupe | 8 | N/A | 24 | 23 | 25 | 5 |
| Litchfield Park | 21 | N/A | 13 | 4 | 3 | 24 |
| Mesa | 19 | 18 | 18 | 19 | 17 | 8 |
| Paradise Valley | 5 | 1 | 4 | 1 | 1 | 6 |
| Peoria | 10 | 7 | 8 | 11 | 10 | 16 |
| Phoenix | 16 | 17 | 14 | 20 | 19 | 9 |
| Queen Creek | 14 | N/A | 9 | 3 | 11 | 20 |
| Scottsdale | 6 | 9 | 6 | 10 | 7 | 23 |
| Surprise | 12 | 3 | 5 | 9 | 13 | 18 |
| Tempe | 17 | 13 | 19 | 16 | 12 | 3 |
| Tolleson | 23 | 8 | 25 | 21 | 20 | 4 |
| Wickenburg | 9 | 5 | 12 | 17 | 15 | 1 |

Source: Authors' calculations

Public health: 15%
Housing: 10%
Transportation: 10%.

The weights are assigned after the overall aggregate ranks for each of the six categories are assessed. The indicators within categories receive similar (or no) weights. That is, for example, ranking on various AIMS scores and NCE scores are not weighted individually although the final "education" ranking is weighted as described above.

Managing missing information on public safety for some metro Phoenix communities was the remaining problem in arriving at a composite and aggregated ranking of QOUL. Seven of the 25 communities included in this study did not have a separate police department and contracted with the Maricopa County Sheriff's Office to administer public safety. These communities could not be ranked with appropriate weights. Table 6.11 shows the individual rankings of the communities

**Table 6.12** The final weighted QOUL rank for each community

| City/town | Overall QOUL rank |
|---|---|
| Paradise Valley | 1 |
| Gilbert | 2 |
| Scottsdale | 3 |
| Surprise | 3 |
| Wickenburg | 3 |
| Peoria | 6 |
| Chandler | 6 |
| Goodyear | 8 |
| Apache Junction | 8 |
| Tempe | 10 |
| Avondale | 11 |
| Glendale | 11 |
| Buckeye | 13 |
| Phoenix | 13 |
| Youngtown | 15 |
| Tolleson | 15 |
| Mesa | 15 |
| El Mirage | 18 |

Source: Authors' calculations

included by the six attributes described earlier. These rankings do not assume any weights. Paradise Valley receives the highest ranking on three attributes – public safety, housing, and jobs – and it is within the top ten ranks in all categories. Carefree is among the top two communities in all but travel and mobility parameters and was unrated in public safety. At the other end of the QOUL rankings are communities like Guadalupe, Gila Bend, and Youngtown.

The final weighted ranks are presented in Table 6.12. Paradise Valley stands out from the rest of the communities as the best place to live in metro Phoenix. Gilbert follows far behind as the second best city in terms of QOUL in Metro Phoenix. However, the rest of the rankings show clusters of cities holding similar ranks. This is a more realistic assessment given that the aggregation involved some subjective assumptions, and therefore finer distinctions would be less defensible. Scottsdale, Surprise, and Wickenburg are all ranked third followed by Peoria and Chandler. Youngtown, Tolleson, and Mesa bring up the bottom together with El Mirage, which is the last city in the rankings according to the approach adopted for this study.

Some of the cities that were left out of the weighted rankings could very well be part of the top ranked places in metro-Phoenix. For example, without public safety, both Carefree and Cave Creek are near the top of the rankings. Even when an average public safety score is assigned to these two cities, they remain firmly in the top cluster. Hence it is reasonable to expect these two communities to be among the best in metro Phoenix in terms of QOL.

## Summary and Conclusions

All studies of QOL are fraught with inter-subjective criteria in choosing the attributes that define QOL. The availability of data and the contextual peculiarities largely determine the indicators selected, especially for studies that rely upon objective data. In this study of intra-metropolitan QOUL, we based our parameters on earlier research but adapted them to an intra-regional context. The reliability and validity of the measures selected depended on how well the measures performed in small communities and whether the measures were affected by spillover effects across communities. Within a large metropolitan region many of the amenities as well as nuisances are shared across city boundaries. Thus, the spatial entities that are chosen to represent the different parts of a metropolitan region become an important determinant of both the indicators chosen and the ultimate QOL assessments.

This study used towns/municipalities within Maricopa County as the spatial entities of concern because the residents' QOL is, in part, determined by the administrative and budgetary decisions made within each political jurisdiction. Six QOUL attributes were selected based on earlier studies conducted by the Morrison Institute, ASU. These earlier assessments of QOL used a survey instrument to gauge residents' perception on the range of attributes selected. The results of the surveys were used in weighting the objective QOUL attributes chosen for this study in order to develop an aggregate QOUL ranking of places or communities within metropolitan Phoenix.

A weighted aggregation of the six QOUL attributes resulted in a ranked list of communities by overall quality of life. The exclusive, small, and mostly residential communities of Paradise Valley, Carefree, and Cave Creek were at the top of that list. The outer suburbs of Gilbert, Surprise, and Wickenburg are among the communities that followed the top three. Scottsdale is perhaps the only large city that made it to the second tier. The older and larger cities of Phoenix, Tempe, Glendale, and Mesa are in the lower half of the list with Mesa approaching the bottom. The ranking of cities by QOUL reinforces the notion that wealthy, exclusive, and small communities provide the highest quality of life for its residents. But larger and more diverse cities such as Scottsdale, Chandler, and Gilbert, also provide a relatively high quality of life. Thus, size or age of communities does not significantly determine the level of residents' well-being. However, the economic status of residents is important in providing better amenities and services, thereby offering a relatively higher QOL in those communities.

## References

Arizona Department of Education (2009) School Report Cards. Retrieved from. http://www10.ade.az.gov/ReportCard/Research.aspx

Arizona Department of Public Safety (2009). Crime in Arizona (2008). Retrived from http://www.azdps.gov/About/Reports/docs/Crime_In_Arizona_Report_2008.pdf

Diener, E. (1995). A value based index for measuring national quality of life. *Social Indicators Research, 36*(2), 107–127.

Ellaway, A., Macintyre, S., & Kearns, A. (2001). Perceptions of place and health in socially contrasting neighbourhoods. *Urban Studies, 38*(12), 2299–2316.

Flowerdew, R., Manley, D. J., & Sabel, C. E. (2008). Neighbourhood effects on health: Does it matter where you draw the boundaries? *Social Science and Medicine, 66*(6), 1241–1255.

Frank, L. D., Andresen, M. A., & Schmid, T. L. (2004). Obesity relationships with community design, physical activity, and time spent in cars. *American Journal of Preventive Medicine, 27*(2), 87–96.

Haynes, R., Daras, K., Reading, R., & Jones, A. (2007). Modifiable neighbourhood units, zone design and residents' perceptions. *Health & Place, 13*(4), 812–825.

Kahn, M. E. (1995). A revealed preference approach to ranking city quality of life. *Journal of Urban Economics, 38*(2), 221–235.

Kawachi, I., & Berkman, L. F. (2003). *Neighborhoods and health* (1st ed.). New York: Oxford University Press.

Lathey, V., Guhathakurta, S., & Aggarwal, R. M. (2009). The impact of subregional variations in urban sprawl on the prevalence of obesity and related morbidity. *Journal of Planning Education and Research, 29*(2), 127–141.

Marans, R. W., & Rodgers, W. (1975). Toward an understanding of community satisfaction. In A. H. Hawley & V. P. Rock (Eds.), *Metropolitan America in community perspective* (pp. 299–352). New York: Wiley.

Morris, M. D. (1979). *Measuring the condition of the world's poor: The physical quality of life index*. New York: Pergamon Press. Published for the Overseas Development Council.

Morris, M. D. (1996). *Measuring the changing condition of the world's poor: The physical quality of life index*, 1960–1990. Working Paper 23–24.

Morrison Institute for Public Policy (1997) *What matters in greater Phoenix: Indicators of our quality of life*. Tempe: Morrison Institute for Public Policy, School of Public Affairs, College of Public Programs, Arizona State University.

Morrison Institute for Public Policy (1998) *What matters in greater Phoenix: Indicators of our quality of life*. Tempe: Morrison Institute for Public Policy, School of Public Affairs, College of Public Programs, Arizona State University.

Morrison Institute for Public Policy (1999) *What matters in greater Phoenix: Indicators of our quality of life*. Tempe: Morrison Institute for Public Policy, School of Public Affairs, College of Public Programs, Arizona State University.

Morrison Institute for Public Policy (2004) *What matters*. Tempe: Morrison Institute for Public Policy, School of Public Affairs, College of Public Programs, Arizona State University.

Moudon, A. V., Lee, C., Cheadle, A. D., Garvin, C., Johnson, D., Schmid, T., Weathers, R. D., & Lin, L. (2006). Operational definitions of walkable neighborhood: Theoretical and empirical insights. *Journal of Physical Activity and Health, 3*, S99–S117.

Mukherjee, R. (1989). *The quality of life: Valuation in social research*. New Delhi: Sage Publications.

Openshaw, S. (1984). *The modifiable areal unit problem*. Norwich: Geo Books.

Perry, C. A., Heydecker, W. D., & Adams, T. (1929). *Neighborhood and community planning*. New York: Regional Plan of New York and Its Environs.

Roback, J. (1982). Wages, rents, and the quality of life. *The Journal of Political Economy, 90*(6), 1257–1278.

Rosen, S. (1979). Wage-based indexes of urban quality of life. In P. Mieszlow-ski & M. Straszheim (Eds.), *Current issues in urban economics*. Baltimore: John Hopkins University Press.

Sadalla, E., Guhathakurta, S., & Ledlow, S. (2005). Environment and quality of life: A conceptual analysis and review of empirical literature. In *The U.S.-Mexican border environment: Dynamics of human-environment interactions* (SCERP monograph series, pp. 29–79). San Diego: San Diego State University Press.

Schwartz, B. (1994). *The costs of living: How market freedom erodes the best things in life*. New York: W.W. Norton.

Schwartz, B. (1996). *The Formation and Transormation of Values*. In the Philosophy of Psychology, W. O'Donohue and R. F. Kitchner, (eds). London: Sage Publications, 319–338.

Simons, L. G., Simons, R. L., Conger, R. D., & Brody, G. H. (2004). Collective socialization and child conduct problems: A multilevel analysis with an African American sample. *Youth and Society, 35*(3), 267–292.

Stein, C. S. (1966). *Toward new towns for America* (3rd ed.). Cambridge: M.I.T. Press.

U.S. Bureau of Census (2009). American Community Survey (2005–2008). Available at hhp://factfinder.census.gov

U.S. Bureau of Census (2009). American Community Survey (2006–2008). Available at hhp://factfinder.census.gov

# Part III
# Empirical Studies of the Subjective Assessment of Quality of Urban Life Illustrated by a Selection of Case Studies

# Chapter 7
# The Quality of Life in Metro Detroit at the Beginning of the Millennium

**Robert W. Marans and Byoung-Suk Kweon**

## Introduction

In the late 1990s, as the new millennium was approaching, the first author contemplated a research program that would track the quality of urban life (QOUL) in world cities and metropolitan regions throughout the twenty-first century. The idea was triggered in part by the World Values Survey (WVS) administered by a University of Michigan colleague, Ronald Inglehart, who worked with researchers in nearly 100 countries in measuring national values and aspirations.[1] If the WVS, which began in the early 1980s, could develop a set of commonly shared national social indicators, it was reasonable that a similar approach could be used to monitor urban life in world cities. One of the many places where the program could be launched was southeast Michigan, which included the city of Detroit and its surrounding counties. The area is commonly referred to as Metro Detroit. The opportunity to conduct the research was presented when the first author served as the faculty investigator for the 2001 Detroit Area Study (DAS2001).[2]

---

[1]For a discussion of the World Values Survey, see Inglehart and Welzel (2005) and http://www.worldvaluessurvey.org/

[2]The Detroit Area Study (DAS) was established at the University of Michigan in 1952 as a practicum in survey research. DAS was intended to train graduate students from the social sciences and professional schools in quantitative social science techniques and provide a facility for faculty to engage in empirical investigations. For a discussion of the DAS and its history, see Couper et al. (2002) and Freedman (1953).

R.W. Marans (✉)
Institute for Social Research, University of Michigan, P.O. Box 1248, Ann Arbor, MI 48106, USA
e-mail: marans@umich.edu

B.-S. Kweon
Department of Plant Science & Landscape Architecture, College of Agriculture and Natural Resources, University of Maryland, 2102 Plant Sciences Building, College Park, MD 20742, USA
e-mail: kweonb@umd.edu

R.W. Marans and R. Stimson (eds.), *Investigating Quality of Urban Life: Theory, Methods, and Empirical Research*, Social Indicators Research Series 45,
DOI 10.1007/978-94-007-1742-8_7, 

In 2001, a major household survey was conducted throughout Metro Detroit. Planning for the survey began in 1999 and was intended to produce information about urban life as experienced by metro area residents. In addition to survey responses, DAS2001 compiled contextual information or data about the communities and environmental settings of each respondent. Contextual information included housing and demographic characteristics, land use characteristics, and other characteristics of the communities where respondents live, such as growth rates, employment statistics, school information, and so forth.

Besides measuring the QOUL at the beginning of the twenty-first century, DAS2001 had several other objectives. These were:

(a) To determine how public perceptions and salient aspects of community life throughout the region have changed since the mid-1970s.[3]
(b) To provide information on aspects of urban life that could inform government, corporate, institutional, and community decision makers.[4]
(c) To establish a benchmark for assessing changes in community and changes in environmental and community conditions throughout the twenty-first century.
(d) To determine how much public perceptions correspond to community and environmental conditions associated with the places people live.

Prior to presenting selected findings from DAS2001, this chapter gives an overview of Metro Detroit where the survey was conducted. The overview discusses the demographic, geographic, and governmental situation of the region. It then reviews the approach used in carrying out the research. Next, findings covering quality of life (QOL), QOUL, and other topics are presented. Finally, the uses of findings for policy and planning are discussed along with key lessons learned from the project.

## About the Region

Southeast Michigan, with its core city of Detroit, is the home to almost five million people. The region – referred to throughout this chapter as Metro Detroit – contains about half of Michigan's population and is located in the heart of the Great Lakes States.[5] For nearly a half-century, Metro Detroit has been characterized by a core city plagued with a myriad of problems including a declining population base, an increasing number of abandoned older homes and vacant commercial and industrial buildings, vacant parcels of land, a weakening tax base, poor schools and other

[3]Initially, comparisons with findings from a 1967 Detroit metro area survey were planned. These plans were abandoned when it was recognized that the 2001 questionnaire would not be replicating items asked in the 1967 survey.

[4]The rationale for this objective is discussed in Marans, 2003.

[5]According to the 2000 US Census, Michigan's population was slightly less than ten million people (9,938,444). In 2010, the population was lower at 9,883,640 people, half of whom lived in Southeast Michigan.

**Fig. 7.1** Southeast Michigan Counties and the City of Detroit (Source: The authors)

public services, and high levels of unemployment and poverty.[6] At the same time, its sprawling suburbs, smaller cities and towns, and rural communities had grown, and in 2000, were home to more than three-quarters of the region's population. Collectively, the outlying population is located in residential settlements interspersed with shopping malls, office and industrial complexes, parks, lakes, and agricultural lands. Together, Detroit and its environs comprise an area of more than 4,500 square miles (or about 12,000 km$^2$).

Metro Detroit is administratively defined by seven counties containing more than 400 cities, towns, and rural townships (see Fig. 7.1).[7] The seven counties are Livingston, Macomb, Monroe, Oakland, St. Clair, Washtenaw, and Wayne. The largest city, Detroit, is part of Wayne County. In order to address regional issues, the

[6] At the time of this writing, Michigan's unemployment rate is the highest in the USA.

[7] County government is a unit of local government within Michigan and other states. Michigan has 83 counties. The US Census divides Southeast Michigan into 3 metropolitan statistical areas (SMA): one consisting of 5 of the 7 Southeast Michigan counties (Wayne, Oakland, Macomb, Livingston, St. Clair) and one additional county (Lapeer); a second SMA is Washtenaw County; and a third SMA is Monroe County.

seven counties and most local governmental units participate in a membership organization called the Southeast Michigan Council of Governments (SEMCOG). DAS2001 and its survey covered the population in the SEMCOG region.

Throughout the last quarter of the twentieth century, the region experienced modest growth in population (2.6%) and urban development expanding into its rural hinterlands. Most of population gain occurred in the outlying counties, particularly Livingston, while there were significant losses in Detroit and the rest of Wayne County. For instance, there were over 1.3 million people in Detroit in 1975; the city's population dropped to 950,000 by 2000, a decline of 30%.[8] This same period also showed declines in jobs, mainly manufacturing jobs, and a corresponding reduction in household income. While there were few shifts in the racial composition of the region during the last half of the twentieth century, the racial mix in Detroit dramatically changed. In 1950, 22% of the city's population was African American. Beginning in the late 1950s, new housing opportunities in the surrounding suburbs and other factors prompted White residents to move from Detroit, resulting in a steady decline in their numbers; by 2000, the African American population represented 84% of the city.[9]

The year 2000 was a special year in Detroit as in other parts of the world. Like elsewhere, celebrations of the new millennium in Detroit and in its surrounding communities were common in 2000 and in 2001. But Detroit had another reason to celebrate: 2001 marked the 300th anniversary of the establishment of the city. The festivities surrounding this milestone and a political campaign aiming at the election of a new city administration created a sense of optimism among Detroiters. As will be noted later, that optimism was not universally shared by residents living outside the city. In fact, economic conditions prior to and during the period of the survey were deteriorating. Unemployment rates were climbing in Detroit and elsewhere in the region, the major automobile companies were reporting loses in sales (Ford) and decreased earnings (GM), and several Detroit area banks were consolidating their operations or leaving Michigan. These conditions portray the region at the time data were being collected for DAS2001.

## Methodology

DAS2001 used a multi-method approach involving the collection of information though structured questionnaires administered to a sample of residents, from the US census, and from other secondary data sources covering the respondents' communities and their physical surroundings. The content of questionnaires was determined

[8] The 2010 US Census shows the Detroit population fell to 713,777, a 25% drop over the past decade.

[9] For a discussion of the multiple factors leading to the white exodus, see Farley et al. (2000) and Thomas (1997).

**Table 7.1** Quality of urban life topics included in the DAS2001 survey

| | |
|---|---|
| Residential domain satisfaction | Residential mobility |
| Housing | Residential history |
| Neighborhood | Factors influencing residential choice |
| Community | Moving intentions |
| County | Neighborhood preference |
| Social capital – neighboring and community involvement | Government and taxes |
| Public services and facilities – schools, police, road maintenance | Environment problems and conservation of open land |
| Willingness to pay for new/improved services and facilities | Prospect for the future |
| Travel | Other domain satisfactions |
| Public transit use | Job |
| Work trips | Health |
| Shopping trips | Standard of living |
| | Family |
| | Friends |
| | Leisure |
| Health and walking – physical activity | Household and respondent characteristics |

Source: The authors

in part by input from sponsors of the study and other key stakeholders. That is, county planners, other government officials, corporate and nongovernmental organizations, and other decision-making bodies operating in the region were queried as to what information about their constituencies was deemed necessary to better inform their decisions. These informational needs were then combined with the interests of the research team which included a desire to replicate questions asked in past regional surveys. The primary criterion used in the selection of questions was the centrality of each to the quality of community life theme.[10] A summary of the questionnaire content is shown in Table 7.1.

## *Questionnaires*

Two types of questionnaires were used to collect information shown in Table 7.1.[11] One was administered through face-to-face interviews, and a shorter version of the questionnaire was administered by mail.

[10]The theme *quality of community life* was subsequently changed to *quality of urban life* in large part because of the different meaning of *community* in other world cities where the research was being replicated.

[11]In addition to topics shown in Table 7.1, several other topics were examined in greater depth in the questionnaire used in the personal interviews but not in the mail questionnaire. These included schools, crime, safety, and community-police relations.

### Face-to-Face Interviews

As part of DAS2001, trained graduate students and professional interviewers from the University of Michigan conducted face-to-face interviews with adults drawn from a sample of households in three of Metro Detroit's seven counties. Beginning in mid-April 2001 and ending in mid-August, 315 interviews were completed. The average length of the interviews was 60 min. In addition to asking each respondent a series of questions, interviewers recorded data about the respondent's dwelling and the area around it. As an incentive to grant the interview, passes to regional parks were mailed along with a cover letter to half of the households that fell in the sample; the remaining half received five dollars. An additional five dollars was given to respondents who completed the interview. The response rate for the face-to-face survey was 59.8%.[12]

### Mail Questionnaires

In order to cover the remaining counties in Metro Detroit and expand the number of respondents in the initial three counties, a shorter version of the questionnaire used in the face-to-face interview was mailed to a sample of over 7,000 adults throughout the region. The mail questionnaire eliminated about half of the original questions in order to ensure a questionnaire that could be completed in approximately 20 min. As in the case of the face-to-face survey, regional park passbooks and five dollars were used as incentives. The mail survey yielded 4,077 responses representing a 56.4% response rate. Data from the face-to-face interviews and the mail questionnaires were merged and weighted so as to represent the correct population distribution within each county and for the entire region.

## *Secondary Sources of Information*

Several sources were used to measure community and environmental conditions associated with the places where the respondents lived. As a first step in gathering these objective measures, the addresses of each respondent in the survey were geocoded; that is, geographic information systems (GIS) were used to spatially map the addresses of over 4,000 respondents throughout the region. Besides being placed in one of the seven counties, each respondent was also placed in a particular community (that is, a city, village, or township, referred to as Minor Civil Divisions, MCDs). They also were assigned to a school district and a census unit (block, block group, or tract). Accordingly, contextual measures related to communities, neighborhoods, and census units were made and matched with the survey respondents and their answers to questions. The creation of separate data files

---

[12]For a discussion of the approach to incentives and its impact on selected responses, see Ryu et al. (2006).

covering survey data, community data, environmental data, and census data and their consolidation allow the researchers to explore numerous relationships suggested by conceptual models.[13]

Among the particular community or MCD measures that were incorporated in the database were tax rates, indicators of growth, crime statistics, and health data. School data such as expenditures per student and standardized educational test scores associated with school districts were also incorporated in the community data file.

The environmental data file included land use information (for example, percent in each land use category, degree of mix, percent of open space and natural resources, etc.), accessibility measures to recreational land, major employment centers, shopping areas, etc., and various density measures. The density measures using census data cover the number of housing units and the size of the population for blocks, block groups, and tracts.

The census data file used 2000 US Census statistics to determine racial mix, housing tenure, and median income for each block, block group, and tract associated with respondents.

## *Analysis, Feedback, and Ongoing Work*

The files containing census data, environmental data, and community data associated with each respondent were merged with the survey data file (see Fig. 7.2).

Figure 7.2 suggests numerous possibilities for examining relationships between contextual data and questionnaire responses using bivariate and multivariate analyses. For example, an analysis might address the question of how density (as reflected by multiple density measures) influences people's responses to crowding, their knowing the names of neighbors, and their interactions with them. Another question might explore the degree to which objective data covering school districts (that is, student–teacher ratios, test scores, expenditures per student) are associated with people's ratings of their public schools. Using multivariate analysis, an examination could be made of the significance and relative importance of several measures covering school districts in predicting ratings scores for respondents with varying numbers of school-age children living at home.

## Findings

As discussed earlier, the DAS2001 survey addressed a range of topics that reflected the informational needs of study sponsors and key stakeholders as well as the interests of the research team. Many of the topics have been addressed in student papers,

[13] See Chaps. 1 and 3 for a discussion of the conceptual models that guided DAS2001.

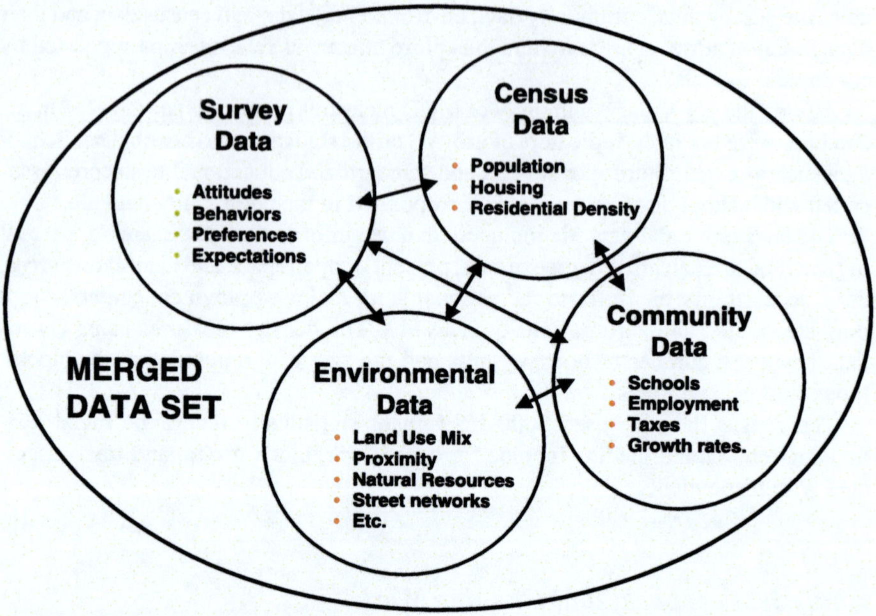


**Fig. 7.2** Merged survey data and objective data (Source: The authors)

technical reports, and conference presentations over the past decade. Data covering other topics are currently being examined or are waiting to be analyzed by other scholars. Rather than presenting all findings from DAS2001, this section presents a limited set that cover QOL, QOUL, neighborhoods and neighboring, transportation, and prospects for the region's future.

## *Quality of Life*

Building on the conceptual framework of Campbell et al. (1976), overall QOL was viewed as a subjective phenomenon and considered a composite of people's assessments of various *domains* of their lives. Accordingly, the metro area survey asked respondents to assess seven specific domains using a 7-point Likert scale measuring satisfaction, ranging from completely satisfied (7) to completely dissatisfied (1). The domains considered were the individual's *family life, health, job, friends, overall standard of living, the use of leisure time, and the amount of time to do the things you want to do*. A final question about satisfaction with *life as a whole* was also asked and was considered a summary measure of QOL. The 7-point satisfaction scale had been used in Campbell et al. (1976) seminal work and in prior Detroit area

**Table 7.2** Overall quality of life: Mean satisfaction scores for Detroit and environs

| Domain | Region | Detroit | Environs |
|---|---|---|---|
| Life as a whole | 5.55 | 5.24 | 5.64 |
| Friends | 5.41 | 5.13 | 5.52 |
| Standard of living | 5.38 | 4.73 | 5.56 |
| Family life | 5.36 | 5.06 | 5.47 |
| Health | 5.18 | 4.81 | 5.26 |
| Leisure | 5.17 | 4.76 | 5.24 |
| Job/school | 5.1 | 4.72 | 5.16 |
| Time to do things | 4.28 | 4.21 | 4.27 |
| Number of respondents (N) | 4,105 | 464 | 3,750 |

Source: The authors
Mean scores on a 7-point scale

surveys; one objective of the 2001 study was to measure change in the QOL of residents of the Detroit region.[14]

Table 7.2 shows the mean satisfaction scores for the entire sample and for the city of Detroit and other parts of the region separately. Overall, Detroit area residents tended to be fairly satisfied with the various domains of their lives. Satisfaction scores for *life as a whole* and for *friends and family* were somewhat higher than satisfaction scores for *health, use of leisure time, and one's job*. People were least satisfied with the *amount of time available for doing things they wanted to do*.[15]

The results showed that:

(a) For the most part, levels of satisfaction among Detroit residents were lower than for those of residents in the surrounding suburbs, smaller cities/towns, and rural areas, that is, the environs.
(b) Both Detroiters and those living outside the city were equally dissatisfied with the *amount of time available to do things they want to do*.

When *satisfaction levels for several domains* were compared with *past satisfaction levels*, there were few and small changes since the mid-1970s. For instance, mean scores for satisfaction with life as a whole and standard of living were virtually identical in 1975 and in 2001. However, satisfaction with *time available for doing things* decreased over the same period (4.89 vs. 4.28). This finding confirmed anecdotal reports that in recent decades, people have had an increasing number of daily obligations, feel increasingly stressed, and have less time for doing the things they wanted to do. The proportions of the population indicating they were *very satisfied*

[14]In earlier studies, the region was defined by three counties, whereas in 2001 the region was defined by seven counties. When comparisons were made between the 2001 and earlier years, only data from the three counties were used.

[15]Satisfaction scores for the domains tend to be strongly associated with one another. The average inter-item correlation is .46, while the alpha value is .87.

also declined when assessing one's leisure, health, job, family life, income, and standard of living between 1975 and 2001.

Several approaches to assessing QOL among Detroit area residents have been used since 2001. In early analysis, responses to the single question were used: *All things considered, how satisfied are you with your life as a whole?* In other instances, an index was created consisting of the combined scores of individual domain satisfaction responses. Finally, more recent analyses view QOL as a latent variable defined by several of the domains. In each situation, QOL was considered as an outcome or dependent variable, and different analytical models produced similar if not identical results.

## *Quality of Urban Life*

Conceptually, urban life implies the "places" where people carry out their daily lives, most of which occur within the residential environment. In DAS2001, the residential environment was considered at three levels[16]:

- The individual home or dwelling
- The neighborhood within which the dwelling was located
- The community (city, town, village, or township) in which the neighborhood was located

Consideration was also given to attributes of each of these *place domains*. For instance, questions about environmental attributes (traffic, noise, upkeep), social attributes (family, friends, safety), and public service (police protection, schools, parks, transportation) were asked about the respondents' neighborhoods. At the community level, consideration was given to shopping, accessibility and transportation, and recreation while people's assessments of housing costs, space, and size of property were examined for the individual dwelling.

At each level, respondents were asked to evaluate several attributes and then give a summary satisfaction score to a single question. Using the same 7-point response categories, they were asked to express their overall satisfaction with their dwelling, their neighborhood, and their community.

Table 7.3 shows overall satisfaction scores for housing, neighborhood and community for the entire regional sample, for Detroit residents, and for residents living elsewhere in the region. Levels of satisfaction are moderately high, in line with other domain satisfaction scores, and interrelated (alpha=.79). That is, a person's feelings about his/her dwelling are associated with feelings about the neighborhood and the community in which the person lives.

[16]To a lesser extent, quality of urban life in the Detroit area considered two additional levels of place: the county in which the respondent lived and the region as a whole. Findings covering these levels are not reported in this chapter.

**Table 7.3** Dwelling, neighborhood, community mean satisfaction scores for Detroit and environs

| Domain | Region | Detroit | Environs |
|---|---|---|---|
| House/dwelling | 5.27 | 4.84 | 5.38 |
| Neighborhood | 5.32 | 4.25 | 5.56 |
| Community | 5.24 | 4.2 | 5.48 |
| Number of respondents (N) | 4,294 | 596 | 3,790 |

Source: The authors
Mean scores on a 7-point scale

As in the case of other domains, Detroiters view the QOUL less favorably than residents in other parts of the region. Satisfaction scores for housing, neighborhood, and the community (Detroit) are consistently lower than corresponding scores for people living elsewhere in the metro area.

To what extent has the QOUL in Metro Detroit *changed over time*? Survey data clearly indicate that the ratings of housing, neighborhoods, and community among Detroiters declined since the mid-1970s. To a lesser extent, declines in the satisfaction scores were also reported by those living outside the core city. Indeed, objective conditions in Detroit and many of its older suburbs did not fare well between 1974 and 2001 since much of the infrastructure was not adequately maintained while the core city experienced a dramatic loss in population and an accompanying deterioration in its housing stock.

Another key question addressed by the research was the *extent to which the three place domains contribute to one's QOL*. A related question was whether these contributions were the same for both Detroiters and for non-Detroiters living in the other parts of the metro area. These questions were explored using two approaches:

(a) One approach involved *structural equation modeling* and considered *QOL as a latent variable* with housing, neighborhood, and community satisfaction scores as predictors. The model involved a simultaneous analysis allowing a comparison between Detroiters and others.
(b) In a second approach, *regression analyses* considered *life satisfaction* as a summary measure of QOL with the place domains as predictors.

Again, separate analyses were made for Detroiters and for non-Detroiters living in other parts of the region.

The structural equation models show moderate but significant relationships between people's feeling about their home, neighborhood and community, and QOL in both settings. The models are represented in Fig. 7.3 for Detroiters and in Fig. 7.4 for non-Detroiters living in the region. The results indicate the following:

(a) Taken together, the three place domains account for about a quarter (24%) of the variance in the QOL of Detroiters and one-fifth (21%) for non-Detroiters.
(b) Satisfaction with the home is the strongest predictor of QOL, while community satisfaction was the least important predictor. In fact, community satisfaction for Detroiters was not a significant contributor to overall QOL.

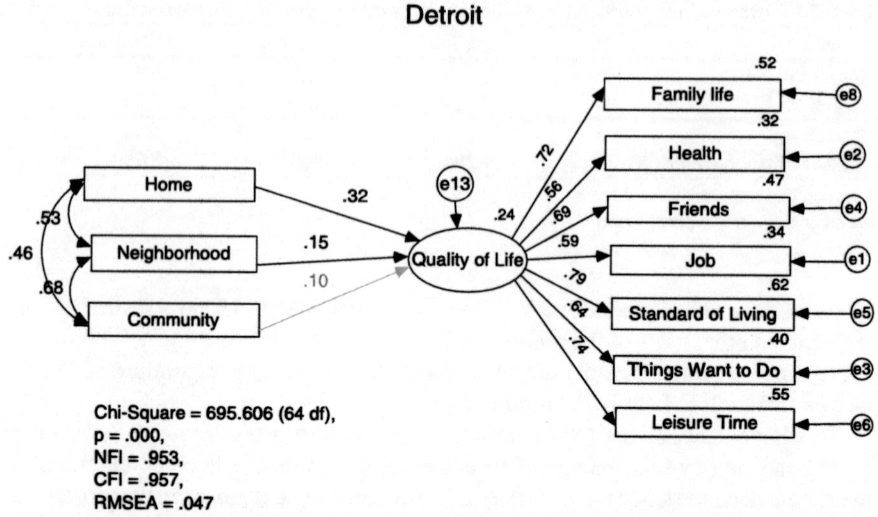


**Fig. 7.3** Structural equation model showing relationships between peoples' feeling about their home, neighborhood and community and QOL for Detroiters (Source: The authors)

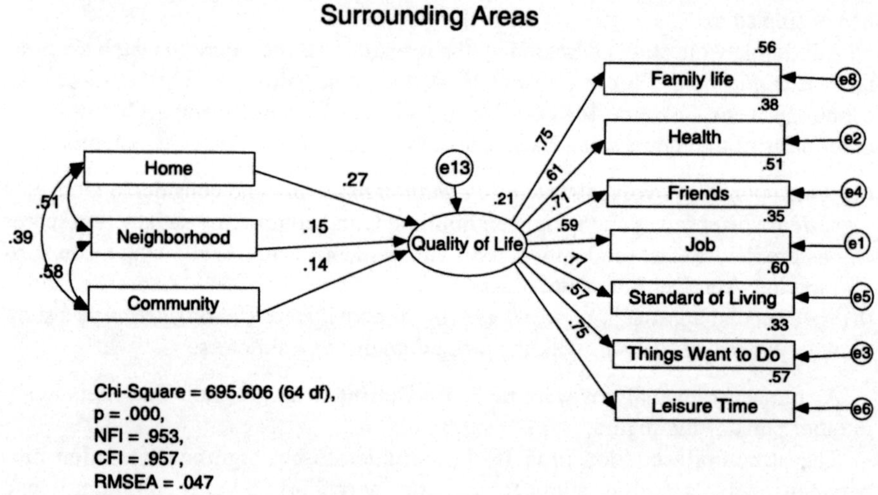


**Fig. 7.4** Structural equation model showing relationships between peoples' feeling about their home, neighborhood and community and QOL for non-Detroiters living in the environs (Source: The authors)

In the second approach, *satisfaction with life as a whole* (that is, life satisfaction) was used as the summary measure of QOL. Regression analysis was initially used with the place satisfaction scores as predictors of life satisfaction. Then, the relative importance as predictors of the place satisfaction measures over and above the other

**Table 7.4** Place domains as predictors of life satisfaction in Detroit and environs

| | Detroit | | | Environs | | |
|---|---|---|---|---|---|---|
| Place domains | *B* | SE *B* | *β* | *B* | SE *B* | *β* |
| Home | .27 | .05 | .28*** | .19 | .02 | .20*** |
| Neighborhood | .06 | .06 | .07 | .11 | .02 | .11*** |
| Community | .11 | .06 | .12 | .11 | .02 | .12*** |
| $R^2$ | | .16*** | | | .12*** | |

Source: The authors
*B* multiple regression coefficient, *SE B* standard error of multiple regression coefficient, *β* (Beta) standardized multiple regression coefficient
*$p \le .05$; **$p \le .01$; ***$p \le .001$

**Table 7.5** The role of place variables in predicting quality of life: Hierarchical regression analysis

| | Detroit | | | Environs | | |
|---|---|---|---|---|---|---|
| | *B* | SE *B* | *β* | *B* | SE *B* | *β* |
| Family | .20 | .05 | .22*** | .20 | .01 | .24*** |
| Health | .11 | .05 | .11* | .08 | .01 | .09*** |
| Friends | .06 | .05 | .05 | .07 | .01 | .08*** |
| Job | .11 | .04 | .12** | .09 | .01 | .11*** |
| Time available | –.16 | .04 | –.18*** | –.04 | .01 | –.06*** |
| Standard of living | .39 | .06 | .37*** | .26 | .01 | .28*** |
| Leisure time | .30 | .05 | .31*** | .25 | .01 | .29*** |
| Home | .15 | .05 | .15** | –.01 | .01 | –.01 |
| Neighborhood | –.00 | .06 | –.00 | .01 | .01 | .01 |
| Community | –.01 | .05 | –.01 | .01 | .01 | .01 |
| $R^2$ | | .68*** | | | .67*** | |
| $R^2$ *change* | | .02** | | | .00 | |

Source: The authors
*$p \le .05$, **$p \le .01$, ***$p \le .001$

domain satisfactions was examined using hierarchical analysis. Again, these analyses covered Detroiters and residents living outside the city.

The first analysis shows that among Detroiters, only satisfaction with housing was significantly related to life satisfaction, while all place satisfaction measures (home, neighborhood, and community) were related to life satisfaction for the non-Detroiters (see Table 7.4).

The second analysis revealed that the non-place domains taken together account for nearly three-quarters of the variance in life satisfaction for both Detroiters and non-Detroiters (see Table 7.5). Adding the place satisfaction variables to the model had no impact on life satisfaction for non-Detroiters and a small but significant impact for Detroiters. That is, housing among Detroiters is a significant factor in their overall assessment of their lives.

Although these analyses produce somewhat different findings depending on how QOL is conceptualized and measured, we conclude that peoples' feeling about housing and, to a lesser extent, their neighborhoods and communities can affect their overall well-being. Furthermore, whether it is in a large urban center like Detroit or in another setting in the metro area, place matters, particularly one's dwelling. However, while satisfaction with housing can be important to overall life satisfaction, it is only marginally important when taking into account the way people feel about other aspects of their lives.

## Neighborhood and Neighboring

DAS2001 data allowed for extensive exploration of the *place domains*. For example, several questions were asked about the neighborhood including people's assessments of public services such as transportation, schools, parks, police protection and street maintenance, interactions with nearby family and friends, involvement in community organizations, and feelings about neighborhood conditions or attributes such as upkeep, crowding, safety, and crime.[17] Besides providing information about specific neighborhood attributes that local planners and service providers might find helpful, the intent was to determine the relative importance of these attributes in predicting neighborhood satisfaction.

The results showed the following:

(a) With respect to specific neighborhood attributes, the pattern of responses was similar to neighborhood satisfaction responses. For the most part, people throughout the metro area were content with their neighborhoods. However, a significant minority, particularly in Detroit, were not happy about many of their neighborhood conditions. For instance, one in three Detroiters said their neighborhoods were noisy compared to less than one in five from other parts of the region. And more than one-fourth of the Detroiters reported living in neighborhoods where dwellings and outdoor areas were poorly kept up. In other parts of the Metro Detroit area, less than one in ten gave poor marks to neighborhood upkeep.

(b) Responses to questions about physical attributes of the neighborhood showed a similar pattern between the counties and Detroit and were examined in relation to one another. Responses were highly correlated, and five of the attributes believed to be stressful to neighborhood residents were combined into a *neighborhood stress index*. The attributes were: noise level, degree of crowding, amount of traffic, and maintenance levels for homes and for yards. This index was subsequently used in modeling determinants of neighborhood satisfaction.

(c) Among other possible determinants of neighborhood satisfaction are various social attributes including perceptions of neighborhood safety, the social cohesiveness of the neighborhood, and neighborhood attachments or social networks.

Once again, these social attributes were examined for the entire region, for the city of Detroit, and for its environs. The results showed that:

---

[17]Neighborhood questions were also asked about the perceived size of the neighborhood, neighborhood factors influencing choice of residence, sense of community, and the likelihood of moving.

(a) In terms of neighborhood safety, Detroiters were less likely than others in the region to describe their neighborhoods as safe.
(b) Social cohesiveness was measured in terms of perceptions of neighbors as similar or dissimilar to respondents, the friendliness of neighbors, and feelings about sense of community. Once again, the views of Detroiters were different than people living in other parts of the region. Detroiters were less likely to describe their neighbors as similar to themselves (31% vs. 56%), as friendly (73% vs. 81%), and less likely to feel a strong sense of community (38% vs. 52%). Residents living in the older suburbs (55%) and in small towns throughout the region (58%) were most likely to feel that there was a strong sense of community in their neighborhoods.
(c) Social networks within the neighborhood were measured by asking about the number of nearby friends and families and attributes of neighboring.[18] Whereas the number of neighborhood friends did not differ significantly between Detroiters and others in the region, Detroiters tended to have more family living nearby. Nonetheless, neighboring patterns between Detroiters and others in the region were similar.

When social networks and patterns of neighboring were examined in relation to neighborhood satisfaction using multivariate analyses, having nearby friends and family, doing favors for one another, and attending neighborhood meetings were significantly related to a satisfying neighborhood. However, when measures of social cohesion, perceptions of crime, and assessment of physical stressors were taken into account, the presence or absence of family and friends and the amount and type of neighboring (that is, doing favors, socializing) had no bearing on overall neighborhood satisfaction.[19]

## *Transportation*

In addition to crime and physical stressors (noisy and crowded conditions, excessive traffic, and poor maintenance) as indicators of neighborhood dissatisfaction, negative ratings of services also contributed to dissatisfaction. Among the key services examined was public transportation, which was prevalent in some parts of the region and nonexistent in many other parts.[20] For the region as a whole, less than one-third of the respondents evaluated public transportation as serving their neighborhoods favorably, while another third evaluated it negatively.[21] Evaluations of the overall

[18] Attributes of neighboring included the number of neighbors known by name, the frequency of visiting and exchanging favors with neighbors, and walking to a neighbor's house.

[19] For a detailed discussion of the predictors of neighborhood satisfaction, see Marans et al. (2005).

[20] Other transportation-related questions covered use of public transit, modes of travel to shopping and work, walking, and willingness to pay for various transportation improvements.

[21] The remaining third gave a neutral response indicating that public transit was either not available, not used if it were available, or the respondent was unaware of its existence.

transportation system covering the entire region were even lower. Nearly half (45%) gave poor marks to everything from public transit to roads and highways.

Irrespective of where people lived, negative assessments of transportation adversely impacted their satisfaction with their particular community and QOL in the region. In fact, many in Metro Detroit believed that life in the region could be improved with better public transit. In response to the statement "Public transit that is safe and reliable would improve the quality of life of metro Detroit residents," a significant number (69%) from all parts of the region strongly agreed or agreed. Whereas most Detroiters (80%) said there would be QOL benefits from improved public transit, at least two-thirds of the respondents from other parts of the region also agreed with the statement.

Although most respondents supported the idea of public transit for the region, their support is tempered when asked if they are willing to pay more in taxes to improve public transportation. While one-third of respondents said they would increase their tax bill for this purpose, another third indicated they would not, while the remaining respondents were noncommittal. It was not surprising to see that Detroiters were most willing to pay more in taxes for public transportation improvements, whereas respondents in the outlying counties show the strongest opposition to a tax increase for this purpose.

Comparing respondents who believed public transit is important and were willing to pay taxes to support it (strong supporters) with those who thought it is important but were not willing to pay taxes for it (weak supporters), the study found that strong supporters significantly outnumbered weak supporters in Detroit and nearby Oakland County. On the other hand, weak supporters outnumbered strong supporters in the outlying counties. For example, public transit proponents were seven times more likely to be found in Detroit than in the outlying counties.

Although there was general support for public transit in the region, questions regarding auto travel revealed stronger sentiments and a greater commitment to private travel modes, especially among non-Detroiters. Whereas Detroiters were as likely to pay taxes for road improvements as on public transit improvements, the wider metro area residents were much more willing to pay for improvements to roads and highways than for public transit improvements (Marans 2008).

## *Prospects for the Future of the Region*

As a way of examining prospects about the region's future, two sets of questions were asked: One set dealt with moving intensions, and the other covered peoples' expectations about the QOL in Metro Detroit 10 years hence.[22]

With respect to moving intentions, almost one-half of the people in the region indicated that they wanted to move from their current residence, and a third said they would "definitely move" or "probably move" within the next 2 years. Detroiters

[22] Another question about people's expectations about the QOL in their particular county is not considered as part of this chapter.

were most likely to say they would move compared to people living elsewhere (48% vs. 35%). And when asked where they would like to move, four in five Detroiters said they intended to leave the city indicating either a preference for another community in Metro Detroit, another part of Michigan, or a move to another state. Only a small proportion of Detroiters (one in five) said they would move elsewhere in their city. Census estimates for 2005 revealed that many Detroit residents acted on their preferences. That is, over 60,000 Detroit residents moved from the city between 2000 and 2005. With the myriad of problems facing the city, the population continued to decline by another 60,000 throughout the remainder of the decade. Similarly, a significant number of respondents who lived elsewhere in Metro Detroit in 2001 expressed a preference for moving from the region or from Michigan. Indeed, deteriorating economic conditions throughout the decade prompted many to leave the Metro Detroit area.

For many, thoughts about the future of the places they live were likely to influence their decision to move or remain. In fact, DAS2001 data show that Detroiters who expected the QOL in the region to decline over the next decade were most likely to say they will move. For others in the region, thoughts about the region's future had no bearing on their plans to move or stay.

On average, survey respondents were more optimistic than pessimistic about the region's future. Those who thought the QOL in Metro Detroit would improve in the coming decade outnumbered respondents who said it would get worse by two to one. Nonetheless, about 15% believed the QOL would deteriorate over the decade, while 45% said there would be no change in regional QOL. Detroiters were the most hopeful; optimists in the city outnumbered the pessimists by four to one. Detroiters tended to think about their city and the metro area in the same way. Unlike others in the region, their feelings about Metro Detroit are closely associated with their feelings about the city.[23]

## Uses of Data for Policy and Planning

One objective of DAS2001 was to produce data that could inform planning and policy making in governmental, institutional, and community organizations throughout the region. At the governmental level, the primary recipients were the counties and the Southeast Michigan Council of Governments (SEMCOG), one of the study's key sponsors.[24] The extent to which the findings have been used is largely unknown. Since the initial data tabulation was completed in early 2002, findings covering all

[23]There is a strong association between community satisfaction ratings and the QOL in Metro Detroit among Detroiters ($r=.51$); for respondents in the rest of the metro area, the association was modest ($r=.19$).

[24]Other sponsors included: Daimler-Chrysler Fund, Macomb and Washtenaw counties, Ann Arbor Transportation Authority, Huron-Clinton Metropolitan Authority, Michigan Economic Development Council. and the USDA-Forest Service.

the questions in the survey have been posted on a DAS2001 website and widely publicized.[25] Posted findings are presented statistically showing the percentage distribution of responses to each question for the entire region as well as for each county and for the city of Detroit. The intent was to have findings easily accessible for target audiences including sponsors, stakeholders, and other decision makers as well as for members of the general public. During the first 3–4 years, there were numerous occasions to speak to governmental, civic, and academic groups about the research and its findings. As a result of the website and presentations, a number of requests for more specific information were made. For example, the data have been used to gage the support for mass transit in the region, to estimate visitations to metroparks from residents in each of the counties, and to tap people's views on the importance of downtown Detroit and the Detroit riverfront to Detroit's revitalization and the QOL in the region.

We have been disappointed that there have not been more requests for study findings or questions posed that could be answered through further data analysis. Nor have we seen evidence that data emanating from the research have impacted planning or policy formation in Detroit or elsewhere in the region. We suspect several factors have contributed to this situation:

(a) First, the release of the data and the establishment of the website did not occur in a timely manner. More than a year transpired between the time planners and policy makers identified issues and questions they wanted the survey to address and when the answers were made available. Other pressing issues and questions for policy makers and planners may have emerged during the interim, or responses to questions posed in 2000 may have been viewed as out-of-date in early 2002.
(b) Second, the findings were not as widely publicized as they might have been. Despite notices about the website and the availability of findings, many governmental units, organizations, and community groups were unaware of their existence.
(c) Third, funding to support dissemination and other outreach activities was not available, prompting the study team to focus efforts on other project issues such as generating objective measures and incorporating them into the data files or turning attention to other research projects. Finally, the follow-up survey that was anticipated in 2006–2007 and that would have produced change measures did not materialize.[26]

Had the follow-up survey been conducted, findings from the initial survey, as well as measures showing changes in the QOUL, would have generated considerable interest among policy makers and in the media.

As noted earlier, an objective of DAS2001 was determining the degree to which perceptions (and evaluations and behaviors) correspond to the community and environmental conditions (objective measures) associated with the places people live.

---

[25] http://sitemaker.umich.edu/das2001/home

[26] The termination of the Detroit Area Study by the University of Michigan in 2004 and deteriorating economic conditions in southeast Michigan were key factors preventing the follow-up survey from occurring.

The fulfillment of this objective required the creation of a comprehensive database incorporating both survey responses and different types of objective measures for each respondent. Conceptually, these measures could be socio-demographic, economic, and/or environmental and associated with one or more geographic units. The units would cover the minor civil division (that is, community) within which each respondent lives, census units (tract, block, block group), and predefined areas, such as a half-mile buffer surrounding the respondent's address. A measure could also be represented by a vector indicating the distance between the respondents address and a designation place such as a hospital, transit stop, or park.

Although there was modest success in developing a database with objective measures and then merging it with survey responses, the process was time-consuming and encountered challenges. These challenges dealt with data quality and data access. Data quality was mostly a question of when the desired data were collected and if the data collection occurred concurrently with period of the survey. Data quality also related to specificity. For instance, the land use data covering places where respondents lived throughout the seven counties were based on aerial photographs from 2000. This was considered close to the time of the survey. However, the areal land use was broad in coverage and not as detailed as the research team hoped to obtain. Consequently, attempts were made to supplement the broad coverage with land use data for parcels in governmental units having a large number of respondents. In the case of Detroit, the city with the largest number of respondents, the problem was twofold. First, the parcel land use was collected in 1998. We suspected considerable land use change between that time and 2001, and therefore, land use was unlikely to correspond to the questionnaire data collection. Second, it was anticipated that obtaining the detailed land use data from Detroit (and from other governments in the region) would be time-consuming. Consequently, the decision was made to incorporate the generalized 2000 land use in the database. Similar problems were encountered in accessing school records and crime statistics for governmental units; therefore, the database did not include this information.

## Continuing Work

Despite disappointments of not being able to systematically monitor the QOUL in Metro Detroit beyond 2001, ongoing activities associated with DAS2001 have taken place over the past decade.

As mentioned earlier, the database has enabled researchers at the University of Michigan to explore various topics based on the questionnaire responses and on questionnaire responses vis-à-vis objective conditions.[27] At the same time, widespread discussions of DAS2001 and the need to measure the QOUL in the twenty-first century

[27] See Marans (2008) for use of data showing the influence of how far respondents live from a transit stop and their use of public transit.

have resulted in parallel studies in other urban settings (see Chaps. 8–14) around the world. While the number of such studies is limited, studies in other cities and/or metro areas have been or are being considered. While the studies noted in the other chapters have replicated questionnaire items used in DAS2001, they also address topics of particular interest to local researchers and their sponsors. Similarly, studies that may occur at some future date will undoubtedly replicate DAS2001 questions and add new questions representing the interests of local researchers, stakeholders, and sponsors. It is anticipated that the collection of future QOUL studies will enable a new generation of researchers to make cross-cultural comparisons on the role of the neighborhoods, cities, and regions in the QOL experience of area residents.

## Summary

This chapter has presented an overview of a study designed to provide systematic data covering the QOUL of residents of Metro Detroit in Southeast Michigan. The primary source of information was a household survey of over 4,000 residents living in seven contiguous counties, one of which contained the largest city, Detroit. Corresponding information about the environment associated with the respondents was also collected.

Among its many objectives, the study was intended to inform policy makers and planners on conditions in the region at the beginning of the twenty-first century and establish a benchmark for measuring social and environmental changes that were expected to occur in the decades to come. At the same time, it aimed at improving our understanding of relationships between perceptions, evaluations, and behaviors and the urban conditions to which people were responding.

While the study was able to achieve several of its objectives, it is not clear whether the study made significant contributions to planning and policy formulation in Southeast Michigan. Nor has it yet to be used as a benchmark for assessing changes in the region throughout the twenty-first century. Presumably, the lessons from DAS2001 will benefit the design, implementation and utilization of other QOUL projects in Metro Detroit and elsewhere.

## References

Campbell, A., Converse, R., & Rodgers, W. (1976). *The quality of American life: Perceptions, evaluations and satisfactions*. New York: Russell Sage.

Couper, M. P., Clemens, J., & Powers, K. (2002). *Detroit Area Study 1952–2001: Celebrating 50 years*. Ann Arbor: The University of Michigan.

Farley, R., Danzinger, S., & Holzer, H. J. (2000). *Detroit divided*. New York: Russell Sage.

Freedman, R. (1953). The Detroit Area Study: A training and research laboratory in the community. *The American Journal of Sociology, 19*, 30–33.

Inglehart, R., & Welzel, C. (2005). *Modernization, cultural change, and democracy: The human development sequence*. New York: Cambridge University Press.

Marans, R. W. (2003). Understanding environmental quality through quality of life studies: The 2001 DAS and its use of subjective and objective indicators. *Landscape and Urban Planning, 65*, 73–83.

Marans, R. W. (2008). Quality of urban life research: Travel findings from the Detroit Area Study. *IAPS Bulletin of People-Environment Studies, 33*(Spring-Summer), 299–352.

Marans, R. W., Carter, G., & Grocman, Z. (2005) *Detroit Area Study 2001: Neighborhoods and neighboring*, Unpublished report (Available from first author).

Ryu, E., Couper, M., & Marans, R. W. (2006). Survey incentives: Cash versus in-kind: Face-to-face versus mail; response rate versus non-response bias. *International Journal of Public Opinion Research, 18*, 89–106.

Thomas, J. M. (1997). *Redevelopment and race: Planning a finer city in Postwar Detroit*. Baltimore: The Johns Hopkins Press.

# Chapter 8
# The Brisbane-South East Queensland Region, Australia: Subjective Assessment of Quality of Urban Life and Changes over Time

**Robert Stimson, Rod McCrea, and John Western†**

## Introduction

In this chapter, we provide an overview of research investigating the subjective assessment of a wide range of quality of urban life (QOUL) domains in the Brisbane-South East Queensland (SEQ) region, Australia. The research is based largely on data collected in two SEQ Quality of Life (QOL) surveys conducted in 1997 (SEQQOL1997)and 2003 (SEQQOL2003). It includes an analysis of the stability and change in those subjective assessments of aspects of QOUL over the 6 year period between the two surveys. The information collected in the surveys generated measures of residents' subjective assessments or evaluations of QOL as a whole and of a number of QOUL domains at both the aggregate level covering the SEQ metro-region and at the disaggregated level of the local area or neighborhood. This chapter reports on findings relating to QOUL domains, as opposed to other QOL domains less related to urban environments (satisfaction with personal relationships, health, independence, money, etc.).

In those surveys, a "satisfaction-based measurement" of QOUL was adopted because – as argued by Campbell et al. (1976) and Marans (2003) – it implies judgmental or cognitive evaluations. That approach was used rather than adopting

R. Stimson (✉)
Australian Urban Research, Infrastructure Network (AURIN), Faculty of Architecture, Building and Planning, University of Melbourne, VIC 3010, Melbourne, Australia
e-mail: rstimson@unimelb.edu.au

R. McCrea
Institute for Social Science Research, The University of Queensland, Level 4, GPN3 Building, Campbell Road, St Lucia, Brisbane, QLD 4072, Australia

† J. Westem (deceased)
School of Social Science, The University of Queensland, Michie Bldg. St. Lucia Campus, Brisbane, Australia

R.W. Marans and R. Stimson (eds.), *Investigating Quality of Urban Life: Theory, Methods, and Empirical Research*, Social Indicators Research Series 45,
DOI 10.1007/978-94-007-1742-8_8, 

one seeking to measure people's "happiness," which is less focused on evaluating "targets" and more influenced by relatively short-term moods of, for example, elation or gaiety.

Thus, QOUL as investigated in the two SEQQOL surveys was conceptualized as being a *multi-layered* and a *multi-dimensional* concept (as proposed by Marans and Rodgers 1975; Marans 2003; McCrea et al. 2005). It is multi-layered in the sense of its representation at various levels: the individual, family, local community, and the wider metropolitan environment in which people live, work and recreate. It is also multi-dimensional in its reflection of various aspects of urban life.

## About the Study Region and the Survey

### *The SEQ Region*

The SEQ region is Australia's third largest metropolitan region. This "sun-belt" metropolis has been experiencing rapid growth and socio-economic transformation for several decades. The region has been referred to as a space of "post-modern urban consumption" (Mullins 1991). That is particularly discernible among the region's coastal communities that are destinations for international and domestic tourists and display an urban lifestyle that characterizes consumerism and fragmented culture, eclecticism, and an aspiration for variety, choices and freedom of expression.

The region is characterized both by low-density urban sprawl development and a multi-centered urban structure. The Queensland state capital, Brisbane City, is at the center of the region and connects to other urban agglomerations via three growth corridors: south to the Gold Coast, north to the Sunshine Coast (both of which are tourism regions), and a less rapidly growing inland corridor west to Ipswich, a long-established industrial and coal mining city (see Fig. 8.1). The hinterland regions beyond include a mix of rural-urban residential acreage developments, farmlands and a number of small towns. Since the early 1990s, a considerable amount of urban consolidation has been occurring in the inner city areas of the region through a mixture of in-fill and urban renewal development. The coastal strip of the Gold Coast has a lot of high-density development with high-rise apartment living.

The SEQ region's population increased from 1.8 to about 2.4 million over the decade (1991–2001), and it is forecasted to reach about 3.2 million by 2011. It is characterized as a "sun-belt growth metropolis," with in-migration accounting for the greatest proportion of the growth in population. The in-migration stream has a diverse age structure and is no longer dominated by retirees as it used to be several decades ago.

The SEQ regional economy has been dominated by employment in the services sectors with economic growth being largely population and construction led, particularly through household sector consumption. Employment growth is heavily concentrated in the retail, education, health and community service sectors. Tourism plays

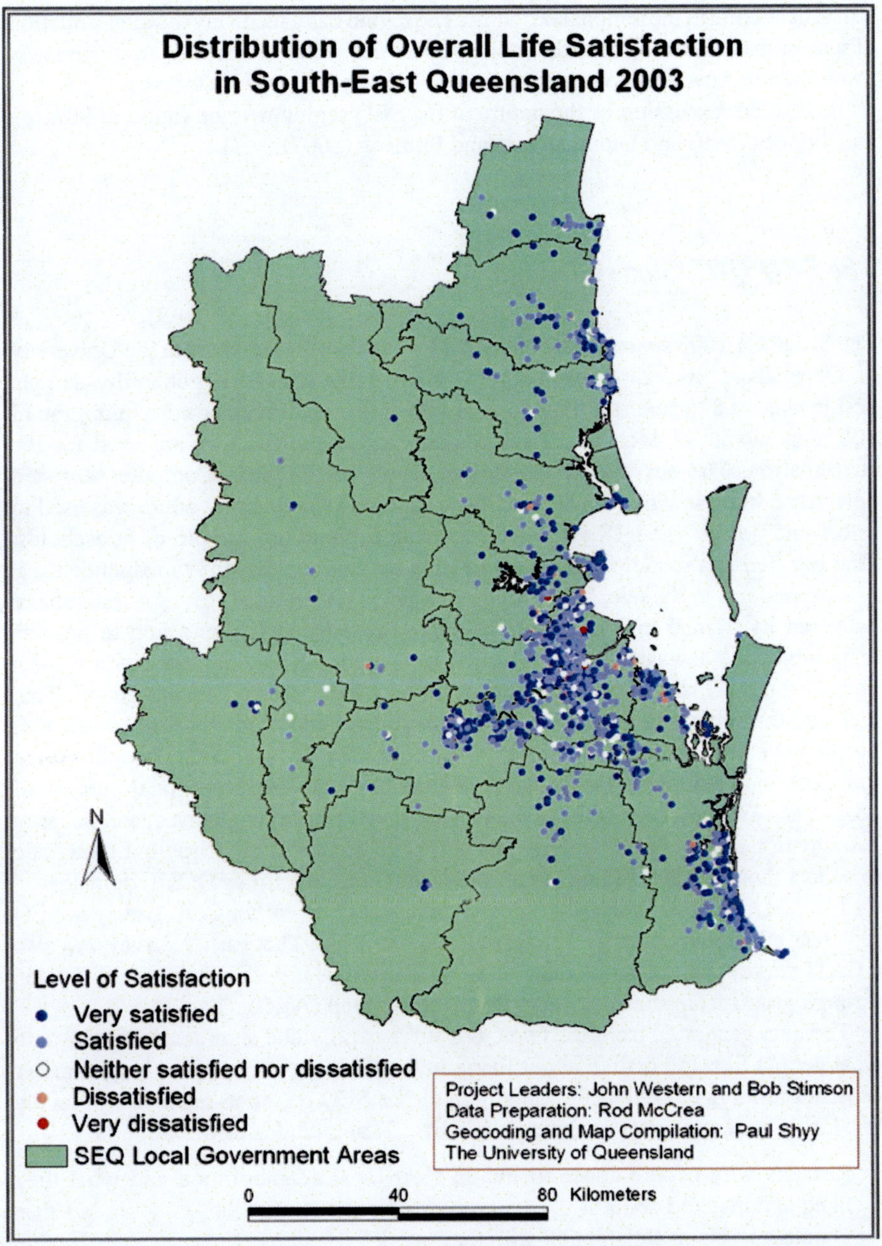


**Fig. 8.1** The Brisbane-South East Queensland region (SEQ) and the location of respondents to the 2003 QOL survey (Source: The authors)

a significant role in the regional economy. The region has a relatively low concentration of jobs in manufacturing, and the producer services sector of business employment is lower than in Australia's two largest metro-cities, Sydney and Melbourne.

A detailed discussion of the nature of the SEQ region may be found in Stimson and Taylor (1999) and Guhathakurta and Stimson (2007).

## *The SEQ QOL Survey Design*

The SEQQOL2003 survey was conducted by a team of researchers at the University of Queensland, including the authors. A spatially stratified probability sample design was used to generate a random sample of households ensuring a minimum of 100 respondents in each of 10 sub-regions across the SEQ region used for the stratification. The survey was conducted using the UQSRC Computer Assisted Telephone Interviewing (CATI) facility. A random selection procedure was used to select one person aged 18 years or over from the random sample of households. That resulted in a random sample of 1,610 individuals as the survey respondents.

Respondents to the telephone survey completed a structured survey questionnaire delivered by trained interviewers. The survey respondents were asked to provide their subjective assessment or evaluation of specific QOUL domains with respect both to the SEQ region as a whole and to their local area or neighborhood. That included the subjective assessment of life in general and of the level of satisfaction people have with access to services, the performance of local councils, attitudes to transport and environmental issues, the nature of work, the recreational pursuits of people, assessment of attributes of the neighborhood where people lived, and people's participation in community affairs. The survey instrument used a 5-point Likert scale to collect those subjective assessments or evaluations of various QOUL domains.

The SEQQOL2003 survey was largely a repeat of an earlier QOL survey in SEQ region conducted by the research team leaders in 1997. That earlier survey was also a CATI survey with a sample size of 1,347 individuals. Both surveys were funded through grants from the Australian Research Council (ARC).

The data generated from the two QOL surveys provided significant insights into what people like and dislike about living in the SEQ region and on a large number of issues relating to QOUL focusing both on the SEQ region as a whole and on the local area or neighborhood where people live. That included information on:

- Reasons why people chose to live in their local neighborhood and what they liked and disliked about it
- People's levels of satisfaction with their work
- The nature of people's work and their journey to and from work
- People's patterns of consumption and their use of local and regional level services and facilities
- What people think are key planning issues for governments to address in making SEQ a better place to live

**Table 8.1** SEQQOL2003 survey data distribution on key demographic and socioeconomic variables of respondents compared with 2001 census data for SEQ region

| | SEQQOL2003 | 2001 census |
|---|---|---|
| Median age of those aged 18 and over | 46 | 43 |
| Percentage female of those aged 18 and over | 49 | 51 |
| Percentage married or in a de facto relationship | 60 | 56 |
| Percentage divorced, separated or widowed | 18 | 18 |
| Percentage born in Australia | 77 | 73 |
| Percentage indigenous | 1.4 | 1.5 |
| Percentage with post-school qualifications | 78 | 78 |
| Percentage with a bachelors degree or higher qualification | 25 | 14 |
| Median income of those aged 20 and over | 26,000 | 23,700 |
| Median household income | 50,100 | 43,700 |
| Percentage employed of those aged 18 and over | 65 | 59 |
| Percentage of total persons aged 18 and over who are employed full-time | 37 | 37 |
| Percentage of dwellings that are separate houses | 84 | 75 |
| Percentage of dwellings that are townhouses, units, flats or semi-detached housing | 15 | 22 |
| Percentage of employed persons working from home | 9 | 5 |

Source: The authors

Because of cost constraints and to keep the time needed to complete the telephone survey questionnaire to a maximum of about 40 min, for the 2003 survey, it was decided to split the sample so that half the sample completed some sets of questions while half the sample completed other sets of questions.

From the data collected in the two SEQQOL surveys, it is possible to investigate the degree to which there has been stability or change over the 6 year period in people's subjective assessments of specific QOUL domains asked in both the 1997 and 2003 surveys.

Summary information on the demographic and socio-economic characteristics of the respondents to the 2003 SEQQOL survey is provided in Table 8.1, which also gives measures for those variables from the 2001 Census of Population and Housing, thus indicating the degree to which the characteristics of the respondents in sample vary from those of the population of the SEQ region. The sample was representative of the SEQ population on most characteristics.

## Assessing Quality of Urban Life "Domains" in the SEQ Region

In this section of the chapter, we focus on the subjective assessment of QOUL "domains" at the aggregate level of the SEQ region using data from the SEQQOL2003 survey only.

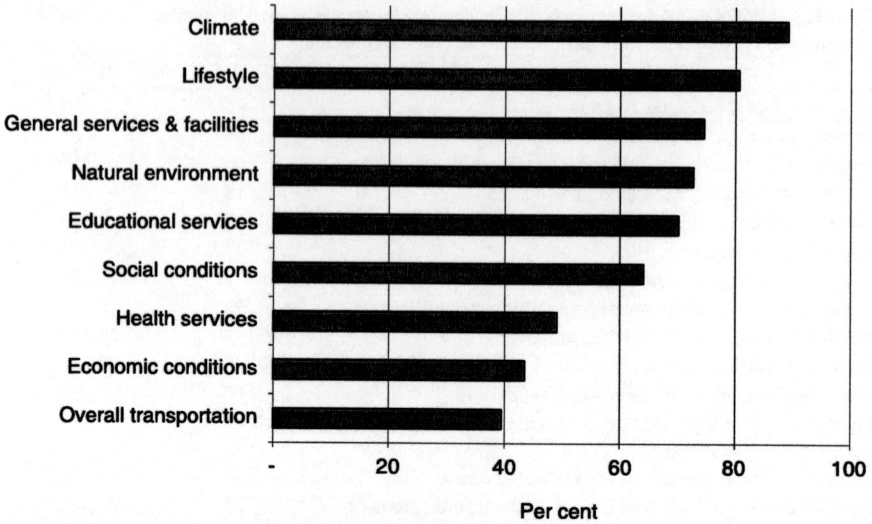


**Fig. 8.2** Rating aspects of QOUL in the SEQQOL2003 survey: Percent of survey respondents giving "good" and "very good" ratings (Source: The authors)

## *Rating "Domains" of QOUL in the SEQ Region*

Survey respondents were asked to rate a number of attributes of the SEQ region that might identify aspects of QOUL. These range from the region's climate and lifestyle to the provision of services, the natural environment, social and economic conditions, and transportation. Figure 8.2 shows the percentage of respondents giving "good" and "very good" ratings. The top ranked domain was *climate* with 89% ranking it "good" or "very good". It was followed by: *lifestyle*, rated "good" or "very good" by 80%; *general services and facilities*, rated by 74% as "good" or "very good"; *natural environment*, 73% and *educational services* (schools) and *social conditions* both on 64%. Rated lowest were: *health services*, rated "good" or "very good" by only 48%; *economic conditions*, 42% and *overall transportation*, 39%. It is noteworthy that the region's *transportation*, *economic conditions*, and *health services* were rated as "good" or "very good" by fewer than 50% of survey respondents. For those three domains, 25%, 15% and 19% respectively rated them as "very poor" or "poor." Of course it is possible that the assessment of some of these domains may have been influenced by factors exogenous to the SEQ region.

## *Overall Satisfaction with Life and with Living in SEQ*

In response to the question asking about *overall levels of satisfaction with life as a whole*, 89% expressed "satisfaction," with 38% indicating "great satisfaction."

The SEQQOL1997 survey found similar results with around 90% of the sample expressing overall satisfaction with life as a whole. These findings are consistent with those reported elsewhere (Cummins 1996). Similar findings were observed for overall *satisfaction with living in the Brisbane-SEQ region.* In this instance, a slightly higher 91% were either "satisfied" (44%) or "very satisfied" (47%). Thus, overall the residents of SEQ had very high levels of satisfaction with their life as a whole and also with living in the region.

It is evident that natural attributes of SEQ (for example, climate, lifestyle and natural environment) contributed more to satisfaction with living in the region than did attributes, such as health services, economic conditions and transport, that were more under government influence to initiate or provide. Therefore, even though the overall QOL in the region was seen by most people as "very good" or "good," there was potential to improve some aspects of QOUL, especially in those domains which could be influenced directly through action by governments.

## *Variations in Assessment of QOUL "Domains" Based on Sub-regions in Which People Live*

Variations in the ratings given to QOUL domains depended frequently on where residents live.

While *climate* was rated as "good" or "very good" by around 85% of respondents in all domains, it obtained its highest rating in the Sunshine Coast (97%) and its lowest (less than 85%) in Ipswich. Generally speaking, those living closer to the coast were most likely to rate the climate highly.

*Lifestyle* was also rated highly fairly consistently across the region, although those living on the Gold and Sunshine Coasts and in Brisbane City's Outer Suburbs gave slightly higher ratings than those in other areas, particularly Ipswich City, Logan-Redland and the inner suburbs of Brisbane City.

*General services and facilities*, such as those relating to retail and entertainment, were less likely to be rated "good" or "very good" by people living in the Rural Hinterland and in Ipswich City (fewer than 60%), while residents of the Gold Coast Brisbane's Middle and Outer Suburbs gave these domains the highest rating.

*Social conditions*, such as friendliness and sense of community, were rated noticeably more highly by residents of Brisbane City's Inner Suburbs, the Rural Hinterland areas and the Sunshine Coast than they were by residents of the Gold Coast Outer areas (away from the coastal strip) and Brisbane's Outer Suburbs, where fewer than 60% gave social conditions "good" or "very good" ratings.

*Economic conditions* were rated as "good" or "very good" by 63% of residents in Brisbane City's Inner Suburbs, while fewer than 40% of residents on the Sunshine Coast, in the Rural Hinterland, and the coastal strip of the Gold Coast rated economic conditions similarly. Over 80% of residents of the Gold Coast coastal strip rated the natural environment highly compared to Ipswich City where the ratings were just over 60%. The other regions came in between these two extremes.

*Overall transportation* was rated poorly by people in all parts of the region, with the highest ratings of "good" or "very good" by around 50% in the Gold and Sunshine Coasts. Residents of Brisbane City's Inner Suburbs, Brisbane City's Middle Suburbs, and Caboolture-Pine Rivers-Redcliffe also gave this domain a rating of over 40%. But for those living elsewhere, the ratings were consistently lower.

*Educational services*, such as public schools, were rated highest by respondents from the Rural Hinterland where over 80% described them as "good" or "very good" compared to no more than around 60% of residents for other areas.

*Health services* were rated highest by residents of Brisbane City's Inner Suburbs (where most of the hospitals are located), with over 60% giving this attribute a "good" or "very good" rating. However, fewer than 50% of residents in other regions gave health services similar ratings.

## Summary QOUL Assessment Profiles for the Sub-regions

Profiles of geographic areas were developed based on QOUL domains perceived "better than" or "worse than" the average, as rated by the sample as a whole. This was achieved by calculating the *z*-scores for the QOUL domains discussed above:

(a) The assessment of health services, economic conditions and social conditions by people living in *Brisbane City's Inner Suburbs* was considerably higher than average, while it was lower than average for educational services (schools), lifestyle, climate and the natural environment. While several attributes rated below average, this sub-region had the highest proportion of residents giving a high rating for the "overall" QOL in the region.

(b) For those living in the *Middle Suburbs of Brisbane City*, assessments were slightly higher than average on all but one of the aspects of QOUL. Respondents *in the Outer Suburbs of Brisbane City* also tended to provide slightly higher than average ratings for the majority of QOUL attributes.

(c) A very different picture emerged for those respondents living in the *Logan City-Redland Shire* sub-region. Here, less than average assessments were given for all the QOUL factors.

(d) An even greater pattern of negativity was shown by those living in *Ipswich City* and most markedly for the natural environment, lifestyle, general services, educational services, overall transportation, climate, and social conditions.

(e) Survey respondents living in the *Caboolture-Pine Rivers-Redcliffe Shires* sub-region showed very little variation from the mean scores on all domains.

(f) A very different picture emerged for those living on the *Gold Coast*. Along the coastal strip, the assessments were considerably better than average with respect to lifestyle, natural environment, general services and facilities, and climate. However, social conditions were perceived to be worse than average. Those living in the *Gold Coast Hinterland* tended to rate general services and facilities and overall transportation more highly than the average, while educational services and social conditions were rated worse than average.

(g) With the exception of economic conditions which were rated much worse than average, *Sunshine Coast* residents tended to have above average ratings for most QOUL domains. Finally, residents living in the *Rural Hinterland* of SEQ tended to give well above average ratings for educational services and social conditions, but well below the average for overall transportation, economic conditions, and general services and facilities.

## Problem Issues in the SEQ Region

Information was collected on the degree to which 11 issues were seen to be a problem for urban living. They included issues to do with the environment, traffic, crime, costs of living, population growth and urban sprawl. The results are summarized in Fig. 8.3.

(a) The greatest problem was *traffic congestions*, seen as a "very great problem" by 23% of respondents, a "great problem" by a further 33%, and "somewhat of a problem" by 26%. Only 18% saw it as "not a problem" or as only a "small problem." Of the 11 issues on which assessment was sought, it was only on this one where more than half the sample reported it as a "great" or a "very great problem."

(b) Of the environmental issues, the *loss of natural places for fish and wildlife to live* and *water pollution*, in the form of the discharge of waste into the rivers, bay and sea, were seen as the greatest problems with just under half (49%)

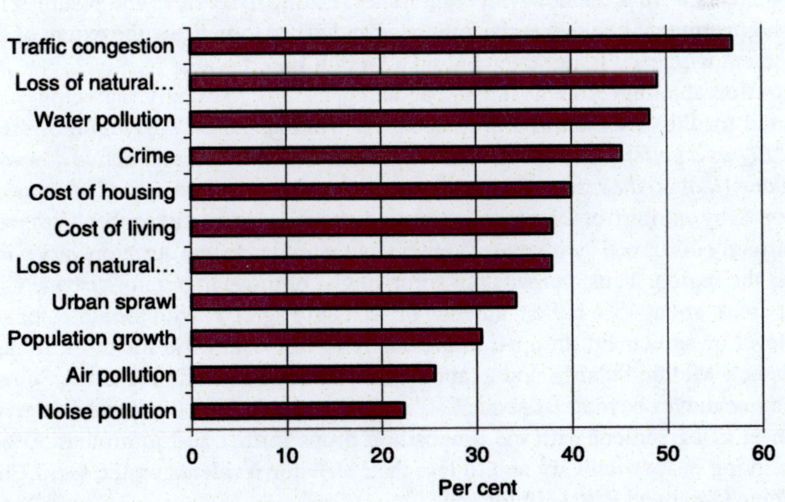


**Fig. 8.3** Rating issues as a "great problem" or "very great problem" in the SEQ region (Source: The authors)

reporting them as "great" or "very great" problems. The *loss of natural areas, urban sprawl*, *air pollution* and *noise pollution* were not as big issues. The loss of natural areas generally was a "great" or "very great" problem to nearly 40%, urban sprawl a problem to around a third, and air and noise pollution to no more than a quarter of the sample.

(c) The social issues of *crime*, *housing costs*, *general cost of living*, and *population growth* were seen as greater problems than air and noise pollution. Crime was high on the list, seen as a "great" or "very great" problem by nearly half (45%) the sample. Cost of housing and cost of living was similarly seen by slightly smaller percentages (40% and 38%, respectively), and population growth was even less of a problem, rated as a "great" or "very great" problem by only around 30% of the sample.

In summary, traffic congestion was the most significant problem these residents of the SEQ region reported experiencing, and it was noted by over 50% of the population studied. Next came the impact of development on the physical environment (loss of natural resources, water pollution and loss of natural areas) noted by between 38% and 48% as a problem. Crime topped the list in the social category, a problem to around 45%, while costs of living and housing were problems to around 40%. These are commonly reported problems of urban living and suggest that the SEQ region may not be too different from many other sun-belt regions.

## Attitudes to Services and Planning Issues in the SEQ Region

Respondents were presented with eight issues relating to services and planning over which government could exercise influence and asked to indicate the extent of their agreement with each issue statement on a 5-point Likert scale.

Considerable agreement (43% "agreeing"; 32% "strongly agreeing") was reported for the statement *A strong economy will depend on developing Brisbane and SEQ as a better place to live.* Although, *Having a vibrant and active downtown Brisbane (that is, the city center) is important to my overall quality of life* was only supported by one third of the survey population. A similar proportion either "disagreed" or "strongly disagreed" with that statement, but attitudes to this question vary widely across the region. Thus, perhaps not surprisingly, of those living in Brisbane City's Inner areas, about 65% either "agreed" or "strongly agreed" with the statement, but this level of agreement dropped dramatically to just over 40% for those living in Brisbane's Middle Suburbs and to about 35% for those living in the Outer Suburbs. When one moves beyond Brisbane City and in particular into the Rural Hinterland, the level of agreement with the proposition drops further still to around 30% for those living in Ipswich City and to less than 30% for residents on the Gold Coast, Sunshine Coast and Rural Hinterland.

In response to the question *Are Casinos good for the Brisbane and SEQ region*?, more than half (55%) of the survey respondents believed not, while only around

16% thought they were. Thus, community-wide support for the proposition that casinos are good – a view current in the late 1990s and early 2000s as the state government pursued a pro-casino policy – was actually quite weak. These findings raised an important issue concerning the lack of public sentiment in support of governments promoting casino developments.

*We need better public transport in the SEQ region was a view supported by a large majority.* Of the 76% who "agreed" or "strongly agreed" with this statement, over 40% (41%) "strongly agreed." Public transport that is reliable and safe is important to the quality of life of residents in the Brisbane-SEQ region was also a view supported by a majority with 60% of survey respondents strongly agreeing and a further 35% agreeing. However, while there was obviously strong support for the proposition, the survey also found that only 9% of survey respondents used public transport "daily or almost daily," while 11% used it "one to three times a week," or only "one to three times a month." The great majority (69%) used public transport "less than once a month" (34%) or "never" (35%). Thus, expressions of strong support for the need for better public transport provision was not matched by high levels of use; but that might also have reflected its inadequacy at the time of the survey.

*The SEQ region has lots of recreational opportunities for me and my family* was a statement supported by around three quarters (73%) of the sample. This finding accorded with the 70% of respondents who said they were "satisfied" or "very satisfied" with their overall recreation opportunities and activities – the most popular of which were eating out or going to the beach, club or pub.

There was almost a universal support for the statement that *the State Government should play an active role in restricting uncontrolled urban growth.* Nearly two thirds (60%) expressed *strong agreement* and a further 30% *agreement.* However, only about one third thought that urban sprawl was a "great" or "very great" problem. Evidently, government is coming up to expectations.

Nevertheless, only around a third (36%) thought that *Local government is doing a good job in managing growth and development.* A further 29% "disagreed" or "strongly disagreed," while the remaining third were unsure. Thus, across the SEQ region there was a very high level of public sentiment suggesting that the state government needed to be actively engaged in growth management and planning, and there was also a perception that local councils could do much better in this regard.

Finally, rural sustainability was also seen as an important issue with 80% of the survey population believing that *Preserving farmland will improve the quality of life for future generations.* Most respondents were also concerned with the loss of natural habitat, and few would like to see more medium-density housing in their neighborhood. Therefore, given the high population growth rate in the region (which around 30% of survey respondents saw as being a significant problem), many of the survey respondents did appear to hold inconsistent attitudes while they wanted to preserve natural and farming tracts of land, but they were reticent about supporting higher-density housing developments which would be necessary if urban sprawl was to be contained. Many people thus wanted to "have their cake and eat it too" – not an uncommon human phenomenon!

## Focusing on Local Level or Neighborhood Issues

An important consideration in addressing QOUL issues in a large scale urban area concerns the local environment or neighborhood in which people live and the judgments they make about it on local issues.

The issues range from general satisfaction with the neighborhood to helping others when the need arises and the sense of community in the respondent's neighborhood. In addition, questions focus on the level of crime, the need for a private motor vehicle to move around the community and the desirability for more medium density housing. With the exception of the last, all questions were asked in both the SEQQOL1997 and SEQQOL2003 surveys, with a 5-point Likert scale ranging from "strongly agree" to "strongly disagree."

(a) The great majority of the sample (87%) stated that they *Feel satisfied living in this neighborhood.* Across the SEQ region, the average neighborhood satisfaction levels were similar in both the 1997 and 2003 surveys. Smaller proportions, however, have a great deal of contact with *people in this neighborhood* although the degree of contact increased between 1997 and 2003 to 46% indicating an increase in interaction with people living in the same neighborhood.

(b) *People in this neighborhood are willing to help each other out.* 61% of the 2003 survey respondents either "agreed" or "strongly agreed," with 21% "strongly agreeing." Only 16% either "disagreed" or "strongly disagreed" with this statement. The average level of agreement with this statement was similar in the 1997 survey.

(c) *There is a strong sense of community in this neighborhood.* Fifty percent of the 2003 survey respondents either "agreed" or "strongly agreed," while 19% either "disagreed" or "strongly disagreed."

(d) *My neighbors are friendly people.* Forty-two percent of the 2003 survey respondents "strongly agreed," and 39% "agreed," while only 7% either "disagree" or "strongly disagree." Generally speaking, in the 2003 survey, the respondents seemed to have favorable attitudes toward their neighbors.

(e) *Vandalism is a problem in this neighborhood.* In the 2003 survey almost 30% of respondents either "agreed" or "strongly agreed," while 49% "disagreed" or "strongly disagreed."

(f) *Breaking and entering is a problem in this neighborhood.* The 2003 survey data show that 35% of respondents either "agreed" or "strongly agreed" with this statement, while 41% either "disagree" or "strongly disagree" with this statement, and 24% neither "agreed" nor "disagreed." Between 1997 and 2003, the average level of agreement with those statements was similar, indicating that residents generally did not perceive any significant improvement in those crimes.

(g) *I feel safe walking around this neighborhood after dark.* In the 2003 survey, 56% of survey respondents either "agreed" or "strongly agreed" with this statement, though 27% of respondents either "disagreed" or "strongly disagreed" with the statement.

(h) *Living in this neighborhood would be difficult without a car.* Fifty-six percent of respondents either "agree" or "strongly agreed" with this statement, with 33% "strongly agreeing." But 34% either "disagreed" or "strongly disagreed," which represented an increase over the 1997 survey data.

(i) *Public transport is adequate for my needs in this neighborhood.* Forty-four percent of the 2003 survey respondents either "agreed" or "strongly agreed" with this statement, but only 17% "strongly agreed," which represented a decline since 1997, although it was not significant. The proportion of the 2003 survey respondents who either "disagreed" or "strongly disagreed" with this statement is 40%, with 21% "strongly disagreeing." Those results suggested that reliance of SEQ residents on the private motor vehicle for transport was due in part to perceived inadequacies in the provision of public transport. In 2003 nearly 70% of the survey respondents aged 18 years or over indicated they used public transport less than once a month or never, and more than 80% of respondents who were employed indicated they used a private motor vehicle to travel to work.

(j) *My neighborhood is conveniently located for people to walk to stores and parks.* In the 2003 survey, 30% of respondents "strongly agreed" and 35% "agreed" with this statement, while 21% either "disagreed" or "strongly disagreed."

(k) *I would like to see more medium-density housing (like townhouses) built in this neighborhood.* As a regional issue, the question of urban sprawl and density is a significant planning concern in the SEQ region, and in some parts of the region, considerable public policy efforts have been encouraging urban consolidation, with the proportion of new dwelling construction that is in the form of medium- and higher-density housing now being considerable. Thus, the attitude of residents to that form of urban redevelopment is an important issue. But in response to this question, only 8% either "agreed" or "strongly agreed" with the statement, and the mean (average) level of agreement with the statement was significantly lower in 2003 than it was in the 1997 survey. The proportion of 2003 survey respondents who "strongly disagreed" with this statement was 55%, with a further 27% "disagreeing." The lack of support for diversification through medium-density housing across the SEQ region was thus very strong, at least with respect to it occurring in people's local neighborhood. These findings tell us that the NIMBY ("not-in-my-backyard") syndrome was strong in the SEQ region.

(l) *Satisfaction with housing.* While not strictly a neighborhood evaluation, survey respondents also had very high levels of satisfaction with their housing situation. The 2003 data tells us that 85% of respondents were either "very satisfied" (40%) or "satisfied" (45%) with their housing. Fewer than 5% expressed any level of dissatisfaction. Satisfaction with housing had increased since 1997.

## Sub-regional Variations in Neighborhood Evaluations

The 2003 survey results revealed considerable variations across the SEQ region in how people living in different geographic areas evaluated the neighborhood issues discussed above. We can get an idea of that variation by examining the mean (average) scores on each 5-point Likert scale for the sub-regions compared with the mean score for the total region. Again, these are expressed as *z*-scores.

In *Brisbane City's Inner Suburbs*, people reported a relatively high level of satisfaction with living in their neighborhood. As might be expected, they were likely to think that their neighborhood was conveniently located for walking to stores, parks, etc. And, they were less likely to think that living in their neighborhood would be a problem without a car. However, given the high overall satisfaction with their neighborhood, it was surprising that so many attributes were rated below average. People living in Brisbane's Inner Suburbs were less likely to: have much to do with people in their neighborhood, think neighbors were friendly people, help each other out, think there was a strong sense of community in the neighborhood and feel safe walking about after dark. But they were more likely to think that breaking and entering was a problem. Those findings suggested that people choosing to live in the Brisbane City's Inner Suburbs highly valued the locational amenities of their neighborhoods and were willing to trade off against other aspects of their neighborhood. In addition, they were less likely to agree with the statement that they would like to see more medium-density housing in their neighborhood. This was somewhat surprising given that these inner city areas comprise relatively high concentrations of medium-density housing development and in which urban consolidation had been occurring. Perhaps close experience of or proximity to this form of housing had soured attitudes or perhaps current densities were considered sufficient.

People living in *Brisbane City's Middle Suburbs* were more likely to have little to do with people in their neighborhood and to think that: vandalism was a problem, their neighborhood was conveniently located to walk to stores and parks and that they would like to see more medium-density housing developed in their neighborhood. But they were less likely to think: their neighbors are friendly people, people were willing to help each other out, there was a strong sense of community in their neighborhoods, and it was safe to walk about after dark.

Moving to *Brisbane City's Outer Suburbs*, survey respondents living here were more likely to: feel safe walking around their neighborhood after dark, think public transport in their neighborhood was adequate for their needs and have little to do with people in their neighborhood. In addition, they were somewhat less likely on average to think: vandalism was a problem; people in their neighborhood were willing to help each other out.

Moving further out into the growth corridors of the SEQ region, survey respondents living in *Logan City-Redland Shire* were more inclined to: think their neighbors were friendly people, think people were willing to help each other out, have things to do with their neighbors, think that public transport in their neighborhood was adequate for their needs and feel safe walking around at night. However, they were considerably more likely to: be less satisfied with living in their neighborhood, think that breaking and entering was a problem and want more medium-density housing in their area.

Residents of *Ipswich City* in the western corridor tended to have below average perceptions that: vandalism was a problem in their neighborhood, breaking and entering was a problem, their neighborhood was conveniently located for people to walk to stores and parks, and public transport was adequate in their neighborhood for meeting their needs. They were more likely to favor more medium-density housing in their neighborhood. They tended to have above average interaction with people in their neighborhood, think their neighbors were friendly, and think that people were willing to help each other out.

In the northern growth corridor area covering *Caboolture, Pine Rivers and Redcliffe Shires*, people were above the average in thinking that their neighborhood was a safe place to walk in after dark and that their neighborhoods comprised friendly people who were willing to help each other out. But, they tended to think that breaking and entering was a problem in their neighborhood and were less likely to think that their neighborhood was conveniently located to stores and shops. Residents of these areas were also less likely to support the notion of more medium-density housing in their neighborhood.

On the Gold Coast in the south of the SEQ region, there were some marked differences between the residents of the coastal strip Gold Coast Inner and the Gold Coast Outer areas further away from the coast. In *Gold Coast Inner*, residents were above average in thinking that: breaking and entering was a problem, there was a strong sense of community in their neighborhood, and their neighbors were friendly people, but they did not have much to do with them. It seems a strong sense of community was not dependent on friendly neighborly interactions. These respondents were less likely to think that living in their neighborhood would be difficult without a car, they tended to think that their neighborhood was conveniently located to stores and parks, they tended to agree that there should be more medium-density housing in their neighborhood (which was not surprising as residents of those areas lived in some of the highest density housing environments in the SEQ region). they considered public transport was adequate for their needs. and they felt safe walking about after dark. Overall, they had above average levels of satisfaction with living in their neighborhood.

In comparison, the residents of *Gold Coast Outer* areas had below average perceptions that breaking and entering was a problem in their neighborhood, and they were more likely to think that public transport was adequate for their needs. However, they were below average in considering that there was a strong sense of community in their neighborhood, they had less than average to do with people in their neighborhood, and they did not want more medium-density housing.

People living on the *Sunshine Coast* in the far north of the SEQ region tended to be above average in thinking that their neighbors were friendly people and willing to help each other out. They were less inclined to think that vandalism and breaking and entering were problems in their neighborhood, but they were also less inclined to think that it was safe to walk outside at night. They were above average in thinking that public transport in their neighborhood was adequate for their needs. And, they were favorably disposed to increasing medium-density housing in their neighborhood.

People living in the diverse *Rural Hinterland* areas of the SEQ region tended to have more extreme positions with respect to their perceptions of neighborhood issues. They were well below average in levels of satisfaction with where they live. They did not think: their neighborhood was well located to walk to stores and parks and that it was safe to walk at night. Vandalism was seen as a problem, but breaking and entering was not. They thought: there was a strong sense of community where

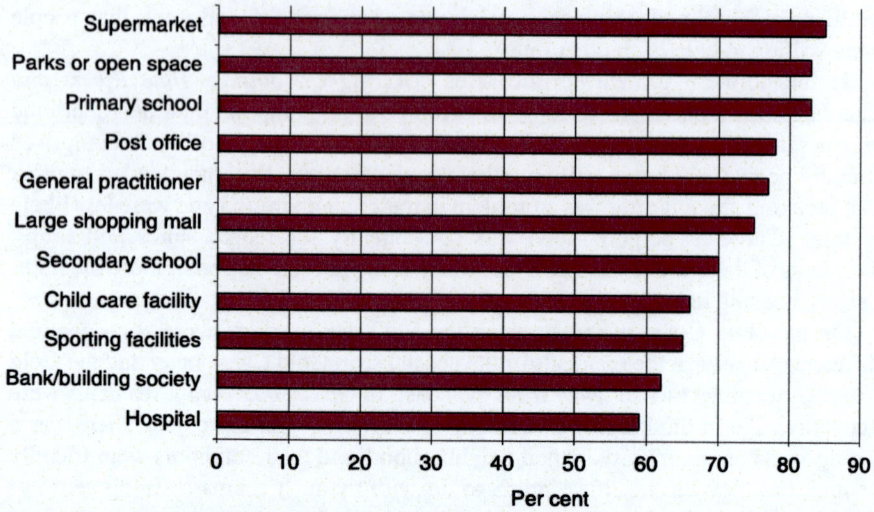


**Fig. 8.4** Level of satisfaction in the SEQ region with access to local services and facilities in 2003: Percent "satisfied" or "very satisfied" (Source: The authors)

they live, their neighbors were friendly people, that they interact with them, and that they help each other out. They were more likely to think that living where they do would be difficult without a car, and they were more inclined to support more medium-density housing in their area. It was surprising that these rural hinterland residents claimed that public transport was satisfactory for their needs, while the reality is that public transport services are either very limited or non-existent; thus, these residents who are highly or totally car-dependent may simply perceive that public transport is not relevant for them. Overall, neighborhood satisfaction was below average even though many social aspects were rated more favorably.

## *Satisfaction with Access to Local Urban Services and Facilities*

The SEQQOL2003 survey asked respondents to rate their level of satisfaction with their *access* to a range of *services and facilities* using a 5-point "satisfaction "/"dissatisfaction" scale. The results are summarized in Fig. 8.4.

*High levels of satisfaction.* In general, across the SEQ region, the survey respondents reported high levels of satisfaction with their access to services and facilities. For example, more than 70% were either "very satisfied" or "satisfied" with their access to: a supermarket (85%); a primary school (84%); parks and open space (83%); local parks (78%); a medical general practitioner (77%); a post office (76%); a large shopping mall (75%) and a secondary school (70%). A similar level of satisfaction was reported by approximately 60% or so of the respondents for their access to: sporting facilities (65%); a child care facility (66%); a public library (62%); a bank/building society (62%) and a hospital (59%).

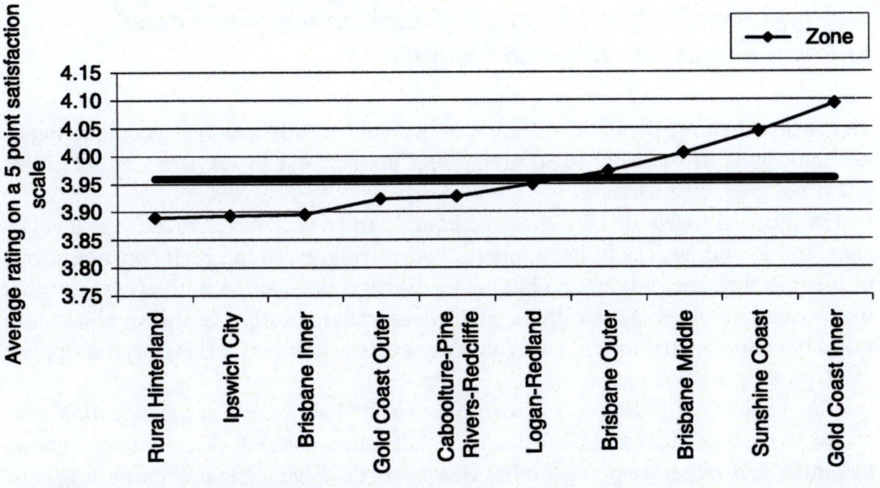


**Fig. 8.5** Variation across the SEQ region in ratings of access to services and facilities (mean scores on a 5-point scale) (Source: The authors)

*Dissatisfaction.* There were, however, some services and facilities with which more than 10% of the survey respondents were either "dissatisfied" or "very dissatisfied" with their access to: a large shopping mall (10%); a public library (11%); a hospital (18%); a bank/building society (20%) and a swimming pool (26%).

*Changes 1997–2003.* With respect to changes in levels of satisfaction, between 1997 and 2003 the results of the two surveys indicated that changes had, on average, *increased*. For example, we find an *increased* level of satisfaction with access to: post offices, supermarkets, large shopping malls, hospitals, primary schools, secondary schools and sporting facilities, while level of satisfaction had *remained about the same* for access to: bank/building society, general practitioners and child care services.

*Spatial variations.* Across the SEQ region, considerable variations exist in the overall levels of satisfaction with access to those services and facilities. This is shown in Fig. 8.5, which plots the composite mean (average) score on a 5-point "satisfaction"/"dissatisfaction" scale for access to the full range of services and facilities.

The *highest levels of satisfaction* were in fact for residents of Gold Coast Inner, the Sunshine Coast, and Brisbane City's Middle and Outer Suburbs, all of which had scores above the average for the region as a whole. In other parts of the region, *below average levels of satisfaction* were recorded, with the lowest level of satisfaction reported by people living in the Rural Hinterland (not surprising given that they lived in the outer geographic margins of the region), Ipswich City, and Brisbane City's Inner Suburbs. These latter two findings were more surprising since both areas were rated relatively well with walking distances to stores and parks. So, satisfaction with access in these two areas is influenced by other factors of accessibility. Overall, there was no systematic relationship between the variation in mean scores and distance from the Brisbane City CBD.

## *Sub-regional Profiles of Levels of Satisfaction with Access to Local Urban Services and Facilities*

To consider in more detail the differences in levels of satisfaction of people living in different parts of the SEQ region regarding their access to services and facilities, again *z*-scores were used.

For those in *Brisbane City's Inner Suburbs*, there was an above average level of satisfaction with access to hospitals (not surprising given the high concentrations of them in the inner-city areas) but below average satisfaction with access to most of the other services and facilities, and in particular sporting facilities, child care, primary schools, post office, bank/building society, a supermarket, large shopping mall, parks and open space.

The residents of *Brisbane's Middle Suburbs* had above average satisfaction with access to most services and facilities, especially supermarkets, primary care doctors, hospitals and large shopping malls. However, they had below average levels of satisfaction with access to parks and open space and child care facilities.

Survey respondents living in *Brisbane's Outer Suburbs* varied in levels of satisfaction with access to services and facilities. In particular, they were above average with respect to satisfaction with access to supermarkets, parks and open space, a bank/building society, and a large shopping mall, but had below average scores with access to child care, doctors and hospitals.

Respondents living in *Logan-Redland* in the southern growth corridor had well above average levels of satisfaction with access to secondary schools, while they had below average levels of satisfaction with access to doctors and child care facilities.

To the west in *Ipswich City*, the survey respondents tended to have below average scores for most issues. This was particularly the case with respect to access to a post office, secondary schools, sporting facilities, parks and open space, supermarkets, and doctors. The only service and facility for which they had a markedly above average score was for access to child care facilities.

In *Caboolture-Pine Rivers-Redcliffe* (the northern growth corridor), survey respondents reported above average levels of satisfaction with access to parks and open space, sporting facilities, child care and secondary schools. But, they had markedly lower than average levels of satisfaction with respect to access to a post office, a bank/building society, doctors, and a hospital.

Residents living on the *Gold Coast Inner* coastal strip expressed above average levels of satisfaction with access to all the services and facilities. But for those living in the *Gold Coast Outer* areas, there was below average satisfaction levels with access to primary schools, parks and open space and supermarkets but above average levels with access to a large shopping mall, child care facilities, sporting facilities, a supermarket, and a bank/building society.

On the *Sunshine Coast*, residents tended to have above average levels of satisfaction with access to almost all the services and facilities, most noticeably with respect to an availability of a doctor, sporting facilities, a primary school, parks and open space, and child care facilities.

The survey respondents living in the *Rural Hinterland* areas tended to have marked below average levels of satisfaction with their access to most services and facilities, and this was particularly so with respect to a large shopping mall, a supermarket, child care facilities, a secondary school, sporting facilities, a bank/building society, and a doctor. However, they had well above average levels of satisfaction with access to a primary school and a post office.

## Rating the Local Provision of Public Services and Facilities and the Role of Local Councils

The SEQQOL2003 survey asked respondents to rate on a 5-point "goodness" scale the provision of a number of public services and facilities in their neighborhood. Most of those services and facilities are the responsibility of local councils, which are the third tier of government, although two protective services (the police) and public schools are a state government responsibility. The services and facilities, and the ratings they received are summarized in Fig. 8.6.

More than three quarters of survey respondents across the SEQ region rated as "very good" or "fairly good": the collection of garbage (88%), local parks (75%), water and sewerage services (77%), the cleanliness of streets and public areas (75%) and public schools (75%). Less than half rated as "very good" or "good" the provision of: swimming pools (48%), police protection (46%), bus services (32%) and community centers (45%).

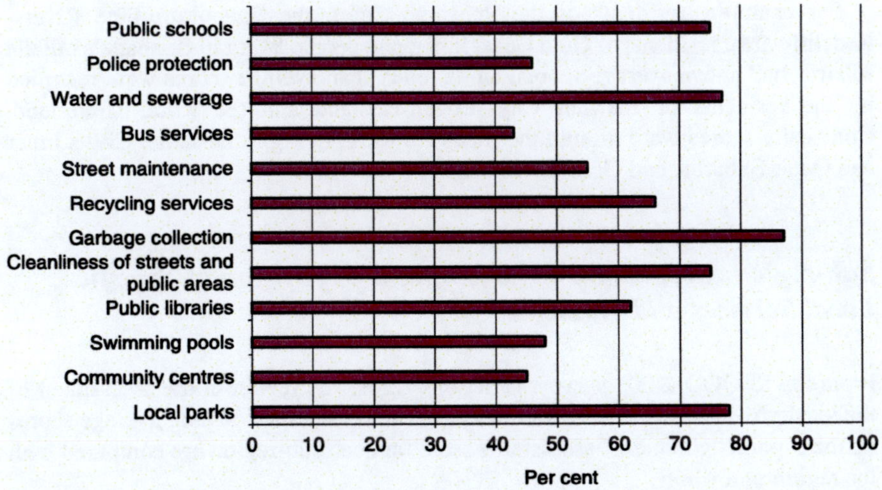


**Fig. 8.6** Assessment of provision of public services at the local area level in 2003: Percent giving a "fairly good" or "very good" rating (Source: The authors)

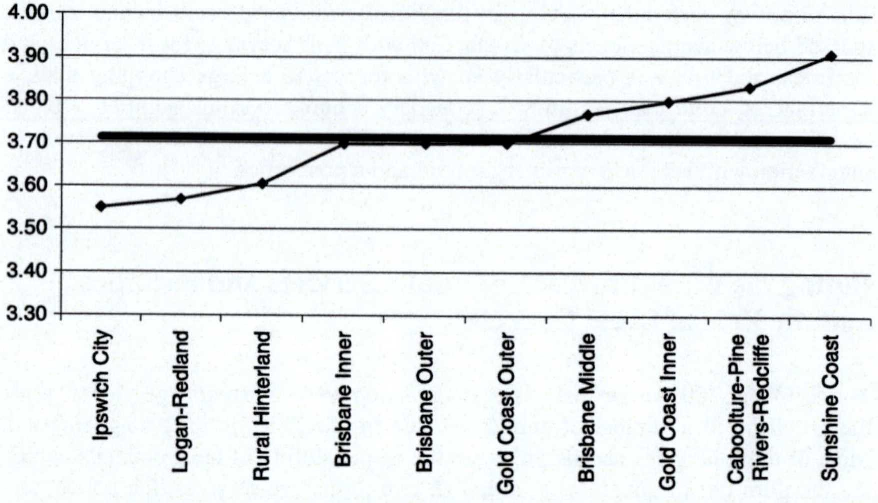


**Fig. 8.7** Variations across the SEQ region in mean scores on a composite 5-point scale ratings of provision of local public services and facilities (Source: The authors)

## *Sub-regional Variations in Overall Assessment of the Provision of Local Services and Facilities*

In Fig. 8.7, the thick horizontal line shows an overall mean of 3.71 on the 5-point "goodness" scale. But as can be seen, there were substantial differences across the region.

For example, residents of the Sunshine Coast, the Caboolture-Pine Rivers-Redcliffe area, residents of Gold Coast Inner, and people living in Brisbane's middle suburbs had above average scores, while lower than average scores were recorded by the residents of Ipswich City, Logan-Redland and the Rural Hinterland. Composite scores were around the mean for those living in Brisbane City's Inner and Outer Suburbs, and Gold Coast Outer.

## *Sub-regional Variations in Assessing the Provision of Specific Local Services and Facilities*

From the SEQQOL2003 survey, variations across the region in the assessment of the local provision of services and facilities can be obtained. Again, average scores on the 5-point "goodness" assessment scale for each sub-region are compared with the region as a whole.

Survey respondents living in *Brisbane City's Inner Suburbs* gave above average ratings on the provision of bus services, water and sewerage services and local parks. But, they gave substantially below average ratings on the provision of public libraries, the cleanliness of streets and public areas, garbage collection, community centers, public schools, police protection and recycling services.

Those living in *Brisbane's Middle Suburbs* rated as above average the provision of bus services, water and sewerage services, police protection and public schools. They gave below average ratings to the provision of community centers and garbage collection.

The residents of *Brisbane's Outer* Suburbs gave well above average ratings to the provision of police protection and also tended to rate a little above average the provision of bus services and recycling services. But, they gave well below average ratings to the provision of public libraries, community centers and swimming pools and were likely to rate less than average the cleanliness of streets and public areas and garbage collection.

Moving to the growth corridors, the survey respondents in *Logan and Redland* in the south and *Ipswich City* in the west consistently rated below average the provision of all of the services and facilities provided by councils and government, with one exception. *Ipswich City* rated the provision of police protection slightly above average. The respondents living in *Caboolture-Pine Rivers-Redcliffe* in the northern corridor rated above average: the provision of recycling services, community centers, public schools, the provision of swimming pools, local parks, garbage collection, street maintenance and the cleanliness of streets and public services.

On the Gold Coast, the survey respondents living on the coastal strip – *Gold Coast Inner* – gave well above average ratings to the provision of local parks and street maintenance. They also rated above average the cleanliness of streets and public areas and the provision of bus services and recycling services. However, they rated well below average the provision of public schools and were also below average in rating police protection. Those living in the *Gold Coast Outer* areas rated street maintenance and the provision of community centers a little above average. However, they gave well below average ratings to the provision of local parks, water and sewerage services, police protection and public schools.

On the *Sunshine Coast*, residents rated highly virtually all the services and facilities, while in the *Rural Hinterland*, well above average ratings were only given to the provision of community centers and public schools, while marginally above average ratings were given to the provision of police protection, public libraries and swimming pools and to the cleanliness of streets and public areas. However, markedly below average ratings were seen for the provision of bus services, water and sewerage services, recycling services, local parks and street maintenance.

## *Assessing the Performance of Local Councils*

Neighborhood problems were discussed with respect to both local government officials and neighbors. The SEQQOL2003 survey revealed that 28% of the respondents across the region had contacted government officials or their council about a local area or neighborhood problem over the past year, and 40% had met with neighbors to discuss such a problem.

When asked to rate their level of satisfaction with the job being done by their local council and councilors, the survey participants responded as follows. In assessing their level of satisfaction with *the job being done by the local council officials in my city or shire*, only 6% said they were "very satisfied" a further 33% stated they were "satisfied," but 26% were "dissatisfied" or "very dissatisfied." Across the region, there were considerable variations.

The most satisfied residents with the job being done by their local council officials were those living in Brisbane City's Inner Suburbs, Brisbane City's Outer Suburbs, Logan and Redland, and the Caboolture-Pine Rivers-Redcliffe areas where nearly half (43%) expressed satisfaction. That level of satisfaction dropped as low as 33% for residents of Ipswich City, and it was also low for those living in the Rural Hinterland where more than 40% were either "dissatisfied" or "very dissatisfied" with the job being done by their local council officials. On the Gold and Sunshine Coasts, this was more than 25%.

With respect to the statement *Local councilors in my community pay attention to what people think*, only 6% of the respondents "strongly agreed," while a further 27% "agreed." In contrast, 27% either "disagreed" or "strongly disagreed" (only 4%). Across the region, there was again considerable variation in these assessments: The incidence of people "strongly agreeing" or "agreeing" with this statement was lowest at 28% for those living in Logan-Redland and on the Sunshine Coast. That level of agreement topped out at only 31% for people living in Brisbane's Middle Suburbs. It was in the Rural Hinterland where the highest proportion of the survey respondents (43%) either "strongly disagreed" or "disagreed" with this proposition, but the lowest proportion expressing that level of negative sentiment was about 34% for the survey respondents living in Gold Coast Outer.

The statement *Local government is doing a good job in managing growth and development* fared little better: only 5% "strongly agreed" with the statement, while a further 30% "agreed". In contrast, 29% disagreed. Again, there was variation in assessment across the region. Only in Caboolture-Pine Rivers-Redcliffe did over 40% of the survey respondents either "agree" or "strongly agree" with this statement. It was a little less for those living in Brisbane's Inner City Suburbs, Logan-Redland, and Brisbane's Outer Suburbs. In the Gold Coast, fewer than 30% of the survey respondents gave that level of rating. In the Rural Hinterland on the Gold Coast, Logan-Redland and Ipswich City, more than 30% of the respondents either "strongly disagreed" or "disagreed" with the statement. In sum, these data suggest that at best, only moderate levels of satisfaction with the performance of the local Council and Councilors were expressed by residents of the region

However, when it comes to levels of *Satisfaction with living in their local council area*, the situation differs. In this case 23% of respondents indicated they were "very satisfied," while a further 44% said they were "satisfied"; only 10% were either "dissatisfied" or "very dissatisfied." But again, there were variations across the region: more that 70% were either "satisfied" or "very satisfied" in Caboolture-Pine Rivers-Redcliffe and across the City of Brisbane from the Inner to the Middle and Outer suburbs. However, that level of satisfaction dropped to 60% or less for those living in Ipswich City, Gold Coast Outer, and in the Rural Hinterland.

## Conclusions

The SEQQOL1997 and SEQQOL2003 surveys revealed that residents have a high *Overall satisfaction with living in the Brisbane-SEQ region*. The survey data provided insights into the subjective assessments people had across a wide range of QOUL issues and at a number of levels of scale, and it has been possible to identify the degree those assessments were stable or had changed over the 6 years between the two surveys. In addition, it has been possible to identify variations in the satisfaction ratings of people living in different parts of the SEQ region to a range of QOUL domains and issues that related to the provision of services and facilities.

The type of QOUL data generated by the two surveys may be of use in public policy, especially with respect to identifying for government agencies levels of satisfaction with specific region-wide and local services and facilities that people use, particularly in helping to identify where deficiencies might exist and where improvements might be made.

**Acknowledgments** The authors are indebted to the people living in the SEQ region who participated in both the 1997 and the 2003 SEQ surveys for providing the information reported in an aggregated form in this paper. We also thank the teams of interviewers who conducted the interviews for those surveys using Social Research Centre's CATI system at The University of Queensland. The surveys were funded by grants from the Australian Research Council Discovery Projects #C595301132 and #DP0209146.

## References

Campbell, A., Converse, P. E., & Rogers, W. L. (1976). *The quality of American life: Perceptions, evaluations and satisfactions*. New York: Russell Sage.

Cummins, R.A. (1996). The domain of life satisfaction: An attempt to order chaos. *Social Indicators Research, 38*, 303–332.

Guhathakurta, S., & Stimson, R. J. (2007). What is driving the growth of new "sunbelt" metropolises?: Quality of life and urban regimes in greater Phoenix and Brisbane-South East Queensland region, Australia. *International Planning Studies, 12*(2), 129–152.

Marans, R. W. (2003). Understanding environmental quality through quality of life studies: The 2001 DAS and its use of subjective and objective indicators. *Landscape and Urban Planning, 65*(1), 73–83.

Marans, R. W., & Rodgers, W. (1975). Toward an understanding of community satisfaction. In A. Hawley & V. Rock (Eds.), *Metropolitan America in contemporary perspective* (pp. 299–352). New York: Halsted.

McCrea, R., Stimson, R., & Western, J. (2005). Testing a moderated model of satisfaction with urban living using data for Brisbane-South East Queensland, Australia. *Social Indicators Research, 72*(2005), 121–151.

Mullins, P. (1991). Tourism urbanisation. *International Journal of Urban and Regional Change, 15*, 326–342.

Stimson, R. J., & Taylor, S. P. (1999). City profile: Brisbane. *Cities, 16*(4), 285–295.

# Chapter 9
# Measuring Quality of Urban Life in Istanbul

**Handan Türkoğlu, Fulin Bölen, Perver Korça Baran, and Fatih Terzi**

## Introduction

In 2005, together with a number of experts and academics from the Istanbul Technical University and other institutions, the authors were invited by the Istanbul Metropolitan Planning and Design Center (IMP) to assist in preparing the 2020 Strategic Plan of Istanbul. IMP was established by Istanbul's mayor under BIMTAS (a public–private partnership company attached to the municipality) as a multidisciplinary research, planning, and design center. Nearly 400 planners, architects, and academics were organized into teams responsible for addressing different components of the plan including demographic projections, the natural environment, housing, transportation, commerce, industry culture, and quality of life (QOL) issues.

The authors were asked to carry out surveys dealing with the physical conditions of housing and residential areas and the overall QOL in the Istanbul Metropolitan Area. Survey results would be used in preparing the Strategic Plan. Two working teams were formed with the first author leading the "Quality of Urban Life" (QOUL) study team and the second author leading the "Physical Analysis of Residential Areas" (PARA) study team. The teams were to report findings dealing with basic housing problems and critical residential neighborhoods and offer alternative strategies to solve these problems.

---

H. Türkoğlu (✉) • F. Bölen • F. Terzi
Department of Urban and Regional Planning, Faculty of Architecture,
Istanbul Technical University, 34347 Istanbul, Turkey
e-mail: turkoglu@itu.edu.tr; bolen@itu.edu.tr; terzifati@itu.edu.tr

P.K. Baran
North Carolina State University, Campus Box 7106, Jordan Hall,
Raleigh, NC 27695, USA
e-mail: perver_baran@ncsu.edu

R.W. Marans and R. Stimson (eds.), *Investigating Quality of Urban Life: Theory, Methods, and Empirical Research*, Social Indicators Research Series 45,
DOI 10.1007/978-94-007-1742-8_9, 

The QOUL study followed an approach similar to that used in the Detroit Area Study (see Chap. 7). The aim of the study was to produce information about the households and housing environments, including residents' perceptions of their neighborhoods and the city.

For the Istanbul study, a single database was created in order to integrate the work of the QOUL study team and the PARA study team (that is, housing densities, building types, structural quality, quality of pavements, accessibility to various facilities, etc.) so as to explore relationships between actual conditions and people's perceptions.

This chapter summarizes findings from the QOUL study covering the Istanbul Metropolitan Area. It begins with a background of the region and then outlines the methodology used in the research. Next, three sets of findings are presented. First, measures of overall QOL domains including three place domains – the house, the neighborhood, and the community – are summarized. That is followed by residents' subjective assessments of selected aspects of urban living, including transportation, recreational facilities, such as parks, schools, and security. Finally, findings used in developing housing strategies within the Strategic Plan are presented. The chapter concludes with a brief discussion of implications of the research for the future of Istanbul Metropolitan Area.

## The Study Area

Istanbul is Turkey's largest city and the dominant center of manufacturing, services, education, and culture. The Istanbul Metropolitan Area, including the city of Istanbul, spreads across both the Anatolian and European sides of the Bosphorus to the coasts of the Sea of Marmara and the Black Sea and covers about 5,200 km$^2$ (Fig. 9.1). The metropolitan city represents less than 1% of the land area of Turkey but accommodates nearly 15% of the country's population.

Since 1950, Istanbul has attracted millions of migrants from the other regions of Turkey. The city's population increased from about 1 million in 1950 to over 12 million by 2006 (Fig. 9.2). Since the 1990s, about a half million people migrated to Istanbul annually, while there was an out-migration of nearly 200,000. This population growth has created a demand for new housing in the metropolitan area.

The metropolitan area is characterized by a linear pattern of urban development in an east–west direction paralleling the Marmara coast, the railroad, and the E5 highway, which joins the European with Anatolian parts of Turkey. Industrial development, transportation facilities, and the shoreline have attracted urban development along this axis. The extent of Istanbul and the distribution of different forms of housing areas include planned and squatter neighborhoods.

Most of the population is concentrated in the central area of Istanbul. The central area contains both a historic and a modern section. The historic core of Istanbul is made up of three ancient settlements separated by the two strips of water, the

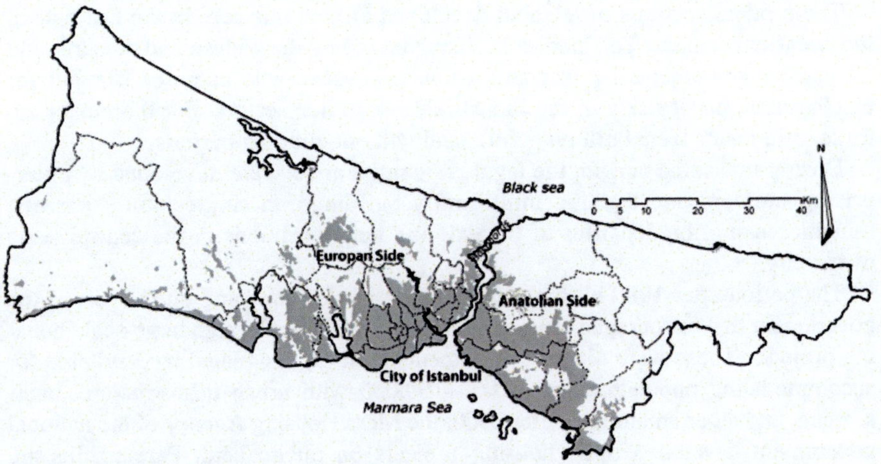


Fig. 9.1 Istanbul metropolitan area (Source: The authors)

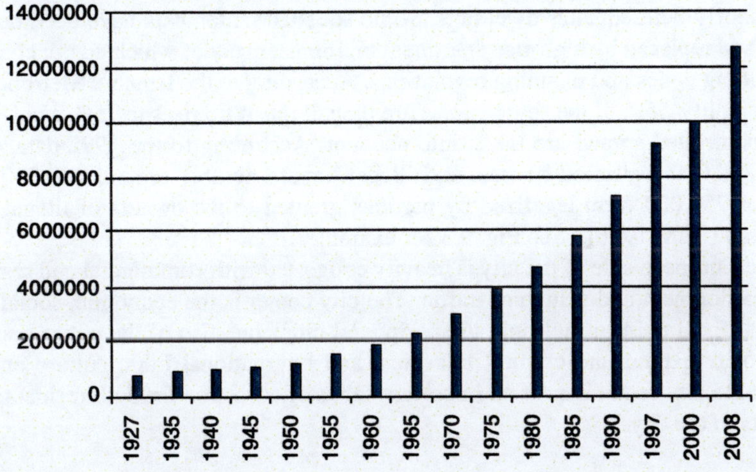


Fig. 9.2 Annual population change in Istanbul, 1927–2008 (Source: Statistical Institution of Turkey 2008)

Bosphorus and the Golden Horn. These are the "historic peninsula": the old city of Constantinople, Galata to the north of the Golden Horn, and Kadikoy to the east of the Bosphorus. These sections of the city contain old residential neighborhoods, some with timber houses and mansions, and some with European style multifamily apartment buildings from the nineteenth century. Many of the historical neighborhoods are designated as conservation areas and are targeted for restoration and rehabilitation. The inhabitants are generally older low-income residents or newly arrived migrants, many of them homeless.

The modern sections of Istanbul developed around this core in the last half of the twentieth century. This period is characterized by the widespread construction of squatter settlements by migrants from rural areas who came to Istanbul for employment, particularly in the industrial and service sectors. For the most part, these settlements were built on public land without building permits.

During this same period, the legal residential areas were developed by enterprising builders constructing multifamily buildings on single family parcels, thus increasing the densities of the existing neighborhoods in the central parts of the city.

The period after 1980 is characterized by a market economy on the part of the government that encouraged mass production of housing through large-scale housing projects. These were planned settlements built on designated land suitable for accommodating populations of 60,000–100,000 with urban infrastructure, landscaping, and other amenities. After 2000, the Mass Housing Agency of the national government developed mass housing projects on public land. Partnerships are formed between the Agency and private construction firms for new housing developments on public land.

Today, a large part of the housing stock in metropolitan Istanbul consists of low-rise, poorly built squatter dwellings. Some squatter settlements have been demolished and replaced by high-rise apartment buildings, many of which do not conform to building codes and planning regulations. According to the Istanbul Metropolitan Municipality, 55% of the settled areas are illegal; this ratio rises to 75% when areas of uncontrolled sprawl are taken into account. According to the 1992 data, there were 850,000 registered houses, and illegal houses number around 400,000, and another 750,000 were legalized by pardons granted to the owners of illegal construction (IMM 1996). (See Fig. 9.3 for examples.)

The European side of the city is heavily congested with commercial and residential development and vehicular traffic. The city center is the economic, social, and cultural focal point of not only the peripheral cities but also of the entire country. Due to its natural and cultural treasures, and international fairs, culture and art events the city center is now an important focus of attraction for both residents and visitors to the city.

## Methodology

The purpose of the research was to focus on two aspects of QOUL:

(a) The first involved an *objective assessment* of the *physical environment in residential areas* (Bolen et al. 2006). Information was collected via a physical survey of neighborhoods across Istanbul. The information was based on a number of objective measures of neighborhood attributes and was used to illustrate problems with the physical environment.

**Fig. 9.3** Examples of different urban characteristics in Istanbul (Source: The authors)

(b) The second involved the *subjective assessment* of the *quality of community life* with respect to social and economic domains and the *satisfaction of the residents* (Turkoglu et al. 2006). Information was collected through a social survey. The survey focused on levels of satisfaction with aspects of urban living, residents' perceptions, and their behaviors and experiences in their living environment.

Data covered neighborhoods defined as small administrative units (known as "mahalles"). There are a total of 941 mahalles in Istanbul, and they display different characteristics based on their residential densities and land values. Therefore, it was necessary to group them with respect to their net residential densities and average land values. We assumed that density indicated the character of the physical environment, and land values indicated the attractiveness of the neighborhood for housing. Thus, the existing neighborhoods were grouped into three categories, first, based on their net densities, and second, based on their land values. These groups were:

- Low density, medium density, and high density
- Low-land value, medium-land value, and high-land value neighborhoods

A comparable strategy has been employed in similar studies (Frank et al. 2005). When combined, we had nine different neighborhood categories with different densities and land values (see Fig. 9.4). These were:

1. Low-density, low-land value neighborhoods
2. Low-density, medium-land value neighborhoods
3. Low-density, high-land value neighborhoods
4. Medium-density, low-land value neighborhoods
5. Medium-density, medium-land value neighborhoods
6. Medium-density medium, high-land value neighborhoods
7. High-density, low-land value neighborhoods
8. High-density, medium-land value neighborhoods
9. High-density, high-land value neighborhoods

A database was created for the nine categories and combined with an ArcGIS database containing all residential buildings and the number of dwelling units in each building. This combined database represented the sampling frame for the survey. From this frame, a random sample of dwelling units was selected proportional to the number of units in the Anatolian and European sides of Istanbul. That allowed us to draw conclusions for either side separately as well as for Istanbul as a whole.

Within each of the nine neighborhood categories, 100 residential buildings were randomly selected by colleagues at North Carolina State University's Centre for Earth Observation. In the first part of the study involving the objective assessments of the selected residential areas, 900 buildings were used.

Subsequently, a sample of the 423 buildings from the 900 buildings were selected, and within each category, six dwelling units were randomly selected for the social survey. Of the 2,538 dwelling units, 1,635 face-to-face interviews were conducted representing a response rate of 66%. Figure 9.5 illustrates the spatial distribution of the clusters for the QOUL sample survey.

## *Questionnaire for the QOUL Survey*

Assessments of QOUL domains were collected through a questionnaire. Experienced interviewers were employed and given detailed sampling instructions in order to ensure their understanding and execution of procedures to follow prior to visiting selected households. Additionally, the questionnaire was pilot tested and modified prior to its administration to the sample of households. The interviewers were given the interview pack, which formed the basis of the research procedure. The pack consisted of a study guide (training material explaining the research, information on completing the cover sheet and questionnaire, persuasion techniques – how to engage with the respondent in conducting the interview, the cover sheet, including guidelines for listing household members and selecting an

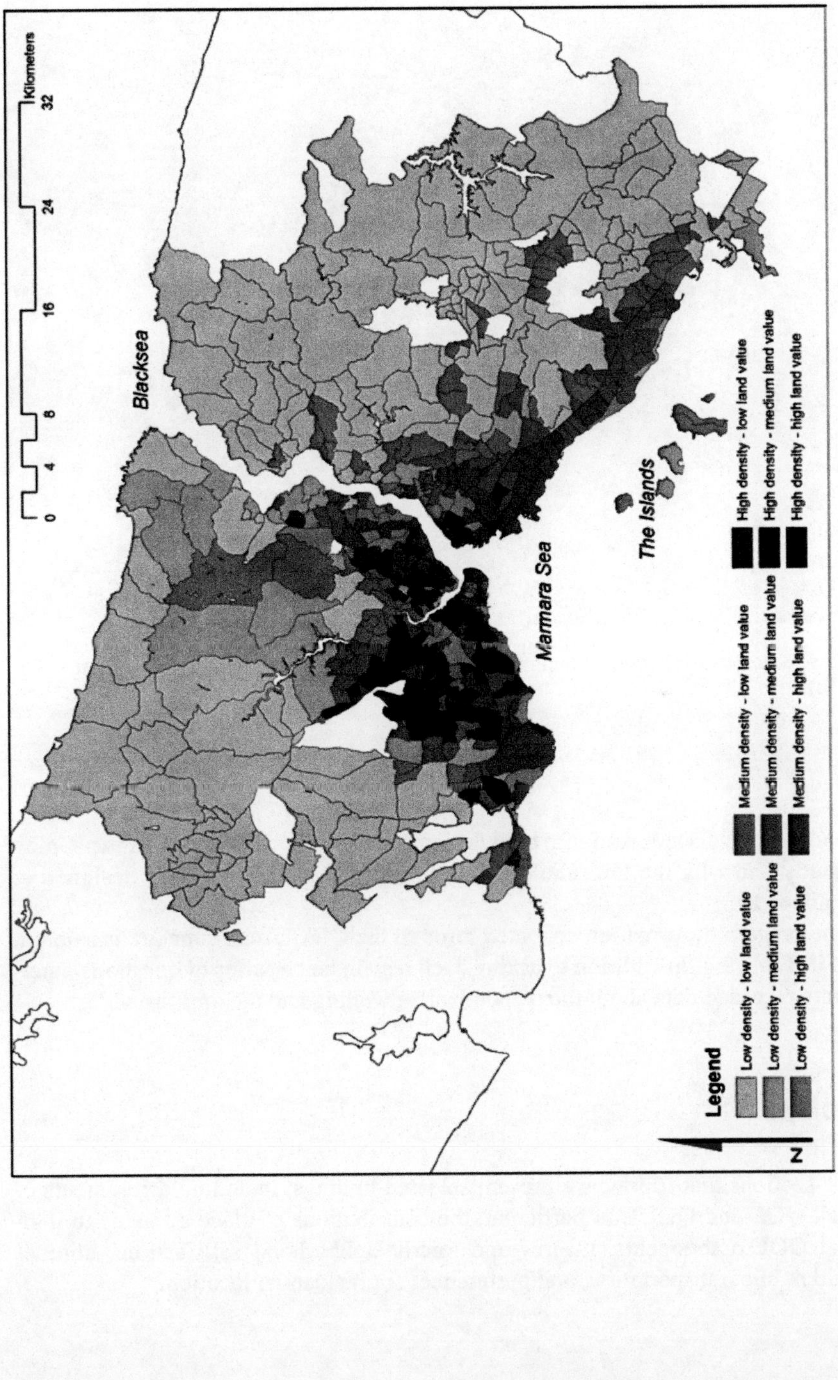


**Fig. 9.4** Density and land value categories according to mahalle groups (Source: The authors)

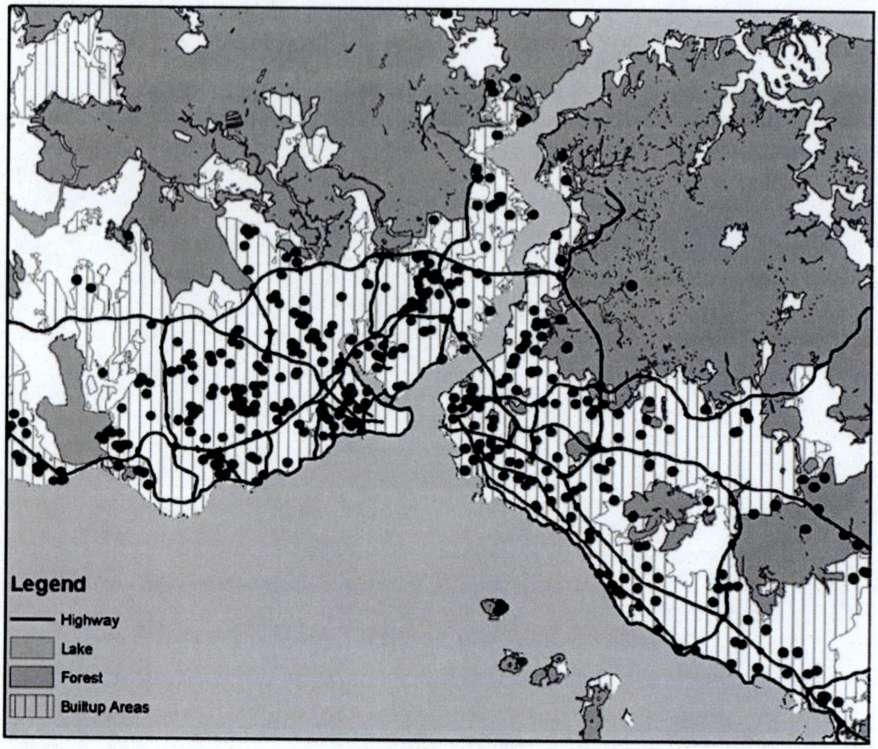


**Fig. 9.5** Sample distribution of building cluster locations for the social survey (Source: The authors)

individual to interview, response booklets, questionnaires, and brochures explaining the study and why the household was selected). The interview was designed to take about 1 h.

The range of information collected through the QOUL questionnaire is summarized in Table 9.1. In addition to asking each respondent a series of questions, interviewers recorded data about the respondent's dwelling and the area around it.

## Findings

In the sections that follow, we present selected findings, including assessments of overall QOL and QOUL in particular, the contributions of place domains to individual QOL assessments, macro- and microneighborhood satisfaction, attitudes toward public transportation, and preferences for residential location.

**Table 9.1** Information collected in QOUL survey

| Quality of life survey characteristics | |
|---|---|
| Heading | Indicator |
| Residential history | How long lived at property |
| | Lived where prior to moving to property |
| | Reasons why living in property |
| Public services and transportation | Cleanliness of streets and public areas |
| | Maintenance of public areas |
| | Quality of public transport |
| | Usage of public transport |
| Taxes | Overall satisfaction in tax payment |
| | Involvement in decisions relating to tax |
| | What they would be prepared to pay more for |
| Schools | Type of school attended |
| | Choice of schools |
| | Transport to school |
| Parks and recreation | How often park are visited |
| | Usage of park space |
| | Importance of access to parks |
| | Types of areas children play |
| | Overall satisfaction with facilities |
| Shopping and entertainment | Main location for shops |
| | Transport used |
| | Satisfaction of shops |
| | Usage of spare time |
| | Satisfaction of neighborhood |
| Community participation and involvement | Attendance to community meetings |
| | Attendance to clubs |
| | Resolving neighborhood disputes |
| Neighborhood and neighboring | Definition of neighbor |
| | Identification of neighborhood problems |
| | Measuring friends and family ties |
| | Measuring cohesion/social capital |
| Housing and residential mobility | Quality of physical living space |
| | Quality of residential building |
| | Status and cost of tenure |
| | Use of apartment (for work) |
| | Overall satisfaction of living space |
| | Likelihood of moving |
| | Assessment of 3x models of neighborhood |
| Safety | Perception of crime |
| | Safety of neighborhood |
| Employment and journey to work | Work status, occupation, and sector |
| | form of transport used |
| | Length of travel in time and distance |
| | Satisfaction in work |
| | Marital status |

(continued)

**Table 9.1** (continued)

| Quality of life survey characteristics | |
|---|---|
| Heading | Indicator |
| Environment | Hazardous waste disposal |
| | Noise/air pollution |
| | Trash disposal |
| | Loss of natural spaces |
| Health and health care facilities | Identification of health related issues |
| | Quality of health care facilities |
| | Walking activities for health |
| | Chosen area to walk |
| | Disability in the family |
| Other domain satisfactions | Social networks |
| | Standard of living |
| Regional issues | Places visited in Istanbul |
| | Identification on how to improve quality of life |
| | Future thoughts on future of neighborhood |
| Demographics | Educational attainment |
| | Household income |
| | Material possessions |

Source: The authors

## *Quality of Life*

There is general agreement that QOL has both a subjective dimension as well as an objective reality. Campbell et al. (1976) measured people's perceptions, evaluations, and satisfactions in their seminal study. Satisfaction was considered a more plausible and realistic objective for policy makers than that of creating happiness, and the researchers were interested in generating data that could potentially influence public policy. Campbell, et al. measured and compared people's assessments of several domains of their lives as well as "life as a whole," and determined the degree to which each domain such as health, family, community, housing, and leisure explained the quality of life experience (Marans 2003).

Accordingly, the Istanbul Metropolitan Area survey asked respondents to assess several domains using a 7-point satisfaction scale ranging from completely satisfied (7) to completely dissatisfied (1). The domains considered were *family life, health, job, friends, standard of living, leisure activities, and* satisfaction with *life as a whole*. A 7-point satisfaction scale had been used by Campbell et al. (1976) in their seminal work and was also used in the 2001 Detroit Area Study (see Chap. 7).

Table 9.2 shows the mean satisfaction scores for the entire sample and for the city of Istanbul. Overall, Istanbul area residents tended to be fairly satisfied with the various domains. Satisfaction score for *friends, family, health, and job* were somewhat higher than satisfaction scores for *life as a whole*. People were least satisfied

**Table 9.2** Overall QOL: Mean scores on a 7-point scale

| Domain | Istanbul | European side | Anatolian side |
|---|---|---|---|
| Life as a whole | 4.59 | 4.59 | 4.59 |
| Friends | 5.48 | 5.47 | 5.51 |
| Standard of living | 4.34 | 4.29 | 4.43 |
| Family life | 5.8 | 5.84 | 5.75 |
| Health | 5.02 | 5.08 | 4.92 |
| Leisure | 4.34 | 4.29 | 4.44 |
| Job/school | 5.36 | 5.42 | 5.26 |
| Number of respondents | 1,635 | 957 | 678 |

Source: The authors

with *use of leisure time and standard of living*. Differences between Istanbul residents living on the European side of the Bosporus and the Anatolian side were negligible.

## *Quality of Urban Life*

Conceptually, urban life means the residential environment where people live. Following the lead of researchers at the University of Michigan (Marans and Rodgers 1975), the residential environment in the Istanbul metropolitan area was considered at three levels:

- The individual's home or dwelling
- The microneighborhood within which the dwelling was located
- The macroneighborhood.

The microneighborhood is defined as the respondent's immediate surroundings that are visible from their immediate home. That is, the immediate street and buildings within it. The macroneighborhood is defined as the larger, overall neighborhood. By using a combination of macro- and microdata, views about overall neighborhood quality can be assessed.

Consideration was given to attributes of place domains related to the individual's home and macro- and microneighborhood. For instance, questions about environmental attributes (traffic, noise, upkeep), social attributes (family, friends, safety), and public services (police protection, schools, parks, transportation) were asked about the respondents' microneighborhoods. At the macroneighborhood level, consideration was given to shopping, transportation, and recreation, while people's assessments of housing costs, space, and size of dwelling were considered for the individual home.

At each level of domain, the respondents were asked to evaluate several attributes and then give a summary satisfaction score to a single question as in the

**Table 9.3** Assessment of dwelling neighborhood and community QOUL: Mean satisfaction scores on a 7 point scale

| Place domain | Istanbul | European side | Anatolian side |
|---|---|---|---|
| House/dwelling | 5.25 | 5.22 | 5.29 |
| Microneighborhood | 4.83 | 4.74 | 5.00 |
| Macroneighborhood | 4.81 | 4.72 | 4.98 |
| Number of respondents | 1,635 | 957 | 678 |

Source: The authors

Detroit Area Study (see Chap. 7). Using the same 7-point scale response categories, they were asked to express their overall satisfaction with their dwelling, their microneighborhood, and their macroneighborhood.

Table 9.3 presents overall satisfaction scores for housing, microneighborhood, and macroneighborhood for the entire sample, for Istanbul residents. Levels of satisfaction are similar in line with other domain satisfaction scores. That is, a person's feelings about his/her dwelling are associated with feelings about the neighborhood in which the person lives. When we compare European and Anatolian sides, it seems the residents who live in the Anatolian side found slightly more satisfaction with their home and neighborhood.

## *Modeling How Place Domains Contribute to Individual Subjective QOL*

Another key question addressed by the research was the extent to which the place domains contribute to one's QOL. A related question was whether these contributions applied to all survey respondents. These questions were explored using two approaches. One approach involved structural equation modeling and considered QOL as a latent variable with housing, microneighborhood, and macroneighborhood satisfaction scores as predictors. In a second approach, regression analyses considered *life satisfaction* as a summary measure of quality of life with the place domains as predictors. This approach followed the process used in analyzing data from the Detroit Area Study (see Chap. 7).

The contributions of the three place domains to one's overall QOL were examined for the metropolitan area, and the question of whether these contributions were the same for both sides of Istanbul was explored using regression analysis.

The first analysis (stepwise regression analysis) considered QOL as a dependent variable with housing, neighborhood, and community satisfaction scores as predictors. The results of the model indicate that all place domains have a low, but significant, impact on life satisfaction. On the European side, the predictors with an $R^2$ of 0.096 have more impact on life satisfaction than on the Anatolian side and Istanbul. In general, home is the strongest predictor of quality of life, while community satisfaction was the least important predictor (see Table 9.4). Nevertheless, the low $R^2$

**Table 9.4** Place domains as predictors of life satisfaction in Istanbul and in the European and Anatolian sides: regression analysis

| | Istanbul | | | European side | | | Anatolian side | | |
|---|---|---|---|---|---|---|---|---|---|
| Place domains | *B* | SE *B* | *β* | *B* | SE *B* | *β* | *B* | SE *B* | *β* |
| Home | .186 | .000 | .170*** | .213 | .001 | .199*** | .118 | .001 | .101*** |
| Neighborhood | .095 | .001 | .092*** | .098 | .001 | .095*** | .099 | .001 | .095*** |
| Community | .062 | .001 | .061*** | .085 | .001 | .085*** | .009 | .001 | .008*** |
| $R^2$ | .074 | | | .096 | | | .032 | | |

Source: The authors
*B* multiple regression coefficient, *SE B* Standard error of multiple regression coefficient, *β* (Beta) Standardized multiple regression coefficient
****p*≤.001

**Table 9.5** The importance of place variables in predicting QOL: Hierarchical regression analysis

| | Istanbul | | | European side | | | Anatolian side | | |
|---|---|---|---|---|---|---|---|---|---|
| Domains | *B* | SE *B* | *β* | *B* | SE *B* | *β* | *B* | SE *B* | *β* |
| Friends | .174 | .001 | .145*** | .187 | .001 | .151*** | .155 | .001 | .136*** |
| Standard of living | .131 | .001 | .139*** | .156 | .001 | .162*** | .104 | .001 | .114*** |
| Family life | .093 | .001 | .072*** | .044 | .001 | .033*** | .176 | .001 | .144*** |
| Health | .066 | .001 | .058*** | .076 | .001 | .064*** | .069 | .001 | .067*** |
| Leisure | .265 | .001 | .294*** | .250 | .001 | .275*** | .274 | .001 | .309*** |
| Job/school | .077 | .001 | .079*** | .097 | .001 | .100*** | .027 | .001 | .027*** |
| Home | .046 | .001 | .044*** | .042 | .001 | .041*** | .056 | .001 | .049*** |
| Neighborhood | .040 | .001 | .040*** | .029 | .001 | .029*** | .052 | .001 | .051*** |
| Community | .032 | .001 | .032*** | .089 | .001 | .091*** | −.111 | .001 | −.102*** |
| $R^2$ | 0.317 | | | 0.336 | | | 0.304 | | |

Source: The authors
**p*≤.05; ***p*≤.01; ****p*≤.001

values suggest that the place domains of home, neighborhood, and community make only modest contributions to life satisfaction, particularly, on the Anatolian side of the Bosphorus.

The second analysis revealed that the nonplace domains taken together, along with the three place domains, account for about a third of the variance in life satisfaction for Istanbul and its parts (Table 9.5). These changes clearly indicate that nonplace variables are most important for life satisfaction in all parts of the Istanbul area.

Leisure and friends are the most important nonplace variables for Istanbul and its parts. The third most important variable is standard of living for the Istanbul area and for residents of the European side, while for those living on the Anatolian side, family life is next in importance.

## *Neighborhood and Neighboring*

The Istanbul survey data allowed for further exploration of place domains. For example, several questions were asked about the macroneighborhood, including people's assessments of neighborhood services (such as public transportation), their interactions with nearby family and friends, involvement in community organizations, and their feelings about neighborhood conditions, such as upkeep, crowding, and safety. Besides providing information about specific neighborhood attributes that local planners might find helpful, the intent was to determine the relative importance of these attributes in predicting neighborhood satisfaction.

In order to measure overall neighborhood quality, a combination of questions such as how satisfied with your neighborhood as "an attractive place," "a nice place to live," and "a nice place for raising children" were used, and the respondents were asked to evaluate them on a scale of 1–7, where 1 is "completely dissatisfied" and 7 is "completely satisfied." The data were then merged to create a Macroneighborhood Quality Index. The spatial distribution of index scores is shown in Fig. 9.6 and is subsequently used in modeling determinants of macroneighborhood satisfaction. The average value was 4.3 out of 7, where 7 represents the most satisfactory situation.

In assessing microneighborhood attributes as defined by the respondent's immediate surroundings that were visible from their home, problems related to the physical environment such as noise, sense of overcrowding, and heavy traffic were considered. Respondents were read a list of problems that exist in some neighborhoods in Istanbul and asked to indicate whether they thought they were "a big problem," "somewhat of a problem," or "not a problem at all." In terms of the most pressing physical neighborhood problems, concerns over car vandalism and theft were most problematic with 73% of the respondents indicating it was a big problem.

The second most pressing physical neighborhood problem was that of poorly kept outside areas followed by bad smells from garbage. In terms of noise in their neighborhood, about half the respondents felt it was a big problem. When respondents were asked about day care in their neighborhood, about two-thirds felt that there was not enough day care for children. Responses were combined into Microneighborhood Stress Index, the spatial distribution of which is shown in Fig. 9.7.

Social cohesiveness was measured by determining respondent's perceptions of their neighbors as similar or dissimilar to respondents, the friendliness of their neighbors, and their feelings about the sense of community. Istanbul residents were more likely to describe their neighbors as being similar than being dissimilar (30.4% vs. 37.6%) and as friendly rather than unfriendly (47.4% vs. 13.9%). They were also likely to report a strong sense of community rather than indicating no sense of community (55% vs. 22%).

Finally, social networks within the neighborhood were measured by asking about the number of nearby friends and families and attributes of neighboring. Responses

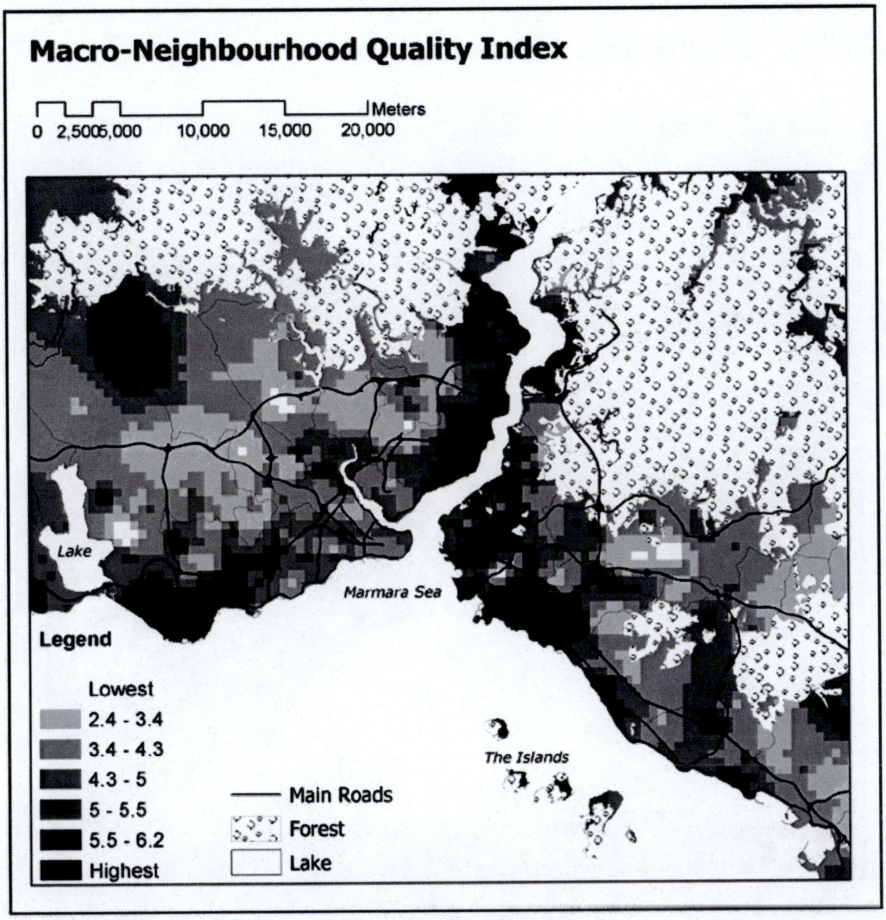


**Fig. 9.6** Spatial pattern for the macroneighborhood quality index (Source: The authors)

were combined into a Neighboring Index, the spatial distribution of which is shown in (Fig. 9.8) Attributes of neighboring included the number of neighbors known by name, the frequency of visiting, and the frequency of exchanging favors with neighbors. Residents tended to have more family living nearby. The mean value of the Neighboring Index was 2.9 on a 5-point scale.

Two important indicators of neighborhood cohesiveness that influence how people feel about their immediate neighborhood are "connections with neighbors" and "similarity with one another." "Similarity and friendliness" are useful building blocks for future neighborhood development. The extent to which a community "engages" in neighborhood affairs is not only based on peoples willingness to participate but is also regulated by the existing channels of engagement offered. For example, if channels of engagement are too formal, they can be foreboding

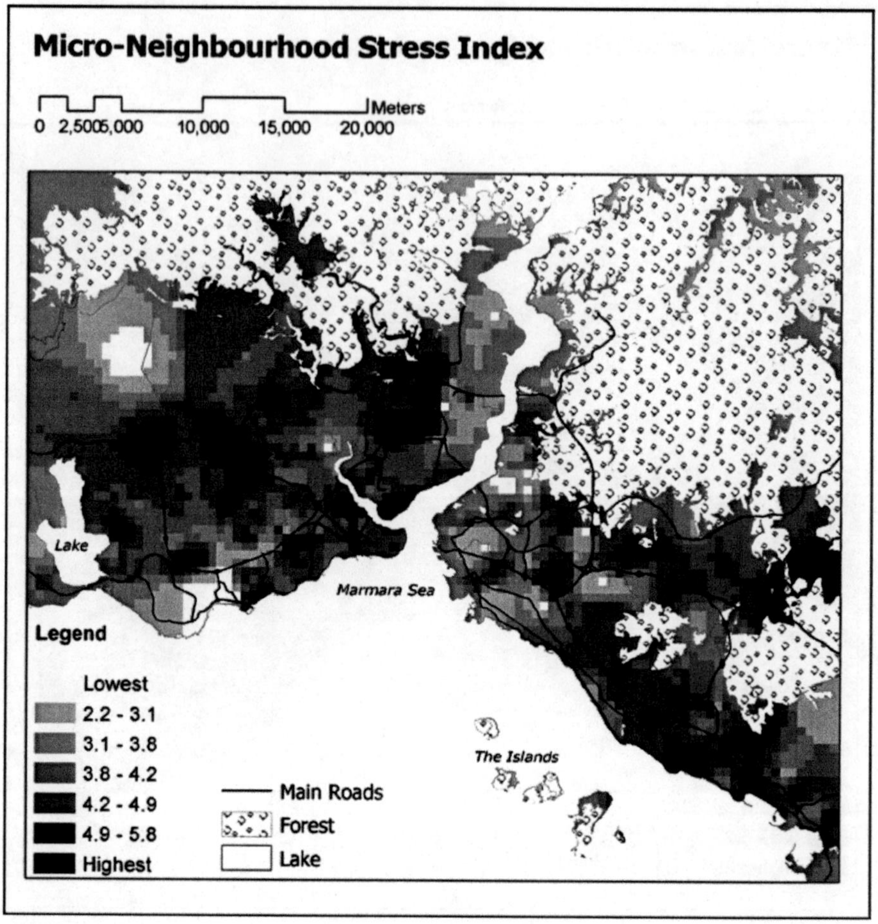


**Fig. 9.7** Spatial pattern for the microneighborhood stress index (Source: The authors)

environments where people avoid going for fear of not belonging and not understanding the "language" and etiquette taking place within a formal meeting. In addition, if "previous experiences in neighborhood participation" have been negative, manifestations of mistrust in the existing channels of engagement may be in existence reinforcing neighborhood disengagement. It could also be possible that people are happy with their environment and do not feel the need to become involved in "neighborhood politics." However, most neighborhoods come with their share of neighborhood problems.

Respondents were presented with a number of forms of engagement for addressing community and neighborhood problems and asked, "During the last year, have you or any of your family taken part in any of the following activities?" Overall, there was relatively little formal engagement among Istanbul area residents.

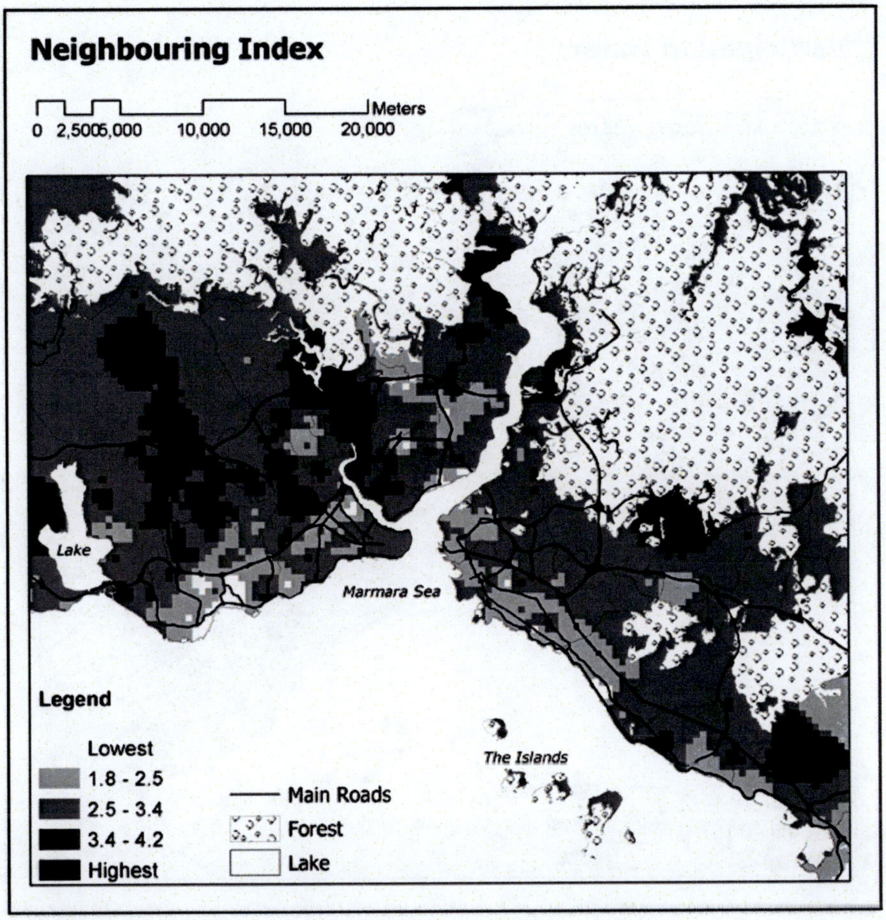


**Fig. 9.8** Spatial pattern for the neighboring index (Source: The authors)

Whereas the most common activity was that of attending a meeting of a neighborhood association or block club (29% responded affirmatively); attendance at a meeting of a local government board or commission was very low at 6%. Attendance at a meeting of the Muhtarlik (Local Administration Office) was even lower at only 4%. The most common indicators were then merged to create a Participation Index in order to measure the overall levels of community involvement throughout the region. The levels of participation are shown in Fig. 9.9. Respondents who answered all questions positively were designated as activists. Conversely, those responding negatively to all four questions were categorized as disengaged. The higher the index score, the greater is the level of engagement. Since overall, 62% of respondents were disengaged, the engagement score was very low for the entire sample.

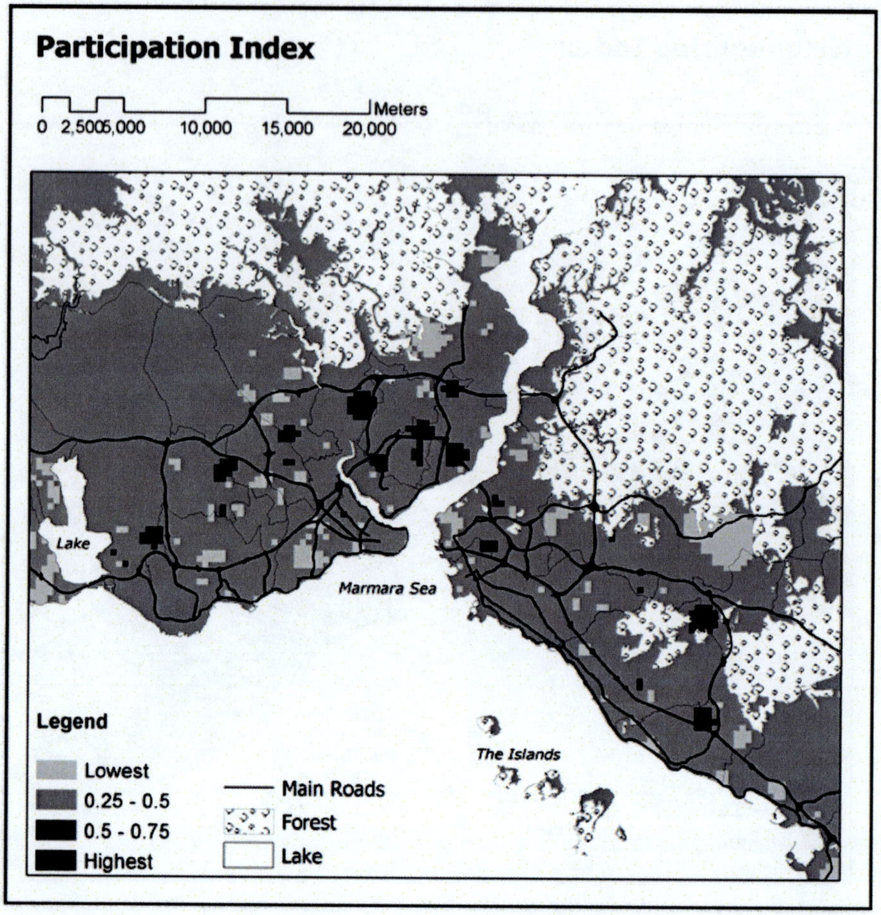


**Fig. 9.9** Spatial pattern for the participation index (Source: The authors)

## *Transportation*

In addition to physical stressors (noisy and crowded conditions, excessive traffic, and poor maintenance) as indicators of neighborhood dissatisfaction, negative ratings of services also contributed to dissatisfaction. One of the key services examined was public transportation. In Istanbul, 93.7% of the respondents said they had public transportation at their place of residence. Public transportation was equally available in all parts of the Istanbul region (Table 9.6).

Given that public transportation was readily available to area residents, we wanted to know how often they used it. Somewhat more than one in five respondents (22.3%) said they used public transportation everyday and 17% stated that

**Table 9.6** Availability of public transportation (Percentage)

| Availability of public transportation | Istanbul | European side | Anatolian side |
|---|---|---|---|
| Available | 93.7 | 94.6 | 92.4 |
| Not available | 6.3 | 5.6 | 7.6 |
| Number of respondents | 1,635 | 957 | 678 |

Source: The authors

**Table 9.7** The frequency of using public transportation (%)

| The frequency of using public transportation | Istanbul | European side | Anatolian side |
|---|---|---|---|
| Everyday | 22.3 | 22.4 | 22.1 |
| 1–3 times a week | 22.2 | 21.9 | 22.9 |
| 1–3 times a month | 20.3 | 22.2 | 16.8 |
| Less than 1 time in a month | 18.2 | 17.7 | 19.2 |
| Never | 17.0 | 15.9 | 19.0 |
| Number of respondents | 1,515 | 906 | 609 |

Source: The authors

**Table 9.8** Main mode of transportation to work or school (percentage)

| Main mode of transportation | Istanbul | European Side | Anatolian side |
|---|---|---|---|
| Walking | 23.8 | 26.1 | 19.3 |
| Car | 19.3 | 19.5 | 19.2 |
| Service Bus | 22.1 | 23.1 | 20.3 |
| Public bus | 25.3 | 23.1 | 29.7 |
| Dolmus/minibus | 4.5 | 3.2 | 7.2 |
| Subway/train | 2.5 | 3.6 | 0.4 |
| Ferry | 1.1 | 0.0 | 3.2 |
| Other | 1.4 | 1.6 | 0.8 |
| Number of respondents | 588 | 347 | 241 |

Source: The authors

they never use it. Nearly half (44.5%) of the respondents used public transportation one to three times a week. Anatolian respondents were most likely to side to never use public transportation (Table 9.7).

Another key issue is the mode of transportation to work or school. When asked how they most often traveled to work or school, one quarter (25.3%) said they typically traveled by public bus, followed by walking (23.8%), service bus (22.1%), and private car (19.3%). More people on the European side walk to work/school compared to walkers on the Antolian side, while the later are more likely to use a public bus to get to work or school. The usage of the subway/train or ferry is dramatically low everywhere in the region. Nearly half of the respondents on both sides of the Bosporus use a public bus or a service bus (provided by companies and schools for their workers or students) (Table 9.8).

**Table 9.9** Public transportation assessment

| Public transportation assessment | Istanbul | European side | Anatolian side |
|---|---|---|---|
| Very bad | 5.3 | 3.7 | 8.4 |
| Bad | 16.5 | 15.7 | 18.0 |
| Neither bad nor good | 30.8 | 29.7 | 32.9 |
| Good | 38.7 | 40.8 | 34.6 |
| Very good | 8.7 | 10.1 | 6.0 |
| Mean score (1–5) | 3.29 | 3.38 | 3.12 |
| Number of respondents | 1,263 | 762 | 501 |

Source: The authors

**Table 9.10** Overall transportation system assessment

| Transportation system assessment | Istanbul | European side | Anatolian side |
|---|---|---|---|
| Very bad | 6.1 | 5.1 | 8.0 |
| Bad | 19.6 | 18.8 | 21.1 |
| Neither bad nor good | 35.9 | 33.1 | 41.1 |
| Good | 31.7 | 34.3 | 26.9 |
| Very good | 6.7 | 8.8 | 2.9 |
| Mean score (1–5) | 3.13 | 3.23 | 2.96 |
| Number of respondents | 1,635 | 957 | 678 |

Source: The authors

For Istanbul, less than half of the respondents (47.4%) evaluated public transportation serving their neighborhoods as good or very good, while one-fifth (21.8%) evaluated it negatively (Table 9.9).The European side is viewed as having better public transportation than the Anatolian side. For the neighborhood, the average value of public transportation assessment was 3.3 out of 5, where 5 is the most satisfactory score. Evaluations of the overall transportation system covering the entire Istanbul metropolitan area was even lower (Table 9.10).

## *Prospects for the Future of the Region*

As a way of examining prospects for the future of the Istanbul metropolitan area, two sets of questions were asked: one dealt with moving intensions, and the other dealt with peoples' expectations about the quality of life in Istanbul in 10 years. These questions were identical to those asked as part of the Detroit Area Study (see Chap. 7).

With respect to moving intentions, only 28% of the people wanted to move from their current residence, and 29% said they would "definitely move" or "probably move" within the next 2 years. Most of them would like to move either within the same neighborhood (40%) or to another neighborhood in Istanbul (50%). Only a small proportion of residents (10%) would like to move somewhere else in Turkey indicating that Istanbul is an attractive place for people to live.

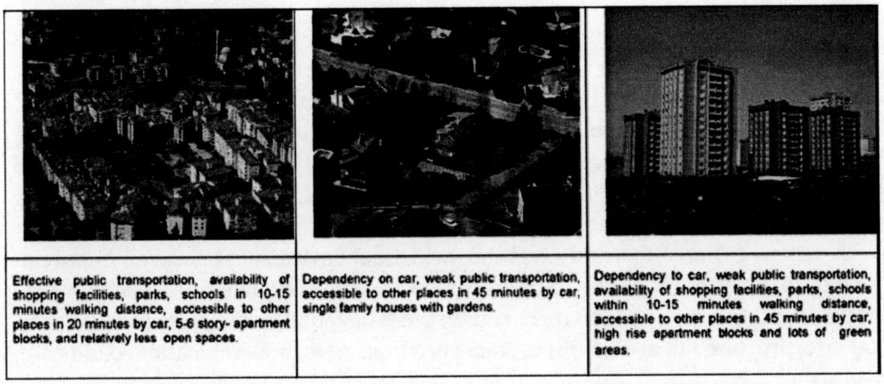


**Fig. 9.10** Housing type preferences. (**a**) Effective public transportation, availability of shopping facilities, parks, schools in 10–15 min walking distance, accessible to other places in 20 min by car, 5–6 story-apartment blocks, and relatively less open spaces. (**b**) Dependency on car, weak public transportation, accessible to other places in 45 min by car, single family houses with gardens. (**c**) Dependency to car, weak public transportation, availability of shopping facilities, parks, schools within 10–15 min walking distance, accessible to other places in 45 min by car, high rise apartment blocks and lots of green areas (Source: The authors)

The respondents who said they were planning to move in the next 2 years were presented with three typical new residential developments and asked to make a choice of which one they would prefer to live in. Interviewers showed pictures of the three developments (Fig. 9.10) and read a brief description of each.

The first development was described as a neighborhood which had effective public transportation; availability of shopping facilities, parks, and schools in 10–15-min walking distance; accessibility to other places in 20 min by car; 5–6-story apartment blocks; and relatively little open spaces. The second development was described as a neighborhood which had a dependency on car, weak public transportation, accessibility to other places in 45 min by car, and a predominance of single family houses with gardens. The third development was described as a neighborhood which was dependent on car; had weak public transportation; had shopping facilities, parks, and schools within 10–15-min walking distance; and was accessible to other places up to 45 min by car. It contained high-rise apartment blocks with lots of green areas. About 40% of the respondents chose the first type of development and a similar proportion chose the second development. The remaining, about one in ten, preferred the third development which had heavy reliance on the automobile.

On average, survey respondents were more optimistic than pessimistic about the region's future. About half of all respondents thought the QOL in the city would improve in the coming decade. Nonetheless, about 27% believed the QOL would deteriorate, while 23% said there would be no change in regional QOL. Residents were more hopeful about their neighborhoods than for the region. Nearly two-thirds (61%) of residents thought the quality of life in their neighborhood would improve in the coming decade.

## Conclusion

This chapter has presented an overview of a study examining the QOUL in the Istanbul Metropolitan Area. The primary source of information was a household survey. The study was carried out as part of a strategic planning process and intended to inform decision makers and planners about the residents' perceptions of urban life in a large and rapidly growing region.

A central focus of the study was the residential environment since an important goal of the strategic plan was to improve housing and the quality of Istanbul neighborhoods. Urban renewal and regeneration were intended to address not only physical infrastructure but also issues of transportation, education, recreation, child care, and social life.

Furthermore, the study was designed to produce baseline data so that future changes in residential conditions as perceived by the residents of Istanbul could be measured with subsequent QOUL.

Data from both the social and physical surveys were analyzed and led to a set of recommendations covering housing in the metropolitan area. In addition to identifying those neighborhoods that needed redevelopment or improvement, infill developments were proposed to curtail future urban growth outside the urbanized area on forest land and on land suitable for agricultural purposes. The proposal to increase the density of the loosely developed areas instead of opening up agricultural land to urban development was in line with the "sustainable development" policy of the Istanbul municipality. However, the proposal was not openly emphasized in the Strategic Plan which was subsequently challenged in court by the Chambers of Architects and Planners. Chambers argued that in addition to their objections to the planning procedure, the plan contained proposals that were often inconsistent and some contradictory to the "sustainable development" policy.

The authors are continuing to make recommendations based upon ongoing analysis of the two surveys. These analyses cover not only peoples' perceptions of housing and neighborhood conditions but also environmental stressors, parks, cultural amenities in urban area, and various municipal services, including public transportation and garbage collection. Further attempts to examine relationships between data from the social survey and the physical survey data are also being explored.

At present, prospects for a follow-up study on QOUL in Istanbul Metropolitan Area are not promising. What is promising is the interest generated by this research examining the quality of life in other metropolitan cities of Turkey.

## References

Bolen, F., Turkoglu, H. D., Ergun, N., Yirmibeşoğlu, F., Kundak, S., Terzi, F., & Kaya, S. (2006) *İstanbul'da Konut Alanlarında Fiziksel Çevre Kalitesi Analizi* (Konut ve Yaşam Kalitesi Grubu Raporu, Cilt I). Istanbul: Istanbul Metropolitan Planning and Design Center.

Campbell, A., Converse, P. E., & Rodgers, W. L. (1976). *The quality of American life: Perceptions, evaluations, and satisfactions*. New York: Russell Sage.

Frank, L. D., Schmid, T. L., Sallis, J. F., Chapman, J., & Saelens, B. E. (2005). Linking objectively measured physical activity with objectively measured urban form: Findings from SMARTRAQ. *American Journal of Preventive Medicine, 2*(Sup. 2), 117–125.

Istanbul Metropoliten Municipality. (1996). *Istanbul master plan report*. İstanbul: Istanbul Metropoliten Municipality.

Marans, R. W. (2003). Modeling residential quality using subjective and objective indicators. In D. U. Vestbro, Y. Hurol, & N. Wilkinson (Eds.), *Methodologies in housing research* (pp. 312–326). Newcastle Upon Tyne: Urban International Press.

Marans, R. W., & Rodgers, W. (1975). Toward and understanding of community Satisfaction. In A. Hawley & V. Rock (Eds.), *Metropolitan America in contemporary perspective* (pp. 299–352). New York: Halstead Press.

Statistical Institution of Turkey. (2008). *Statistical year book*. Ankara: Statistical Institution of Turkey.

Turkoglu, H. D., Bölen F., Baran P. K., & Marans, R.W. (2006). *İstanbul'da Yaşam Kalitesinin Ölçülmesi* (Konut ve Yaşam Kalitesi Grubu Cilt II). Istanbul: Istanbul Metropolitan Planning and Design Center.

# Chapter 10
# The Quality of Urban Life and Neighborhood Satisfaction in Famagusta, Northern Cyprus

**Derya Oktay and Ahmet Rustemli**

## Introduction

The idea of carrying out a quality of urban life (QOUL) survey in Famagusta (Northern Cyprus) was suggested in 2005 when the first author served as a visiting scholar at the University of Michigan. During the visit, she learned of the Detroit Area Study (DAS) – discussed in Chap. 7 – involving a household survey covering urban life in Metro Detroit. Since Famagusta was a unique city in terms of its population mix, its rapid growth, and its unclear future due to the political situation in Cyprus, such a study had the potential of being significant and valuable to both political leaders and local and national planners.

Following a preliminary literature review, an analysis of existing demographic data, interviews with the local governmental officials, and a research grant from TUBITAK (The Higher Council for Scientific and Technological Research in Turkey) were obtained in 2006 to launch a household survey of Famagusta residents.[1] The survey was conducted in 2007 and was intended to produce information

[1] The Famagusta Study, titled "Measuring the Quality of Community Life in Gazimagusa" and directed by the first author, is one of the partner cities included in the International Program of Research on Quality of Life coordinated at the University of Michigan, USA, under the coordination of Robert Marans. The study was funded by the Scientific and Technical Research Council of Turkey (TUBITAK) for the period of September 2007–May 2008 and is placed under the auspices of the Urban Research & Development Center of Eastern Mediterranean University (EMU).

D. Oktay (✉)
Urban Research & Development Center, Eastern Mediterranean University,
Famagusta, North Cyprus, via Mersin 10, Turkey
e-mail: derya.oktay@emu.edu.tr; de.oktay@gmail.com

A. Rustemli
Faculty of Business and Administration, Eastern Mediterranean University,
Erdemkent Funda Sokak No 26, Çankaya, Ankara, Turkey
e-mail: rustem.ahmet2@gmail.com

R.W. Marans and R. Stimson (eds.), *Investigating Quality of Urban Life: Theory, Methods, and Empirical Research*, Social Indicators Research Series 45,
DOI 10.1007/978-94-007-1742-8_10, © Springer Science+Business Media B.V. 2011

covering various aspects of urban life among local residents. In addition to survey responses, the Famagusta Area Study (FAS) aimed at compiling contextual information or data about the community and environments associated with each respondent. Contextual information included housing and demographic characteristics, land use, selected physical environmental characteristics (traffic density, etc.), and socioeconomic information about the local areas where respondents lived (that is, employment, school information, etc.).

This chapter first provides an overview of Famagusta including its demographic and geographic characteristics and its urban context. It then reviews the methodology used to carry out the research. Next, selected findings covering overall QOUL, neighborhood satisfaction, the assessment of neighborhood attributes, and other community issues are presented. Finally, the uses of the findings from the study for policy, planning, and design are discussed, along with opportunities for further analysis of the data.

## About Famagusta

Famagusta (Gazimagusa) is the second largest city of Northern Cyprus with a population of 35,381 (TRNC 2006 Population and Dwelling Census). The city is located on the eastern coast of the island of Cyprus and has a historic core including a harbor (see Figs. 10.1–10.3). The history and urban development of Famagusta date back to the first century AD, and the contemporary city has been developed over

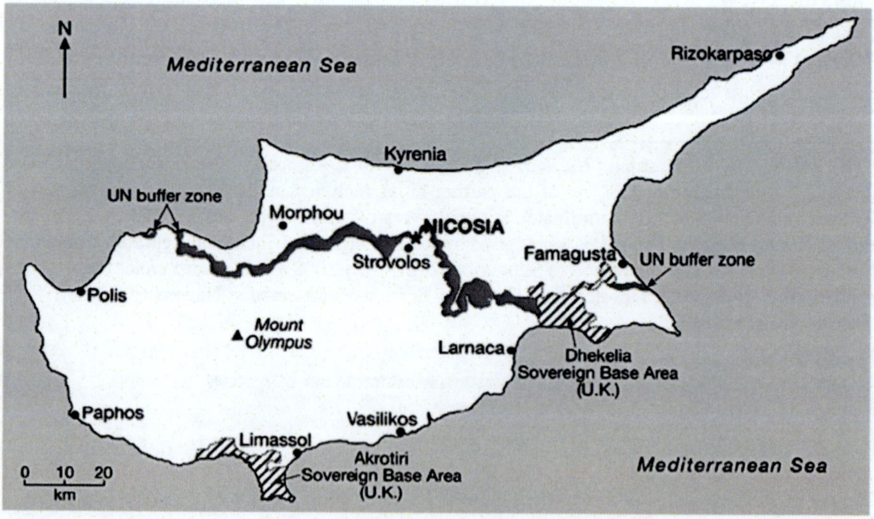


**Fig. 10.1** Map of the island of Cyprus showing main cities and the north–south divide (Source: Oktay 2007, 231)

**Fig. 10.2** The air view of the walled city and the Famagusta harbor (Source: www.emu.edu.tr)

**Fig. 10.3** The air view of the Eastern Mediterranean University campus and surrounding neighborhoods (Source: www.emu.edu.tr)

**Fig. 10.4** A view of the mass housing complex in peri-urban areas (Source: D. Oktay archive)

seven periods: the early period (648–1192 AD – the foundation period), the Lusignan period (1192–1489), the Venetian period (1489–1571), the Ottoman period (1571–1878), the British period (1878–1960), the period between 1960 and 1974 (the Greek and Turkish period), and the period after 1974 (the Turkish period).

The city was an important trade and tourism destination and served as a regional center before Cyprus was divided in 1974. Today, despite some restrictions on its capacity owing to the new circumstances of the island, the harbor still plays an important part in the trade activities of the northern Turkish region. However, changes in the last two decades in Famagusta result largely from the establishment and growth of the Eastern Mediterranean University (EMU).

The development of the university, with a student population of nearly 14,000 who come from 67 different countries (in addition to the de facto population), has led to remarkable changes in the sociodemographic makeup of the city. It has also benefited property owners throughout the city as the demand for rental housing has increased.

Famagusta has experienced uncontrolled and rapid urban development in the form of multistory housing, haphazard additions to existing houses, and incompatible land uses. The construction and commercial sectors, and new housing in the form of multistory apartments or "villas," are shaped in a spontaneous way without following a coordinated master plan. Housing is often built on flat sites with no trees and stand as isolated concrete units on separate lots (see Fig. 10.4). The most negative design aspects in these settlements are the site layout creating unusable spaces between buildings and poor transition between indoor private spaces and outdoor public spaces (Oktay 2002). Traditional family and kinship patterns that led to lively and well-connected neighborhoods have therefore broken down, and social life has been deteriorated (Oktay 2005).

The decaying character of the historic walled city is another problem in Famagusta. Although the walled city, with its organic form, fortifications, and moat was declared a conservation area under the new Town Planning Law (55/1989), the measures undertaken for its conservation and revitalization have not led to the attainment of a satisfactory state in terms of cultural or economic sustainability. Consequently, the walled city is functionally isolated from the other parts of Famagusta. Finally, the uncertain future of the unoccupied Varosha (Maras) district has affected urban development by preventing the city from growing towards the south. This district was a prosperous tourism and residential area but was vacated by Greek inhabitants in line with the United Nations demarcation decision in 1974.

Because Famagusta has a dynamic socioeconomic structure, a problematic pattern of urban development, and an uncertain future due to the political situation of Cyprus, evaluating the quality of life (QOL) of its residents is crucial for policy makers and planners who will shape the city's future. Whether the residents are aware of the city's problems and how they might respond to change can be determined through a comprehensive survey.

## Methodology

The collection of survey information was achieved through a structured questionnaire administered to a sample of housing units selected from Famagusta neighborhoods. Within each housing unit, a resident was selected and interviewed by trained graduate students. The interviews were conducted between spring and fall 2007.

### *Sampling Approach*

The survey was conducted within housing units using a multistage sampling procedure.

First, the total number of housing units (13,455) within the city limits was determined by counting the parcel plots. Eight neighborhoods of the city were identified, and the number of households in each neighborhood was determined.

A random sample of 540 households was selected from the neighborhoods in proportion to its overall size. With each selected household, an adult, 18 years old or older, was asked to participate in the survey. Of the total number of households contacted, 398 occupants completed the interview, representing a response rate of 74%. Nonresponse was largely due to no one being home when the interviewers made their visit.

The distribution of the households in the sample is shown on the map of Famagusta in Fig. 10.5.

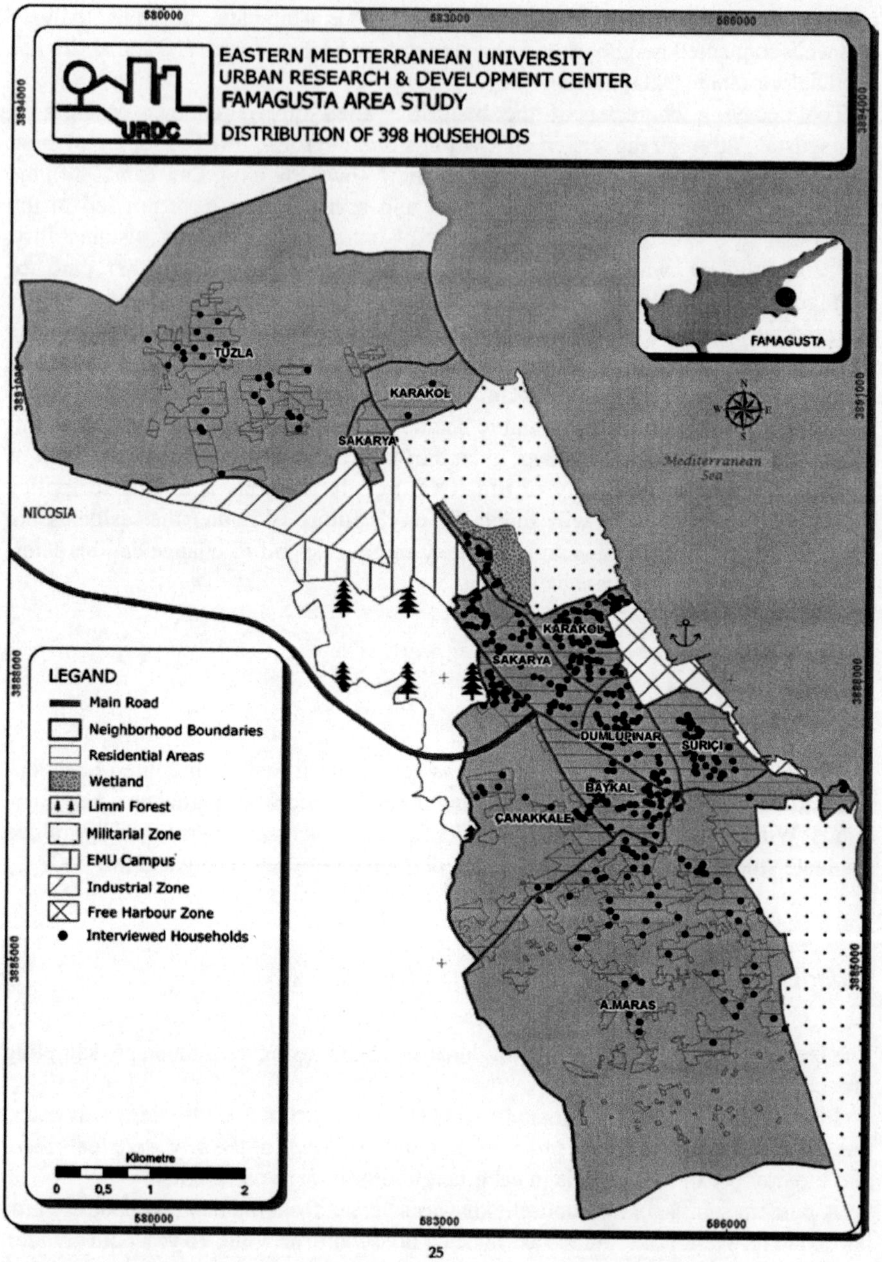


Fig. 10.5 Distribution of the 398 households in the sample (Source: The authors)

**Table 10.1** Famagusta quality of urban life topics

| | |
|---|---|
| *Residential history/mobility* | *Satisfaction* |
| Factors influencing residential choice | City |
| Moving intentions | Neighborhood |
| Neighborhood preference | Community |
| *Neighboring and community involvement* | *Public services and facilities* |
| Social ties | Schools |
| Identification of neighborhood problems | Public spaces and arteries |
| Sense of belonging and attachment | Public transportation |
| | Shopping |
| | Maintenance of streets and open spaces |
| *Government and taxes* | *Travel* |
| General satisfaction with taxes | Travel behavior/type |
| Willingness to pay for new/improved services and facilities | Public transportation use |
| | Work/school trips |
| | Shopping trips |
| *Parks and recreation areas* | *Regional issues* |
| Frequency of use | Participation in urban activities |
| Type of use | Town and historic core relationship |
| Accessibility | Town and university relationship |
| Places where children play | General evaluation of the city |
| Relationship with sea | Environment problems |
| Satisfaction with facilities | |
| *Safety* | *Prospects for the future* |
| Perception of crime | |
| Neighborhood safety | |
| *Other domain satisfaction* | *Household and respondent characteristics* |
| Job/health/standard of living/family/friends/leisure | |

Source: The authors

## Data Collection

The interviews were administered face-to-face by trained graduate students and were conducted in the respondents' homes. The interview schedule included questions that examine people's feelings and behaviors with reference to their households and dwelling. Many of the questions were drawn from the questionnaire used in the DAS (see Chap. 7). A summary of the QOUL topics included in the questionnaire is shown in Table 10.1.

However, as QOL considerations are likely to vary from one city to another (Mazumdar 2003), many of the DAS questions were modified, and new questions were added to reflect the local situation. In the next section, we report findings covering only a portion of the questionnaire. These deal with the overall quality of urban life in Famagusta, neighborhood satisfaction, and assessments of neighborhood social attributes including safety, social ties, neighbors, and neighborhood attachment. Several attributes of the surrounding environment (accessibility, attractiveness, greenery, public transportation), recreational facilities (parks, playgrounds) along with selected physical attributes (traffic density, noise level, crowding, and street maintenance) are analyzed as factors that potentially influence the neighborhood satisfaction.

## Findings

As outlined above, the survey addressed a range of topics that met the potential informational needs of local governmental units as well as the interests of the research team. Many of the topics have been addressed in a technical report, articles, and conference presentations (Oktay 2010; Oktay and Marans 2008). Data covering other topics are currently being examined or will be analyzed in the future. Rather than presenting all findings, this section presents a limited set that covers overall QOL and its determinants, neighborhood satisfaction and the attributes associated with it, and respondents' views about the future of their neighborhoods and Famagusta.

### *Overall Quality of Life in Famagusta*

Overall QOL was viewed as a subjective phenomenon and considered a composite of people's assessments of various domains of their lives as first discussed by Campbell et al. (1976). Accordingly, the Famagusta survey asked respondents to assess seven specific domains using a 5-point satisfaction scale ranging from "completely dissatisfied" (1) to "completely satisfied" (5). The domains considered were the individual's overall standard of living, job/school, family life, friends, health, leisure, and the amount of time to do the things they want to do.

Table 10.2 shows the mean satisfaction scores for the overall sample. Satisfaction scores for family life and friends were somewhat higher than satisfaction scores for

**Table 10.2** Overall quality of life in Famagusta: domain satisfaction scores

| Domain | Mean score | Standard deviation |
|---|---|---|
| Family life[a] | 4.24 | .63 |
| Friends | 4.09 | .58 |
| Health | 3.78 | .90 |
| Job/school[b] | 3.70 | .97 |
| Standard of living | 3.69 | .79 |
| Life as a whole | 3.46 | .93 |
| Leisure | 3.29 | .97 |
| Time to do things | 3.05 | 1.01 |

Source: The authors
$N=398$
[a]The relevant question was responded by those who live with their families ($N=242$)
[b]The relevant question was responded by those who have work or who are students ($N=372$)

health, job/school, standard of living, and life as a whole. People were least satisfied with the amount of time to do the things they want to do and the way they spend their spare time.

## *Quality of Urban Life in Famagusta*

In the Famagusta area study, the residential environment was considered at three levels:

- The individual home or dwelling
- The immediate (microscale) neighborhood
- The overall (macroscale) neighborhood.

Consideration was also given to attributes of each of these place domains.

At the city level, respondents were asked to evaluate several attributes and then give a summary score for *satisfaction with life in the city of Famagusta today?*'. Using the same 5-point response categories, they were also asked to express their overall satisfaction with their dwelling, their immediate neighborhood, and their overall neighborhood.

Table 10.3 presents overall satisfaction scores for the Famagusta respondents' dwellings, their immediate neighborhood, their overall neighborhood, and urban life. The mean scores are fairly comparable for the first three domains but somewhat lower for the overall QOUL (3.51 vs. 3.29). Indeed, while 40% of the respondents were either satisfied or very satisfied with the overall QOL in Famagusta, 14% were either dissatisfied or very dissatisfied. The remaining respondents said they were indifferent (neither satisfied nor dissatisfied).

However, when the respondents were asked about specific attributes of Famagusta, the city was seen as a place with serious environmental problems such as lack of effective public transportation, unprotected natural resources, not having a "green city" image, ineffective environmental problem solving, a lack of cultural amenities, and little sense of historical values. On the other hand, most of the respondents agreed that "Famagusta is a safe city," that having "a strong economy will depend on developing the city as a better place to live and work," and that "public transit that is reliable and safe can be important to the quality of life of Famagusta residents."

**Table 10.3** Dwelling, immediate neighborhood, overall neighborhood, urban life mean satisfaction scores in Famagusta

| Domain | Mean score | Standard deviation |
|---|---|---|
| House/dwelling | 3.50 | 1.002 |
| Immediate neighborhood | 3.48 | .90 |
| Overall neighborhood | 3.54 | .87 |
| Urban life | 3.29 | .77 |

Source: The authors
$N$=398

Findings indicate that the overall QOL scores are comparable among local Cypriots, immigrants from Turkey, as well as MEU international students. Similarly, whether respondents were working, unemployed, retired, or students had no bearing on their QOL scores.

## *Overall Satisfaction with Neighborhood*

An important area of exploration in the Famagusta study was neighborhood satisfaction. For more than a quarter of a century, neighborhood satisfaction has been studied within the field of housing research (Campbell et al. 1976; Hall and Ring 1974; Michelson 1979; Marans and Wellman 1978; Marans 2003; Galster and Hesser 1981; Francescato 1998) and is seen as influencing the overall QOL of people.

In the Famagusta survey, neighborhood satisfaction was measured by a single question, "How satisfied are you with the overall neighbourhood quality?" with respondents given the opportunity to indicate they were "very dissatisfied" (1), "very satisfied" (5), or some value in between. Nearly two-thirds (63%) said they were satisfied, 25% were neither satisfied nor dissatisfied, and 12% were dissatisfied.

Empirical studies suggest that a number of specific attributes of a neighborhood contribute to residents' overall satisfaction. Accordingly, three sets of satisfaction questions were asked covering the social environment, urban environmental attributes, and attributes of the physical environment. Table 10.4 presents the mean scores and the standard deviations covering the responses to these sets of questions.

Table 10.4 reveals that the respondents tended to be more satisfied than dissatisfied with the attributes; scores above 3.0 represent some level of satisfaction while scores below 3.0 represent a level of dissatisfaction. Attributes with the highest average scores were satisfaction with safety (mean = 4.08) and satisfaction with friendliness of neighbors (mean = 3.90). The mean scores for satisfaction with recreational facilities, greenery, maintenance of streets, and traffic density reflect a level of dissatisfaction with these aspects of urban life. However, an important point needs to be attended when interpreting the values in Table 10.4. Considering the limited range of responses, the standard deviations are high, indicating wide variations in responses.

In order to determine the degree to which assessments of these attributes taken together influence the overall quality of life measure, a regression analysis was employed. In this analysis, the attribute variables were entered in blocks as shown in Table 10.5. Overall, the set of variables explain 23.5% of the total variance in QOUL measure (multiple $r=.485$). Two of the five social attributes had significant effects on the QOUL measure. Safety and sense of belonging contributed positively to overall QOUL, $t$ values being 3.58 ($p \leq .001$) and 2.26 ($p \leq .03$), in that order.

Satisfaction scores for the city's environmental attributes were also related to the QOUL measure. Among these attributes, satisfaction with public transportation contributed most to QOUL ($t=4.80$, $p \leq .001$), but negatively. Satisfaction with

**Table 10.4** Satisfaction scores for social, environmental, and physical attributes

| Attributes | Mean[a] | Standard deviation[a] |
|---|---|---|
| *Social* | | |
| Safety | 4.08 | 0.95 |
| Social network | 3.18 | 1.43 |
| Friendliness of neighbors | 3.90 | 0.93 |
| Perceived similarity of others | 3.43 | 1.14 |
| Sense of belonging | 3.09 | 1.35 |
| *Urban Environmental* | | |
| Accessibility | 3.62 | 1.47 |
| Attractiveness of place | 3.19 | 1.40 |
| Greenery | 2.34 | 1.24 |
| Public transportation | 3.62 | 1.06 |
| Recreational facilities | 2.35 | 1.07 |
| *Physical* | | |
| Traffic density | 2.79 | 1.07 |
| Level of noise | 3.34 | 1.41 |
| Crowding | 3.30 | 1.34 |
| Maintenance of Streets | 2.69 | 1.35 |

Source: The authors
$N=398$
[a] The higher the mean value, the greater the level of satisfaction

**Table 10.5** Zero-order correlation coefficients between QOUL and environmental attributes and the results of the regression analysis

| Domain | Zero-order correlation coefficient | Standardized beta coefficient | *t* | Prob |
|---|---|---|---|---|
| *Satisfaction with social attributes* | | | | |
| Safety | .19** | .19 | 3.58 | .000 |
| Social support (no. of rel. and fr.) | .10 | .09 | 1.61 | |
| Friendliness of neighbors | .11* | .06 | 1.17 | |
| Perceived similarity of others | .08 | .03 | 0.58 | |
| Sense of belonging | .15** | .13 | 2.26 | .024 |
| *Satisfaction with urban/environmental attributes* | | | | |
| Accessibility | .05 | .00 | 0.06 | |
| Attractiveness of place | .08 | –.04 | –0.65 | |
| Greenery | .22** | .16 | 2.99 | .003 |
| Public transportation | –.34** | –.26 | –4.80 | .000 |
| Recreational facilities | .08 | .08 | 1.29 | |
| *Satisfaction with physical attributes* | | | | |
| Traffic density | .05 | –.02 | –0.35 | |
| Level of noise | .21** | 19 | 2.71 | .007 |
| Crowding | .11* | –.02 | –0.28 | |
| Maintenance of streets | .24** | .14 | 2.50 | .010 |

Source: The authors
*Significant at alpha .05
**Significant at alpha .01

greenery, on the other hand, had a positive effect on QOUL ($t=2.99$, $p \leq .003$). On the other hand, ratings of accessibility, attractiveness of the environment, and recreational facilities had no impact on QOUL.

Table 10.5 also shows that satisfaction with noise level and maintenance of streets scores were associated with QOUL evaluations, the correlation coefficients being +0.21 and +0.24, in that order. The higher the satisfaction with noise level and maintenance of streets, the higher was the QOUL. Satisfaction scores for traffic density and crowding, on the other hand, did not relate to the QOUL measure.

## Prospects for the Future

As a way of examining prospects about the future of residents' neighborhood and QOUL in Famagusta, three questions were asked: one dealt with moving intentions, another with expectations about the future quality of their neighborhood, and the third with QOUL in Famagusta in the next 10 years.

### *Moving Intentions*

With respect to moving intentions, more than half of the respondents (61%) indicated that they did not want to move from their current residence, and a half (52%) said they would "definitely not move" within the next 2 years. When the respondents were asked where they would like to move, 45% said they would prefer another neighborhood in Famagusta, and a comparable number (40%) mentioned the new low-rise residential estates that were being built outside the city.

### *Preferences for Alternative Types of Residential Environment*

In response to the question about a desirable place to live, almost half (49%) of the respondents preferred to live in an environment which had effective public transportation, available shopping, parks, and schools within a 5–10-min walk distance and accessible to other places within 5–10 min by car. They also said that an environment with 4–5-story apartment blocks and less urban open spaces was desirable (Fig. 10.6 – Type 1). On the other hand, more than a third (38%) of the respondents preferred a residential environment where a car was a necessity, where there was no public transportation, and where there were single family houses with gardens and access to natural areas (Fig. 10.6 – Type 3). The remaining 13% said they preferred an environment with row houses and single family housing, shopping facilities, entertainment, parks, and schools within 10–20-min walking distance, and the opportunity to walk to nearby open spaces (Fig. 10.6 – Type 2).

*TYPE 1*

- Effective public transportation and convenient walking
- Shopping, entertainment, parks, and school are within a 5-10 minutes walk
- Commute locations are about 5-10 minutes
- 4-5 storeys apartment blocks
- Relatively less urban open spaces
- **Preferred by 49 percent**

*TYPE 2*

- Convenient driving opportunities but weak public transportation
- Availability of shopping facilities, entertainment, parks, and school within a 10-20 minutes walk distance
- Commute locations are about 15-25 minutes
- Row houses / single family houses
- Opportunity of walking in open spaces
- **Preferred by 13 percent**

*TYPE 3*

- Dependency to car
- No public transportation
- Commute locations are about 30-45 minutes
- Single family houses with gardens
- Access to natural areas
- **Preferred by 38 percent**

**Fig. 10.6** Preferences for alternative types of residential environment (Source: D. Oktay archive)

In order to understand how people felt about their neighborhood and Famagusta in the future, they were asked whether the quality of life would get better, stay about the same, or get worse over the next 10 years. On average, respondents were more optimistic than pessimistic about the future of their neighborhoods. Those who believed the QOL in their neighborhood would improve outnumbered respondents who said it would get worse by three to one (64% vs. 21%). Only 15% said their neighborhood would not change over the next decade.

Survey respondents were less optimistic about the city's future. Nearly half (48%) said the QOUL in Famagusta would get better over the next decade while 12% felt it would deteriorate; the remaining 40% believed there would be no change. These findings are somewhat surprising in that nearly half of the respondents were optimistic about the city's future despite the numerous environmental problems mentioned by most of them.

## Summary and Conclusions

Urban life revolves around the places where people carry out their daily lives, and for most people, the residential environment is the central environmental setting where this occurs. The residential environment often provides a haven of security and a comfortable, supportive milieu in which individuals organize their daily activities (Oktay 2001). When the congruence between people and their surroundings is impaired, however, emotional disturbances, health problems, and social disorder may occur (Vliet 1999). Accordingly, creating a satisfactory housing environment is the most important social objective in urban development.

This chapter has presented an overview of a study designed to produce empirical data covering the QOUL of residents in Famagusta, Northern Cyprus. A key focus of the study was Famagusta neighborhoods and how they were viewed by their residents. The primary data source was a household survey that was designed to meet the interests of local governmental officials and researchers dealing with various aspects of the city. The information obtained was initially presented as descriptive statistics, while regression analyses were used to examine relationships between the different factors.

The findings reveal that compared to satisfaction with an individual's dwelling, the immediate neighborhood, and several neighborhood attributes, satisfaction with the overall QOUL in Famagusta was low. While almost two-thirds of the residents were satisfied with their neighborhoods, just 40% expressed satisfaction with the QOUL.

In general, people were most likely to be dissatisfied with recreational facilities, greenery, maintenance of streets, and traffic in Famagusta. However, the standard deviations and the mean scores were large, indicating that city dwellers differed widely in their assessments of the various domains of the city and the urban life.[2]

The results of the regression analysis (see Table 10.5) on QOUL showed that safety and sense of belonging are the two social attributes of importance to residents. The other social attributes, namely, satisfaction with social support, friendliness of neighbors, and perceived similarity of others, did not relate to QOUL. Similarly, only two of the four physical attributes, noise level and maintenance of streets, had significant effects on QOUL evaluations. Satisfaction scores with traffic density and crowding as the other two physical attributes did not relate to QOUL measure.

Concerning the prospects for the future, respondents were more optimistic about the future of their neighborhood than they were about the city's future. Nonetheless,

---

[2] A preliminary study by the authors has proved the existence of these differences across the neighborhoods (Oktay et al. 2009).

nearly half of the respondents were optimistic about the city's future despite various environmental problems recognized by most residents. This might be explained by cognitive dissonance (Festinger 1957), that is, residents recognized that their neighborhood and the city have problems; at the same time, they realized that they reside there and, in all likelihood, will continue to live there in the future. Their dissonance is reduced by expressing satisfaction with the neighborhood and the city and being optimistic about the future.

Famagusta residents tended to be unhappy with their neighborhoods, and when asked if they would like to move, most said they would move under the right conditions. Of those expressing a desire to move, about half mentioned another Famagusta neighborhood with high-rise buildings, and the other half preferred a move to one of the low-rise residential estates outside the city. Many said they wanted an environment which has effective public transportation, available shopping facilities, parks, and schools within a 10–15-min walk distance, but accessible by car to other places. Surprisingly, they also indicated a preference for living in apartment blocks with relatively little open spaces between them. Others preferred an environment which was auto-dependent, characterized by single family houses with gardens and accessibility to natural areas, while a small minority preferred row houses or single family houses where shopping facilities, entertainment, parks, and schools were within a 10–20-min walk from their homes.

These findings suggest that the future development policies should consider the creation of urban housing schemes with full services (shopping, parks, and other services) that are easily accessible. Developing new multistory mixed-use housing schemes and rehabilitating existing residential areas with supporting activity nodes would also support the idea of defragmentation as a strategy for combating urban sprawl. This strategy would also contribute to urban sustainability in Famagusta. Considering the central employment location in Famagusta, an urban regeneration scheme would make great sense if the proposed scheme is designed to meet the needs of local residents and students.

On the other hand, a significant number of respondents were interested in moving to housing estates outside Famagusta. At the same time, it was shown that views on public transportation had no bearing on QOUL. These findings suggest that a master plan is needed that aims at controlling urban sprawl and providing better living environments in both urban and peri-urban areas. The master plan should seek to provide more greenery, better recreational facilities, improved street maintenance, and reduced traffic density. In particular, the provision of greenery and the maintenance of streets need serious attention as they were rated poorly by residents and are strongly associated with QOUL.

## Future Work

Further analysis of the survey data is planned and will focus on other QOUL topics such as public transportation, shopping, community involvement, parks and recreation areas, and relationship between Famagusta's historic core and other areas of the city. Furthermore, objective environmental measures covering the residential setting of respondents will be compiled using a geographic information system (GIS) and analyzed together with survey data.

The initial findings from the Famagusta survey have produced information that can inform governmental, institutional, and community leaders as they plan for and implement programs designed to enhance the QOUL of their constituents. At the same time, the study has produced indicators that represent baseline measures for assessing changes in people's attitudes and behaviors in the future. Finally, once the objective measures are compiled, the study will determine how much they correspond to residents' perceptions and behaviors about QOUL. From a research perspective, the measurement of people's opinions and actions on the one hand and the environments in which they live on the other presents opportunities to explore the meaning of social and physical changes that take place in the city.

## References


Campbell, A., Converse, P. E., & Rodgers, W. L. (1976). *The quality of American life: Perceptions, evaluations, and satisfactions*. New York: Russell Sage.

Festinger, L. (1957). *A theory of cognitive dissonance*. Standford, CA: Stanford University Press.

Francescato, G. (1998). Residential satisfaction. In W. V. Vliet (Ed.), *Encyclopedia of housing* (pp. 484–486). London: Sage.

Galster, G. C & Hesser, G. W. (1981). Residential satisfaction: Compositional and contextual correlates. *Environment and Behavior, 19*(5), 539–558.

Hall, J., & Ring, J. (1974). *Indicators of environmental quality and life satisfaction: A Subjective approach*. London: Survey Unit, Social Science Research Council.

Marans, R. W. (2003). Understanding environmental quality through quality of life studies: The 2001 DAS and its use of subjective and objective indicators. *Landscape and Urban Planning, 65*, 73–83.

Marans, R. W., & Wellman, J. D. (1978). *The Quality of nonmetropolitan living: Evaluations, behaviors, and expectations of northern michigan residents*. Ann Arbor: Institute for Social Research, University of Michigan.

Mazumdar, S. (2003). Sense of place considerations for quality of urban life. In N. Z. Gülersoy, N. Esin, & A. Özsoy (Eds.), *1st International Conference on Quality of Urban Life: Policy Versus Practice* (pp. 83–97). Istanbul: Istanbul Technical University, Urban and Environmental Planning Research Centre.

Michelson, W. H. (1979). *Environmental choice, human behavior and residential satisfaction*. New York: Oxford University Press.

Oktay, D. (2001). *Planning housing environments for sustainability: Evaluations in cypriot settlements*. Istanbul: Building and Industry Center (YEM) Publications.

Oktay, D. (2002). The quest for urban identity in the changing context of the city: Northern Cyprus. *Cities, 19*(4), 31–41.

Oktay, D. (2005). Cyprus: The South and the North. In R. Van Kempen, M. Vermeulen, & A. Baan (Eds.), *Urban issues and urban policies in the new EU countries* (pp. 205–231). Aldershut: Ashgate.

Oktay, D. (2007). An analysis and review of the divided city of Nicosia, Cyprus, and new perspectives. *Geography, 92*(3), 231–247.

Oktay, D. (2010). *Gazimağusa'da Kentsel Yaşam Kalitesi (Quality of urban life in Famagusta)*. Famagusta: Urban Research and Development Center Publications, EMU Press.

Oktay, D., & Marans, R. W. (2008, May 28–June 1). Assessing the quality of life in Famagusta Neighborhoods: Initial findings from an ongoing research program. *EDRA 39th Conference*, Veracruz, Mexico.

Oktay, D., Rüstemli, A., & Marans, R. W. (2009). Neighbourhood satisfaction, sense of community and attachment: Inital findings from famagusta (Gazimagusa) Quality of life study. *ITU A/Z Journal, 6*(1), 6–20.

Vliet, V. V. (1999). *The encyclopedia of housing*. New York: Sage.

# Chapter 11
# The Quality of Life in Dhaka, Bangladesh: Neighborhood Quality as a Major Component of Residential Satisfaction

**Abul Mukim Mozammel Haque Mridha and Gary T. Moore**

## Introduction

Quality of life (QOL) is complex concept that includes a number of contributing factors, including housing, education, work, and environment (Blanco and Chacon 1985). Quality implies the degree of excellence of a characteristic, but the concept of the QOL means different things to different people (Das 2008).

There are three different approaches to the study of QOL in the context of housing and environment (García-Mira et al. 2005). First, QOL studies have focused on subjective well-being or life satisfaction (Donovan and Halpern 2002). Second, QOL has been perceived by governments and some researchers to be synonymous with standard of living (Jackson 2002; see also Chap. 3). The third interpretation of QOL has been to link the concept directly to sustainable development such that the two terms are used almost interchangeably (García-Mira et al. 2005).

In the first approach, researchers are concerned with individuals' subjective experience of well-being in their lives. The underlying assumption is that well-being can be defined by people's conscious experiences in terms of hedonic feelings or cognitive satisfaction (Diener and Suh 1997). It is therefore appropriate to examine directly how they feel about life in the context of their own standards and life experiences, including their experience and satisfaction with housing and

A.M.M.H. Mridha (✉)
Urban Research Centre, University of Western Sydney, 34 Charles Street, Parramatta, NSW, Australia
e-mail: M.Mridha@uws.edu.au

G.T. Moore
Faculty of Architecture, Design and Planning, The University of Sydney, 148 City Road, NSW, Australia
e-mail: gary.moore@sydney.edu.au

R.W. Marans and R. Stimson (eds.), *Investigating Quality of Urban Life: Theory, Methods, and Empirical Research*, Social Indicators Research Series 45, DOI 10.1007/978-94-007-1742-8_11, neighborhood environment. The Dhaka study discussed in this chapter uses this approach. It was the first study of residential satisfaction conducted in Bangladesh.

## The Context: Dhaka, Bangladesh

Dhaka is one of the oldest and largest cities in South Asia. As a formal city, its history extends over 400 years. Before that, a number of urbanized settlements evolved from the seventh century in the area that is now Dhaka. During the past 400 years, it has been a capital city on four occasions (Hossain 1990). First, Dhaka was the capital of the Bengal province from 1610 to 1713 under the Mughal Empire (Fig. 11.1). In 1905, with the beginning of British rule, it became the capital of East Bengal and Assam (Fig. 11.2). In 1947, after independence from two hundred years of British rule and the subsequent division of the Indian subcontinent, Dhaka

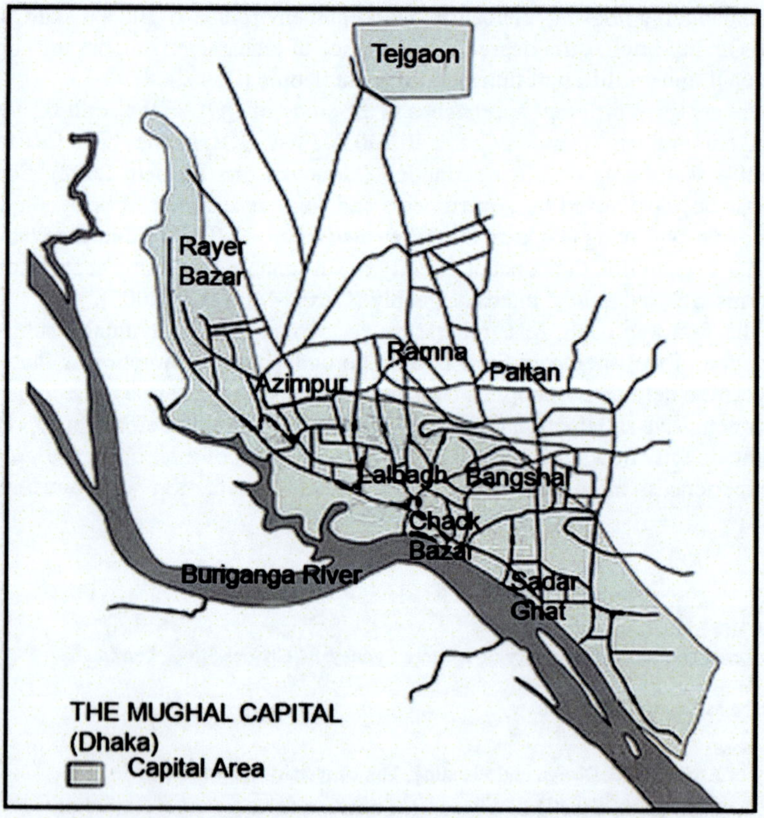


Fig. 11.1 Dhaka under the Mughal emperor (Source: geology.com 2009)

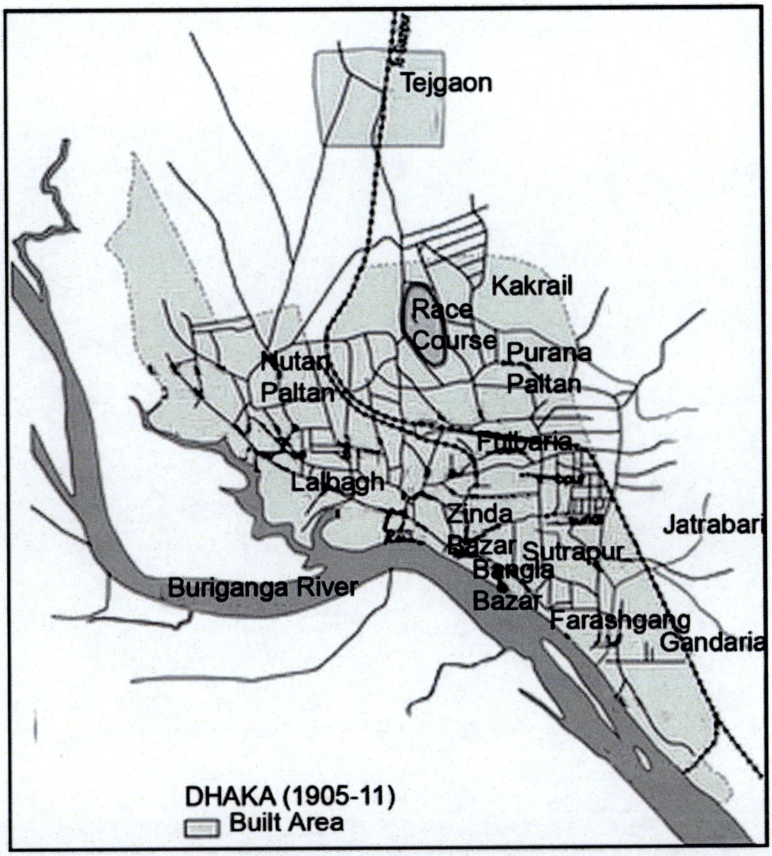


**Fig. 11.2** Dhaka as the capital of East Bengal and Assam (Source: Ahmed 2009)

became the capital of East Pakistan, a province of the newly created Pakistan. In 1971, following a bloody 9-month liberation war, Dhaka became the capital of Bangladesh (Fig. 11.3), the land of the Bangla-speaking people (Mamun 1991).

Dhaka is situated between latitudes 24°40′N and 24°54′N and between longitudes 90°20′E and 90°30′E and is defined by the Tongi Canal in the north, the Buriganga River in the south, the Shitalakhya and the Balu Rivers in the east, and the Turag River in the west.

The area of Dhaka under the jurisdiction of the Dhaka City Corporation (DCC) stands at approximately 145 km$^2$ (DCC 1999), constituting 90 wards.

Dhaka has a rich background of urbanization with various phases of cultural, social, and political transformation. The dynamics of these changes have had a deep influence on the people and their life style. Since the 1990s, rapid urbanization has lead to sharp changes in living style. The social acceptance of these changes, especially in the formal and spatial concepts of housing, has lead to a major dependency on building developers, technology, and various regulatory measures.

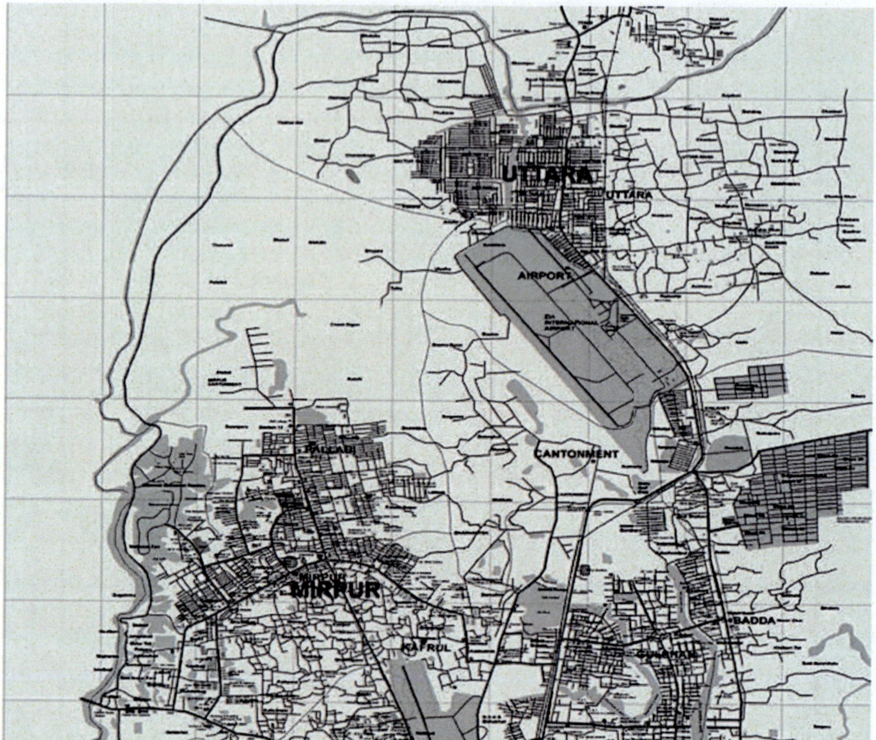


Fig. 11.3 Dhaka in its present context (Source: Ahmed 2009)

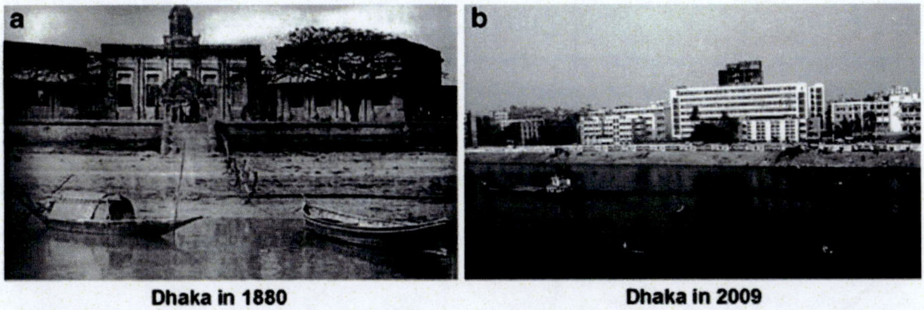


**Fig. 11.4** Riverfront developments in Dhaka in 1880 versus 2009 (**a**) Dhaka in 1880 (**b**) Dhaka in 2009 (Source: (a) British Library (2009); (b) Ahmed (2009))

The urban infrastructure service system has had difficulties in matching the pace of rapid transformation (Fig. 11.4), resulting in stressful conditions, for example, the large-scale and frequent failure of the electricity supply, impacts directly on housing satisfaction.

The population of Dhaka has risen rapidly. The UN Development Program (UNDP) has reported an annual growth rate of 10.8% during the period 1974–1981 (UNDP 1994). By 2001, Dhaka had over five million people, the result of a 14-fold increase in urban population since from 1980 to 2001 (BBS 2003). The population continues to grow at an estimated 4.2% per annum, one of the highest rates amongst Asian cities (McGee 2006). It is now the ninth largest city in the world. As per the UNDP, 90% of urban housing (apartments) was developed by the private sector, while only 10% is public housing, including that made available publicly to government employees.

## The Study Area

### *Six Different Housing Areas in Dhaka*

The study was undertaken in six major residential areas of Dhaka (see Fig. 11.5). Those areas contain the majority of medium- to high-density, mid- to high-rise housing in the city. In Fig. 11.6, housing that is typical of that found in each of those areas is illustrated.

*Dhanmondi* is one of the most crowded yet planned areas in Dhaka. Its origins date back to the late 1950s when the Dacca Improvement Trust (DIT) developed the area for purely residential purposes. Plots of land were allotted to ministers, government officials, public leaders, professionals and executives. A grid pattern of wide roads was built, and vegetation was planted along roadsides and in other public places. Dhanmondi began as an affluent residential area and primarily consisted of two-storey houses fostering a quiet neighborhood environment. In the

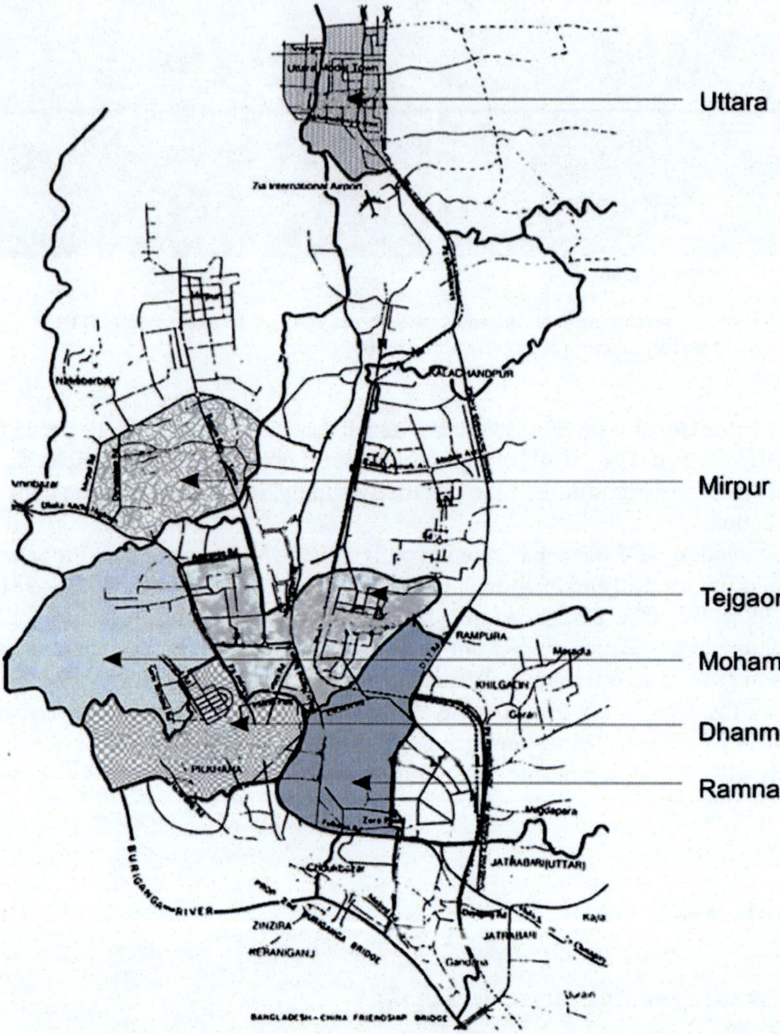


**Fig. 11.5** Location of the six areas of Dhaka chosen for the study (Source: The authors)

early 1970s, in response to tremendous pressures on land, that rule was relaxed, and over the decades, the area has evolved into a miniature city of major apartment buildings, plus hospitals, malls, schools, banks, offices and universities. Dhanmondi has now become a multipurpose area. The non-residential elements of Dhanmondi are now much larger and more important to the area than the owner-occupied and rental apartment buildings.

*Mirpur* is one of the most populous areas of the Dhaka City Corporation. It evolved rapidly after 1962. In this area, the majority of land has been used for

**Fig. 11.6** Typical apartment buildings surveyed in each of the six study areas (Source: All photographs by the first author, unless otherwise indicated)

residential purposes followed by commercial activities. Most of the governmental housing complexes are located in Mirpur. The area is situated at a distance from the CBD but is well connected with important areas of the city.

*Mohammadpur* had been developed primarily as a moderate-density residential area for middle-income people. Now, massive urbanization has turned Mohammadpur into another miniature city of major apartment buildings and has resulted in the loss of natural environment, including swamps and wetlands. Everything is available, and the transportation system connects the area well to other parts of the city. One of the largest apartment blocks in Dhaka is located here along with a number of housing societies. This has resulted in a real estate construction boom accompanied with the erection of different malls and shopping complexes. The development of most of the parts of Mohammadpur was planned in the 1950s with relatively broad streets and avenues. This has resulted in relative ease of traffic movement. However, with the increase of population (over 27,000 per km$^2$ in 2010), traffic congestion in the streets and roads has increased. This area also accommodates a number of refugee camps, of which one is the largest in Bangladesh. These camps lack in every socio-environmental quality and have considerable negative impact on the area.

*Ramana* is a very important area of Dhaka as many government, semi-government, autonomous, and private institutions are located here. It is adjacent to the CBD and therefore experiences a very high demand for residential accommodation. Many of the taller apartment buildings in the city are located in this area.

*Tejgaon* is a historic center of industrial activity in the city. The area contains numerous plants and factories, including such industries as garment manufacturing, food processing, metal works, and pharmaceuticals. Many government, semi-government, and autonomous organizations and institutions are also located in the area. Most banks operating in Bangladesh have a branch in Tejgaon. Being near the city center and close to civic facilities, Tejgaon has emerged as an important business district of Dhaka as well becoming a densely residential populated area (over 25,000 per km$^2$).

*Uttara* is a very large area measuring 36.91 km$^2$. It is primarily a residential suburb, and is divided into several sectors. The majority of the residents are from upper- and higher- middle-class backgrounds and prefer to live far from the congestion and pollution of the city. In recent years, with the increasing influx of people moving out from the city, Uttara has evolved into a bustling community, similar to other areas of Dhaka, thus losing its quiet, leafy, suburban character. The majority of the residents in this area do office jobs in the public or private sector. The area has several shopping complexes as well as governmental and private offices. Population density is the lowest of the six study areas, being only approximately 3,000 per km$^2$.

## Demographics of the Study Area

The study focused on investigating the relationships between apartment living, neighborhood, and residential satisfaction. The population studied was middle-income families living in predominantly mid- to high-rise apartment buildings in these six areas of Dhaka.

Dhaka has recently emerged as a rapidly growing megacity. According to the 2001 census, the size of Dhaka's population (under the Dhaka City Corporation) was 5.3 million (BBS 2003). The population is growing by an estimated 4.2% per annum. The estimated population of Dhaka in 2008 was 12.7 million (BBS 2009). The most recent census figures show that approximately one million people or 20% of the population live in apartment buildings (BBS 2003).

Survey results in 2005 (Hackenbroch and Gruebner 2005) indicated that 3.4 million people of very low socioeconomic status lived in slums of poor housing, high population density, and poor environmental services, especially water and sanitation. In 2005, this was 37% of Dhaka's population and increased from 3,000 slum communities in 1996 to 5,000 by 2005. This study was not about residential satisfaction for that third of the population but for the 20% of the population who live in medium- to high-density, mid- to high-rise apartment buildings.

As of 2001, the gender balance in Dhaka as a whole was 3.0 million (56%) male and 2.3 million (44%) female. The gender ratio (males per 100 females) of the population is calculated as 123.4 based on the population census (BBS 2001). The gender ratio of Dhaka has decreased over the years due mainly to the reunion of females with their male partners living in the city and the increase in the number of single females in the urban work force (Siddiqui et al. 1993).

The proportion of Dhaka's population that is young is relatively high due to age-selective rural–urban migration (Afsar 2000). About 40% of the city's total population is in what is considered "unproductive age groups" of 0–14 years and over 60 years, indicating a high dependency burden on the working age population (BBS 1997).

The city population is composed of people from virtually every region of Bangladesh. Population density is 222 persons per hectare within the central Dhaka City Corporation area and 71 persons per hectare for the metropolitan area as a whole.

The main occupations are in service (31.5%), commerce (23%), transport (8.5%), labor (9%), agriculture (7.6%), and construction (2.8%) (Islam 2004). Per capita income for the city as a whole is approximately US$500, and the average literacy rate is 53.9%, evenly divided between male and female.

## *Sociocultural Characteristics of the Study Areas*

Three major historic-cultural phases of Dhaka have marked influences on its urban morphology: first, an indigenous culture before the arrival of central Asians when the majority of the population represented different Hindu castes; second, a traditional culture of central Asia and north India; and third, the western culture of the Europeans.

Largely, the Aryans shaped the indigenous culture of the subcontinent. They invaded from Central Asia before the Christian era and drove the Dravidians south. They introduced a caste system to aid in their survival and to suppress the native agrarian people. Subsequent Central Asians played a similar role in breaking down custom and in the transmission and adaptation of new techniques. However, they were not altogether new people for the indigenous people in terms of habits and way of thinking, mode of production, and consumption. The English people who came from an entirely different course of life and different attitude towards life, mode of production, and consumption, were more alien to the indigenous and Central Asian populations. The initial intent of the English migration was for business and exploitation. They rapidly introduced sophisticated techniques for commercial benefit, tried to reorganize production and consumption behavior of the population, and tried to force their surplus products into this vast market.

Although since the beginning of the early twentieth century, Dhaka's recent growth rate has dropped to 4.2% per annum, as compared to 6.2% in the 1980s; resources available to local government have not been adequate to provide for the minimal level of services with which they are charged. Dhaka is experiencing a number of urban problems (for example, traffic congestion and malfunctioning of the overall transit system) due to the inability of its existing urban tissue to meet the requirements of rapid urbanization and population growth. In addition to its continuous expansion, the built environment of the city continues to change its texture predominantly by the rapid development of shopping centers and office and apartment buildings, meeting the needs of ongoing changes in social, cultural, and economic factors.

The urban population living in high-density housing has limited open space and an antiquated road infrastructure on which to move. The rapidly growing population in Dhaka and in its outskirts is also contributing to environmental pollution, posing problems to health and threatening the quality of life (Mashreque 2009).

## The Problem

As a physical setting, the residential environment is one of the factors critical for human well-being and quality of life. Typically, people spend most of their working time in buildings and most of their leisure time at home or close by in their neighborhood (García-Mira et al. 2005).

Neighborhoods are ideal units to study and assess QOL as they combine physical and social aspects that impact on daily life (Romice 2005). Social networks tend to overlap on spatial arrangements, and issues of territoriality, identity, and well-being become attached to location (Morrison 2003; Moudon and Ryan 1994). Neighborhoods act as important sources of opportunity and provide a sense of identity. On the other hand, neighborhoods can also act as a constraint on personal life chances (Madanipour et al. 1998). These facts justify the need to determine the role that neighborhoods play in the quality-of-life experience of individuals in communities.

The study reported here explored QOL in six major residential neighborhood areas in Dhaka as part of a larger research project on residential satisfaction. It placed quality of urban life (QOUL) as an important focus while taking into account the interaction between people and their neighborhood.

The quality of the urban environment throughout the world has emerged as an issue of fundamental concern for researchers, policy makers, and citizens (Das 2008). However, there is little research done on quality of life in Dhaka. One study considered the QOL, mental health, and nutritional status of adolescents (Izutsua et al. 2006). Other studies examined the quality of life of older people in rural areas (Nilsson et al. 2005, 2006; Rana et al. 2009) and QOL of workers (Khaleque 1995, 1999). Nevertheless, to date, no study has explicitly investigated the quality of *urban* life, focusing on the physical environment or on residential satisfaction as a component of QOUL.

## Methodology

### Research Design

The study examined housing satisfaction in its socio-physical context. The research defined independent and dependent variables within a new model of residential satisfaction.[1] The central questions of interest were to what degree

[1] Presentation of the new model of residential satisfaction goes beyond the scope of this chapter and will be presented by the authors in a subsequent paper.

residents were satisfied with their housing, including their neighborhood, and what aspects of the housing and neighborhood contributed to overall residential satisfaction or dissatisfaction.

Although we wished to study residential satisfaction *in situ,* control of the independent and dependent variables was neither possible nor desirable. Therefore, the basic research design used survey research. In comparison to laboratory experiments, which involve examination of a phenomenon in a controlled setting by manipulating independent variables and observing their effect on dependent variables, this study did not attempt to make causal conclusions. Instead, this study makes correlational conclusions. This is the approach of the vast majority of housing studies, including those dealing with residential satisfaction.

Survey research is the most appropriate method in this context and enabled the collection of information from a broad and representative sample of inhabitants regarding residential satisfaction in Dhaka. The survey research design and the large sample also allowed broad generalization to the rest of Dhaka's population. To the best our knowledge, this research was the first of its kind conducted in Dhaka or anywhere in Bangladesh.

The study surveyed respondents using a pilot-studied, structured questionnaire. The study involved two stages:

- First, pilot studies were conducted in a Bangladeshi community in Sydney and later in Dhaka to develop and test the research instrument and to acquire practical experience with the potentials and constraints of the survey process.
- Second, using the refined instrument, the main study was conducted in Dhaka households with the purpose of addressing the research questions.

## Sampling and Sample Size

The survey used stratified sample design to generate a random sample of Dhaka households.

To ensure that a representative and random sample of the population living in mid- to high-rise housing, several steps were taken:

- First, six areas of the city with the majority of medium- to high-density, mid- to high-rise housing were selected (the six described above).
- Second, a subset of buildings was chosen randomly from within each area.
- Third, as previous research has suggested that residential satisfaction is influenced by distance of the housing unit from the ground floor, sub-populations were selected based on floor levels (two upper floors versus two lower floors of each building).
- Fourth, from among the upper and lower floors, apartment units were then randomly selected.
- Finally, one adult respondent 18 years of age or older was selected at the time of the interview in the sampled units.

The sample size was based on published tables, which provided necessary guidelines to determine the actual sample size from established sampling criteria (e.g. percentage of sampling error, confidence level, and split-level). A further consideration for sample size was the minimum number needed for data analysis, as the sample size needs to be appropriate for the planned analysis. Due to the objectives of the research and the nature of the research questions, the study employed multivariate statistical techniques. Accordingly, a sample size of at least 200 respondents was required according to tables in de Vaus (2002) in order to achieve no more than a 7% sampling error with 95% confidence level, assuming a 50/50 split. In anticipation of non-responses, a total of 236 people were contacted to obtain the final sample of 204 interviews.

## *Measurement Instrument*

Based on previous residential satisfaction studies and an informal reconnaissance in the study areas in Bangladesh, a structured questionnaire – called The Residential Satisfaction in Dhaka Questionnaire – was developed. It was refined through a two-part pilot study, first in a Bangladeshi residential area in Sydney and second by interviewing 30 residents living in apartment buildings in Dhaka. The questionnaire was designed to measure resident perceptions and evaluations about a number of aspects of their housing and neighborhood environment.

The final questionnaire comprised of 33 questions in five sections covering the apartment unit, the building, management and maintenance, neighbors, and the neighborhood. Responses to each question were measured on a linear numeric version of a Likert-type scale. When items are to be judged on a single dimension and arrayed on a scale with equal intervals, a simple, linear numeric scale with the extremes labeled appropriately is the most statistically appropriate method of scaling. Whereas the traditional Likert-type scale produces only ordinal data and is thus inappropriate for parametric statistics, linear-numeric scales lead to equal-interval data that may be analyzed using the most powerful parametric statistics (Alreck and Settle 1995).

For the majority of questions, respondents were requested to place a tick in one of five boxes to indicate their level of satisfaction with each item on a five-point linear numeric version of a Likert scale, "1" standing for strong level of dissatisfaction and "5" representing a strong level of satisfaction. To give more options to respondents and to achieve precision consistent with reliability, at the end of each section, questions were asked about peoples' overall satisfaction, and responses were rated on a seven-point linear numeric scale. Examples of both types of questions are given in Figs. 11.7 and 11.8 – one of the 29 specific residential satisfaction questions is given in Fig. 11.7, and an example of one of the four overall questions of satisfaction, as the predictor dependent variables, in Fig. 11.8.[2]

As the primary language of the respondents was Bangla, the state language of Bangladesh, the questionnaire was translated into Bangla.

[2] The full "Residential Satisfaction in Dhaka Questionnaire" is available on request from the first author.

| Q.11. How satisfied are you with the following features of your neighborhood? | | | | | |
|---|---|---|---|---|---|
| **Neighborhood** | Strongly dissatisfied | | | | Strongly satisfied |
| | 1 | 2 | 3 | 4 | 5 |
| Physical appearance of the neighborhood buildings | □ | □ | □ | □ | □ |
| Children's recreation facilities (park, play field) | □ | □ | □ | □ | □ |

Fig. 11.7 Example of one of the satisfaction questions regarding neighborhood on a five-point linear numeric scale (Source: The authors)

| **Overall Neighborhood** | Strongly dissatisfied | | | | | | Strongly satisfied |
|---|---|---|---|---|---|---|---|
| | 1 | 2 | 3 | 4 | 5 | 6 | 7 |
| Q.12. Overall, how satisfied are you to live in your neighborhood? | □ | □ | □ | □ | □ | □ | □ |

Fig. 11.8 Example of an overall satisfaction question on a seven-point linear numeric scale (Source: The authors)

## Findings

The survey data were analyzed using a number of statistical tool to identify QOL by considering satisfaction with the apartment unit, the apartment building, and the neighborhood. The following summarizes the major results regarding the components of residential satisfaction in Dhaka.

### *The Quality of Neighborhood: A Major Component of Residential Satisfaction*

To examine empirically the components of residential satisfaction, all continuous variables used to measure residential satisfaction were analyzed using factor analysis. Principal component analysis revealed six components of residential satisfaction:

- Management and maintenance
- Architectural features
- Neighborhood
- Neighbors
- Recreation facilities
- Ambient environment

**Table 11.1** Factor analysis of the component "neighborhood"

| Item description | Factor loading |
|---|---|
| Neighborhood cleanliness | .81 |
| Traffic safety | .79 |
| Neighborhood parking facilities | .74 |
| Noise level as created by traffic | .71 |
| Street condition in the neighborhood | .66 |
| Crime and vandalism situation | .65 |
| Garbage disposal of neighborhood | .59 |
| Open space condition in the neighborhood | .52 |
| Physical appearance of the neighborhood buildings | .49 |
| Neighborhood building maintenance | .46 |

Source: The authors

Three of these relate to the quality of the socio-physical neighborhood – Neighborhood, Neighbors, and Recreation facilities – the third including satisfaction with parks, playing fields, and libraries. Taken together, the factors Neighborhood, Neighbors, and Recreation facilities clearly indicate the importance of the socio-physical neighborhood environment as a major contributor to residential satisfaction in medium- to high-rise apartment living in Dhaka.

The component Neighborhood included ten items of residential satisfaction (shown in Table 11.1). The factor loading of each of these ten items was higher than on the other five components of residential satisfaction. Overall, the items all relate to the quality of Neighborhood. This component had an eigenvalue of 3.85 and accounted for 6% of the total variance in residential satisfaction across the 204 respondents. Correlation values of the items were very strong, ranging from 0.227 to 0.629. The reliability coefficient of 0.869 showed strong internal consistency of this item set.

## *The Impact of the Quality of Neighborhood on Overall Residential Satisfaction*

To examine how well the six components of residential satisfaction predicted overall residential satisfaction and to identify the best predictor of overall residential satisfaction, multiple regression analysis was employed.

The model was composed of the six components (as independent variables) of residential satisfaction. Each component was evaluated to ascertain which of the variables contributed to the prediction of overall residential satisfaction. As seen from Table 11.2, Neighborhood had the second largest beta coefficient (0.28), indicating that after Management and maintenance, this component was not only the

**Table 11.2** Coefficients of the components of residential satisfaction

| | Standardized coefficients | | Correlations | | | Collinearity statistics | |
|---|---|---|---|---|---|---|---|
| Component | Beta | Sig. | Zero-order | Partial | Part | Tolerance | VIF |
| Management and maintenance | .48 | .001 | .70 | .62 | .42 | .79 | 1.26 |
| Neighborhood | .28 | .001 | .50 | .45 | .26 | .90 | 1.11 |
| Architectural features | .22 | .001 | .49 | .35 | .20 | .85 | 1.18 |
| Neighbors | –.19 | .001 | –.38 | –.33 | –.17 | .94 | 1.06 |
| Ambient environment | –.14 | .001 | –.40 | –.23 | –.13 | .87 | 1.15 |
| Recreation facilities | .08 | .05 | .13 | .14 | .08 | .99 | 1.01 |

Source: The authors
Dependent variable: overall residential satisfaction

second factor extracted but also made the second strongest contribution to overall residential satisfaction. Taken together with the factor of Recreation facilities (which included satisfaction with parks, playing fields, libraries and other neighborhood facilities), these results clearly indicate the importance of the neighborhood environment as a major component of residential satisfaction in medium- to high-rise apartment living in Dhaka.

A further analysis focused on the location of neighborhoods. As the survey was administered in various neighborhoods in Dhaka, it was of interest to ascertain if there were differences in levels of reported satisfaction between neighborhoods. Figure 11.9 shows these differences in terms of percentage of satisfied and dissatisfied residents.

Descriptive statistics suggest that percentages of satisfied residents were highest in Uttara and Tejgaon and lowest in Mirpur and Mohammadpur. Two thirds of the respondents in Uttara (65%) were satisfied with "living here," while the fewest respondents (17.3%) were satisfied in Mohammadpur.

Taking Uttara and Mohammadpur as extreme cases, an ANOVA with planned comparisons was run to investigate whether or not there was a statistically significant difference between these two locations and extremes of satisfaction.

The statistical analysis showed that Levene's test was significant, implying that it was not possible to assume equal variances. Therefore, the second row (that is, do not assume equal variances) of Table 11.3 was considered for this analysis. The row indicates that the level of contrast is highly significant as $p \leq .001$. Therefore, there is a statistically significant difference in the perception of satisfaction between residents living in Uttara and Mohammadpur.

A frequency analysis was performed on the neighborhood items (see Table 11.1) for Uttara and Mohammadpur. That analysis showed that despite the significant difference in overall neighborhood satisfaction between the two areas, respondents in both locations were most *satisfied* with the low levels of crime and vandalism in

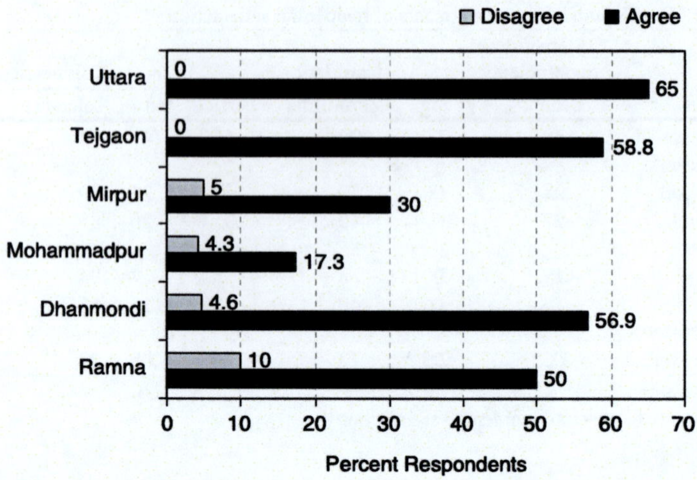


**Fig. 11.9** Percentage of agreement or disagreement of satisfaction with "living here" by location (Source: The authors)

**Table 11.3** Contrast test between Uttara and Mohammadpur for comparison of "overall satisfaction in living here"

| | | Contrast | Value of contrast | Std. error | t | df | Sig (2-tailed) |
|---|---|---|---|---|---|---|---|
| Overall satisfaction | Assume equal variances | 1 | 1.43 | .37 | 3.86 | 199 | .001 |
| | Do not assume equal variances | 1 | 1.43 | .29 | 4.84 | 43.52 | .001 |

Source: The authors

their neighborhoods, followed by the quality of garbage disposal, and then the three variables related to traffic: noise, safety and parking. Conversely, residents of both areas were most *dissatisfied* with neighborhood recreational facilities for children and adults (Table 11.4).

What differentiates between the areas the most is:

- The relative levels of neighborhood satisfaction with the physical appearance of the neighborhood (for example, 70% satisfaction in Uttara versus only 13% satisfaction in Mohammadpur)
- Building maintenance, street condition and neighborhood cleanliness (all with greater than 30% differences in satisfaction between the two areas)

**Table 11.4** Comparative assessment of neighborhood by respondents of low (Mohammadpur) and high (Uttara) satisfaction area

| | Mohammadpur | | Uttara | |
|---|---|---|---|---|
| | Satisfied (%) | Dissatisfied (%) | Satisfied (%) | Dissatisfied (%) |
| Physical appearance of the neighborhood buildings | 13.00 | 39.10 | 70.00 | 15.00 |
| Open space | 17.40 | 43.50 | 42.50 | 27.50 |
| Street condition | 21.70 | 39.10 | 57.50 | 12.50 |
| Children's recreation facilities | 13.00 | 65.20 | 27.50 | 50.00 |
| Adult recreation facilities | 13.00 | 65.20 | 5.00 | 55.00 |
| Building maintenance | 4.30 | 56.50 | 42.50 | 15.00 |
| Noise level as created by traffic | 43.40 | 39.10 | 50.00 | 25.00 |
| Neighborhood parking facilities | 39.10 | 30.40 | 50.00 | 15.00 |
| Neighborhood cleanliness | 17.30 | 60.90 | 50.00 | 15.00 |
| Traffic safety | 30.40 | 26.10 | 60.00 | 10.00 |
| Crime and vandalism | 56.50 | 8.70 | 75.00 | 15.00 |
| Garbage disposal | 43.40 | 26.10 | 67.50 | 12.50 |
| Overall satisfaction with neighborhood | 8.70 | – | 50.00 | 2.50 |

Source: The authors

## Discussion

Neighborhood satisfaction is one of the most frequently researched topics in neighborhood studies (Hur and Morrow-Jones 2008) and is also a typical component of residential satisfaction studies. In other parts of the world, high neighborhood satisfaction has been found to increase households' sense of community and vice versa (Brower 2003; Mesch and Manor 1998). Neighborhood dissatisfaction also influences people's decisions to relocate to another area (Brower 2003; Lee et al. 1994; Marans and Rodgers 1975; Marans and Spreckelmeyer 1981). In Malaysia, studies by Salleh (2008) and by Husna and Nurizan (1987) found that neighborhood factors influence residents' needs and expectations and were very important in determining residential satisfaction. Those findings are confirmed in the current study, the first of its kind for Bangladesh. The quality of neighborhood, including neighborhood cleanliness, a number of factors about traffic, parking and noise, low crime and vandalism, and good garbage collection and disposal, are also critical to residential satisfaction in Dhaka.

The value attached to a neighborhood is expressed in both internal and external image-forming processes. The perception of the public and the opinions of stakeholders make up a neighborhood's external reputation; however, the image that the residents have is its internal reputation (Adriaanse 2007). A growing proportion of people in Dhaka have access to and ability to pay for housing of reasonable quality. It appears, therefore, that other aspects of the residential environment have become more important to overall residential satisfaction, namely, the perception and evaluation of their neighborhood. For instance, as found in this study, people are more concerned about neighborhood cleanliness and traffic safety and about the presence of children's and adult recreation facilities, as well as the condition of surrounding streets and open space in the neighborhood, than about the physical appearance of their own apartment building or layout of their own apartment unit.

In the quest to identify the components of residential satisfaction, the results from the present study have revealed that neighborhood is a very strong predictor of residential satisfaction. It is second only to management and maintenance. The latter has been found in many studies to be the most important predictor of residential satisfaction, going all the way back to the now classic studies by Francescato et al. (1979). Other than physical appearance, other parts of overall neighborhood satisfaction are satisfaction with the other people living in the area.

Satisfaction clearly relates to an absence of or at least low levels of crime and vandalism. Clean neighborhoods are one of the prime concerns of respondents for satisfaction. However, the majority of respondents in the current Dhaka study were dissatisfied with the cleanliness of their neighborhoods. Furthermore, respondents were not happy with parking in the neighborhood, they were bothered by the noise generated by traffic, and they were dissatisfied with overall traffic safety. Finally, there was concern about the lack of parks and open space for recreation, indicating the necessity for a more sympathetic approach to planning.

The relationships between overall residential satisfaction with neighborhood and traffic safety and neighborhood street conditions indicated that safety issues associated with location are significant in predicting satisfaction. Since the overall rates of crime and vandalism are relatively low within the neighborhoods studied and are generally satisfactory for the residents, the major safety concern was with traffic. Improved transportation infrastructure to relieve residential areas of traffic noise and traffic safety concerns, combined with street improvements (for example, adequate lighting, signage, maintenance), would upgrade traffic conditions for the neighborhood.

Neighborhood garbage disposal is also important. Proper, effective, and regular garbage collection and disposal combined with other aspects of neighborhood cleanliness are important to residents.

Location of the neighborhood in the larger urban conglomeration is also an important aspect of residential satisfaction. For the most part, this finding agrees with those of other researchers. For example, in their study of US new towns, Burby and Weiss (1976) found that housing and the immediate neighborhood character were major factors in families' decisions to move to both new and conventional communities. Similarly, Michelson (1977) reported that 14% of wives interviewed in residential environments in Toronto and 13% of husbands mentioned neighborhood factors as reasons for moving away from their home. Francescato et al. (1979) pointed out that in the open market, people who can make a choice tend to place great importance on locational factors, particularly in the presence of good schools (if they have children), the general upkeep of the neighborhood, and the type of people living in the immediate environment. In 1980, Chi and Griffin (1980) conducted a social indicator study in two squatter settlements and one public housing area in Limon, Costa Rica, and measured residential satisfaction. They found that the proximity of a squatter settlement to the center of the city contributed to high levels of residential satisfaction. Conversely, residents in public housing projects were less satisfied with the location of their community because it was not easily accessible to churches, schools, and the market. These findings stressed the importance of location in predicting residential satisfaction.

In the research reported here, the greatest difference in residents' expectations was between the two neighborhoods of Mohammadpur and Uttara. The majority of the residents in the former were dissatisfied, while those in the latter were satisfied. What makes them different? As previously shown in Table 11.4, although a high percentage of residents of Mohammadpur were satisfied with safety and security issues, however, they were less satisfied with the physical appearance of neighborhood buildings, building maintenance, open space conditions, and cleanliness. These items appeared to be the main source of dissatisfaction. Earlier literature supports the concept that appearance is a predictor of residential satisfaction (Francescato et al. 1979). The majority of the residents of Uttara expressed satisfaction with the physical appearance, safety and security, and street conditions in their neighborhood. These were dominant predictors of satisfaction. In both locations, the fact that people were dissatisfied with recreational facilities was not a strong influence on their overall satisfaction.

However, despite the importance of locational factors, this finding should not be interpreted to mean that housing development could not be successful if located in an undesirable neighborhood. Success in both the developed world and the developing world can be achieved as long as undesirable locational factors are offset by a combination of residents whom others find compatible, a pleasant and healthy physical environment, and effective and friendly management system.

## Conclusions

This chapter has reported on the first study conducted on residential satisfaction in Bangladesh. From a stratified random sample of 204 middle-income households living in medium- to high-density, mid- to high-rise housing in Dhaka, a number of conclusions have arisen.

It is clear from the factor analysis and the regression analyses that the physical quality of the neighborhood is a dominant predicator of overall residential satisfaction (second most important in Dhaka behind the management and maintenance of the housing itself). This finding mirrors numerous findings from elsewhere in the developing and developed world, both that the quality of neighborhood is critical in residential satisfaction and that it falls behind the management and maintenance of the housing itself.

This finding is buttressed by two of the other factors found in this study – neighbors and recreation facilities for children and adults, including parks, playing fields, and libraries. The latter relates to the socio-physical qualities of the neighborhood.

The importance of neighborhood is likely due to its being the basic environmental unit in which social life occurs, therefore affecting residents' overall QOL. High neighborhood satisfaction contributes to a strong sense of community, while low satisfaction influences people's desire and ultimately their decision to move elsewhere. Most respondents in the current study have access to housing of reasonable quality (as reported by respondents) and are more concerned about neighborhood cleanliness, traffic, parking and noise, low crime and vandalism, and good garbage collection and disposal than about the physical appearance of their own buildings or the architectural layout of their apartment units. Therefore, like many other cities in the world, satisfaction with neighborhood is one of the most important elements of overall residential satisfaction in Dhaka.

The current study found that the social, economic, and physical environments all contribute to the quality of urban life. How in particular do these dimensions of the environment affect the quality of life for Dhaka residents? The answer is through the mediating effects of neighborhood and community on residential satisfaction and thus onto overall QOL. Specifically, satisfaction with the physical environment of the neighborhood, neighbors, and neighborhood social life affects life satisfaction through what we might call community satisfaction. That is, satisfaction with one's interactions with neighbors, the compatibility of people living in one's neighborhood, low crime and vandalism, and the availability of neighborhood resources like outdoor recreation space for children and adults all contribute significantly to Dhaka residents' overall satisfaction with the community. These overall feelings about the community, in turn, likely play a significant role in life satisfaction. Similarly, neighborhood socioeconomic features, such as satisfaction with one's neighbors, cost of living in the neighborhood, and neighborhood improvements can affect life satisfaction through overall feelings about the house, home, and community.

The Dhaka study findings are similar to those of other studies in other parts of the world, that the overall socio-physical features of the neighborhood and community influence life satisfaction more than the physical features of the individual dwelling. That is, in Dhaka, as elsewhere in the world, community and neighborhood are more central to people's lives than is the quality of their immediate dwelling unit.

**Acknowledgements** This chapter reports on the part of the research from an unpublished PhD thesis submitted by the first author under the supervision of the second author to the University of Sydney, 2010. The research was supported by Endeavour International Postgraduate Research Scholarship to the first author. Acknowledgement is extended also to members of the Environment, Behaviour and Society Research Group for critical feedback, to five research assistants in Dhaka, and to the residents of Dhaka who participated as respondents in this research.

## References

Adriaanse, C. C. M. (2007). Measuring residential satisfaction: A residential environment satisfaction scale (RESS). *Journal of Housing and Built Environment, 22*(3), 287–304.

Afsar, R. (2000). *Rural-urban migration in Bangladesh: Causes, consequences and challenges*. Dhaka: Dhaka University Press.

Ahmed, E. (2009). *Dhaka: Visual records of its colourful past*. Retrieved September 25, 2009, from http://dhakadailyphoto.blogspot.com/2007/06/maps-dhaka-and-bangladesh.html

Alreck, P. L., & Settle, R. B. (1995). *The survey research handbook* (2nd ed.). Chicago: Irwin.

BBS. (1997). *Population census 1991: National report provisional*. Dhaka: Bangladesh Bureau of Statistics (BBS), Government of Bangladesh.

BBS. (2001). *Population census 2001: National report provisional*. Dhaka: Bangladesh Bureau of Statistics (BBS), Government of Bangladesh.

BBS. (2003). *Population census 2001: National report provisional*. Dhaka: Bangladesh Bureau of Statistics (BBS), Government of Bangladesh.

BBS. (2009). *Statistical pocket book 2009*. Dhaka: Bangladesh Bureau of Statistics (BBS), Government of Bangladesh.

Blanco, A., & Chacon, F. (1985). The evaluation of the quality of life. In J. F. Morales, A. Blanco, & C. H. y. Fernandez (Eds.), *Applied social psychology*. Bilbao: Desclee de Brouwer.

British Library. (2009). Dhaka circa 1880s: Buriganga River (photograph from the Asia, Pacific & Africa collections). *British Library Online Gallery*. London, British Library. Retrieved September 2009, from http://www.bl.uk/onlinegallery/onlineex/apac/index.html

Brower, S. (2003). *Designing for community*. College Park: University of Maryland Press.

Burby, R. J., III, & Weiss, S. F. (1976). *New communities USA*. Lexington: Lexington Books.

Chi, P. S. K., & Griffin, M. D. (1980). Social indicators for measuring residential satisfaction in marginal settlements in Costa Rica. *Social Indicators Research, 8*(4), 453–465.

Das, D. (2008). Urban quality of life: A case study of Guwahati. *Social Indicators Research, 88*, 297–310.

DCC. (1999). *Dhaka City Corporation (DCC) profile*. Dhaka: Dhaka City Corporation. Retrieved February 21, 2009, from http://www.dhakacity.org/Page/Menu_Profile/Profile/Category/1/Profile_info.

de Vaus, D. A. (2002). Surveys in Social Research (5th ed.). Crow's Nest, Australia: Allen and Unwin.

Diener, E., & Suh, E. (1997). Measuring quality of life: Economic, social, and subjective indicators. *Social Indicators Research, 40*, 189–216.

Donovan, N., & Halpern, D. (2002). *Life satisfaction: The state of knowledge and implications for government*. London: Cabinet Office/Prime Minister's Strategy Unit.

Francescato, G., Weidemann, S., Anderson, J. R., & Chenoweth, R. E. (1979). *Residents' satisfaction in HUD-assisted housing: Design and management factors*. Washington, DC: US Department of Housing and Urban Development.

García-Mira, R., Uzzell, D. L., Real, E., & Romay, J. (2005). Housing, space and quality of life: Introduction. In R. García-Mira, D. L. Uzzell, E. Real, & J. Romay (Eds.), *Housing, space and quality of life* (pp. 1–5). Aldershot: Ashgate.

Geology.com. (2009). *World map: Political*. Retrieved September 25, 2009, from http://geology.com/world/world-map.shtml

Hackenbroch, K., & Gruebner, O. (2005). *Land use and land cover mapping in informal settlements: The case of Dhaka, Bangladesh*. ITC expert meeting on slum mapping, Technical University Dortmund, May 2008. Retrieved November 12, 2010, from http://www.ciesin.org/confluence/download/attachments/34308102/Hackenbroch_presentation_Dhaka.pdf?version=1

Hossain, S. (1990). *Echoes from old Dhaka*. Dhaka: Hakkanni.

Hur, M., & Morrow-Jones, H. (2008). Factors that influence residents' satisfaction with neighborhoods. *Environment and Behavior, 40*(5), 619–635.

Husna, S., & Nurizan, Y. (1987). Housing provision and satisfaction of low-income households in Kuala Lumpur. *Habitat International, 11*(4), 27–38.

Islam, S. (Chief Ed.). (2004). *Banglapedia: National encyclopedia of Bangladesh*. Dhaka: Asiatic Society of Bangladesh.

Izutsua, T., Tsutsumi, A., Islam, A. M., Kato, S., Wakai, S., & Kurita, H. (2006). Mental health, quality of life, and nutritional status of adolescents in Dhaka, Bangladesh: Comparison between an urban slum and a non-slum area. *Social Science and Medicine, 63*, 1477–1488.

Jackson, T. (2002). Quality of life, economic growth and sustainability. In M. Cahill & A. Fitzpatric (Eds.), *Environment and welfare: Towards a green social policy* (pp. 97–116). London: Palgrave Macmillan.

Khaleque, A. (1995). Industrial pollution and quality of life of workers in Bangladesh. *Journal of Human Ergology, 24*(1), 13–23.

Khaleque, A. (1999). Sleep deficiency and quality of life of shift workers. *Social Indicators Research, 46*(2), 181–189.

Lee, B. A., Oropesa, R. S., & Kanan, J. W. (1994). Neighborhood context and residential mobility. *Demography, 31*(2), 249–270.

Madanipour, A., Cars, G., & Allen, J. (Eds.). (1998). *Social exclusion in European cities*. London: Jessica Kingsley Publishers.

Mamun, M. (1991). *Dhaka: Tale of a city*. Dhaka: Dhaka City Museum.

Marans, R. W., & Rodgers, W. L. (1975). Toward an understanding of community satisfaction. In A. H. Hawley & V. P. Rock (Eds.), *Metropolitan America in contemporary perspective* (pp. 299–352). New York: Wiley.

Marans, R. W., & Spreckelmeyer, K. F. (1981). *Evaluating the built environment: A behavioural approach*. Ann Arbor: University of Michigan Institute for Social Research and the Architectural Laboratory.

Mashreque, M. S. (2009). Dhaka's environment suffers urban impact [electronic version]. *The Daily Star Online (Bangladesh)*. Retrieved December 24, 2009, from http://www.thedailystar.net/newDesign/news-details.php?nid=107999

McGee, T. (2006). Urbanization takes on new dimensions in Asia's population giants [electronic version]. *Population Reference Bureau*. Retrieved February 10, 2009, from http://www.prb.org/Articles/2001/UrbanizationTakesonNewDimensionsinAsiasPopulationGiants.aspx

Mesch, G. S., & Manor, O. (1998). Social ties, environmental perception, and local attachment. *Environment and Behavior, 4*, 504–519.

Michelson, W. (1977). *Environmental choice, human behavior, and residential satisfaction*. New York: Oxford University Press.

Morrison, N. (2003). Neighbourhoods and social cohesion: Experiences from Europe. *International Planning Studies, 2*(8), 115–138.

Moudon, A. V., & Ryan, M. (1994). Reading the residential landscape. In S. J. Neary, M. S. Symes, & F. E. Brown (Eds.), *The urban experience: A people-environment perspective* (pp. 183–313). Edmonds, Suffolk: St Esmondsbury Press.

Nilsson, J., Grafström, M., Zaman, S., & Kabir, Z. N. (2005). Role and function: Aspects of quality of life of older people in rural Bangladesh. *Journal of Aging Studies, 3*, 363–374.

Nilsson, J., Rana, A. K. M. M., & Kabir, Z. N. (2006). Social capital and quality of life in old age. *Journal of Aging and Health, 3*, 419–434.

Rana, A. K. M. M., Wahlin, A., Lundborg, C. S., & Kabir, Z. N. (2009). Impact of health education on health-related quality of life among elderly persons: Results from a community-based intervention study in rural Bangladesh. *Health Promotion International, 1*, 36–45.

Romice, O. (2005). Neighbourhood quality of life-global and local trends, attitudes and skills for development. In R. García-Mira, D. L. Uzzell, E. Real, & J. Romay (Eds.), *Housing, space and quality of life* (pp. 71–79). Aldershot: Ashgate.

Salleh, A. G. (2008). Neighbourhood factors in private low-cost housing in Malaysia. *Habitat International, 32*(4), 485–493.

Siddiqui, K., Qadir, S. R., Alamgir, S., & Haq, S. (1993). *Social formation in Dhaka city*. Dhaka: Dhaka University Press.

UNDP. (1994). *UNDP 1994 report on human development in Bangladesh: Empowerment of women*. Dhaka: United Nations Development Programme (UNDP).

# Chapter 12
# The Salzburg Quality of Urban Life Study with GIS Support

**Alexander G. Keul and Thomas Prinz**

## Introduction

Contemporary studies about quality of urban life (QOUL) outlined and presented in this book usually deal with communities, towns, cities, megacities, or urban regions. As quality of life (QOL) and, more specifically, QOUL, is a subjective, context-dependent, transactional phenomenon, it will show different domain-specific features when – to use Bronfenbrenner's terminology (1977) – studied in the environment of a social-behavioral micro-, meso- or exo-system. For the meso- or exo-system of a whole urban area, methodology and evaluation will involve highly aggregated data, whereas QOUL analysis on the microlevel of individual households can include phenomena of local geography and small-group sociology. This was the reason why the Salzburg QOUL study which is discussed in this chapter took a bottom-up instead of a top-down approach. It started with a 1-km² test area, and was later widened to include six city districts for purposes of comparison.[1]

[1]Further details of the Salzburg project are available in Prinz et al. (2007) and Keul et al. (2007a, b).

A.G. Keul (✉)
Psychology Department, Salzburg University, Hellbrunnerstr.34, 5020 Salzburg, Austria
e-mail: alexander.keul@sbg.ac.at

T. Prinz
Research Studio iSPACE, Schillerstrasse 25, 5020 Salzburg, Austria
e-mail: thomas.prinz@researchstudio.at

R.W. Marans and R. Stimson (eds.), *Investigating Quality of Urban Life: Theory, Methods, and Empirical Research*, Social Indicators Research Series 45,
DOI 10.1007/978-94-007-1742-8_12, 

## Context for the Study

In European research, QOL is a colorful topic. Noll (2004) lists seven main avenues of research related to QOUL, including the EC Eurobarometer, the International Social Survey Program, and the European Values Study. Maderthaner (1995, 1997) expressed methodological caution since QOL and well-being have short- and long-term components that are triggered rationally and emotionally, with heuristics and errors well known in decision theory. In his QOL Atlas for Germany, Korczak (1995) included 44 objective QOL indicators relating to six dimensions. In total, he listed nine QOL dimensions: Environment, welfare, culture, safety, provision, health, leisure, nutrition, and freedom.

As discussed in Chap. 4, the investigation of QOUL can be enhanced by making use of the Geographic Information Systems (GIS) technology, which is already widely used to enhance data in studies of public health (see, for example, Pearce et al. 2006), biology, criminology, political science, and in environmental protection. In environmental psychology, GIS has only been recognized more recently as providing an opportunity to enhance investigations in that field (called "open door" by Golledge 2002), but its use seems to have been limited to studies investigating spatial orientation and environmental cognition. However, in QOUL studies the use of GIS is very recent and is still relatively rare. In a U.S. study of QOUL in the Detroit region, Marans (2003, 2004, 2007) has promoted an interface of unit record based subjective measures of QOUL with GIS-based spatial objective measures of QOL.

In investigating subjective QOL, the various domains that make it up may assume different levels of importance for different people and it has received varying emphasis by researchers in different disciplines. For example, in Austria, housing satisfaction has been studied separately by psychologists, sociologists, economists and geographers, but without GIS support. The latest and biggest QOUL survey studied Linz, Upper Austria (Stadtforschung Linz 2005), with a sample of 21,069 respondents. But it did not geocode the data set so the visualization of results gives no spatial information, which limits the usefulness of the data for city planning and for political purposes.

The study discussed in this chapter evolved in the following way. After a pilot study in Salzburg and Linz in 2004, which involved the questionnaire used in the Detroit region study (referred to above), and following a 2006 visit to the Institute for Social Research at the University of Michigan by the first author, the authors tried to incorporate geo-informatics analysis into their research design and data analysis. GIS research had already created a set of quantitative spatial indicators for the urban area of the Salzburg City Council. A QOUL-GIS project with student interviewers was started in 2006 in the implementation phase of the new Salzburg Spatial Development Concept (Magistrate Salzburg 2007). It followed a bottom-up approach exploring the subjective assessment of QOUL seen by local residents. A project with the Salzburg City Council then commenced in 2007 investigating QOUL in six Salzburg urban districts (Keul 2008).

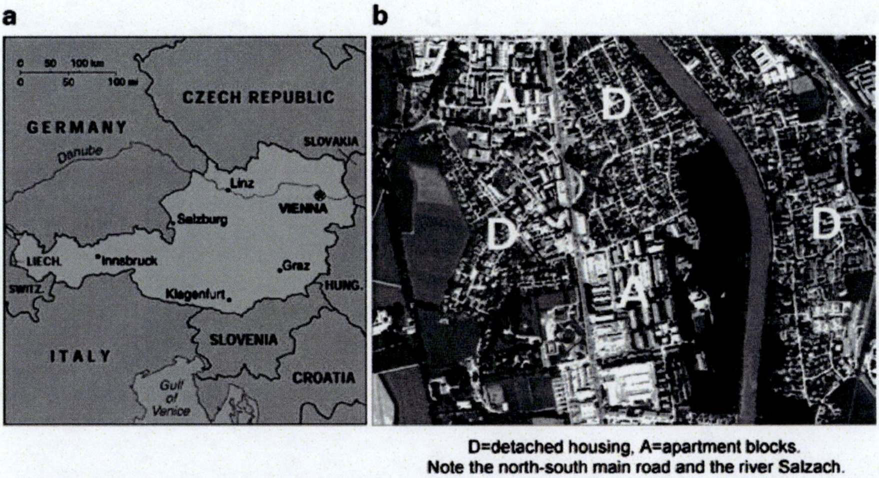


**Fig. 12.1** The location of Salzburg in Austria (**a**), and a satellite image of south east Salzburg (**b**). *D* detached housing, *A* apartment blocks. Note the north-south main road and the river Salzach (Source: CIA The World Factbook (a) and iSPACE (b))

## The Study Area

Austria is situated mostly in the Alpine region of Central Europe (see Fig. 12.1a) and consists of nine Federal Provinces. It has long had a rather constant population of about eight million inhabitants – 16% aged under 15 years, 62% 15–60 years, and 22% 60 years and over. Life expectancy for Austrians born in 2000 was 75 years. The mean net annual income in 2003 for employees was 17,148 Euros (Statistik Austria 2009).

Salzburg City, the capital of the Federal Province of Salzburg, covers 65.6 km$^2$ at 424 m altitude. In 2009, it had a population of 149,108 (Magistrate Salzburg 2009), and the population density was 2,273 per km$^2$ – which is about half the density of Vienna (the Austrian capital) with 1.7 million inhabitants. Salzburg City has 68,693 households. In Salzburg Province, 30% of the population is single and 30% are two-person-households. In Salzburg City, there are more single households. The mean number of children was 1.1 in 2007 for the Salzburg Federal Province, and 37.3% had no children. Salzburg City residents have fewer children. In 2007, 20.5% foreign citizens lived in Salzburg City. The unemployment rate for Salzburg Province was 6.0% in 2009 (Salzburger Landesregierung 2009). In a consumer survey on household budgets in 2000, it was found that 23.5% of expenditure was used for housing, 15.0% for transportation, 13.2% for food, and 12.3% for leisure (Statistik Austria 2009).

First mentioned in the year 1120, Salzburg is probably Austria's oldest city. Known internationally for its classical composer Wolfgang Amadeus Mozart, the Salzburg Festival, and the scenery of the movie "The Sound of Music," it is organized

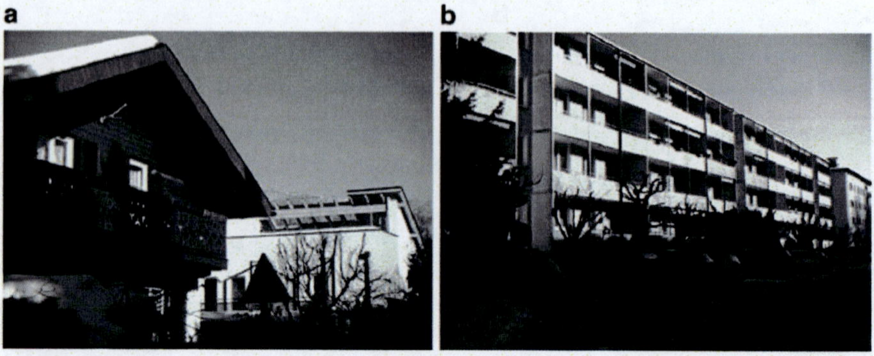

**Fig. 12.2** (a) Typical traditional and modern detached housing, and (b) multistory in the Salzburg-South study area (Source: The first author)

in 24 urban districts, some old, some new, such as Liefering which joined the city in the twentieth century (Medicus 2009). The Salzburg outskirts are a mixture of green spaces (wood, meadows), detached housing areas, and apartment blocks.

The study area location in Salzburg-South is shown in Fig. 12.1, and typical examples of housing in the area are illustrated in Fig. 12.2.

## Methodology

The study was conducted in two phases:

(a) The first phase (Study 1) followed a bottom-up research strategy and focused on a 1-$km^2$ area in southern Salzburg, representing about 3.6% of the city's population. A random sample design was used for the QOUL survey, and socio-demographic data and spatial indicators should be matched on a micro-system level with local geographical features.

(b) The second phase (Study 2) continued with the sample survey approach conducted in six districts of Salzburg (out of a total number of 24 city districts), representing one-third of the city's population. But this QOUL survey moved from the micro- to the meso-system, comparing social and spatial features at the district level.

### *Study 1: QOUL and GIS Analysis*

The QOUL-GIS test area situated in Salzburg City is referred to as Salzburg-South, and its center is defined by the map reference N 47°47 E 13°04, east and west of the southern main road Alpenstrasse. It covers an area of 1 $km^2$ (0.386 square miles) and has 100 grid squares of 100 m×100 m. The grid structure is identical with the new GIS system of the Salzburg spatial indicators. Aggregated socio-demographic data (the number of inhabitants, gender, age structure, etc.) and quantitative spatial indicators (including housing structure, child welfare, leisure areas, and greenery) already existed for the grid squares. The total population is 5,307 people. Of the 100 grid squares, 56 have a population density of more 30 inhabitants.

From an address-based master list, a random sample design was used to select dwellings in which interviews were conducted. The survey questionnaire focused on 22 QOUL items mostly derived from the Marans Detroit region QOUL 2001 study (which has been discussed in Chap. 7). That included questions collecting information on the socio-demographic characteristics of respondents (gender, age, profession, education, household size, housing type), and qualitative data on their subjective assessment of QOUL domains. Information on the residential address was later used for geocoding. Fourteen questions collected information relating to the following:

- Years of residence at the address
- Subjective assessment of QOUL
- Public transport quality
- Distance to public transport
- Housing satisfaction
- Food shops and food shop quality
- Use of green space
- Leisure space accessibility
- Neighborhood quality
- Assessment of safety and threats

A 5-point Likert scale was used to collect data on peoples' subjective assessments. In addition, the interviewer rated the quality of the visible residential building on a 5-point scale.

With the address data list and the instructions for the random sampling of dwellings, 293 personal doorstep interviews were conducted, representing a sample of 5.5% of the residential population. Those interviews were conducted in the mild winter of 2006/2007 in 55 $ha^2$.

The interview address data were later geocoded, and the QOUL data were analyzed using SPSS as well as a GIS map (a satellite-based orthophotograph).

Methodically, geocoded psychological field data are a relatively new field without traditional rules of scientific data processing. From everyday knowledge it is obvious that people or families living in detached houses will appear as single dots on a GIS map whereas people from a multi-story housing block – although possibly with different QOUL values – occupy almost the same location on the GIS map.

In geocoding spatial positions of QOUL values by local residents (see Fig. 12.6a), multiple occupants of the same geographical position were indicated by the total number of all residents per grid square. A different way of visualizing this might

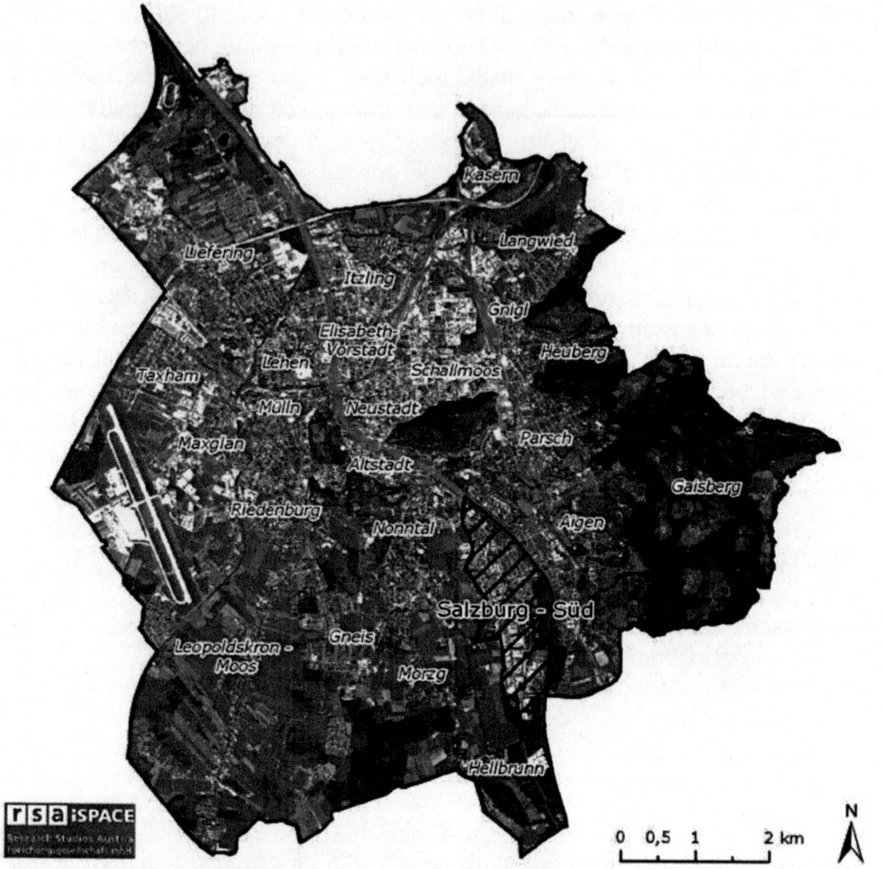


**Fig. 12.3** Position of Salzburg districts, Salzburg-South highlighted (Source: iSPACE)

have been to enlarge the respective position dots. In a next step, QOUL values per resident were transformed into mean grid (hectare) values by dividing the sum of QOUL values through the number of grid (hectare) residents.

The two main (ad hoc) research hypotheses for the QOUL-GIS test were:

- Local QOUL is a transactional product of person *and* environment. Person- and environment-based effects can be distinguished.
- Geocoded QOUL follows environmental "quality gradients" (for example, QOUL has a maximum in high-greenery areas and gets lower with decreasing green environment).

**Fig. 12.4** Superimpositions of mental maps of Salzburg-South inhabitants, age 30–50 (Source: First author, result of a 2007 student's survey; background map SAGIS)

On the official City Council map, the district of Salzburg-South is a compact block along the river Salzach and along the main road Alpenstrasse (see Fig. 12.3). Drawn as a "mental map" (that is, as represented in the heads of its residents), the same area can look quite different to different people and may vary from the spatial objective reality. In a 2004 seminar, students asked 26 inhabitants to draw mental map sketches of their "home area" on an unmarked map of southern Salzburg. The results are shown in Fig. 12.4. The superposition of the individual mental maps does not merge into a well-defined structure, but falls apart into totally different shapes from a line around the whole city quarter to isolated activity islands. A generally accepted "edge" (as in the terminology of Lynch 1960) is the river Salzach. At its western edge, the city district merges into green spaces.

## *Study 2: QOUL by Districts*

In collaboration with the Department of City Planning in the Salzburg City Council, six of 24 Salzburg districts were selected for this phase of the study. The intent was to collect information from a survey approach to potentially support the discussions about the new Salzburg Spatial Development Concept (REK; Magistrate Salzburg 2007). The city districts Gnigl, Gneis, Leopoldskron-Moos, Liefering, Parsch, and Salzburg-South (the latter being familiar from Study 1) are located on the city outskirts to the northwest, northeast, and south. Figure 12.5a shows their positions on a city map, and Fig. 12.5b overlays a GIS hectare cell population density map.

**b**

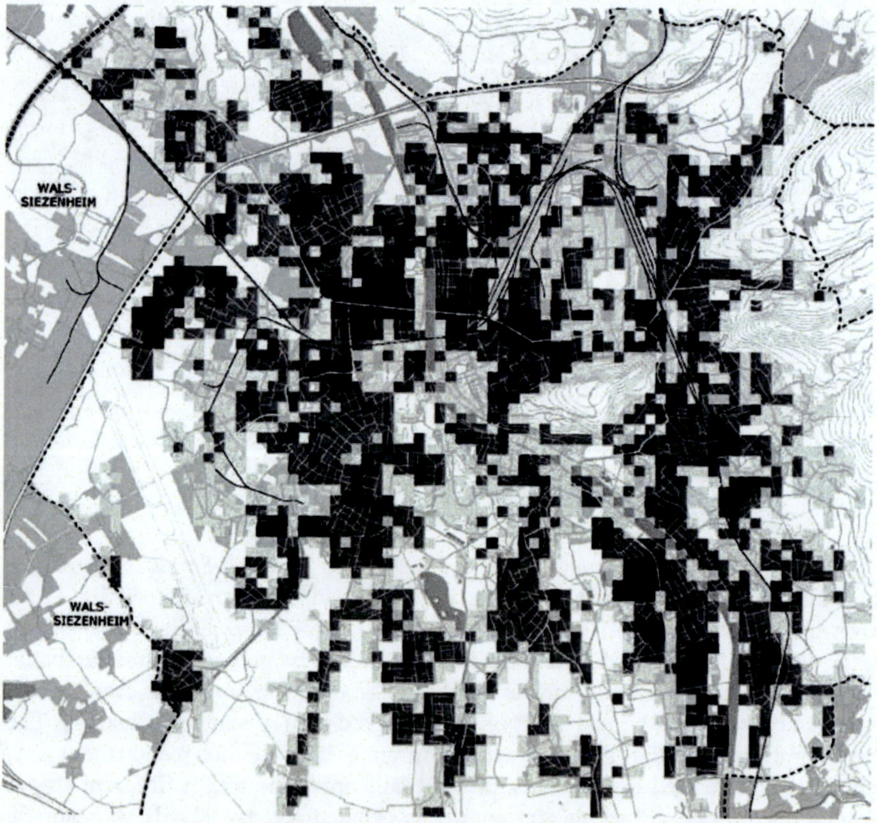


**Fig. 12.5** Grid overlay of population density per hectare within Salzburg (Source: iSPACE)

Parts of Gnigl, Liefering, Parsch and Salzburg-South are densely populated, whereas Gneis and Leopoldskron are not. As "new" districts from the historical point of view, the social borderlines of the six city districts are not so pronounced as those of the old Salzburg city center.

To compare local QOUL and its domains and characteristics in the six different areas, a questionnaire was compiled, pre-tested and delivered by interviewers face-to-face with respondents. For the sampling, no address-based master list was used but interviewers were instructed to select inhabitants of single houses in a street or of tenants in multi-story housing according to a random number list. A sample of 372 was thus selected and the questionnaires were administered in late 2007. For an area with a total population of about 49,000 inhabitants (33% of the city of Salzburg population) this represents a 0.75% sample.

The questionnaire gathered the following information:

- Address of survey respondent for geocoding
- Eleven socio-demographic variables (gender, age, education, profession, household size, number of children, family status, housing square meters, housing type, birthplace location, and residence time in district)
- A sketch of the subjective outline of the home district on an unmarked Salzburg 1:50,000 SAGIS map
- Four qualitative variables (benefits/drawbacks of life in the district, elements of the district center, cultural events/supply)
- Fifty-seven quantitative variables, each on a 5-point Likert scale (personal selection of district or not; modal mix of personal mobility walking/bicycle/ public traffic/car; 6 on overall district quality; 3 on image, 9 on social issues; 2 on social infrastructure; 2 on social plus culture; 5 on quality for children/ youth; 2 on quality for old people; 6 on environment; 4 on infrastructure - for example, shopping-; 2 on culture; 1 on future of district; 5 on place attachment; 5 on meeting points)

This gives a total of 74 items/variables on the respective city district.

The three (ad hoc) research hypotheses/propositions to be tested were:

- Mean Salzburg QOUL values, when different for the six city districts (as defined by the City Council), follow objective geographical differences (such as maximum building density).
- QOUL in the six Salzburg city districts studied is a combination of image (e.g. safety) and behavioral components (e.g., leisure quality) which can be distinguished.
- Variables of statistical influence on mean district QUOL are person- and environment-based and may be analyzed separately.

## Results for Study 1: QOUL and GIS Analysis

On the small scale of Study 1, a test field of 1 $km^2$, QOUL phenomena can be shown in fine grain within hectare cells of $100 \times 100$ m. With the aid of GIS, the transactions environment-QOUL were visualized as being changes in building density, green spaces, and infrastructure.

### *Mean Scores for QOUL*

Figure 12.6a shows the locations of the 293 field interviews geocoded onto the iSPACE GIS map. A mean QOUL value was computed for every grid hectare square ($100 \times 100$ m). Fifty five grid squares with a population over 30 inhabitants were sampled.

On the 5-point Likert scale, the resulting QOUL values have a left-skewed distribution between 1.0 and 4.0. Thirty six squares are QOUL $1<2$, 17 are $2<3$, and 2 are 3–4. In Fig. 12.6b, all QOUL hectare areas are marked with white dots and are coded as follows:

- "+" represents a mean QOUL value of 1.0
- "O" represents a QOUL value between 1.1 and 1.9
- "-" represents a mean QOUL value between 2.0 and 4.0
- The lowest QOUL between 2.5 and 4.0 is indicated with "- -."

Nine squares have top QOUL (1.0) ratings; 27 squares have an average QOUL rating (1.1–1.9); and 19 squares have a reduced QOUL rating (2.0–4.0). Of the latter, five squares have the lowest QOUL rating (2.5–4.0). The minimum rated QOUL square (mean 4.0) is in the left hand upper corner. Low and highest rated QOL hectare cells are clustered irregularly.

### *Variations in QOUL Mean Scores*

It was found that there are QOUL differences related to type of housing and to some environmental factors. The following is evident from the QOUL mean scores:

(a) There is no QOUL gradient from negative to positive away from the main road Alpenstrasse which intersects the area almost vertically down the middle.
(b) Apartment houses have a tendency to have a more negative QOUL and detached houses a more positive QOUL, but with some exceptions. A negative QOUL square with a mean of 3.0 is found in a green detached housing area east of Alpenstrasse, and the most negative square (4.0) is a low-rise apartment housing, not high-rise blocks.

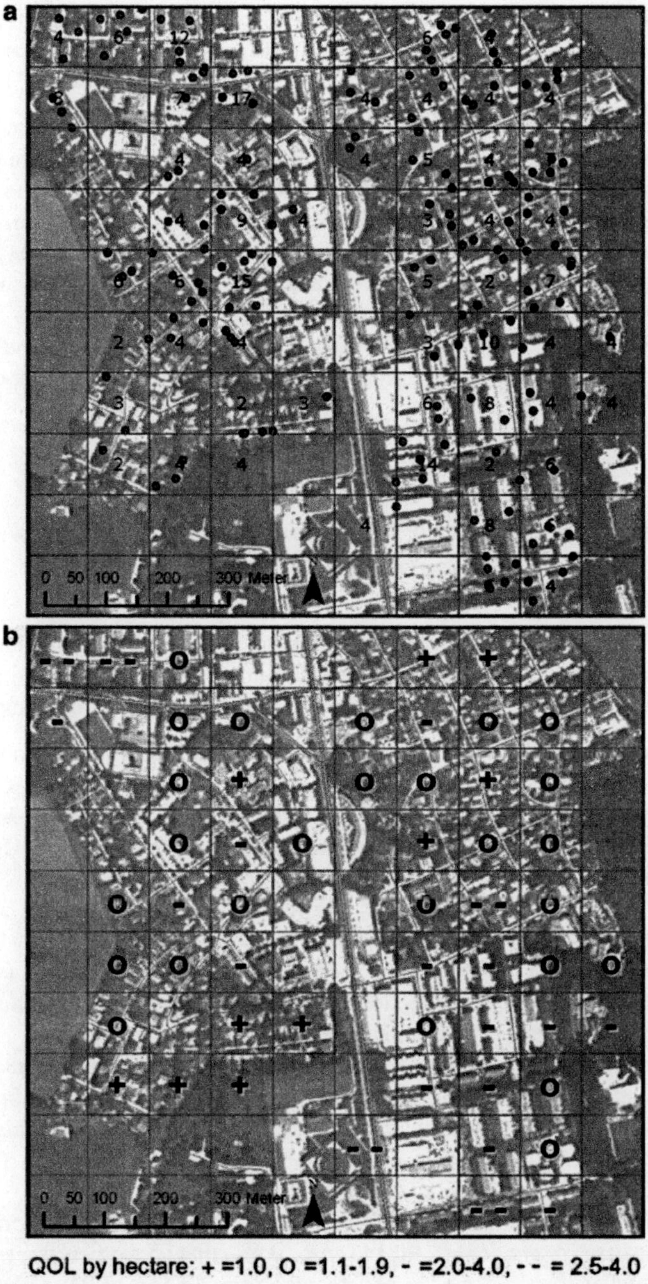


**Fig. 12.6** **(a)** Location of QOUL interviews in Salzburg-South; and **(b)** the mean scores on QOUL for *grid squares* (QOL by hectare: +=1.0, O=1.1–1.9, -=2.0–4.0, - -=2.5–4.0; Source: The authors; aerial map by iSPACE)

(c) A GIS overlap (not reproduced here) of geocoded satisfaction with the green environment versus percent of categorized green space per hectare does not show a gradient of rising dissatisfaction from "garden landscape" towards the main, treeless road.

As a consequence, subjective QOUL seems to be a function of more than geography, not only reflecting the actual (static) spatial situation, but also the (dynamic) time-aggregated experiences with the surrounding housing landscape. In a housing quality map of a real estate agent, you will find a quality gradient from positive to negative when approaching a main road and proportional to the local greenery. In the subjective QOUL of the resident, the distinctions are not so pronounced and the resulting values are – like housing satisfaction – rather a compromise product of cognitive dissonance than a product of sharp local quality analysis.

## *Using Correlation Analysis to Explore Relationships Between Assessed QOUL and Environmental and Person Factors*

The relationships between spatial QOUL mean scores and environmental and personal characteristics were analyzed using the non-parametric correlation analysis (Kendall's *t*). That revealed further interrelations, including the following:

(a) For respondent ratings of their satisfaction with six factors – housing, food shops, greenery, leisure, neighbors and safety, four of the ratings – housing ($\tau$=+0.314), greenery ($\tau$=+0.310), neighbors ($\tau$=+0.307), and safety ($\tau$=+0.193) – correlate positively and highly with the general QOUL rating, whereas leisure quality and food shops do not.
(b) There were positive correlations between greenery and leisure satisfaction ($\tau$=+0.427), and between neighbors and safety ($\tau$=+0.232).
(c) Housing satisfaction significantly increases with dwelling duration ($\tau$=+0.235) and age ($\tau$=+0.279) – a reflection of rootedness to the local community – and is inversely related to the distance to public transport ($\tau$=–0.214).
(d) Population density is relevant for QOUL. Hectare cell QOUL mean scores correlate positively with the number of inhabitants ($\tau$=+0.280, $p$<.006). Different housing types have a different population density, and high-density multistory housing needs a compensatory leisure environment for a good QOUL.
(e) Significant correlations were found for some person-based characteristics. This was the case for gender ($\tau$=–0.141, $p$<.012) – that is, women assessed QOUL to be higher than did men – and for age ($\tau$=–142, $p$<.002) – that is, older people assessed QOUL to be higher.

Thus, QOUL in Salzburg-South is clearly both person- and environment-based.

Also, environmental esthetics play a role. A Likert scale value for shabby/well-kept buildings (coded by the interviewers) correlates positively with inhabitant-reported QOUL ($\tau$=+0.335, $p$<.000).

## *Hypotheses Tested*

Summing up, for the Salzburg QOUL Study 1, the two research hypotheses have been tested with the following results:

(a) Local QOUL is a product of person-based *and* environment-based effects that can be identified/analyzed independently. This proposition was found to be *true* as both gender and age are significant, as are housing, greenery, population density.
(b) Geocoded QOUL follows environmental "quality gradients" – for example, QOUL has a maximum rating in high-greenery areas and gets a lower rating with a decreasing green environment. This proposition was found to be *false*, and such a simple relation is not present as QOUL is a function of more than geography.

# Results for Study 2: QOUL by Districts

As a first step, in Study 2, the QOUL data and the 74 socio-demographic and district-related image and behavior variables were statistically evaluated using the SPSS package. Some means, percentages and nonparametric correlation results are shown in Tables 12.1–12.3 for the six city districts.

## *QOUL, Socio-Demography and Housing*

Table 12.1 shows that city district QOUL mean scores on the Likert scale varied from between 1.24 in the Gneis district to 1.81 in the Salzburg-South district. For comparison, the results of the test area in Salzburg-South (project 1) were also put into Table 12.1.

The results indicated the following:

(a) Participant mean age and age ranges do not show large differences for the six city districts.
(b) High QOUL/housing satisfaction and percentages of people born in Salzburg City do not match. Only Salzburg-South, the lowest-rated-QOUL and housing satisfied district, has the lowest local birth rate.

**Table 12.1** QOUL, socio-demography and housing for six Salzburg city districts

| Criteria/measure | Gneis | Leopoldskron-Moos | Liefering | Gnigl | Parsch | Salzburg-South | GIS area Sbg-South |
|---|---|---|---|---|---|---|---|
| District population | 6,500 | 2,500 | 14,500 | 6,500 | 10,000 | 11,000 | 11,000 |
| *Sample size n=* | *50* | *50* | *75* | *50* | *73* | *74* | *293* |
| QOUL mean | 1.24 | 1.48 | 1.56 | 1.58 | 1.71 | 1.81 | 1.69 |
| QOUL sd | 0.45 | 0.71 | 0.76 | 0.76 | 0.74 | 0.80 | 0.80 |
| HS mean | 1.22 | 1.58 | 1.35 | 1.50 | 1.49 | 1.76 | 1.69 |
| HS sd | 0.42 | 0.76 | 0.60 | 0.77 | 0.73 | 0.92 | 0.80 |
| Age years | 18–87 | 21–74 | 20–82 | 20–84 | 20–87 | 20–84 | 14–88 |
| Mean age | 45.5 | 41.0 | 48.6 | 50.4 | 48.0 | 41.8 | 47.7 |
| Born Salzburg-city | 46% | 36% | 50% | 46% | 36% | 16% | – |
| HH no children | 80% | 84% | 58% | 64% | 79% | 88% | 73% |
| Mean square meter | 83.6 | 82.4 | 96.0 | 89.2 | 76.0 | 68.2 | – |
| Residence mean | 16.0 years | 12.5 years | 25.1 years | 24.6 years | 17.7 years | 11.7 years | 15.5 years |
| Residence range | 1–60 | 1–35 | 1–70 | 1–84 | 1–48 | 1–45 | 1–67 |
| Detached housing | 10% | 12% | 35% | 24% | 6% | 3% | 21% |
| Apartment blocks | 76% | 60% | 35% | 52% | 82% | 85% | 62% |
| Max. building density[a] | 1.1 | 0.5 | > 1.1 | 1.1 | > 1.1 | > 1.1 | > 1.1 |

Source: First author, 2007 students' project
*sd* standard deviation, *HS* housing satisfaction, *HH* household
[a]GFZ values from REK 2007, Salzburg City Council

**Table 12.2** QOUL and residential image/behavior variables for six Salzburg city districts

| Criteria/measure | Gneis | Leopolds-kron-Moos | Liefering | Gnigl | Parsch | Salzburg-South | GIS area Sbg-S |
|---|---|---|---|---|---|---|---|
| *Sample size n=* | *50* | *50* | *75* | *50* | *73* | *74* | *293* |
| QOUL mean score | 1.24 | 1.48 | 1.56 | 1.58 | 1.71 | 1.81 | 1.69 |
| District self-selected | 56% | 42% | 28% | 46% | 44% | 45% | – |
| District has center | 48% | 42% | 72% | 56% | 51% | 52% | – |
| Bike use | 56% | 30% | 52% | 54% | 36% | 41% | – |
| Car use | 42% | 54% | 49% | 42% | 47% | 32% | – |
| Very friendly | 52% | 56% | 68% | 62% | 38% | 34% | – |
| Very green | 67% | 78% | 56% | 49% | 36% | 34% | 1.3[a] |
| Very safe | 48% | 71% | 49% | 48% | 23% | 30% | 1.7[a] |
| Very good leisure | 27% | 20% | 24% | 20% | 8% | 12% | 1.2[a] |
| Very good for kids | 52% | 54% | 62% | 48% | 30% | 25% | – |
| Very good shopping | 12% | 18% | 60% | 24% | 41% | 47% | 1.5[a] |
| Social problems | 25% | 12% | 32% | 30% | 80% | 73% | – |
| Very positive development | 25% | 17% | 48% | 33% | 16% | 15% | – |
| District a part of the city | 20% | 42% | 77% | 62% | 51% | 55% | – |

Source: First author, 2007 students' project
[a]Scale 1–5

**Table 12.3** QOUL interrelations (non-parametric correlations, Kendall's $\tau$) for six city districts

| Criteria | Gneis | Leopolds-kron-Moos | Liefering | Gnigl | Parsch | Salzburg-South | GIS area Sbg-South |
|---|---|---|---|---|---|---|---|
| *Sample size n=* | *50* | *50* | *75* | *50* | *73* | *74* | *293* |
| QOUL mean score | 1.24 | 1.48 | 1.56 | 1.58 | 1.71 | 1.81 | 1.69 |
| Feeling at home | .001 | .000 | .000 | .000 | .005 | .000 | – |
| Housing satisfaction | .002 | .000 | .008 | .000 | .012 | .000 | .000 |
| Friendly district | n.s. | .000 | n.s. | .001 | .000 | .002 | – |
| Green district | n.s. | .000 | .000 | n.s. | n.s. | n.s. | .000 |
| Safe district | .002 | n.s. | n.s. | n.s. | n.s. | n.s. | .000 |
| Good leisure quality | .001 | .000 | n.s. | n.s. | n.s. | n.s. | .011 |
| Good for old people | n.s. | .000 | .000 | .002 | n.s. | n.s. | – |
| Good neighbors | n.s. | .000 | .000 | n.s. | n.s. | n.s. | .000 |
| Gender | n.s. | n.s. | n.s. | n.s. | n.s. | n.s. | .012[a] |
| Age | n.s. | n.s. | n.s. | n.s. | n.s. | n.s. | .002[b] |
| Building quality rating | – | – | – | – | – | – | .000 |

Source: First author, 2007 students' project
[a]QOL higher for female residents
[b]QOL higher for older people

(c) Mean housing floor spaces are higher in districts with higher QOUL means and smaller in the two districts with lower QOUL means.
(d) The length of residence and the percentage of detached houses show no relation with QOUL.
(e) Apartment blocks were more frequent (over 80%) in the lower rated QOUL districts.
(f) Maximum building density (measured as GFZ, which is the sum of built space for every story divided through the ground space of the site) also shows no simple relation, however, for two of the three highest-GFZ districts, mean QOUL was lowest.

## *QOUL and Residential Image/Behavior Variables*

Table 12.2 shows the mean QOUL ratings and several residential image/behavior variables for the six districts under study.

The results showed the following:

(a) Gneis, the district with the highest QOUL mean score, self-selected as their homeplace by 56% of the survey respondents, had 56% bike use, and was rated "very good for leisure" by 27%.
(b) Other image variables – district center, friendly, green, safe, good for kids/shopping, social problems, very positive development, part of the city – were not the highest at Gneis.
(c) Gneis had the lowest mean percent values for good shopping and being part of the city.
(d) Nevertheless, Salzburg-South, the district with the lowest QOUL mean score, is also assessed lowest for being friendly, green, good for children, and very positive development.

## *QOUL Interrelations*

Non-parametric correlations (Kendall's $\tau$) were run to test whether QOUL is statistically related in all six districts to main image and socio-demographic variables. Results are listed in Table 12.3.

Interpretations of the results are as follows:

(a) It becomes obvious that –when comparing district QOUL rating mean scores– feeling at home and housing satisfaction are the only key factors in relation to QOUL, as they correlate highly with QOUL in all six city districts.
(b) Other important image items (friendly, green, safe, leisure, for old people, neighbors) interrelate with QOUL only in some or few districts, not in all six.

(c) For the six districts with their sample sizes of 50–75, gender and age do not correlate with QOUL, but for the GIS grid test area and its big sample of 293, they do.
(d) The interviewer-coded assessment of building quality (which was only collected in the Salzburg-South GIS grid test area) also correlates with the QOUL values.

The highly significant correlations (Kendall's $\tau$) for the mean score ratings of QOUL for the six districts the following image items were found to be significant:

(a) *Gneis district*: safe, safe at night, leisure, knowing people, getting news quickly
(b) *Leopoldskron district*: friendly, green, leisure, church, for children, environment, beautiful, for old people, neighbors
(c) *Liefering district*: positive development, for old people, finding ones way in the district, neighbors, green
(d) *Gnigl district*: friendly, good environment, parks, walking, beautiful, for old people
(e) *Parsch district*: friendly, for children/youth, beautiful
(f) *Salzburg-South district*: friendly, church, positive development, beautiful

## Hypotheses Tested

Summing up the Salzburg QOUL project 2, the three research hypotheses have been tested with the following results:

(a) QOUL rating mean values for six different districts in Salzburg follow objective geographical differences (such as maximum building density). This proposition is shown to be *false*. For the meso-systems of the districts, no geographical relations were noticeable.
(b) The six Salzburg districts show different image and behavior components resulting in different QOUL ratings. This proposition is shown to be *partly true and partly false*. Feeling at home and housing satisfaction is related to QOUL rating in all six districts (*false*). QOUL shows no general (that is, is present in all six districts) relation to other image and behavior variables (*true*).
(c) Variables of statistical influence on district QUOL rating means are person-based *and* environment-based and may be identified/analyzed separately. This proposition is shown to be *false*.

## Conclusion

Contemporary spatial planning faces the challenge of simultaneously taking account of ecological, social and economic issues in the planning process. QOUL needs to be tested and spatial development should meet sustainability criteria. Socially responsible urban planning requires a feedback on environmental and infrastructural resources, and this may be enhanced through having access to subjective QOUL data collected from local populations.

What makes up housing areas with a high QOUL? What are the right infrastructure decisions to enhance QOUL? These are important questions. Public discussions on housing and residential quality and on residents' subjective evaluations of QOUL typically run on a number of levels (citizens, planners, politicians, etc.), and are characterized by different interests and languages.

In Salzburg City in Austria, an evaluating GIS tool has already been designed and tested in conjunction with the City Council. It uses a set of quantitative spatial indicators operated by methods of Applied Geo-informatics. This indicator set has been matched in Study 1 reported in this chapter with qualitative data collected through a sample survey of 293 geocoded respondents interviewed as part of a local QOUL survey in the same grid (that is, in high resolution). The QOUL spatial indicator set used at Salzburg City – in an analogy with the Detroit Area Study conducted by Marans (see Chap. 7) – contained variables of basic everyday life value like play areas, public transport, green leisure areas, shops, among and others. Also, QOUL and its constituent domains have been tested, but not geocoded, in six different Salzburg districts in Study 2. With 46 district image items and 15 district knowledge/familiarity/transport items, the 372 field interviews tested local QOUL using a total of 61 variables.

From the two studies involving 665 field interviews with Salzburg residents collected by 25 trained psychology students for their seminar work, the following conclusions can be drawn:

(a) In a GIS fine grain analysis of Salzburg-South, the rating of local QOUL revealed a close relationship with population density and housing type (detached house versus apartment blocks). However, no spatial interrelations with "good" versus "bad" environmental conditions (green areas versus main road) were found. QOUL reflects aggregated experiences with the landscape and its resources, and, like housing satisfaction, seems to depend on the personal level of expectation and habituation.
(b) Variations in the rating of QOUL in the areas of Salzburg that were studied are both person- and environment-based. Housing satisfaction, greenery, neighbors, and safety influence QOUL rating, as do the gender and age of the survey respondents.
(c) In a comparison of local QOUL rating in six Salzburg outer districts, QOUL mean scores were closely correlated with housing satisfaction and with "feeling at home," but not with other important image and behavior items.

For the user-oriented evaluation of local residential attraction, the combination of different indicators is important. One research possibility is a top-down approach; that is, an expert-rating of the overall importance of singular indicators. A bottom-up approach looks more promising as local residents have to express subjective QOUL using different aspects of residential quality. This is an innovative view of the city through the eyes of local inhabitants as true location experts and it is able to improve the diagnostic value of top-down GIS spatial indicator systems. The full potential of the comparative use of spatial indicators from geo-informatics and QOUL rating has yet to be unfolded, and synergies with other fields – like urban ecology bio-indicators – have only been pointed out in discussions. In a century of online citizen participation, sustainable planning goals and sophisticated data mining, the investigation of QOUL must incorporate GIS technology to merge environmental and health psychology with the research on urban well-being (Keul 1995). Evaluating subjective QOUL and overlapping it with GIS QOUL indicators is of vital political interest as Guideline 5 of the Salzburg Spatial Development Concept (Magistrate Salzburg 2007) demonstrates, the aim of all urban planning actions is the promotion of overall quality of life. The manifold urban functions are to be further developed and improved.

## Future Work

During the planning time of the QOUL book project, geocoding of the 2007 Study 2 data set was realized in cooperation with iSPACE Salzburg. Compared with the Study 1 $km^2$ and its 293 data points, the six Study 2 test spaces were each of about the same general size, but only filled with 50–70 data points (i.e. population samples of 0.5–2.0%). With the smaller population samples, in some districts (Leopoldskron-Moos, Gneis, and Parsch) there were not enough hectare cells left for a statistical evaluation. In the best GIS sample at district Gnigl, 26 of 65 population hectare cells with population size over 10 per hectare showed a non-significant correlation (Kendall's $\tau = -.145$, $p < .430$) of QOUL with population density, contrary to Study 1. What has to be concluded from this is not the invalidity of Study 1 findings but the inability to replicate a fine-grain sample phenomenon (293 data points per $km^2$) with a coarse-grained sample. In future QOUL studies, close attention has to be paid to the spatial data density.

Fine-grain micro-system (Bronfenbrenner) surveys demand plenty of field working time and energy, similar to Roger Barker's *behavior setting* approach (1968), and it is economically unlikely that they will be realized very often outside academic testing grounds. Nevertheless, a system theory perspective of environmental psychology asks the researcher to keep in mind that phenomena unfolding through empirical field research will always depend on the complexity level of the system studied – which is different for a street, a city district, a whole city and geographical units beyond that. As in social networks theory, more complex systems are governed by other rules as simpler systems. Realizing this will help to use QOUL as a calibrated tool for innovative social research.

**Acknowledgments** The field interviews of study 1 were undertaken by Salzburg psychology students Susanne Bachler, Anna Enz, Rene Hackstock, Julia Hochfilzer, Lars Larsen, Gerald Moser, Judith Schider, Eva Strasser, Sonja Tejkl, and Melanie Trieb. The field interviews of study

2 were undertaken by Stefanie Spoeth and Silvia Steingruber (Gneis), Melanie Bachinger and Sabine Hemmers (Leopoldskron-Moos), Tugba Doenmez, Tamara Plass and Daniela Udovicic (Liefering), Claudia Biehl and Martin Beinsteiner (Gnigl), Anita Koeberl, Christina Liesinger and Claudia Reslhuber (Parsch), as well as Sylvia Hametsberger, Karolina Horngacher and Isabella Uhl (Salzburg South). Study 1 was conducted in cooperation with RSA Research Studios Austria iSPACE Salzburg; project 2 in cooperation with the Salzburg City Council, Department of Townplanning. No financial grants were involved.

## References

Barker, R. (1968). *Ecological psychology: Concepts and methods for studying the environment of human behavior*. Stanford: Stanford University Press.

Bronfenbrenner, U. (1977). Towards an experimental ecology of human development. *American Psychologist, 32*, 513–532.

Golledge, R. G. (2002). The open door of GIS. In R. B. Bechtel & A. Churchman (Eds.), *Handbook of environmental psychology* (pp. 244–255). New York: Wiley.

Keul, A. G. (Ed.). (1995). *Wohlbefinden in der Stadt: Umwelt- und gesundheitspsychologische Perspektiven* [German: Well-being in the city: Perspectives of environmental and health psychology]. Weinheim: Beltz, Psychologie Verlags Union.

Keul, A. G. (2008). *Bericht ueber das Empirische Seminar Sozialpsychologie* WS 2007/08 [German: Report on the empirical seminar in social psychology, winter term 2007/08]. Salzburg: Psychology Department.

Keul, A. G., Prinz, T., & Zocher, D. (June, 2007a, June 29–30). *GIS-Darstellung staedtischer Lebensqualität (QOL) – Das Planquadrat Salzburg-Sued* [German: GIS-based visualization of urban QOL – The Salzburg South Interface]. Poster, Internationales Seminar Stadtoekologie, Salzburg, Austria.

Keul, A. G., Prinz, T., & Lang, S. (2007b). The Salzburg QOL-GIS interface – Spatial analysis of 293 urban QOL interviews. In *ERSA 2007, Local governance and sustainable development*, August 29–September 2, Book of Abstracts II. Special Sessions (pp. 32–33). Paris: CEE.

Korczak, D. (1995). *Lebensqualitaets-Atlas: Umwelt, Kultur, Wohlstand, Versorgung, Sicherheit und Gesundheit in Deutschland* [German: QOL-Atlas: Environment, culture, welfare, supply, safety and health in Germany]. Opladen: Westdeutscher Verlag.

Lynch, K. A. (1960). *The image of the city*. Cambridge: MIT Press.

Maderthaner, R. (1995). Soziale Faktoren urbaner Lebensqualitaet [German: Social factors of urban quality of life]. In A. G. Keul (Ed.), *Wohlbefinden in der Stadt* (S.172–197). Weinheim: Psychologie Verlags Union.

Maderthaner, R. (1997). Wohlbefinden und Lebensqualitaet [German: Well-being and quality of life]. *Psychologie in Oesterreich, 17*, 62–65.

Magistrate Salzburg. (2007). Raeumliches Entwicklungskonzept 2007 [German: Spatial development concept 2007]. Salzburg: Magistrat, Stadtplanung und Verkehr.

Magistrate Salzburg. (2009). Statistische Daten. Available at: www.stadt-salzburg.at

Marans, R.W. (2003, September 22–24). *Modeling residential quality using subjective and objective indicators: Opportunities through quality of life studies*. Presentation at IAPS Methodologies in Housing Research Conference, Stockholm, Sweden.

Marans, R.W. (2004, July 7–10). *The 2001 Detroit Area Study (DAS) on the quality of community life*. Presentation, Board Session, 18th IAPS Conference, Vienna, Austria.

Marans, R.W. (2007, August 29–September 2). *Theory and policy driven quality of urban life studies: Examples from the Detroit metropolitan area*. Paper for the session, Analysing Urban Quality of Life, ERSA, Paris.

Medicus, R. (2009). *SalzburgWiki*. Available at: www.salzburg.com/wiki/

Noll, H.-H. (2004, December 2–3). *Subjektives Wohlbefinden im Europaeischen Vergleich* [German: Subjective well-being in a European comparison]. Presentation, Workshop Subjektives Wohlbefinden und Subjektive Indikatoren: Konzepte, Daten und Analysen, ZUMA Mannheim, Germany.

Pearce, J., Witten, K., & Bartie, P. (2006). Neighborhoods and health: A GIS approach to measuring community resource accessibility. *Journal of Epidemiology and Community Health, 60*, 389–395.

Prinz, T., Keul, A. G., & Lang, S. (2007). Raeumlich differenzierte Lebensqualität – GISgestuetzte Analyse von 293 QOL-Interviews in der Stadt Salzburg [German: Spatially differentiated QOL – GIS-based analysis of 293 QOL interviews in Salzburg City]. In J. Strobl, T. Blaschke, & G. Griesebner (Eds.), *Angewandte Geoinformatik 2007* (pp. 589–594). Heidelberg: Wichmann.

Salzburger Landesregierung. (2009). *Statistische Daten*. Available at: www.salzburg.gv.at

Stadtforschung Linz. (2005). *Analyse Buergerbefragung 2004* [German: Analysis of the 2004 citizen survey]. Linz: Magistrate Linz.

Statistik Austria. (2009). *Statistisches Jahrbuch 2009* [German: Statistical yearbook 2009]. Vienna: Statistik Austria.

# Chapter 13
# Subjective Quality of Life in Queensland: Comparing Metropolitan, Regional and Rural Areas

**Rod McCrea, Mark Western, and Tung-Kai Shyy**

## Introduction

Little research in subjective quality of urban life (QOUL) examines differences between metropolitan cities, regional cities and towns, and rural areas. Research usually focuses on one urban size or the other (for example, Jongudomkarn and Camfield 2006; Talen 2001; Walter-Busch 2000). This chapter focuses on variations in subjective QOUL in the State of Queensland, Australia, by comparing and contrasting it within the state's metropolitan and non-metropolitan areas, the latter consisting of regional urban centers and rural areas. It divides the non-metropolitan (or what are referred to in Australia as regional urban centers) into cities and towns to examine and differentiate between those categories. The chapter is descriptive, using both univariate statistics and multivariate discriminant function analysis to distinguish between these different urban size categories in terms of both overall subjective QOUL and four main attributes of urban environments, namely:

- Access to services and facilities
- Noise pollution
- Incivilities
- Social capital.

R. McCrea (✉) • M. Western
Institute for Social Science Research, The University of Queensland,
Level 4, GPN3 Building, Campbell Road, St Lucia, Brisbane, QLD 4072, Australia
e-mail: r.mccrea@uq.edu.au; m.western@uq.edu.au

T.-K. Shyy
School of Geography, Planning and Environmental Management,
The University of Queensland, Brisbane, QLD 4072, Australia
e-mail: t.shyy@uq.edu.au

R.W. Marans and R. Stimson (eds.), *Investigating Quality of Urban Life: Theory, Methods, and Empirical Research*, Social Indicators Research Series 45,
DOI 10.1007/978-94-007-1742-8_13, 

Two main research questions are addressed in this chapter:

RQ1: How do levels of subjective QOUL in regional cities and towns differ from those in metropolitan and rural areas?

RQ2: Which main attributes of urban environments best distinguish regional cities and towns from metropolitan and rural areas?

## Overall Subjective QOUL

Overall subjective QOUL can be conceptualized in two broad ways (as discussed in Chap. 3) and can be viewed as:

- All aspects of QOL experienced within an urban environment; or
- Only those aspects of QOL derived from an urban environment.

The chapter takes the latter view of QOUL and focuses on the four main attributes of the urban environment mentioned above. Within the latter view, there are also different geographic levels of urban environment such as housing, neighborhoods, local communities and regions (see Campbell et al. 1976; Marans and Rodgers 1975). This chapter examines subjective QOUL and urban attributes associated with subjective evaluations of residents' local areas and neighborhoods, even though each resident is also situated within the broader classification of metropolitan cities, regional cities, regional towns and rural areas. Subjective urban QOL in this paper does not include subjective evaluations of a resident's housing or broader region.

Differences in overall subjective urban QOL for residents living in metropolitan, regional and rural areas may be limited. As discussed in Chaps. 1 and 3, various psychological mechanisms create a tendency for residents to be satisfied with the place in which they live (for example, the satisfaction treadmill and psychological adaptation). Also, during the residential relocation process, people tend to move to local areas which satisfy them on aspects important to them. This applies to both intra-urban migration (McCrea 2008) and migration between major urban, regional and rural areas (Williams and Jobes 1990). Thus, differences in overall subjective QOUL may not be great between metropolitan, regional and rural areas. However, we expect evaluations in these differently sized urban areas to still be different on main dimensions of urban environments.

Residential location preferences may also reflect compositional differences in socioeconomic and demographic characteristics of local areas. For example, higher socioeconomic status residents tend to place more importance on lifestyle characteristics afforded by local areas in choosing where to live (Malecki and Bradbury 1992; Williams and Jobes 1990). Thus, variations in subjective urban QOL between areas may be partly accounted for compositional differences in socioeconomic and demographic characteristics of the areas. However, this chapter does not aim to control for compositional differences (which would tend to explain differences in evaluations between local areas in compositional effects). We aim simply to describe differences in subjective QOUL between metropolitan, regional and rural areas.

## *Attributes of Subjective QUOL*

RQ2 aims to distinguish between metropolitan, regional and rural areas using four main attributes of urban environments: access to services and facilities, noise pollution, incivilities and social capital. These are discussed in turn.

### Access to Services and Facilities

This is an important predictor of subjective QOUL (Glaeser et al. 2000; Rogerson et al. 1989, 1996) and can be expected to vary between local areas in metropolitan, regional and rural areas. The *theory of optimal centrality* (Archibugi 2001; Cicerchia 1999; see also Chap. 3) suggests that the level of access to services and facilities available in an urban center depends on the size of an urban center. This is particularly so for access to higher level services and facilities (such as medical specialists, universities and theaters) which are unlikely to be available or viable in smaller urban centers or rural areas. Even perceptions of whether a place is rural or urban can depend more so on perceived access to services and facilities than perceived community size (Pagliccia et al. 1995).

However, optimal centrality theory also suggests that as the size of urban centers becomes large (for example, metropolitan cities), they can become "overloaded." Then, negative attributes of QOUL become more prevalent, such as pollution, congestion and social incivilities. So optimal centrality theory asks, what is the optimal size to which an urban center might grow before the onset of urban overload outweighs any added benefits of access to services and facilities? Cities can grow too large, and smaller regional cities can potentially offer higher subjective QOUL than larger metropolitan cities. In this study, residents evaluate their access to services and facilities as well as two aspects of urban overload: noise pollution and incivilities.

### Noise Pollution

This is commonly associated with traffic (for example, car, train and plane noise) as well as noise from industry and entertainment venues. Traffic is also associated with a range of interrelated urban problems such as air pollution, congestion, accidents and parking problems (Proost and Van Dender 2001; Romilly 1999). So noise pollution in a local area can be indicative of urban overload and is a significant predictor of neighborhood dissatisfaction (Parkes et al. 2002).

### Incivilities

Incivilities may be both physical and social in nature. Physical incivilities include things like vandalism, littering, vacant housing, abandoned cars and buildings, and untidy allotments. Social incivilities include things like gang activities, loud parties, homelessness, drunkenness and loitering. These incivilities are linked to subjective QOUL because incivilities indicate a breaking down of both norms of behavior and social control in neighborhoods (Perkins and Taylor 1996; Skogan and Maxfield 1981).

Incivilities associated with the breaking down of social norms and social control are expected to occur more in large, rapidly growing areas with high in-migration, according to social disorganization theory (Burgess 1967 [1925]; Shaw and McKay 1942). High rates of residential growth, mobility inflows and ethnic heterogeneity are hypothesized to disturb the functioning of family and other social organizations (for example, schools, churches, neighborhood and civic associations) that help establish social norms and exert informal social control. Areas that experience rapid population growth and changing population composition are therefore hypothesized to have higher levels of urban incivility than areas that are not growing or becoming demographically more diverse. Accordingly, more incivilities may be expected in metropolitan South East Queensland (SEQ), which is a rapidly growing region, compared to more residentially stable regional cities, towns and rural areas in Queensland.

## Social Capital

Social capital is also examined in this chapter and is expected to be inversely related to incivilities in neighborhoods or local communities. Social capital is a community attribute that is underpinned by relationships which exhibit mutual trust and reciprocity (a willingness to assist each other) and facilitate cooperation and action for mutual benefit (Coleman 1988, 1990; Putnam 1993, 1995). Consequently, neighborhood social capital should facilitate pro-social norms and informal social control which inhibit neighborhood incivilities (Bursik 1999).

While more incivilities are expected in larger urban centers, more social capital may be expected in smaller urban centers. In smaller urban centers, residents are much more likely to know each other and interact with each other, and are more easily able to monitor each other's activities. This should lead to more trust, reciprocity and social capital (Besser 2009). From a functionalist perspective, more social capital may also arise in smaller urban centers to compensate for fewer services and facilities provided in these centers. Fewer services and facilities may encourage the formation of voluntary and informal groups to cover gaps in services and facilities (Halseth and Ryser 2007; Labonte 1999). Conversely, residents in larger urban centers, with greater access to services and facilities, may have less need to foster social capital.

Apart from the usefulness of social capital in inhibiting incivilities and meeting gaps in services and facilities, social capital is associated with favorable evaluations of neighbors and a sense of community(for example, Farrell et al. 2004; Prezza et al. 2001). These in turn contribute to overall subjective QOUL (Davidson and Cotter 1991; Farrell et al. 2004; Sirgy and Cornwell 2002; Western et al. 2007).

### *This Study*

Rather than predicting overall subjective QOUL from evaluations of attributes of the urban environment, as is commonly done in subjective QOUL studies, this study aims to describe the differences in overall subjective QOUL and various attributes of urban environments between metropolitan SEQ, regional cities, regional towns and rural areas. As stated earlier, the main aim is to be able to describe subjective QOUL in regional areas by contrasting it with that in metropolitan and rural areas.

The study reported here begins with *univariate* analyses of differences in overall subjective QOUL and four main attributes of urban environments (access to services and facilities, noise pollution, incivilities and social capital). It then uses *multivariate discriminant function* analysis to identify underlying dimensions in the four main attributes, which maximally distinguish between metropolitan SEQ, regional cities, regional towns and rural areas. These attributes of urban environments are all expected to vary across the group urban size categories, so differences can be expected in univariate analyses for each attribute. Access to services and facilities, noise pollution and incivilities are expected to be higher in larger urban centers, while social capital is expected to be higher in smaller urban centers. No predictions are made as to which underlying dimensions may maximally distinguish between the urban centers in each urban size category since we have no strong theoretical expectations about the relative importance of these dimensions. Discriminant function analysis is an exploratory technique that is appropriate in this situation.

## Method

### *The Sample*

The sample is wave 1 of the "Living in Queensland" study, a longitudinal state-based household panel survey investigating social well-being and quality of life (Boreham et al. 2008). A geographically stratified random sample of residents aged over 18 years was selected across the State of Queensland, Australia. The state was divided into five large regions and randomly sampled, ensuring a minimum of 300 residents in regions outside South East Queensland (SEQ), the most densely populated region (which is discussed in Chap. 8). This enabled a random selection of residents living in metropolitan SEQ, regional cities, regional towns and rural areas throughout Queensland.

Residents were contacted by telephone using random digit dialing and invited to participate in the survey. Of those residents, approximately 69% agreed to complete either a hardcopy or an Internet-based questionnaire. Of the initial sample who agreed to participate, 3,959 respondents (or 61%) completed the questionnaire. The residential addresses of these respondents were subsequently geocoded to determine

**Table 13.1** Distribution of respondents by size of urban size category

| Urban size category | Urban center population | Number of respondents | % in 2008 sample | % in 2006 population[a] |
|---|---|---|---|---|
| Metropolitan SEQ | 250,000 or more | 1,674 | 49.9 | 53.4 |
| Regional cities | 20,000 to less than 250,000 | 632 | 18.8 | 19.9 |
| Regional towns | Less than 20,000 | 627 | 18.7 | 14.7 |
| Rural areas | Not applicable | 424 | 12.6 | 12.0 |
| Total | | 3,357 | 100.0 | 100.0 |

Source: The authors and the Australian Bureau of Statistics (2006a)
[a]Persons by place of usual residence

whether they lived in metropolitan SEQ, regional cities, regional towns or rural areas. This involved locating each resident's address on a digital street dataset (MapInfo StreetPro 2007 version) within a Geographic Information System (GIS), overlaying digital census boundaries (Australian Bureau of Statistics 2007).

Approximately 85% of respondents provided sufficient address information to geocode their residential location to a Census Collection District (CCD). CCDs are relatively small areas designed for the collection and processing of population census data, consisting of approximately 225 dwellings on average (Australian Bureau of Statistics 2006b). The CCDs were then assigned to differently sized urban centers based on the Australian Bureau of Statistic's Section of State classification (see Australian Bureau of Statistics 2006c). This provided a sample size of 3,357 geocoded respondents.

## *Measures*

### Urban Size Category

Respondents were categorized as living in one of the four categories of settlement:

- Metropolitan SEQ
- Regional cities
- Regional towns
- Rural areasaccording to the size of the urban centers in which they lived.

Metropolitan SEQ is the largest conurbation in Queensland, consisting of the merging Brisbane and Gold Coast cities. Of the state's 3.9 million population in 2006, 53.4% lived in this conurbation. By contrast, regional cities had between 20,000 and 250,000, and regional towns had less than 20,000 residents. Residents in rural areas lived outside urban centers. Table 13.1 shows the distribution of respondents by urban size category compared to the population in 2006. The table indicates that the sample distribution was similar to the population distribution, though by design, it was under-sampled in metropolitan SEQ and over-sampled in regional towns.

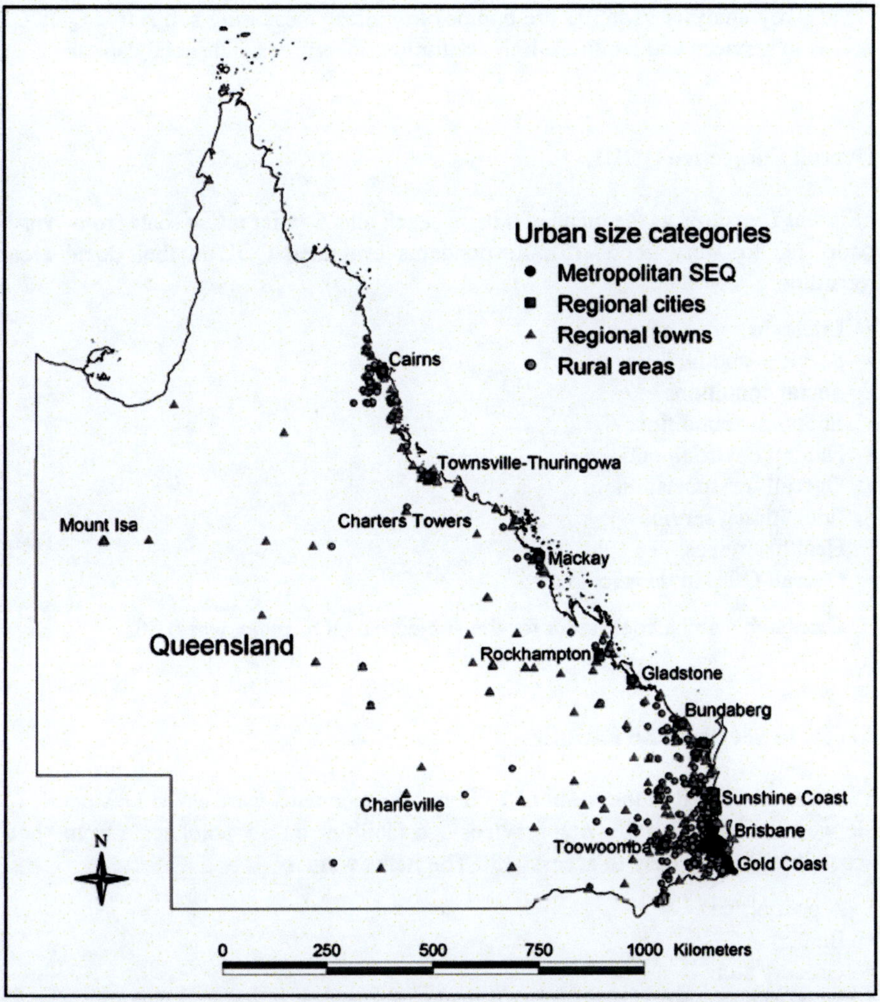


Fig. 13.1 Spatial distribution of survey respondents by urban size category (Source: The authors)

Figure 13.1 shows the spatial distribution of respondents by the size of the urban center in which they lived. Respondents living in metropolitan SEQ were in either the Brisbane or Gold Coast major urban centers. Those living in regional cities were scattered intermittently along the coast, except for respondents in Toowoomba, which is a regional city inland west of Brisbane. Those living in regional towns and rural areas were more dispersed along the coast as well as scattered across inland Queensland, though most were still living in the south-east corner of Queensland, in accord with the general population distribution.

The key analysis variables are composite indices measuring subjective QOUL, access to services and facilities, noise pollution, incivilities and social capital.

### Overall Subjective QOUL

This was measured as the mean of 9 items, each on a 5-point rating scale from "very poor" (1) to "very good" (5). Respondents evaluated QOL in their local area regarding:

- Lifestyle
- Services and facilities
- Social conditions
- Economic conditions
- Natural environment
- Overall transportation
- Educational services
- Health services
- Overall QOL in their local area.

Cronbach's alpha coefficient for the subjective QOL index was 0.86.

### Access to Services and Facilities

This was measured as the mean of 13 items. Respondents were asked how accessible were each of the following services and facilities on a 5-point scale from "not accessible" (1) to "easy to access" (5). The items were:

- A post office
- Bank
- Grocery store
- Shopping center
- Hospital/health center
- General practitioner (or doctor)
- Sporting facilities
- Child care facilities
- Primary school
- Secondary school
- Local police station
- Leisure and entertainment facilities
- Public transport.

Cronbach's alpha for the index was 0.93.

### Noise Pollution

Noise pollution in the neighborhood was a self-report index computed as the mean of two items on a 5-point scale from "never happens" (1) to "very common" (5). Respondents were asked how common in their neighborhood were:

- "Loud traffic noise"
- "Noise from airplanes, trains or industry."

Cronbach's alpha for the index was 0.64.

### Incivilities

The level of incivilities in the neighborhood was measured as the mean of 7 items on the same 5-point scale from "never happens" (1) to "very common" (5). Respondents were asked the frequency in their neighborhood of:

- Homes and gardens in bad condition
- Rubbish and litter lying around
- Teenagers hanging around on the streets
- People being hostile or aggressive
- Vandalism and deliberate damage to property
- Burglary and theft
- People being noisy or inconsiderate.

Cronbach's alpha for this index was 0.90.

### Social Capital

Social capital in the neighborhood was measured with an index based on the mean 3 items. Respondents were asked to rate their level of agreement on a scale from "strongly disagree" (1) to "strongly agree" (7) to the following statements:

- "This is a close-knit neighborhood";
- "People around here are willing to help their neighbors";
- "People in this neighborhood can be trusted."

Cronbach's alpha was 0.86.

## Results

### *Descriptive Statistics*

Table 13.2 below shows *descriptive statistics* for subjective QOUL and the four main attributes of urban environments. The number of valid responses (*N*) for each

**Table 13.2** Descriptive statistics for subjective urban QOL and attributes of urban environments

| | *N* | Missing (%) | Mean | Median | Mode | C.V. (%) | Min | Max |
|---|---|---|---|---|---|---|---|---|
| Subjective urban QOL | 3,189 | 5.3 | 3.72 | 3.78 | 4.00 | 17 | 1 | 5 |
| Access to services and facilities | 3,191 | 5.2 | 4.19 | 4.38 | 5.00 | 18 | 1 | 5 |
| Noise pollution | 3,180 | 5.6 | 2.85 | 3.00 | 3.00 | 34 | 1 | 5 |
| Incivilities | 3,185 | 5.4 | 2.53 | 2.43 | 3.00 | 28 | 1 | 5 |
| Social capital | 3,185 | 5.4 | 4.51 | 4.67 | 4.00 | 30 | 1 | 7 |

Source: The authors
The higher the value, the more positive the response

variable was between 3,185 and 3,189. This was less than the 3,357 geocoded respondents because of missing values; however, the percentage of missing values was at an acceptable level for each variable.

The differences in *mean*, *median* and *mode* reflect some skewness in the measures. However, only access to services and facilities was *highly skewed*, with the modal response equal to the maximum value of 5 or "easy access." Looking at other modal values and associated value labels:

- *Overall subjective QOUL* can be described as 4 "good";
- *Noise pollution* and *incivilities* as 3 "not common";
- *Social capital* as 4 "agree" with the statements indicating social capital.

The *coefficient of variation* (C.V.) – which is the standard deviation expressed as a percentage of the mean – shows relatively little variation in overall subjective QOUL and access to services and facilities, compared to more variation in noise pollution, incivilities and social capital. The minimum and maximum values for each variable covered the full range of possible values for each variable.

Table 13.2 shows that respondents tend to report high scores for *subjective QOUL, access to services and facilities*, and *social capital*, and low scores for *noise pollution* and *incivilities*.

Table 13.3 shows *correlations coefficients* between overall subjective QOUL and four attributes of urban environments. Spearman correlations were used to correct for any skewness in variables. All correlations were significant ($p<.01$) except for the correlation between *noise pollution* and *access to services and facilities*:

- *Overall subjective QOUL* was most highly correlated with *access to services and facilities* (+.46);
- *Social capital* and *incivilities* were moderately correlated with *overall subjective QOUL* (+.31 and –.27 respectively);
- *Noise pollution* was least correlated with overall subjective QOL but in an anticipated direction (–.11).

The four attributes of the urban environment were not very highly correlated with each other, except for *incivilities* and *noise pollution* (+.46).

**Table 13.3** Correlations between subjective urban QOL and attributes of urban environments

| | Overall subjective urban QOL | Access to services and facilities | Noise pollution | Incivilities |
|---|---|---|---|---|
| Access to services and facilities. | +.46 | | | |
| Noise pollution | −.11 | +.03 | | |
| Incivilities | −.27 | −.07 | +.46 | |
| Social capital | +.31 | +.10 | −.19 | −.30 |

Source: Boreham et al. (2008)
List-wise $N = 3{,}168$

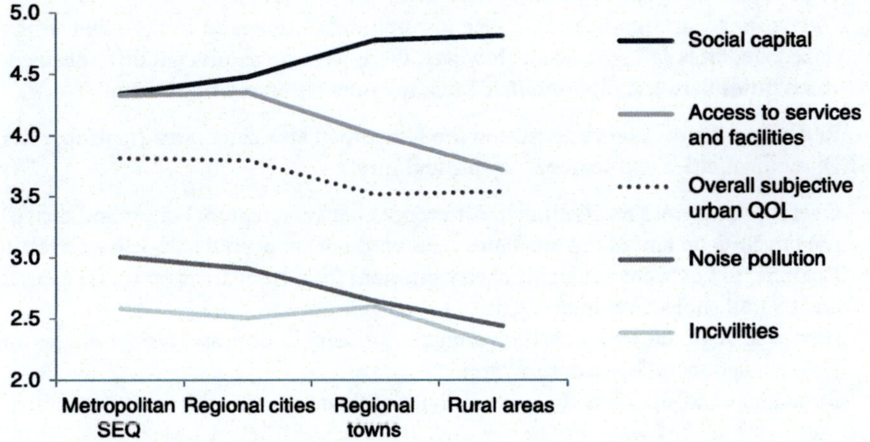


**Fig. 13.2** Overall subjective urban QOL and attributes of urban environments (means) by urban size category. Note: All measures use a scale from 1 to 5 except social capital which uses a scale from 1 to 7 (Source: The authors)

Figure 13.2 shows how the means for subjective QOUL and attributes of urban environments vary by urban size category. The results may be summarized as follows:

(a) *F* tests showed that significant differences existed in *subjective QOUL* and each of the four attributes across the different urban size categories (all $ps < .001$).[1] Follow-up *t*-tests[2] were conducted to identify where these significant differences were:

- Mean overall subjective QOUL was not significantly different between metropolitan SEQ and regional cities nor significantly different between regional towns and rural areas;
- However, regional towns and rural areas had significantly lower subjective QOUL than metropolitan SEQ and regional cities (all $ps < .001$).

[1] Since access to services and facilities was highly skewed, a Kruskal–Wallis test was also conducted using mean ranked scores to confirm that this also varied significantly across the urban size categories ($p < .001$). The pattern of variation was the same.

[2] Bonferroni corrections were used to limit the family-wise error rate to .05.

(b) *Social capital* was not significantly different between metropolitan SEQ and regional cities, nor was it significantly different between regional towns and rural areas. However, social capital was significantly higher in regional towns and rural areas than in either metropolitan SEQ or regional cities (all *ps* < .001).

(c) *Access to services and facilities* was not significantly different between metropolitan SEQ and regional cities. However, it was significantly higher than in regional towns, which in turn was significantly higher than rural areas (all *ps* < .001).

(d) Similarly, *noise pollution* followed the same pattern as access to services and facilities. It was not significantly different between metropolitan SEQ and regional cities, though it was significantly higher than in regional towns, which was significantly higher than in rural areas (all *ps* < .005).

(e) *Incivilities* was significantly lower in rural areas compared to the other urban size categories (all *ps* < .001). However, there were no significant differences in incivilities between metropolitan SEQ, regional cities and rural towns.

Regarding the differences between the four urban size categories (metropolitan SEQ, regional cities and regional towns, and rural):

- Overall, there were no significant differences between regional cities and metropolitan SEQ on any of the attributes measured nor on overall subjective QOUL;
- Regional towns were not significantly different from rural areas on social capital and overall subjective urban QOL;
- However, regional towns did have higher noise pollution and higher access to services and facilities than rural areas;
- Rural areas had significantly lower incivilities than any other urban size category.

## Discriminant Function Analysis

As showed by Fig. 13.2 above, there are similarities in the patterns by which attributes of the urban environment vary across different urban size categories. However, it is difficult to identify the main dimension that distinguishes between the different urban size categories using univariate analyses. Therefore, a multivariate technique called *discriminant function analysis* (DFA) was employed to find underlying dimensions (or functions) that maximally distinguish between the urban size categories using composites of the four main attributes of urban environments. In this way, differences in urban size categories can be described in terms of various combinations of the four attributes of the urban environment.

Differences between urban size categories were predicted by access to services and facilities, noise pollution, incivilities and social capital. DFA assumes multivariate normality, which is unlikely since access to services is skewed. However, DFA is robust to violations of multivariate normality when the sample size is large, and when the violations are associated with skewness rather than outliers (Tabachnick and Fidell 1996). As this is the case with this study, DFA is robust to skewness in access to services and facilities.

**Table 13.4** Attributes of the urban environment loading onto functions 1 and 2

| | Functions | |
|---|---|---|
| | 1. Access versus social capital | 2. Incivilities |
| Access to services and facilities | .75 | .17 |
| Incivilities | .23 | .75 |
| Social capital | –.36 | .23 |
| Noise | .53 | –.01 |

Source: The authors

After list-wise deletion for missing values, 3,172 valid cases were used in the DFA. The overall relationship between the four attributes and urban size was significant ($\chi^2(12)=540.45, p<.001$). This means that living in differently sized urban centers can be significantly differentiated by these four main attributes of the urban environment. Although most of the variation in the attributes was within the four urban size categories (84%) rather than between the categories (16%), three significant dimensions or functions were produced by the DFA. Of the between-group variation, 88.6% was explained by the first function, 10.2% was explained by the second function, and 1.2% was explained by the third function. Since the third function explained so little between group variation and did not have a clear interpretation, only the first two functions were interpreted.

Functions are interpreted by identifying which attributes load with coefficients greater than 0.30 on each function (see Table 13.4).

## Function 1

It is evident that:

- Access to services and facilities loaded very highly onto the first function (0.75);
- Noise pollution also loaded highly (0.53);
- Also loading negatively onto the first function was social capital (–0.36).

So while noise and access to services and facilities were not highly correlated at an individual level, they did combine to distinguish between respondents living in different sizes of urban categories. We describe this first function as "access versus social capital" to reflect the main positively and negatively loading attributes.

## Function 2

On the second function, the only attribute which loaded substantially (0.75) is incivilities, and so, this function was called "incivilities." The second function is also interpreted as controlling for (or "net of") the first function because it is extracted after the first function.

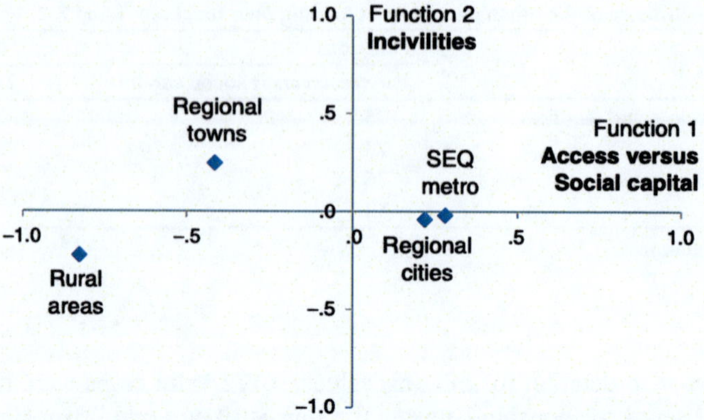


**Fig. 13.3** Mean scores on the first and second functions for each urban size category (Source: The authors)

To explore further how the two functions discriminated between the different urban size categories, the mean scores on each function were plotted for each urban size category (see Fig. 13.3). These mean scores were calculated from the function scores for each resident in each urban size category, expressed as standardized ($z$) scores. Figure 13.3 is interpreted by examining how far apart the mean scores for each category are on each function.

As the discriminant function analysis showed, Function 1 distinguishes between the urban sizes better than Function 2. In DFA, the first function accounts of most of the between-group variance. On Function 1, regional cities and metropolitan SEQ score more highly on access versus social capital, while regional towns and rural areas have lower access versus higher social capital (or higher social capital versus lower access), particularly in rural areas. The pattern on Function 2 reflects the same pattern for incivilities in Fig. 13.3 since only incivilities load onto Function 2. Most of the difference in incivilities is between regional towns and rural areas, with both regional cities and metropolitan SEQ having approximately mean levels of incivilities ($z=0$).

There was surprisingly little difference between the means for metropolitan SEQ and regional cities on either Function 1 or Function 2. Combined with findings from the univariate analyses, this suggests that no distinction should be drawn between metropolitan SEQ and regional cities on overall subjective urban QOL or the four main attributes of urban environments. More of a distinction can be drawn between regional towns and rural areas and between regional town and cities (metropolitan or regional).

## Discussion

Overall, most residents rated their overall subjective QOUL as good, though it was higher on average in metropolitan SEQ and in regional cities than in regional towns and rural areas. Regional cities were similar to metropolitan SEQ in terms of overall subjective QOUL and in terms of the four main attributes of the urban environment measured: access to services and facilities, noise pollution, incivilities and social capital. So although the initial aim in this paper was to contrast subjective QOUL in regional cities and towns with that in metropolitan SEQ and rural areas, the main differences are between cities (metropolitan or regional), regional towns and rural areas.

With respect to the first research question about how levels of subjective QOUL differ between urban size categories, overall subjective QOUL in regional cities was similar to that in metropolitan SEQ on average and is higher than that in regional towns and rural areas. In regional towns (that is, less than 20,000 residents), overall subjective QOUL was similar to that in rural areas on average.

With respect to the second research question about the main attributes distinguishing between differently sized urban categories, discriminant function analysis (DFA) identified two interpretable dimensions (or functions) with significant discriminatory power. The first function was called "access versus social capital" and was associated with higher access to services and facilities (and also higher noise pollution) and lower social capital. It distinguished between rural areas, rural towns and cities generally, though it did not distinguish between regional cities and metropolitan SEQ. Cities had more access versus social capital than rural towns and rural areas, with rural areas having the lowest. Conversely, rural areas were more endowed with social capital versus access compared to rural towns and cities. The DFA findings were consistent with the univariate findings and highlighted the primary importance of access and social capital in distinguishing between the urban size categories compared to noise pollution and incivilities.

The contrast between social capital and access to services and facilities is consistent with the hypothesis that social capital formation is stimulated in regional towns and rural areas to help meet gaps in services and facilities in these areas. However, the levels of both social capital and access to services and facilities were perceived as relatively good in cities, rural towns and rural areas on average.

The second function with reasonable discriminatory power was "incivilities" because this was the only attribute of the urban environment loading onto this factor. Incivilities explained less variation between the different urban size categories than access versus social capital. Like the first function, incivilities did not distinguish between metropolitan SEQ and regional cities, both of which had average levels of incivilities. However, rural areas had slightly lower incivilities on average, while rural towns had slightly higher, after controlling for "access versus social capital." Notwithstanding this, incivilities did not greatly distinguish between local areas in different urban size categories.

## Limitations and Future Research

Overall subjective QOUL did not vary greatly between the different urban size categories on average, and most of the variation in the four main attributes of the urban environment occurred within rather than between the categories (84% and 16%, respectively).

This suggests that future research should focus on examining variations in subjective QOUL within rather than between urban centers of different sizes as that is where most of the variation occurs. Such variation may take place at smaller spatial scales than the aggregate categories we have focused on, or it may be associated with socio-structural and demographic differences between individuals and households, or it may reflect a combination of these two factors. However, cities and towns can be considered as two broad categories of QOUL research as the interplay between access and social capital may operate differently (for example, more privatized versus more community-based lifestyles).

This study does not support a research focus on optimally sized urban centers (Archibugi 2001; Cicerchia 1999). For example, metropolitan SEQ and much smaller regional cities had very similar measures on subjective QOUL and four main attributes of urban environments. Rather than focussing on variations in subjective QOUL between cities of differing sizes, optimal centrality theory may be modified to examine variations in subjective QOUL between different local areas or neighborhoods with varying densities (for example, McCrea et al. 2006). Subjective QOUL may be depend more on where you live in an urban center than the size of the urban center in which you live.

There is still the problem of relating objective densities and other characteristics of local areas with subjective evaluations of those local areas given what is known about mechanisms which weaken their links between objective characteristics and subjective evaluations in urban environments (for example, psychological adaptation and residential relocation processes). If residents move to local areas which satisfy them on criteria important to them and adapt to other less important urban attributes, then variation in subjective evaluations is dampened, and links between objective characteristics and subjective evaluations are weakened. This is a basic problem in relating subjective evaluations of urban environments to objective characteristics of local areas (McCrea 2008).

An alternative approach to using subjective evaluations of urban attributes in future research in subjective QOUL may be to measure the subjective *importance* of various attributes of urban environments in choosing where to live, as opposed to subjective *evaluations* of those attributes. Subjective importance ratings reflect both subjective evaluations and objective characteristics of local areas while being less influenced by residential relocation and psychological adaptation processes. For example, if access to services and facilities is very important in a resident's location decision, it should reflect good objective and subjective access, independent of residential relocation and psychological adaptation processes.

## Conclusion

This chapter set out to contrast regional cities and towns with metropolitan SEQ and rural areas in terms of overall subjective QOUL and four main attributes of urban environments (access to services and facilities, noise pollution, incivilities and social capital).

However, regional cities were not very different from metropolitan SEQ. Univariate and multivariate analyses suggested that the main distinction in urban size categories should be between cities (whether metropolitan or regional), regional towns and rural areas. However, there was much more variation in the four main attributes of the urban environment within urban size categories than between them. So future research should focus on examining variations in subjective QOUL and associated attributes of urban environments within urban centers while keeping a broad distinction between cities, towns and rural areas.

**Acknowledgement** The 2008 Living in Queensland Survey was funded as part of an Australian Research Council Linkage Project (#LP0775040) with additional support from the Queensland Public Sector Union (QPSU).

## References

Archibugi, F. (2001). City effect and urban overload as program indicators of the regional policy. *Social Indicators Research, 54*(2), 209–230.

Australian Bureau of Statistics. (2006a). *2006 Census of population and housing* (customised data extraction). Canberra: Australia Bureau of Statistics.

Australian Bureau of Statistics. (2006b). *Census dictionary* (2901.0). Canberra: Australian Bureau of Statistics.

Australian Bureau of Statistics. (2006c). *Statistical geography volume 1: Australian Standard Geographic Classification (ASGC) 2006* (1216.0). Canberra: Australian Bureau of Statistics.

Australian Bureau of Statistics. (2007). *Census of population and housing: Census geographic areas digital boundaries, Australia, 2006* (2923.0.30.001). Canberra: Australian Bureau of Statistics.

Besser, T. L. (2009). Changes in small town social capital and civic engagement. *Journal of Rural Studies, 25*(2), 185–193.

Boreham, P., Dow, G., Western, M., & Laffan, W. (2008). *The living in Queensland survey* [electronic data file for wave 1 – 2008]. Brisbane: Institute for Social Science Research (ISSR)/ The University of Queensland.

Burgess, E. (1967 [1925]). The growth of the city: Metropolitan structure and violent crime. In R. E. Park & E. W. Burgess (Eds.), *The city*. Chicago: University of Chicago Press.

Bursik, R. J., Jr. (1999). The informal control of crime through neighborhood networks. *Sociological Focus, 32*(1), 85–97.

Campbell, A., Converse, P., & Rodgers, W. (1976). *The quality of American life: Perceptions, evaluations and satisfactions*. New York: Russell Sage.

Cicerchia, A. (1999). Measures of optimal centrality: Indicators of city effect and urban overloading. *Social Indicators Research, 46*, 276–299.

Coleman, J. (1988). Social capital in the creation of human capital. *The American Journal of Sociology, 94*, S95–S120.

Coleman, J. (1990). *Foundations of social theory*. Cambridge: Harvard University Press.

Davidson, W. B., & Cotter, P. R. (1991). The relationship between sense of community and subjective well-being: A 1st look. *Journal of Community Psychology, 19*(3), 46–253.

Farrell, S. J., Aubry, T., & Coulombe, D. (2004). Neighborhoods and neighbors: Do they contribute to personal well-being? *Journal of Community Psychology, 32*(1), 9–25.

Glaeser, E., Kolko, J., & Saiz, A. (2000). *Consumer city* (Working Paper 7790). Cambridge: National Bureau of Economic Research.

Halseth, G., & Ryser, L. M. (2007). The deployment of partnerships by the voluntary sector to address service needs in rural and small town Canada. *Voluntas: International Journal of Voluntary and Nonprofit Organizations, 18*(3), 241–265.

Jongudomkarn, D., & Camfield, L. (2006). Exploring the quality of life of people in North Eastern and Southern Thailand. *Social Indicators Research, 78*(3), 489–529.

Labonte, R. (1999). Social capital and community development: Practitioner emptor. *Australian and New Zealand Journal of Public Health, 23*(4), 430–433.

Malecki, E. J., & Bradbury, S. L. (1992). Research-and-development facilities and professional labor: Labor-force dynamics in high technology. *Regional Studies, 26*(2), 123–136.

Marans, R. W., & Rodgers, W. (1975). Toward an understanding of community satisfaction. In A. Hawley & V. Rock (Eds.), *Metropolitan America in contemporary perspective* (pp. 639–662). New York: Halsted.

McCrea, R. (2008). *Urban quality of life: Linking objective dimensions and subjective evaluations of the urban environment*. Ph.D. dissertation, University of Queensland, Brisbane.

McCrea, R., Shyy, T. K., & Stimson, R. (2006). What is the strength of the link between objective and subjective indicators of urban quality of life? *Applied Research in Quality of Life, 1*, 79–96.

Pagliccia, N., Apland, L., & Kazanjian, A. (1995). The impact of professional and personal satisfaction on perceptions of rural and urban: Some analytic evidence. *Social Indicators Research, 34*(3), 367–376.

Parkes, A., Kearns, A., & Atkinson, R. (2002). What makes people dissatisfied with their neighbourhoods? *Urban Studies, 39*(13), 2413–2438.

Perkins, D. D., & Taylor, R. B. (1996). Ecological assessments of community disorder: Their relationship to fear of crime and theoretical implications. *American Journal of Community Psychology, 24*(1), 63–107.

Prezza, M., Amici, M., Roberti, T., & Tedeschi, G. (2001). Sense of community referred to the whole town: Its relations with neighboring, loneliness, life satisfaction, and area of residence. *Journal of Community Psychology, 29*(1), 29–52.

Proost, S., & Van Dender, K. (2001). The welfare impacts of alternative policies to address atmospheric pollution in urban road transport. *Regional Science and Urban Economics, 31*(4), 383–411.

Putnam, R. (1993). The prospectus community: Social capital and public life. *The American Prospect, 4*, 35–42.

Putnam, R. (1995). Bowling alone: America's declining social capital. *Journal of Democracy, 12*, 65–78.

Rogerson, R., Findlay, A., Morris, A., & Coombes, M. (1989). Indicators of quality of life: Some methodological issues. *Environment and Planning A, 21*(12), 1655–1666.

Rogerson, R., Findlay, A., Paddison, R., & Morris, A. (1996). Class, consumption and quality of life. *Progress in Planning, 45*, 1–66.

Romilly, P. (1999). Substitution of bus for car travel in urban Britain: An economic evaluation of bus and car exhaust emission and other costs. *Transportation Research Part D: Transport and Environment, 4*(2), 109–125.

Shaw, C. R., & McKay, H. D. (1942). *Juvenile delinquency and urban areas*. Chicago: University of Chicago Press.

Sirgy, M. J., & Cornwell, T. (2002). How neighborhood features affect quality of life. *Social Indicators Research, 59*(1), 79–114.

Skogan, W., & Maxfield, M. (1981). *Coping with crime: Individual and neighborhood reactions*. Beverly Hills: Sage.

Tabachnick, B., & Fidell, L. (1996). *Using multivariate statistics* (3rd ed.). New York: Harper Collings College Publishers.

Talen, E. (2001). Traditional urbanism meets residential affluence – An analysis of the variability of suburban preference. *Journal of the American Planning Association, 67*(2), 99–216.

Walter-Busch, E. (2000). Stability and change of regional quality of life in Switzerland, 1978–1996. *Social Indicators Research, 50*(1), 1–49.
Western, J., McCrea, R., & Stimson, R. (2007). Quality of life and social inclusion. *International Review of Sociology, 17*(3), 525–577.
Williams, A. S., & Jobes, P. C. (1990). Economic and quality-of-life considerations in urban rural migration. *Journal of Rural Studies, 6*(2), 187–194.

# Chapter 14
# Comparing Urban and Rural Quality of Life in the State of Washington

**Benjamin L. Messer and Don A. Dillman**

## Introduction

The State of Washington in the Pacific northwest of the USA is heavily urbanized with nearly three quarters of the population concentrated in just seven of the 39 "urban" counties (see Fig. 14.1). These counties contain large metropolitan cities such as Seattle, Tacoma, Olympia, Everett, Spokane, and Vancouver and comprise only 15% of Washington's total land area. The remainder of the state population is dispersed across 32 non-urban counties; most of which are largely "rural" although many also have smaller cities and towns. Urban–rural differences in the population distribution, common throughout the world, often coincide with disparities in demographic, social, and economic population characteristics as well as the perceived overall quality of life (QOL) of the residents and their communities (Economic Profile System 2009; World Bank 2009; Shucksmith et al. 2006; UNCHS 1994).

In the early 1970s, quality of urban life (QOUL) in Washington followed a trend that seemed to be occurring across the USA: Americans living in urban areas were less satisfied overall with their communities than their counterparts in the non-urban, less populated regions (see Dillman and Tremblay 1977; Campbell et al. 1976; Dillman and Dobash 1972). In 1971 in Washington, more than half of the residents living in communities with fewer than 150,000 people reported very much satisfaction with their community, compared to fewer than 40% of residents living

B.L. Messer (✉)
Department of Sociology, Washington State University, P.O. Box 644020, Pullman, WA 99164-4020, USA
e-mail: bmesser@wsu.edu

D.A. Dillman
Social and Economic Sciences Research Center, Washington State University, P.O. Box 644014, Pullman, WA 99164-4014, USA
e-mail: dillman@wsu.edu

R.W. Marans and R. Stimson (eds.), *Investigating Quality of Urban Life: Theory, Methods, and Empirical Research*, Social Indicators Research Series 45,
DOI 10.1007/978-94-007-1742-8_14, 

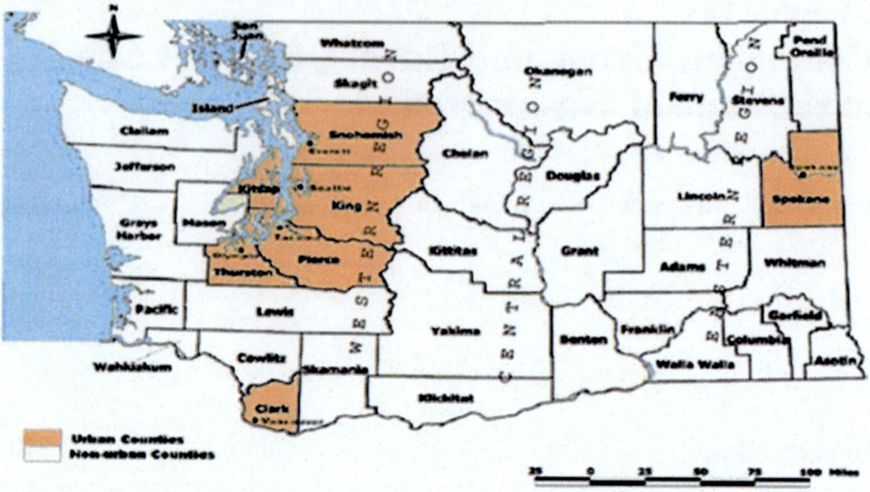


**Fig. 14.1** Urban and non-urban counties in Washington State (Source: The authors)

in communities of 150,000 or more (Dillman and Dobash 1972). That observation pointed to a contradiction in which urban residents, who were less satisfied with their subjective QOL than were rural residents, nonetheless were living in places with better objective indicators of QOL – such as higher income and lower poverty levels, more and higher paying jobs, and higher educational attainment – than their rural counterparts (Dillman 1979).

In this chapter, we revisit the QOL between urban and non-urban residents in the state of Washington in 2008, 37 years after the Dillman and Dobash study (1972):

(a) First, we provide a brief history of Washington up to 1970 and then present data from several federal and state sources to illustrate some of the objective demographic, social, and economic changes that have occurred in Washington since 1970.
(b) Second, using data from the 2008 Washington Community Survey (WCS), we explore the differences in subjective community satisfaction between urban and non-urban residents and employ logistic regression to identify some of the important predictors of community satisfaction in the two areas. The WCS contains measures of QOL, such as overall satisfaction with various QOL aspects of communities, as well as some proximate determinants of QOL, including where and how far Washingtonians have to go for various services and activities, such as work, recreation, and shopping. The WCS also provides information on demographic and technology characteristics, such as age, gender, income, and Internet and cell phone use.
(c) Finally, we compare and contrast the findings and offer a brief explanation of the results with some possible policy implications.

## Background

### *Context for the Study: Washington State*

The State of Washington is located in the northwestern-most corner of the contiguous USA, bordering Canada to the north, Oregon to the south, Idaho to the east, and the Pacific Ocean to the west. The western region of Washington – from the ocean to the Cascade Mountains – has a temperate marine climate with large areas of dense forests and numerous waterways. Most of the major population centers of the state, including Seattle, Tacoma, and the capital, Olympia, are also located in this region (refer to Fig. 14.1). The central region of Washington is dominated by the Cascade Mountains, which run north–south across the state and range upwards to over 14,000 ft elevation. This region is largely rural with tourism, agriculture, and resource extraction industries. Finally, much of the eastern region of Washington is semi-desert with areas of forest in the northeast and large-scale agriculture in the southeast. Most of the region is rural with Spokane – near the eastern border – being the only major urban area and the primary economic hub for the region.

Washington officially entered the USA as part of the Oregon Territory in 1848 and achieved statehood in 1889. During that time, the majority of the population was Native American, with tribes including the Yakima, Walla Walla, Umatilla, Nez Perce, Quinault, Snohomish, and others (Ritter 2003). The first white American settlers in the eastern region were primarily agricultural, with some in the northeast engaged in timber, mining, and the fur trade (Ritter 2003). Settlers in the western region of the state were largely involved in forestry, mining, hunting and trapping, fishing, and shipping (Ritter 2003). The western region also became the destination for many Asian immigrants, particularly Chinese, who were employed in various resource and transportation industries for cheap labor (Ritter 2003).

The Puget Sound area in western Washington, with its easy access to the Pacific Ocean, became the population and economic hub of the state. The region contains Seattle, Washington's largest city and the primary port for trade with Alaska and other areas, Tacoma, the original western terminus of the Northern Pacific Railroad and a center for industry, and Olympia, the capital city. During World Wars I and II, the region became more industrialized with the development of large shipping and aircraft industries, with the Boeing Company, a producer of aviation technologies, becoming the state's largest corporation and private employer (Ritter 2003). The Great Depression era between the wars also brought in funds for several hydroelectric projects along many of Washington's rivers, including the Grand Coulee Dam on the Columbia River – the largest of its kind – to supply electricity throughout the Northwest USA (Ritter 2003; Schwantes 1989).

During and after World War II, the remainder of Washington also experienced significant population growth and economic development. Large-scale agriculture began to predominate in southeast and south central Washington while commercial timber and mining became the leading industries in northeast and north central Washington and in parts of rural western Washington (Ritter 2003).

Following the post–World War II boom, the western, urban areas of the state experienced a severe economic decline. This decline lasted until the early 1960s, with Boeing's entry into the commercial airline market and the arrival of the World Fair in Seattle in 1962 spurring some economic growth in terms of industry and tourism (Ritter 2003).

Moreover, from 1940 to 1970, Washington population nearly doubled from 1,736,191 to 3,409,161 residents. Seattle's population also experienced heavy growth during the 1940s through the 1950s, increasing by 51%, with similar trends occurring in other urban areas (OFM 2009).

## *Community Satisfaction and QOL in 1970s in Washington*

In the early 1970s, the entire state's economy experienced a temporary but precipitous economic decline. Large-scale agriculture and resource industries across the state, and particularly in the rural east, remained relatively profitable until the nationwide energy crisis struck in 1973 (Ritter 2003). Spokane's economy, heavily reliant on agriculture and resource extraction industries, began to diversify in the early 1970s as a result of severe stagnation (Stratton 2005). Much like the economy, Spokane's population also began to decline in the 1970s, shrinking by over 6% during the decade (OFM 2009). On the other side of the state, Seattle was especially hard hit in the early 1970s, particularly after Boeing laid-off over half of its workforce. An iconic Seattle billboard from the time read:

> Would the last person who leaves Seattle please turn out the lights? (Crowley 2001).

During the 1970s, Seattle's population also shrank by nearly 5%, (OFM 2009). Other population centers throughout the state experienced similar economic and population declines (OFM 2009).

It was in this context that Dillman and Dobash (1972) conducted a survey in Washington to better understand Washingtonians' preferences for community living. Their findings pointed to a paradox in which the majority of Washingtonians, those living in urban regions, were less satisfied with their communities compared to Washingtonians in non-urban regions despite the fact that urban communities seemed at an advantage in regards to many objective social and economic indicators of material well-being, including income, jobs, education, health care, housing, and others (Dillman and Tremblay 1977). Table 14.1 shows some of these differences for 1970. Urban Washingtonians were, on average, younger and had slightly higher levels of education, income, and earnings per job compared to their counterparts in non-urban communities despite the greater impact of the economic downturn in these urban areas.

Yet in the Dillman and Dobash study (1972), community satisfaction increased as community size decreased such that those living in communities of 2,500 or less

**Table 14.1** Characteristics of Washington urban, non-urban, and total counties for 1970 and 2007

| | 1970 | | | 2007 (% annual change from 1970) | | |
|---|---|---|---|---|---|---|
| Characteristics | Urban counties | Non-urban counties | State total | Urban counties | Non-urban counties | State total |
| Population | 2,427,458 | 981,711 | 3,409,169 | 4,658,879 | 1,809,545 | 6,468,424 |
| % of Total | 71.20 | 28.80 | 100.00 | 72.02 | 27.98 | 100.00 |
| 10-year change[a] | 24.21 | 12.45 | 19.49 | 9.71 | 9.82 | 9.74 |
| Population density[b] | 240.68 | 16.03 | 47.79 | 461.92 | 29.54 | 90.67 |
| Families | 610,881 | 251,661 | 862,542 | 1,165,713 | 453,824 | 1,619,537 |
| Households | 690,324 | 415,263 | 1,105,587 | 1,825,815 | 675,694 | 2,501,509 |
| Persons/household | – | – | 2.98 | – | – | 2.53 |
| Gender (% female) | 50.36 | 50.22 | 50.32 | 50.17 | 49.98 | 50.12 |
| Age (median) | – | – | 27.50 | – | – | 37.60 |
| % 18–64 | 57.17 | 54.84 | 56.50 | 65.74 | 61.45 | 64.54 |
| % 65+ | 8.91 | 10.60 | 9.40 | 10.77 | 14.09 | 11.70 |
| Race (% non-white) | 5.26 | 3.09 | 4.64 | 20.59 | 16.03 | 19.31 |
| % Foreign born | 5.03 | 3.53 | 4.60 | 12.93 | 10.66 | 12.29 |
| Education[c] | | | | | | |
| % High school degree or less | 71.11 | 76.77 | 72.76 | 32.75 | 45.72 | 36.27 |
| % Some college or more | 28.89 | 23.23 | 27.24 | 67.25 | 54.28 | 63.73 |
| % Married[d] | 64.00 | 65.51 | 64.44 | 51.43 | 52.73 | 51.79 |
| Family income (Median)[e] | – | – | $58,246 | – | – | $66,642 |
| % Below poverty | 6.37 | 10.59 | 7.60 | 6.81 | 10.35 | 7.80 |
| % Less than $25k/year | 14.18 | 22.26 | 16.54 | 11.72 | 18.30 | 13.57 |
| % $75k/year or more | 25.63 | 15.94 | 22.80 | 48.60 | 30.73 | 43.59 |
| Per capita[e] | $22,729 | $19,427 | $21,776 | $41,999 | $28,46 | $38,212 |
| Hourly minimum wage[e] | | | $9.51 | | | $7.93 |

(continued)

**Table 14.1** (continued)

| Characteristics | 1970 Urban counties | 1970 Non-urban counties | 1970 State total | 2007 (% annual change from 1970) Urban counties | 2007 Non-urban counties | 2007 State total |
|---|---|---|---|---|---|---|
| Jobs | 1,068,321 | 422,822 | 1,491,143 | 2,920,466 | 948,347 | 3,868,813 |
| Per capita | 1.10 | 1.14 | 1.11 | 1.17 | 1.11 | 1.16 |
| Mean earnings/job[e] | $42,817 | $35,705 | $40,798 | $53,108 | $35,882 | $48,885 |
| % Employed[f] | 92.13 | 92.17 | 92.14 | 94.48 | 92.64 | 94.01 |
| % Non-labor share of total income[g,h] | 23.0 | 26.0 | 24.0 | 28.7 | 38.1 | 30.7 |
| Physicians/1,000 persons[g] | – | – | – | 2.69 | 1.58 | 2.38 |
| % Uninsured[i] | – | – | – | 13.46 | 16.66 | 14.33 |
| Crimes/1,000 persons | – | – | 48.81 | 42.19 | 47.37 | 43.73 |
| Officers/1,000 persons | – | – | – | 1.42 | 2.08 | 1.60 |

Sources: American Community Survey (2009), U.S. Census Bureau (2009), Economic Profile System (2009), and Northwest Area Foundation (2009)

Notes:

[a] 8-year change for 2007 data (2000–2008)
[b] Persons/sq. mile
[c] 25 years and older population
[d] 1970: 14 years and older population, 2007: 15 years and older population
[e] 2007 dollars
[f] Percent of civilian population in labor force 16 years and older
[g] 2006 data
[h] Non-labor income from money earned on investments or transfer payments
[i] 2005 data

were the most satisfied, with nearly 20% more indicating very much satisfaction than those living in communities of 500,000 or more. Aggregating the communities into urban and non-urban categories, the results demonstrated that over 10% more non-urban Washingtonians (150,000 or less) indicated very much satisfaction with their communities compared to those in urban communities (more than 150,000) (Dillman and Dobash 1972).

Urban communities did experience a higher rate of growth over the previous 10 years and had slightly higher unemployment and fewer jobs per capita than non-urban communities (see Table 14.1). Also, non-urban communities may also have had other advantages, including less traffic and crime, higher environmental quality, and more opportunities for recreational activities (Dillman and Tremblay 1977). However, it remains somewhat counterintuitive for such a large disparity in perceived community satisfaction to exist between the two types of community residents.

## *Thirty-Seven Years of Change*

In the late 1970s, Washington's economy exhibited renewed growth, primarily in the forms of technology and diversification. For example, Spokane hosted the World Fair in 1974 and engaged in a massive downtown revitalization project, which spurred tourism and diversified the local economy with less reliance on resource extraction and processing industries (Stratton 2005). A decrease in fuel prices in the late 1970s and early 1980s helped lower overall prices for agriculture across the state, and viticulture became a prominent industry in the south in the Columbia River Basin (Ritter 2003). The development of Microsoft, Amazon.com, Nintendo, and other technology firms in Seattle, along with coffee (for example, Starbucks) and major sports franchises (for example, Seattle Seahawks, Seattle Super Sonics, etc.), assisted in spurring the region's next economic boom, which has continued through the 1990s and into the 2000s (Moody 2004). Washington's population also experienced renewed growth, with overall growth rates hovering just under 20% from the 1980 to 2000, although rates were much lower in the major cities, which had experienced negative growth rates in the previous decade (OFM 2009). While Washington has been, and still is, largely white in terms of the racial and ethnic characteristics of the population, the 1980s saw dramatic increases in urban minority populations, particularly Asians; many Hispanics also migrated to rural, agricultural regions of Washington for agricultural labor (Ritter 2003).

Thus, 37 years after the Dillman and Dobash (1972) study, the State of Washington seems quite different. Turning again to Table 14.1, in the state as a whole, Washington had a much lower rate of population growth in previous years compared to 1970 although the population and population density has nearly doubled, with an annual growth rate of nearly 2.5% (EPS 2009). There were also more than twice as many families and households in Washington in 2007 but smaller household sizes. In fact, the growth in households outpaced total population growth over the period, a trend related to suburbanization in urban regions such as Seattle and Spokane.

The proportion of the female population has remained fairly consistent, but Washingtonians are nearly 10 years older on average and are less likely to be married than in 1970, which is also consistent with nationwide trends (UNSD 2007). The proportion of minorities and the foreign born has grown substantially since 1970, increasing nearly fivefold and with rates of annual increase greater than that of total population growth. Educational attainment also increased considerably in the state, with the proportion of the population having at least some college education more than doubling.

Washingtonians have made substantial economic gains in the past 37 years as well. For example, Table 14.1 shows that median income increased by about 14%, but per capita income climbed at nearly the same rate as the population at 2.04% per year, or 75.5% over the entire period. Although the proportion of families below the poverty level has remained steady, the proportion making less than US$25,000 has slightly declined, and the proportion earning US$75,000 or more has increased dramatically, nearly doubling. The minimum wage has slightly declined, and percent employed and jobs per capita have remained comparatively constant, but the annual growth rate in jobs did outpace population growth by nearly 2%. Average earnings per job have also kept pace with the overall increase in median income, with a 19.8% overall or a 0.54% annual increase. In addition, most of the growth in jobs has been in the services, construction, finance, insurance, and real estate sectors, while retail trade, agriculture, transportation, and other sectors that were dominant in Washington in the post–World War II period only slightly increased or even declined.

Many changes have also occurred in the daily lives of Washingtonians in the past 37 years due to several technological and economic developments. For example, the advent of cell phones and the Internet have been associated with transformations in the way people communicate, people's personal networks and community participation, satisfaction with community, and the amount of and means through which people receive information (see, for example, Stern and Dillman 2006; Dutta-Bergman 2005). Along with the growth of the auto industry, these changes have enabled Washingtonians to become more mobile, but increased mobility may not translate into greater satisfaction with community and could have an adverse effect. For example, non-urban residents typically must travel further distances for jobs and many amenities and services, while many urban residents have to navigate worse traffic conditions. Developments in communications and the growth of autos and infrastructure could have had counteracting effects on QOL by enabling more mobility necessary to maintain certain aspects of living while also eroding other aspects of local communities, such as the availability of jobs and businesses that provide local services and amenities.

In sum, the data in Table 14.1 demonstrates that Washington has become more dense while its population is older, more diverse, better educated, and somewhat more economically advantaged compared to 1970. The Washington economy has also become more service-oriented than in the past, and the Internet and cell phones, along with other technological developments, have undoubtedly created transformations in Washingtonians' QOL.

Moreover, many of the disparities between urban and non-urban communities in 1970 remain, with many growing wider. For example, Table 14.1 shows that differences in age, race, education, income, and jobs have increased such that non-urban residents in 2007 are even more likely, on average, to be older and white with less income, education, and job opportunities or higher paying jobs compared to urban residents. Urban residents also enjoy more physicians and fewer crimes per capita and a greater variety of choices to perform different activities such as shopping, recreation, etc. On the other hand, in absolute terms, urban regions have become much denser, which has led to traffic congestion and problems associated with crowding, but urban residents do have more public transportation options and shorter distances to travel for work, amenities, etc. Internet and cell phone penetration has been about the same in urban and non-urban Washington (not shown in Table 14.1), with about 71% having Internet in the home and over 80% having a cell phone (Horrigan 2008; Blumberg and Luke 2008).

The data presented in Table 14.1 do not portray a complete profile of all demographic, social, and economic changes that have occurred in Washington, but they do highlight some general trends. Given these trends, in which urban and non-urban populations seem to be diverging in multiple ways, the question remains as to whether the overall perceived QOL between urban and non-urban Washingtonians has also changed and, if so, how? Equally important are questions over which QOL indicators significantly influence perceived community satisfaction and whether these are different in the two regions.

## Methodology

### *The WCS Sample Design and Questionnaire*

We designed the WCS to obtain measures of QOL, including community satisfaction and distance from various activities, as well as information about Internet and cell phone usage and demographic characteristics of Washingtonians. The survey was implemented in two separate waves of data collection in 2008 – one during the summer (June–August) and one during the fall (September–November) – to test some methodological aspects of the survey, as reported in Messer and Dillman (2010). However, the two different waves of data collection resulted in very few differences between respondents to each wave and thus have been combined for this analysis.

To conduct the survey, a total sample of 5,400 residential addresses was obtained from the US Postal Service's Delivery Sequence File (DSF), which provides a near complete listing of all addresses in the USA. After excluding those addresses in the sample that no longer receive mail, the total sample size for the WCS came to 5,062 residential addresses, or 93.7% of the original sample. A total of 2,243 responses were obtained for an overall response rate of 44.3%.

Given the concentration of the Washington population in only a few urban counties, the sample was stratified into urban and non-urban counties.[1] That stratification meant that 50% of the sample came from five urban counties (Clark, King, Pierce, Snohomish, and Spokane) and 50% came from the remaining 34 counties in Washington. This was done to ensure that a significant number of responses were obtained from some of the less populous counties. Post-stratification weights were applied to the data to offset the under-representativeness of the urban counties. In addition, weights were applied to gender and age using data from the 2007 American Community Survey (ACS) for Washington to enhance the representativeness of the sample.

Both mail and Internet survey modes were used to collect data from respondents to the WCS. Four mail contacts were employed to implement both versions of the questionnaire. The first contact was a pre-letter, and the second contact contained a URL letter for Internet respondents and a paper questionnaire with a return envelope for mail respondents and a US$5 incentive for the majority of respondents. The third contact was a reminder/thank you postcard to all sampled respondents, and the fourth contact offered non-respondents a different mode, mail or Internet, depending on which mode was presented first to respondents.

The mail and Internet versions of the questionnaire were also designed using a unified-mode construction, as described by Dillman et al. (2009), in which each mode employs the same graphical features, questions, and question order. This helps reduce differences between the two modes, or the likelihood that people provide different answers to the same questions based on which survey mode was administered.

## *WCS Variables*

Several questions about specific aspects or indicators of community satisfaction were included in the WCS and were designed based on social indicator research considerations (as discussed in Diener and Suh 1997; Land 1983; Liu 1976; Campbell et al. 1976). Other measures of interest, and also related to QOL, include Internet and cell phone usage and demographic characteristics.

Each of the measures is discussed in more detail below.

---

[1] Urban counties were distinguished by total population, population density, and location (refer back to Fig. 14.1 above). The urban counties in Washington – Clark, King, Kitsap, Pierce, Snohomish, Spokane, and Thurston – all have populations greater than 200,000, population densities greater than 250 persons per square mile, and contain or are located adjacent to large cities greater than 150,000 people. (Two counties – Kitsap and Thurston – have relatively smaller populations but have large population densities with relatively large cities; each county also borders other urban counties.) All non-urban counties, on the other hand, fail to meet these criteria and range in population from just a few thousand to nearly 200,000 with population densities from 3 persons per square mile to about 95 persons per square mile. (Most of these counties are comprised of large rural areas, but some also have medium to small size cities of less than 100,000 and/or relatively dense residential areas that serve nearby urban centers.)

## Overall Community Satisfaction

Respondents were asked:

Overall, how satisfied are you with living in this community?

Answers ranged on a 4-point scale from "not at all satisfied" (0) to "completely satisfied" (3). In the following analyses, the community satisfaction variable was constructed dichotomously, with "0" representing "not at all satisfied" and "somewhat satisfied" and 1 indicating "mostly satisfied" and "completely satisfied."

## Past Few Years

Respondents were asked:

During the last few years, how much better or worse do you think this community has become as a place to live?

Response categories were based on a 5-point scale ranging from "much worse" (0) to "no change" (2), to "much Better" (4), and coded from 0 to 4.

## Community Satisfaction Indicators

Respondents were asked:

How would you rate the community where you live on each of the following?

on a 5-point scale ranging from "very poor" (0) to "excellent" (4) and coded 0 to 4.

Respondents rated 14 different indicators:

- Access to good medical care services
- Availability of affordable child care
- Quality of public schools
- Amount of crime in the community
- The availability of good jobs within a reasonable driving distance
- Variety of local businesses
- Quality of community or city government
- Amount of traffic on streets and roads
- Maintenance of streets and roads
- Quality of public transportation
- Quality of parks and recreational areas
- Residents who care about the community
- The cost of living in the community
- The natural environment of the community.

A factor analysis and correlation matrix resulted in the following indices:

- Child care/public education
- Traffic/streets
- Jobs/businesses
- Parks/environment.

The other indicators remained as single measures.

Moreover, about 10% of respondents chose the "not sure" answer category were coded as missing data.

### Distance Factors

The WCS also asked residents to indicate where they go most of the time to do various activities. Respondents chose one of four answer categories for each of 12 activities, coded 0–3: "Does not apply to me" (0), "Your postal address city/town" (1), "Another city or town close by (less than 30 miles away)" (2), and "A distant city or town (more than 30 miles away)" (3).

The activities include:

- Buy groceries
- Buy other household items
- Buy gasoline
- Work
- Get medical care
- Outdoor recreational activities
- Attend religious services
- Participate in clubs or other organizations
- To eat out
- Entertainment (movies, sports, bowling, etc)
- Visit with friends
- Visit with relatives

Several indices were created using factor analysis and a correlation matrix:

- Buy commodities/food (groceries, household items, gasoline, and eat out)
- Outdoor recreation/entertainment
- Clubs/religious services
- Friends/family.

Other distance factors remain as single measures.

### Technology Characteristics

One of the goals of the WCS was to determine what impact the Internet and cell phone usage has on community satisfaction. Respondents were asked whether they use a computer, the Internet, and a cell phone with "yes" and "no" answer categories coded 1 and 0, respectively.

### Communication with Others

A fourth question prompted residents to indicate how they most often communicate with others (face-to-face, phone, Internet, or other) with each successive category denoting further distance away from other persons. These categories were coded 0–3 respectively.

### Personal and Household Characteristics

WCS respondents were asked several questions about their personal and household characteristics. These characteristics include:

- Gender
- Age
- Race/ethnicity
- Marital status
- Employment
- Educational attainment
- Household income
- Years in community
- Where the people they feel closest to live.

The answers were coded as follows:

- Gender, race, marital status, and employment are dichotomous variables, coded 0 and 1.
- Age and years in the community are continuous variables.
- Educational attainment, household income, and closest people are categorical variables.
- Education is measured on a 4-point scale, coded 0–3, in which 0 is "High school diploma or less," 1 is "Some college, no degree," 2 is "2- or 4-year college degree," and 3 is "Graduate or Professional degree."
- Income is measured on a 6-point scale, from 0 to 5, in which 0 is "less than US$10,000," 1 is "US$10,000 to less than US$25,000," 2 is "US$25,000 to less thanUS$50,000," 3 is "US$50,000 to less than US$75,000," 4 is "US$75,000 to less than US$100,000," and 5 is "US$100,000 or more."
- Closest people is measured on a 3-point scale in which 0 is "closest to people who live in the community where you live," 2 is "about the same for people who live in and outside the community where you live," and 3 is "closest to people who live outside the community where you live."

### *Missing Data*

Missing data constituted about 10% of all data but was dispersed across questions so that a deletion method resulted in a significant decrease in the overall number of responses. With sampling error a concern, the smaller *n* without the missing data was considered too low. Thus, we employed multiple imputation methods (Royston 2004) to account for missing data on all variables except the dependent variable and those that received weights (i.e., overall community satisfaction, urban/non-urban county, gender, and age), in which case we used list-wise deletion. After employing these methods, the total *n* was 2,149, of which 1,253 are urban residents (58%) and 856 are non-urban residents (42%).

### *Data Analysis*

Several different statistical methods were used to analyze community satisfaction in Washington:

(a) First, we compare the proportion of urban vs. non-urban residents who indicated high satisfaction on a number of indicators as well as distance traveled to perform various activities.
(b) Second, a comparison of means was used to illustrate some of the relative differences in satisfaction and the characteristics between urban and non-urban residents.
(c) Finally, we constructed ordinal logistic regression models in a vein similar to the integrative model used by Filkins et al. (1999) to determine which characteristics are significantly associated with overall satisfaction for the state and for urban vs. non-urban regions of the state.

## Findings

### *Comparing Urban and Non-urban Overall Community Satisfaction and Perceived Quality of Aspects of the Community*

In light of the 1972 Dillman and Dobash study showing higher community satisfaction in non-urban Washington communities, we first compared the distribution of urban and non-urban residents' overall community satisfaction. The results are illustrated in Fig. 14.2.

Overall, the majority of Washingtonians are mostly or completely satisfied with their communities, but residents in urban counties are slightly more satisfied than their counterparts in non-urban counties. The difference is 5.7 percentage points between urban and non-urban residents indicating "most" or "complete" satisfaction with their community, which is a substantial reversal of results from 1972.

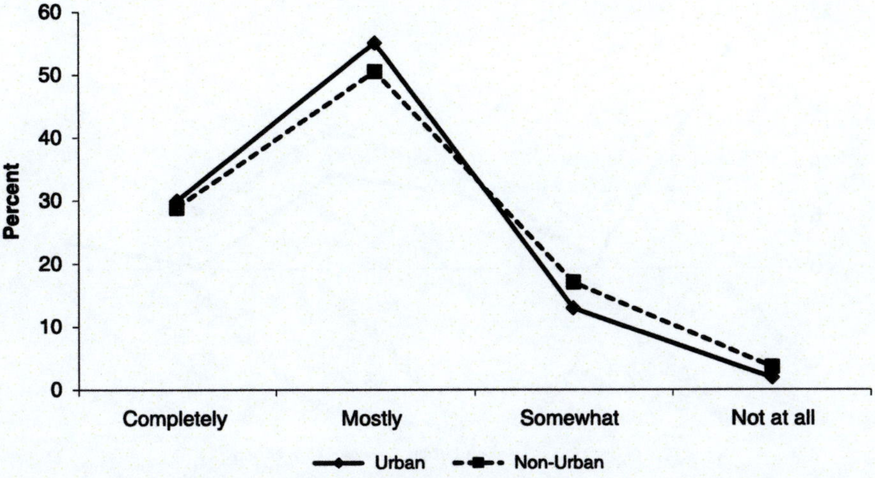


**Fig. 14.2** Percent of urban and non-urban respondents choosing "completely," "mostly," "somewhat," or "not at all" satisfied with their community (Source: The authors)

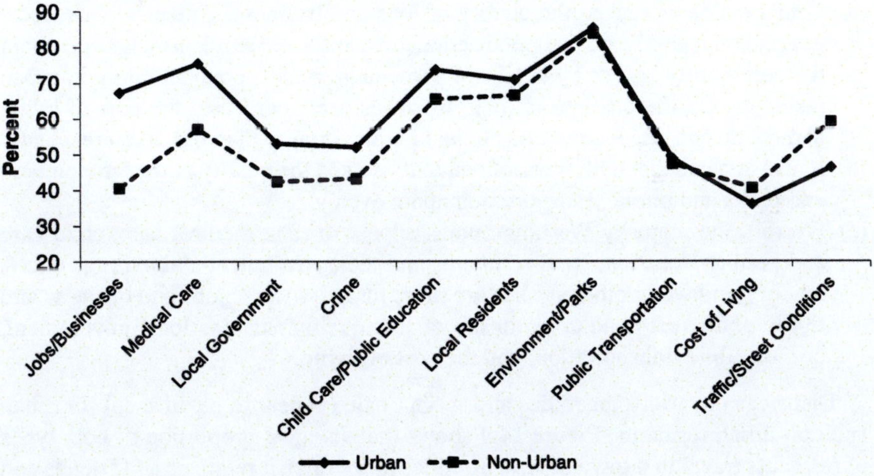


**Fig. 14.3** Percent of urban and non-urban respondents choosing "excellent" or "good" for satisfaction with various community characteristics (Source: The authors)

Figure 14.3 illustrates the differences between urban and non-urban residents' perceived quality of various aspects of their community. The figure shows the percent of residents that ranked each community characteristic "excellent" or "good" (on a 5-point scale), and although both types of residents' perceptions follow a similar trend, many of the differences are substantial.

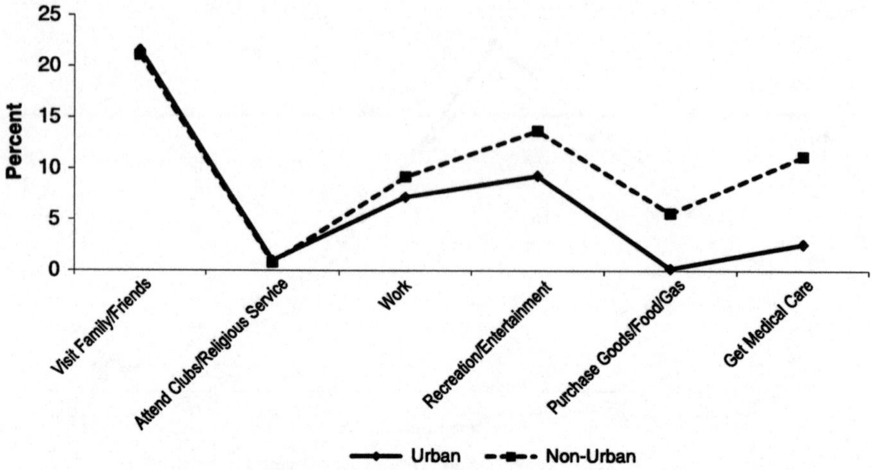


**Fig. 14.4** Percent of urban and non-urban respondents that travel to a town 30 or more miles away from their residence for various activities (Source: The authors)

Examples of the findings are as follows:

(a) Urban residents ranked the quality of jobs and businesses, medical care, local government, child care and public education, and local residents who care about the community higher by at least 4 percentage points, compared to non-urban residents. On the other hand, non-urban residents perceived the cost of living and traffic and street conditions to be of better quality than did urban residents.
(b) It also appears that both types of residents ranked the quality of the environment and parks and public transportation about evenly.
(c) Overall, the majority Washingtonians appear to rank medical care, child care and public education, local residents, and the environment and parks as excellent or good while other aspects are more mixed, such as jobs and business and traffic and street conditions, or are of less quality, such as local government, crime, public transportation, and the cost of living.

Distances to various amenities and services also appear to be different for urban and non-urban residents. Figure 14.4 shows that an equal proportion of both types of residents travel to a town at least 30 miles away to visit relatives and friends and to participate in clubs or religious services. However, compared to urban residents, non-urban residents travel further from their communities for work, for recreation and entertainment, to purchase certain amenities, and to get medical care.

## *Testing Whether Place of Residence Is a Salient Predictor of Community Satisfaction*

While there does not appear to be much variation in overall community satisfaction between urban and non-urban residents, it is important to determine whether place of residence is a salient predictor of community satisfaction in Washington and also whether and in what ways community, distance, and personal characteristics influence community satisfaction in urban and non-urban counties. Substantial differences have been found to exist between the objective characteristics of urban and non-urban residents and communities and the perceived quality of specific community characteristics. However, Table 14.2 shows the comparison of means of these characteristics from the 2008 WCS data to further underscore the differences that exist.

(a) Urban counties have more females and minorities and fewer married residents who are younger and who have lived less time in their communities, on average, compared to non-urban counties.
(b) Urban residents also have higher educational attainment, household incomes, and employment compared to non-urban residents.
(c) Technologically, more urban residents report use of a computer, the Internet, and cell phones and have Internet in the household compared to non-urban residents although the proportions for both types of residents using these technologies are rather high.
(d) Moreover, in support of Fig. 14.3 above, urban residents ranked, on average, more community characteristics of better quality than did non-urban residents.
(e) Medical care, public schools and child care, jobs and businesses, crime, local government, and public transportation all receive higher rankings, on average, from urban Washingtonians compared to non-urban Washingtonians.
(f) Traffic and street conditions appear to be of better perceived quality in non-urban counties, and no differences were significant for the environment and parks, local residents, and cost of living between the two regions.
(g) The means for overall community satisfaction and satisfaction with the past few years are relatively high (on a scale from 0 to 3) for both types of residents, but urban residents expressed more overall community satisfaction, on average, than non-urban residents, with a significant difference of .08 for overall satisfaction and .17 for satisfaction with the past few years.
(h) Finally, urban residents report having to travel farther for work and to attend clubs and/or religious services, while non-urban residents report further travel to buy goods, food, and gas, to get medical care, and for recreation and entertainment.

## *Strategies to Model Factors Impacting on Overall Community Satisfaction*

We used two different modeling strategies to determine if these mean differences in characteristics impact overall community satisfaction in Washington:

(a) First, we constructed several ordinal logistic regression models to illustrate differences in community satisfaction across the state controlling urban and non-urban place of residence and other variables.

**Table 14.2** Urban vs. non-urban means of demographic, distance, and community satisfaction and attachment characteristics

| | Urban counties (mean)[a] | Non-urban counties (mean)[a] | $X^2$ |
|---|---|---|---|
| Gender[b] | **0.52** | 0.49 | 5.70* |
| Age | 45.4 | **48.0** | 44.35** |
| Race[c] | **0.20** | 0.17 | 2.87* |
| Education | **1.65** | 1.35 | 38.2** |
| Married[d] | 0.54 | **0.60** | 5.76* |
| Employed[e] | **0.72** | 0.62 | 23.36** |
| Income | **3.11** | 2.63 | 50.54** |
| Years in community | 13.3 | **16.1** | 20.77** |
| Closest people[f] | 1.00 | 0.93 | 3.30 |
| Use computer[g] | **0.92** | 0.86 | 20.98** |
| HH Internet[g] | **0.90** | 0.82 | 21.98** |
| Use Internet[g] | **0.93** | 0.85 | 38.84** |
| Cell phone[g] | **0.89** | 0.84 | 9.51** |
| Communicate with others[h] | 1.21 | 1.16 | 3.03 |
| Satisfaction with… | | | |
| Medical care[i] | **3.02** | 2.54 | 83.58** |
| Public schools/child care[i] | **2.59** | 2.49 | 6.74** |
| Traffic/streets[i] | 2.13 | **2.40** | 38.2** |
| Jobs/businesses[i] | **2.57** | 1.95 | 195.15** |
| Crime[i] | **2.47** | 2.27 | 14.05** |
| Local government[i] | **2.44** | 2.23 | 19.06** |
| Public transportation[i] | **2.32** | 2.20 | 4.71* |
| Parks/environment[i] | 2.99 | 2.99 | 0.02 |
| Local residents[i] | 2.80 | 2.80 | 0.02 |
| Cost[i] | 2.16 | 2.20 | 0.58 |
| Overall community satisfaction[j] | **2.13** | 2.05 | 5.40* |
| Past few years[j] | **2.41** | 2.24 | 10.51** |
| Distance to… | | | |
| Buy goods/food/gas[k] | 1.30 | **1.46** | 45.01** |
| Work/job[k] | **1.37** | 1.17 | 24.45** |
| Get medical care[k] | 1.45 | **1.54** | 6.63** |
| Recreation/entertainment[k] | 1.50 | **1.56** | 4.12* |
| Clubs/religious services[k] | **0.84** | 0.77 | 5.51* |
| Visit friends/relatives[k] | 1.83 | 1.80 | 0.61 |

Source: The authors
*$p \leq .05$; **$p \leq .01$
[a] Significantly higher value in bold
[b] Coded as 1 = Female
[c] Coded as 1 = Non-White
[d] Coded as 1 = Married
[e] Coded as 1 = Employed
[f] Coded as 0 = In Community, 1 = About same, 2 = Outside Community
[g] Coded as 1 = Yes
[h] Coded as 0 = Face-to-face, 1 = Phone, 2 = Internet, 3 = Other
[i] Coded as 4 = Excellent
[j] Coded as 3 = Completely Satisfied
[k] Coded as 3 = Town 30 or more miles away

(b) Second, we then used several regression models but divide the dependent variable into two dependent variables:
    - Overall urban community satisfaction
    - Overall non-urban community satisfaction

The results of the first approach are shown in Table 14.3, with six models being tested:

(a) *Model 1* shows the results of a bivariate relationship between place of residence and overall satisfaction. The statistically significant odds ratio[2] of 0.82 demonstrates that non-urban residents are 18% less likely to indicate higher satisfaction than are urban residents. In this simple analysis, place of residence appears to be a salient factor.
(b) In *Model 2*, we analyzed the impact of the various community characteristics on overall satisfaction, controlling for place of residence. This model shows that, for all Washingtonians, place of residence and the quality of public transportation and the environment and parks do not impact the odds of overall satisfaction while all other factors have a significant, positive relationship. For example, those residents that express higher satisfaction with the past few years are 28% more likely to report higher overall community satisfaction compared to those who indicated lower satisfaction with the past few years. It is also interesting to note that, although not significant, place of residence changed directions so that non-urban residents appear more likely to be more satisfied with their community than urban residents.
(c) In *Model 3*, we analyzed the relationship between the distance factors and community satisfaction controlling for place of residence. All the distance factors are significant except where residents go to buy commodities and get medical care. Interestingly, those that attend clubs and/or religious services further away from their communities are 19% more satisfied with their community compared to those who perform these activities closer to their community.

---

[2]Odds ratios are a measure of the odds of being in or selecting a certain outcome (for example, mostly/completely satisfied vs. somewhat/not at all satisfied) for every one unit increase in an independent or explanatory variable (e.g., place of residence, distance traveled from residence, satisfaction with aspects of community, etc.). An odds ratio of 1.00 indicates that there is no difference between the odds of being in the outcome category and the odds of the not being in that outcome category. An odds ratio greater than 1.00 indicates that the odds of being in the outcome category are better or greater than the odds of not being in that outcome category while the opposite is true for odds ratios smaller than 1.00. For example, in Table 3, Model 1, the odds ratio of 0.82 demonstrates that the odds of non-urban residents being mostly or completely satisfied with their community is 18% lower than the odds of urban residents being mostly or completely satisfied with their community. Similarly, in Table 3, Model 2, the odds ratio of 1.14 for the "medical care" variable demonstrates that a one-unit increase in satisfaction with medical care increases the odds of being satisfied with the community by 14%. In other words, respondents choosing "somewhat satisfied" with medical care in their community are 14% more likely to be satisfied with their community than those respondents choosing "not at all satisfied" with medical care, controlling for other variables in the model; those choosing "mostly satisfied" with medical care are 14% more likely to be satisfied with their community than those choosing "somewhat satisfied" with medical care, and so on.

Table 14.3 Odds ratios[a] for community satisfaction in Washington under Models 1 through 6

| | Model 1 | Model 2 | Model 3 | Model 4 | Model 5 | Model 6 |
|---|---|---|---|---|---|---|
| Urban/non-urban | 0.82** (.080) | 1.04 (.121) | 0.79** (.077) | 0.80** (.078) | 0.80** (.082) | 0.91 (.110) |
| Past few years | | 1.28*** (.068) | | | | 1.32*** (.072) |
| Medical care | | 1.14** (.074) | | | | 1.12 (.078) |
| Public schools/child care | | 1.70*** (.165) | | | | 1.48*** (.153) |
| Traffic/streets | | 1.15** (.081) | | | | 1.22*** (.091) |
| Jobs/businesses | | 1.23*** (.093) | | | | 1.35*** (.110) |
| Crime | | 1.16*** (.071) | | | | 1.23*** (.079) |
| Local government | | 1.24*** (.095) | | | | 1.28*** (.103) |
| Public transportation | | 1.00 (.051) | | | | 0.98 (.051) |
| Parks/environment | | 1.17 (.117) | | | | 1.21* (.125) |
| Residents | | 1.48*** (.120) | | | | 1.39*** (.118) |
| Cost | | 1.14* (.077) | | | | 1.09 (.079) |
| Distance to… | | | | | | |
| Buy commodities (groceries, HH items, food, and/or gas) | | | 1.21 (.167) | | | 1.33* (.213) |
| Work/job | | | 0.78*** (.044) | | | 0.88* (.064) |
| Get medical care | | | 0.98 (.088) | | | 1.02 (.102) |
| Recreation/entertainment | | | 0.73*** (.062) | | | 1.03 (.099) |
| Clubs/religious services | | | 1.19** (.099) | | | 1.02 (.087) |
| Visit friends/relatives | | | 0.78*** (.062) | | | 0.87 (.080) |
| Use computer | | | | 0.68*** (.124) | | 0.87 (.197) |
| Internet in HH | | | | 1.83*** (.356) | | 1.59** (.334) |
| Cell phone | | | | 0.74* (.116) | | 0.72** (.116) |
| Communicate w/others | | | | 0.90 (.061) | | 0.86** (.068) |

| | | | | | | |
|---|---|---|---|---|---|---|
| Gender | | | | | 1.23** (.115) | 1.24** (.126) |
| Age | | | | | 1.01*** (.004) | 1.01*** (.004) |
| Race | | | | | 1.08 (.143) | 1.22 (.172) |
| Education | | | | | 0.96 (.050) | 0.83*** (.049) |
| Married | | | | | 1.04 (.116) | 1.07 (.126) |
| Employed | | | | | 0.86 (.105) | 0.94 (.128) |
| Income | | | | | 1.17*** (.054) | 1.12** (.054) |
| Years in community | | | | | 1.00 (.004) | 1.01* (.004) |
| Closest people | | | | | 0.55*** (.036) | 0.73*** (.054) |
| LR x2[b] | 1.03 | 783.70*** | 73.24*** | 17.72*** | 167.51*** | 901.25*** |
| F | 3.89** | 43.42*** | 9.15*** | 4.23*** | 12.81*** | 22.18*** |
| N | 2,149 | 2,149 | 2,149 | 2,128 | 2,149 | 2,128 |

Source: The authors
*$p \leq .10$; **$p \leq .05$; ***$p \leq .01$
Notes:
[a]Standard errors in parentheses
[b]Computed without weights

The remaining significant distance factors all have a negative relationship with community satisfaction in which those that travel further away from their residence are less satisfied with their community than those traveling shorter distances from their residence. Moreover, in this model, place of residence is a significant factor in which urban residents, who travel shorter distances on average, are more likely to have higher satisfaction with their community than non-urban residents.

(d) *Model 4*, which controls for technology characteristics, demonstrates that residents who use a computer or cell phone are less likely to have higher satisfaction than those who do not use these technologies. This seems somewhat contradictory because Internet use is positively related to community satisfaction. Place of residence also remains significant and in the same direction, indicating more satisfaction in urban communities.

(e) *Model 5* illustrates the relationship between the personal and household characteristics of Washingtonians and overall community satisfaction. In this model, gender, age, and income all have significant, positive impacts on community satisfaction, and those whose closest networks live further away are less satisfied than those whose closest networks live in the same community. Again, place of residence is significant and in the direction of greater urban satisfaction.

(f) Finally, in *Model 6* we included all the variables. In this model, place of residence is in the same direction but is no longer significant. All other variables are also in the same direction as in previous models, but some lost significance while other became significant. For example, quality of medical care; the cost of living; distance to recreation and entertainment, to clubs and/or religious services, and to visit friends and/or family; and use of a computer are no longer significantly associated with community satisfaction. Concurrently, quality of the environment and parks, distance to buy commodities, method of communication with others, education, and years in the community all gain statistical significance.

From the analysis of this first modeling strategy, it appears that several community, technological, and demographic characteristics are important determinants of community satisfaction in Washington, including place of residence, although it is not significant in the final model, which has the best model fit according to the LR chi-Square statistic. However, these models only indicate which factors are salient predictors of community satisfaction in Washington, controlling for place of residence, and do not determine whether and to what extent differences exist between urban and non-urban community satisfaction. To explore this, we created several regression models similar to those used to generate the results in Table 14.3, but in this second modeling strategy, we divided the dependent variable into two dependent variables: overall urban community satisfaction and overall non-urban community satisfaction. Table 14.4 shows the results of running five groups of models to do this, and the results are discussed below:

(a) *Models 1a and 1b* demonstrate the association between the quality of community characteristics on urban vs. non-urban community satisfaction. The models show that the quality of jobs and businesses is a significant predictor of

Table 14.4 Odds ratios[a] for urban and non-urban community satisfaction

| | Model 1a (urban) | Model 1b (non-urban) | Model 2a (urban) | Model 2b (non-urban) | Model 3a (urban) | Model 3b (non-urban) | Model 4a (urban) | Model 4b (non-urban) | Model 5a (urban) | Model 5b (non-urban) |
|---|---|---|---|---|---|---|---|---|---|---|
| Past 5 years | 1.24*** (.082) | 1.38*** (.119) | | | | | | | 1.27*** (.088) | 1.41*** (.129) |
| Medical care | 1.24** (.108) | 1.01 (.099) | | | | | | | 1.16* (.106) | 1.06 (.117) |
| Public schools/ child care | 1.58*** (.194) | 1.82*** (.282) | | | | | | | 1.35** (.180) | 1.63*** (.284) |
| Traffic/streets | 1.14 (.102) | 1.16 (.131) | | | | | | | 1.17* (.107) | 1.24* (.150) |
| Jobs/businesses | 1.16 (.200) | 1.33*** (.147) | | | | | | | 1.29** (.142) | 1.46*** (.178) |
| Crime | 1.28*** (.098) | 1.03 (.100) | | | | | | | 1.31** (.105) | 1.16 (.121) |
| Local government | 1.21** (.119) | 1.31*** (.159) | | | | | | | 1.24** (.132) | 1.35** (.170) |
| Public transportation | 1.03 (.073) | 0.96 (.070) | | | | | | | 1.02 (.075) | 0.98 (.074) |
| Parks/ environment | 1.18 (.164) | 1.21 (.174) | | | | | | | 1.21 (.174) | 1.25 (.186) |
| Residents | 1.56*** (.166) | 1.39*** (.175) | | | | | | | 1.48*** (.168) | 1.31** (.166) |
| Cost | 1.24*** (.108) | 1.02 (.106) | | | | | | | 1.20* (.111) | 0.97 (.109) |
| Distance to... | | | | | | | | | | |
| Buy commodities (groceries M, HH items, food, and/or gas) | | | 1.39 (.292) | 1.13 (.202) | | | | | 1.44* (.325) | 1.26 (.275) |
| Work/job | | | 0.88* (.065) | 0.67*** (.056) | | | | | 0.96 (.097) | 0.81** (.086) |
| Get medical care | | | 0.83 (.122) | 1.10 (.117) | | | | | 0.83 (.128) | 1.22 (.164) |
| Recreation/ entertainment | | | 0.80** (.090) | 0.69*** (.090) | | | | | 1.04 (.131) | 1.09 (.156) |
| Clubs/religious services | | | 1.06 (.113) | 1.39*** (.180) | | | | | 0.97 (.104) | 1.10 (.171) |

(continued)

**Table 14.4** (continued)

| | Model 1a (urban) | Model 1b (non-urban) | Model 2a (urban) | Model 2b (non-urban) | Model 3a (urban) | Model 3b (non-urban) | Model 4a (urban) | Model 4b (non-urban) | Model 5a (urban) | Model 5b (non-urban) |
|---|---|---|---|---|---|---|---|---|---|---|
| Visit friends/ relatives | | | 0.80** (.082) | 0.74*** (.091) | | | | | 0.91 (.111) | 0.80* (.112) |
| Use computer | | | | | 0.50** (.148) | 0.68 (.200) | | | 0.66 (.307) | 1.48 (.659) |
| Internet in HH | | | | | 1.78** (.489) | 1.92** (.628) | | | 1.26 (.367) | 2.29*** (.746) |
| Cell phone | | | | | 0.95 (.183) | 0.68** (.138) | | | 0.80** (.169) | 0.62** (.152) |
| Communicate w/ others | | | | | 1.00 (.082) | 0.75** (.093) | | | 0.95 (.095) | 0.67*** (.096) |
| Gender | | | | | | | 1.14 (.140) | 1.33** (.193) | 1.19 (.158) | 1.30 (.217) |
| Age | | | | | | | 1.01** (.004) | 1.01** (.006) | 1.01 (.005) | 1.01** (.007) |
| Race | | | | | | | 1.07 (.177) | 1.03 (.235) | 1.12 (.206) | 1.38 (.322) |
| Education | | | | | | | 0.91 (.064) | 1.01 (.077) | 0.80*** (.064) | 0.85** (.079) |
| Married | | | | | | | 0.94 (.127) | 1.20 (.226) | 1.03 (.153) | 1.13 (.213) |
| Employed | | | | | | | 0.81 (.121) | 0.88 (.183) | 0.82 (.154) | 1.07 (.214) |
| Income | | | | | | | 1.33*** (.079) | 0.99 (.073) | 1.26*** (.083) | 0.97 (.070) |
| Years in community | | | | | | | 1.00 (.005) | 1.00 (.006) | 1.00 (.005) | 1.02** (.007) |
| Closest people | | | | | | | 0.50*** (.044) | 0.62*** (.065) | 0.69*** (.068) | 0.78** (.089) |
| LR x2[b] | 440.61*** | 313.93*** | 29.27*** | 50.71*** | 9.12 | 18.09*** | 121.44*** | 63.31*** | 515.61*** | 423.82*** |
| F | 26.55*** | 20.52*** | 2.89*** | 9.71*** | 1.50 | 3.88*** | 9.64*** | 5.21*** | 11.69*** | 11.18*** |
| N | 1,293 | 856 | 1,293 | 856 | 1,281 | 847 | 1,293 | 856 | 1,281 | 847 |

Source: The authors

*$p \leq .10$; **$p \leq .05$; ***$p \leq .01$

Notes:

[a]Standard errors in parentheses

[b]Computed without weights

community satisfaction for non-urban residents and not for urban residents, while quality of medical care, crime, and cost of living is significantly associated with urban satisfaction but not non-urban satisfaction. Many of the characteristics that are significant across the two models also have differential impacts on satisfaction in the two regions, as suggested by the different odds ratios. For example, urban residents who indicated higher satisfaction with the past few years are 24% more likely to have higher overall community satisfaction compared to urban residents who were less satisfied with the past few years while this figure is 38% among non-urban residents. This indicates that satisfaction with the past few years has differential impacts on urban- vs. non-urban-perceived overall satisfaction. Other factors considered more influential for non-urban community satisfaction include public schools and child care, jobs and businesses, and local government. For urban residents, medical care, crime, residents, and cost all appear to be more salient predictors of community satisfaction compared to non-urban residents.

(b) Distance factors also impact urban and non-urban satisfaction, as shown in *Models 2a and 2b*. Most of the significant factors are in the same direction for both types of residents, but differential impacts are present. Distance to clubs and religious services is significant for only non-urban satisfaction, while distance to work, to recreation and entertainment, and to visit friends and family have a greater impact on non-urban compared to urban satisfaction.

(c) *Models 3a and 3b* show the relationship between technological characteristics and urban and non-urban satisfaction. However, in these models there is little overlap in which characteristics significantly influence satisfaction in the two regions. It appears that household Internet service has a positive impact in both regions and is more salient for non-urban residents while use of a computer is significant only for urban satisfaction, and use of a cell phone and method of communication with others are significant only for non-urban satisfaction.

(d) *Models 4a and 4b* show similarities between demographic characteristics and urban vs. non-urban satisfaction. In these models, gender is a significant predictor only for non-urban residents while income is significant only for urban residents. Both age and closest networks are significant for both types of residents and are in the same direction, but the latter is more effectual for urban residents.

(e) The final models, *Models 5a and 5b*, control for all the variables and demonstrate several differences between what influences urban and non-urban community satisfaction. The differences for the community characteristics are similar to those in Models 1a and 1b, with traffic and street conditions significant for both types of residents and slightly more important for non-urban residents. However, distance in these models does not impact urban satisfaction except for buying commodities, which is not significant for non-urban residents. Non-urban residents' distance to work and family and friends remains significant and in the same direction but are no longer significant predictors of urban satisfaction. Technological characteristics remain the same for non-urban satisfaction while use of cell phones becomes significant and use of computer loses significance for urban satisfaction; use of cell phones also seems to have a greater effect on non-urban satisfaction. Finally, among personal characteristics, education is a significant predictor of both urban and non-urban satisfaction and is slightly more effectual for urban residents, and years in community gains significance for non-urban satisfaction while gender loses significance. According to the LR chi-square statistic, these models and Models 1a and 1b provide the best overall fit.

## Discussion and Conclusions

In the early 1970s, non-urban Washingtonians expressed higher overall community satisfaction compared to urban Washingtonians. Thirty-seven years later in 2008, it appears that this trend has reversed. Our data show that, in 2008, urban Washingtonians were more satisfied overall with the QOL in their community than non-urban Washingtonians. Although the difference is small and both types of residents are highly satisfied with their community QOL, the trend is a substantial reversal from the early 1970s.

Many of the community and personal characteristics in these two regions have diverged in the past 37 years, which offers a partial explanation for why the shift in satisfaction has occurred.

(a) For urban residents in 2008:

- A larger percentage are, on average, younger, more educated, and have higher incomes and more and higher paying jobs available compared to non-urban residents than was the case in 1970
- They are also more satisfied with many QOL and institutional aspects of their communities, including jobs and businesses, medical care, local government, police protection (crime), child care and education, and other local residents
- However, they appear to be equally or less satisfied with the natural environment and parks, public transportation, the cost of living, and traffic and street conditions.

These trends are similar to those outlined by Dillman and Tremblay (1977) using data from the 1970s.

(b) Non-urban residents in 2008 on average:

- Traveled further from their communities than urban residents for important services and amenities, including their job and recreation and entertainment, to purchase goods, gas, and food, and to get medical care
- Are less likely to have a computer, Internet access, or a cell phone compared to urban residents

However, our analyses of whether and to what extent these characteristics influence community QOL satisfaction produced insightful but mixed results:

(a) It appears that the urban/non-urban place of residence distinction is not an important indicator of satisfaction among Washingtonians when taking other factors into account. For example, regardless of place of residence, higher satisfaction with community is significantly predicted by:

- Higher satisfaction with community aspects, including medical care, public schools and child care, traffic and street conditions, local jobs and businesses, local government, the natural environment and parks, and local residents
- Technology characteristics such as access to the Internet in the home
- The use of more personal forms of communication with close friends and relatives, and absence of a cell phone
- Personal characteristics like increases in age, higher income and less education, more years living in the community, having friends and family in the community, and being a female.

(b) Although we do not find a significant relationship between overall satisfaction with community QOL and urban/non-urban place of residence, the various community and personal characteristics were found to have differential impacts on the community satisfaction of residents living in the two areas:

- On the one hand, non-urban residents' overall satisfaction was influenced more by satisfaction with community aspects such as medical care, traffic and street conditions, jobs and businesses, and local government; by distance traveled to work and to visit friends and family; by access to the Internet, absence of a cell phone, and use of more personal forms of communication with friends and family; and by age and years lived in the community
- On the other hand, urban residents' overall community satisfaction was impacted more by satisfaction with crime, local residents, and cost of living; by distance to buy commodities; and by income, education, and having close friends and relatives in their community.

These findings suggest that policy aimed to improve the QOL of Washingtonians should not necessarily be crafted to target only urban or non-urban residents but should also not be urban/non-urban "blind" or have a one-size-fits-all approach. Instead, different state policies could have universal aims with differential priorities and implementation in urban and non-urban counties. For example, among all Washingtonians, Internet access in the home significantly and positively impacts overall QOL satisfaction, both net of and controlling for other factors, including place of residence, implying that policies related to Internet access would be well-received across the state. However, net of other factors, Internet access in the home reflects more on non-urban residents' satisfaction and, controlling for other factors, is one of the most influential characteristics for non-urban resident satisfaction tested in this study while it is not significantly associated with urban satisfaction. Thus, when combined with urban/non-urban Internet access figures, the results suggest that policies related to household Internet access in Washington should possibly have a non-urban bias but also apply across the state and not be limited to non-urban counties. The same also applies to crime, except that policies aimed at the state level should have an urban bias since it is found to be more salient for residents in these counties.

The conclusions drawn from this research apply only to data drawn from a sample of the adult population in the state of Washington. That is, the results may not be generalizable to other states or regions of the country. The social, economic, demographic, and technological trends that have occurred and the results on community QOL satisfaction in Washington reported in this study may also be found in other states or across the entire USA. Regardless, it is likely that policy implications will differ widely across states even if these trends persist elsewhere. Moreover, due to the smaller sample size resulting from the WCS, we are unable to conduct a more nuanced analysis using different levels of urbanicity/rurality, such as the rural–urban continuum detailed by the USDA (2009), which could provide results applicable to lower levels of aggregation.

## References

American Community Survey (ACS). (2009). *2007 American community survey 1 year estimates: Washington*. American Factfinder, U.S. Census Bureau, U.S. Department of Commerce. Retrieved July 2009, from http://factfinder.ensus.gov/servlet/DatasetMainPageServlet?_program=ACS&_submenuld=&_lang=en&_ts=

Blumberg, S. J., & Luke, J. V. (2008). *Wireless substitution: Early release of estimates from the National Health Interview Survey, January–June 2008*. Hyattsville: National Center for Health Statistics. Retrieved June 2009, from http://www.cdc.gov/nchs/nhis.htm

Campbell, A., Converse, P. E., & Rodgers, W. L. (1976). *The quality of life in America: Perceptions, evaluations, and satisfactions*. New York: Russell Sage.

Crowley, W. (2001). *Turning point 7: A bumpy ride: Seattle's economic booms, busts, and comebacks*. HistoryLink.org. Accessed August 25, 2009, http://www.historylink.org/index.cfm?DisplayPage=output.cfm&file_id=3224

Diener, E., & Suh, E. (1997). Measuring quality of life: Economic, social, and subjective indicators. *Social Indicators Research, 40*, 189–216.

Dillman, D. A. (1979). Residential preferences, quality of life, and population turnaround. *American Journal of Agricultural Economics, 61*(5), 960–96.

Dillman, D. A., & Dobash, R. P. (1972). *Preferences for community living and their implication for population redistribution* (Agricultural Experiment Station, Bulletin 764). Pullman: Washington State University.

Dillman, D. A., & Tremblay, K. R. (1977). The quality of life in rural America. *The Annals of the American Academy of Political and Social Science, 429*, 115–129.

Dillman, D. A., Smyth, J. D., & Christian, L. M. (2009). *Internet, mail, and mixed-mode surveys: The tailored design method*. Hoboken: Wiley.

Dutta-Bergman, M. J. (2005). The antecedents of community-oriented Internet use: Community participation and community satisfaction. *Journal of Computer-Mediated Communication, 11*(1), 97–113. article 5.

Economic Profile System (EPS). (2009). *County and regional data from 1970 to 2000 for the West*. Sonoran Institute. http://eps.sonoran.org/ and Headwaters Economics, http://headwaters-economics.org/profiles/searchprofile2.php?string=Washington&submit=Search

Filkins, R., Allen, J.C., & Cordes, S. (1999). Predicting community satisfaction among rural residents: An integrative Model. *Center for Applied Rural Innovation*. Journal Series No. 12451. Lincoln: Agricultural Research Division, University of Nebraska.

Horrigan, J. B. (2008). *Home broadband adoption 2008: Adoption stalls for low-income Americans even as many broadband users opt for premium services that give them more speed*. Retrieved December 10, 2008, from http://www.pewInternet.org/pdfs/PIP_Broadband_2008.pdf

Land, K. C. (1983). Social indicators. *Annual Review of Sociology, 9*, 1–26.

Liu, B. C. (1976). *Quality of life indicators in U.S. Metropolitan Area*. New York: Praeger.

Messer, B. L., & Dillman, D. A. (2010). *Using address based sampling to survey the general public by mail vs. 'Web Plus Mail'*. Social and Economic Sciences Research Center (SESRC) Technical Report 10–13, Pullman, WA. http://www.sesrc.wsu.edu/dillman/papers/2010/Messer%20Dillman%20WCSTechReport.pdf

Moody, F. (2004). *Seattle and the demons of ambition: From boom to bust in the number one city of the future*. New York: St. Martin's Press.

Northwest Area Foundation (NWAF). (2009). Indicators website: Washington. http://www.indicators.nwaf.org/DrawRegion.aspx?RegionID=53000

Office of Financial Management (OFM). (2009). *Population, State of Washington*. http://www.ofm.wa.gov/pop/default.asp

Ritter, H. (2003). *Washington's history: The people, land, and events of the far Northwest*. Portland: Westwinds Press.

Royston, P. (2004). Multiple imputation of missing values. *Stata Journal, 4*(3), 227–241.

Schwantes, C. A. (1989). *The Pacific Northwest: An interpretive history*. Lincoln: University of Nebraska Press.

Shucksmith, M., Cameron, S., Merridew, T., & Pichler, F. (2006). *First European quality of life survey: Urban-rural differences*. European Foundation for the Improvement of Living and Working Conditions. Retrieved August 2009, from http://www.eurofound.europa.eu/pubdocs/2006/63/en/1/ef0663en.pdf

Stern, M. J., & Dillman, D. A. (2006). Community participation, social ties, and use of the internet. *City and Community, 5*(4), 409–424.

Stratton, D. H. (2005). *Spokane and the Inland Empire: An interior Pacific Northwest anthology*. Pullman: Washington State University Press.

U.N. Centre for Human Settlements (UNCHS). (1994). *Population, urbanization, and quality of life*. New York: United Nations Publications. 47 pp.

U.N. Statistics Division (UNSD). (2007). *Demographic yearbook*. New York: United Nations. Retrieved August 2009, from http://unstats.un.org/unsd/demographic/products/dyb/dybsets/2007%20DYB.pdf

U.S. Census Bureau. (2009). *1970 census population Washington: Characteristics of the population* (Vol. 1(49)). Washington, DC: U.S. Department of Commerce. Retrieved July 10, 2009, from http://www2.census.gov/prod2/decennial/documents/00496492v1p49_TOC.pdf

U.S. Department of Agriculture (USDA). (2009). *Measuring rurality: Rural-urban continuum codes*. Economic Research Service, USDA. http://www.ers.usda.gov/Briefing/Rurality/RuralUrbCon/

World Bank. (2009). *World development report 2009: Reshaping economic geography* (p. 393). Washington, DC: The World Bank.

# Part IV
# Innovations in Modeling Quality of Urban Life Data

# Chapter 15
# Modeling Determinants of Subjective QOUL at Different Geographic Scales: The Case of the Brisbane-SEQ Region[1]

**Rod McCrea, John Western†, and Robert Stimson**

## Introduction

This study investigates the relative importance of different urban attributes in contributing to people's *satisfaction with urban living*. In particular, it assesses how the importance of those attributes varies between different demographic groups of residents. It focuses on the relative importance of satisfaction with respect to three urban domains at different levels of scale—*housing*, the *neighborhood*, and the wider metropolitan *region*—in predicting the overall life satisfaction of residents in a large, rapidly growing metropolis. Most existing studies measure satisfaction with housing, the neighborhood and the local community (for example, Bruin and Cook 1997; Campbell et al. 1976; Lu 1999; Parkes et al. 2002; Sirgy and Cornwell 2002; Sirgy et al. 2000); however, few studies include

[1]With kind permission from Springer Science+Business Media, this chapter is a condensed version of a paper by McCrea, R., Stimson, R. and Western, J. (2005) Testing a moderated model of satisfaction with urban living using data for Brisbane-South East Queensland, Australia, *Social Indicators Research*, 72, pp. 121–152.

R. McCrea (✉)
Institute for Social Science Research, The University of Queensland,
Level 4, GPN3 Building, Campbell Road, St Lucia, Brisbane, QLD 4072, Australia
e-mail: r.mccrea@uq.edu.au

†Deceased

R. Stimson
Australian Urban Research Infrastructure Network (AURIN), Faculty of Architecture, Building and Planning, University of Melbourne, VIC 3010, Melbourne, Australia
e-mail: rstimson@unimelb.edu.au

R.W. Marans and R. Stimson (eds.), *Investigating Quality of Urban Life: Theory, Methods, and Empirical Research*, Social Indicators Research Series 45,
DOI 10.1007/978-94-007-1742-8_15, © Springer Science+Business Media B.V. 2011

a consideration of satisfaction with the wider metropolitan regional setting in which people live (for example, Turksever and Atalik 2001). Such consideration is important as large metropolitan regions incorporate a large number of diverse neighborhoods or local communities.

The chapter begins with a discussion of the context for investigating QOL considerations in urban settings. This is done at two levels: on *regions* in terms of considering phenomena such as migration, economic viability, and environmental sustainability and on *individuals* in terms of considering people's perceived level of satisfaction with the urban environment and overall QOL. That discussion is then placed within the context of a bottom-up meta-theory which combines a number of perspectives. The segment of the meta-theory model tested relates to satisfaction with three urban domains at different levels of scale—satisfaction with housing, neighborhood, and the region. As well, the moderating effects of selected demographic characteristics of residents are tested. The chapter concludes with a discussion of the limitations and implications of the study.

Quality of urban life (QOUL) has been defined in two broad ways according to two measurement traditions (see Chap. 4). In the *objective measurement* tradition, QOUL has been conceptualized as a weighted average of various objective measures of the urban environment like actual crime rates, pollution levels, and housing costs (for example, Boyer and Savageau 1981; Pierce 1985). In this study, such objective measures are not used. In the *subjective measurement* tradition, urban QOL has been conceptualized as satisfaction in a number of urban domains (e.g. housing and neighborhood satisfaction), which in turn contribute to overall life satisfaction along with satisfaction in other life domains (for example, Marans and Rodgers 1975; Sirgy et al. 2000). This study adopts this latter approach, with most of the analysis and discussion referring to measures of satisfaction in different urban domains.

## *Urban Quality of Life and Broader Implications*

The importance of subjective QOUL extends beyond individual perceptions of satisfaction. It has broader implications for migration patterns, economic growth, and environmental sustainability of communities and regions (for example, Glaeser et al. 2000; Keeble 1990; Ley 1996; Liaw et al. 2002).

Subjective QOUL also has implications for residential relocation processes within regions, for example, the attraction of high-income households to inner city areas by high consumption opportunities (Ley 1996) or the attraction to rural-residential areas by the natural environment (Keeble 1990).

Economic capital also flows to places with high subjective QOUL (Sirgy and Cornwell 2002). For example, subjective QOUL may influence business location decisions (Brotchie et al. 1985; Grayson and Young 1994; Rogerson 1999) and attract skilled labor. Further, both economic and population growth are tied to environmental issues which together contribute to overall subjective QOUL (Kemp et al. 1997).

## *Subjective QOUL at Different Levels of Geographic Scale*

QOUL issues are not only important because they affect population growth, economic growth, and environmental sustainability, but because they affect individual satisfaction with urban living.

Findings in research investigating people's satisfaction with urban living can be organized into satisfaction with different urban domains which relate to different levels of scale within an urban environment (that is, house, neighborhood, community and region). Three commonly studied urban domains are housing satisfaction, neighborhood satisfaction and community satisfaction (for example, Bruin and Cook 1997; Campbell et al. 1976; Lu 1999; Parkes et al. 2002; Sirgy and Cornwell 2002). However, regional satisfaction is not often studied (for example, Turksever and Atalik 2001).

In Campbell et al. (1976) model of residential satisfaction (see Chap. 1, Fig. 1.1), residents make evaluations of community attributes at different levels of geographic scale which in turn affect satisfaction at different levels of urban domain (i.e. housing, neighborhood and community satisfaction). Not surprisingly, studies often use evaluations of attributes relating to one level to predict satisfaction with that level of urban domain. For example:

- Housing satisfaction is predicted by attributes of homes such as dwelling age, size, structure, and tenure (Campbell et al. 1976; Lu 1999).
- Neighborhood satisfaction is predicted by physical, social, and economic attributes of neighborhoods (Sirgy and Cornwell 2002).
- Community satisfaction is predicted by attributes relating to local governance, such as education provision, police relations, local taxes, cost of living, public transport, noise levels, parks, and green areas (for example, Campbell et al. 1976; Turksever and Atakik 2001).
- Regional satisfaction is predicted by climate, travel to work and environmental pollution (Turksever and Atakik 2001).

However, the model by Campbell et al. (1976) also suggests that attributes at one geographic level may predict satisfaction in other levels of urban domain. This has also been found to be the case. For example, housing satisfaction can be influenced by surrounding features such as one's neighbors, characteristics of housing in the local area and community size (Campbell et al. 1976; Lu 1999; Parkes et al. 2002). Also, community satisfaction is predicted by neighborhood social features (Sirgy and Cornwell 2002) and broader regional attributes like climate, leisure opportunities and shopping (Turksever and Atakik 2001).

Finally, the model also suggests that satisfaction in levels of urban domain can influence each other. Support for this has been found by Campbell et al. (1976) and Sirgy and Cornwell (2002) with community satisfaction being related more to neighborhood satisfaction than housing satisfaction.

However, these interrelationships between subjective evaluations and satisfaction at different levels of the urban domain are rarely tested together in one model.

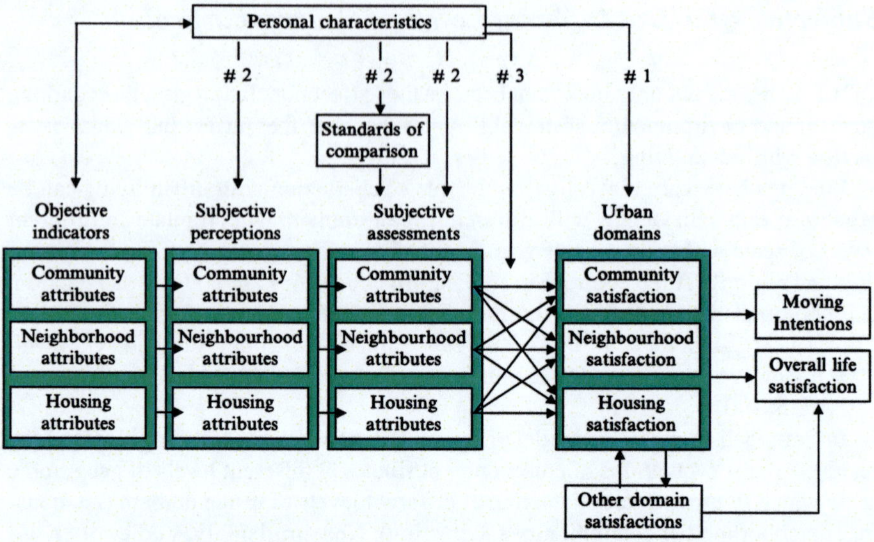


**Fig. 15.1** Model of satisfaction with urban living (Adapted from Marans and Rodgers 1975)

This study models these interrelations in one model and including the less common examined urban domain of regional satisfaction.

## *A Meta-theory of Satisfaction with Urban Living*

One of the most comprehensive models of satisfaction with urban living was first proposed by Marans and Rodgers (1975) (see Fig. 15.1). It is useful both because it is comprehensive and because it incorporates a number of theoretical perspectives (i.e., it is a *meta-theory*). As such, it is useful for conceptualizing the variety of findings on satisfaction with urban living within a broad theoretical framework.

As already mentioned, there are two measurement traditions in QOUL studies. The far left shaded box in Fig. 15.1 reflects the first tradition of measuring objective indicators of urban attributes relating to urban QOL (for example, actual crime rates, pollution levels, and housing costs). The other shaded boxes reflect the second measurement tradition of measuring subjective indicators which usually involves predicting satisfaction in different urban domains from perceptions and assessments of various urban attributes. This study follows the second tradition. However, the meta-theory shows that the two traditions are linked.

In Fig. 15.1, three levels of urban living are reflected within each shaded box—housing, neighborhood and community. However, there is no reason why a regional level cannot be incorporated as an additional layer. Satisfaction in the three urban domains in Fig. 15.1 is predicted by assessments of urban attributes associated with that domain and also by urban attributes associated with other urban domains (that is, cross paths are hypothesized). This is supported by the evidence reviewed earlier, although the meta-theory does not specify which cross paths are important.

Personal characteristics in the meta-theory are hypothesized to have three types of effects on satisfaction in the urban domains: *direct*, *mediated*, and *moderating* effects. *Direct effects* (see #1 in Fig. 15.1) and *mediated effects* (see #2) are more commonly studied (for example, Bruin and Cook 1997; Campbell et al. 1976; Lu 1999). Those personal characteristics mediated by standards of comparison reflect a number of psychological processes involved in satisfaction judgments (for example, Kahneman 1999; Michalos 1985; Schawrz and Strack 1999). However, the less commonly studied *moderating effects* of personal characteristics (see #3) are the focus of this study; that is, the varying importance of urban attributes for different groups of people.

In the meta-theory, satisfactions in the different urban domains have implications for both overall life satisfaction and intentions of people to move. Overall life satisfaction relates more to individual outcomes, while intentions to move have broader regional implications for population growth, economic growth, and environmental sustainability.

The meta-theory is based on a *bottom-up* model (i.e. satisfaction in urban and other life domains predicts overall life satisfaction). The following analyses use a bottom-up model which is common with urban quality of life studies (for example, Marans and Rodgers 1975; Campbell et al. 1976; Sirgy and Cornwell 2002). Top-down models also exist where satisfaction with overall life predicts satisfaction in different life domains (for example, Jeffres and Dobos 1995; Lance et al. 1995) as a result of global personality influences (Diener 1984). However, the model tested in this study of SEQ only considers bottom-up paths because the study focuses on the bottom-up effects of urban domain satisfactions on overall life satisfaction.

## The Model Tested

The study conducted in SEQ focuses on examining the importance of regional satisfaction in predicting overall life satisfaction in comparison with other urban domains and on examining the moderating effects of selected demographic characteristics on the importance of various urban attributes. As such, the model tested here relates to those parts of the meta-theory where *overall life satisfaction* is predicted from *satisfaction in urban domains* (including regional satisfaction), which in turn are predicted from subjective assessments of urban attributes (see later in Fig. 15.2). The model is then tested for moderating effects of gender, age, and family status using *multi-group path analysis*.

In the model tested, assessments of urban attributes predict the urban domain that was referenced in the relevant question asked of survey respondents.

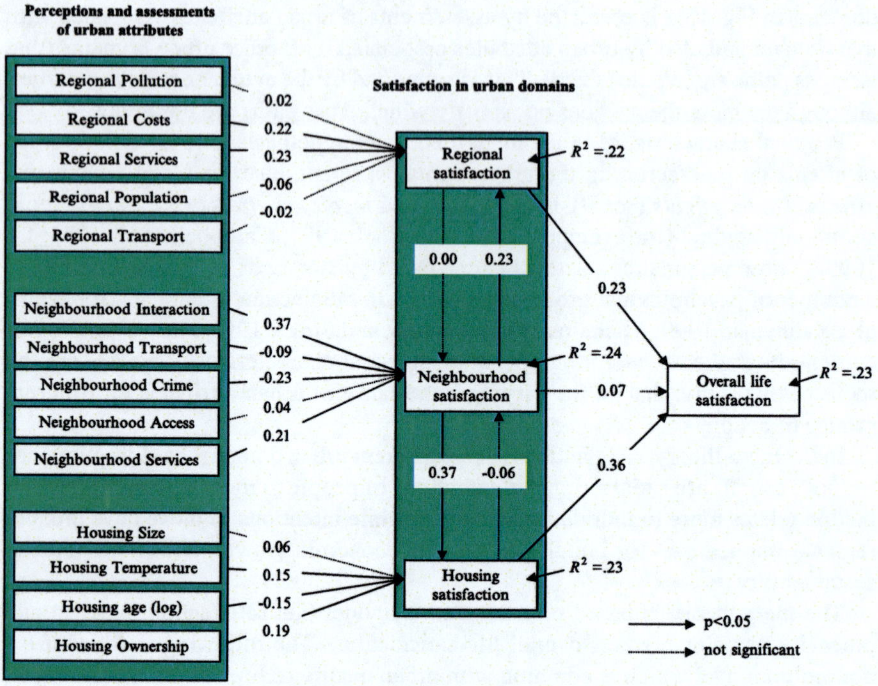


**Fig. 15.2** Model of satisfaction with urban living

For example, questions on air, water, and noise pollution related to the region as a whole and therefore are used to predict regional satisfaction. No cross paths are initially specified in order to minimize model complexity although cross paths are later added using the criteria that they significantly improve the fit of the model.

## Method

### *The Study Region*

The SEQ region has a contemporary urban economy dominated by the services sector and has a multi-centered urban form. Brisbane City (the state capital) is at the center of the region, which has two large growth corridors to the south and north linking Brisbane City to the tourism areas of the Gold Coast and the Sunshine Coast. As well, there is a western corridor incorporating the old coal mining and industrial city of Ipswich. The SEQ region also incorporates a variety of rural–urban shires around the fringe of the urbanized areas which include a number of small towns, farming and hobby-farming areas, and acreage living areas. The SEQ region is growing rapidly. Its population increased from 1.8 to 2.35 million between 1991 and 2001, and it is projected to grow to about 3.2 million over the decade to 2011. SEQ is a sprawling low-density, multi-centered metropolitan region encompassing the Brisbane and Moreton Statistical Divisions.

## *Sample Procedures and Survey Participants*

The data used in this study were collected in a 1997 survey of QOL in the SEQ region. Information was collected from a random sample of 1,347 residents spread across the region using a telephone interview mode. The survey participants were generally representative of the study population and were selected through a random digit dialing process, with the sampling method ensuring distribution of residents across the region roughly in proportion to the distribution of population. A random selection procedure was used to select a person aged 18 and over in each contacted household. The sample comprised 606 males and 741 females; the mean age being 42.3 years (SD = 15.8). Most participants were employed (46.3% wage and salary earners and 13.5% employed in their own business), 15.7% were retired, 5.4% unemployed, 5.4% were undertaking full time education and 2.2% were living on pensions. Most respondents were either married or in *de facto* relationships (63.1%), while 21.4% were single, 10.2% were divorced or separated and 5.2% were widowed. No incentives were given for participation in the survey, and the response rate was approximately 30%. Responses were entered directly onto a computer-assisted telephone interviewing (CATI) system at the University of Queensland.

## *Measures*

The *satisfaction measures* used in the study are single item measures with a 5-point scale where 1 represents "very dissatisfied" and 5 represents "very satisfied" (that is, satisfaction of respondents with their housing, neighborhood, employment situation, money available to them personally, time to do things, relationship with partner, relationship with children, independence or freedom, life as a whole). A satisfaction measure was not available directly for *regional satisfaction*; thus, a proxy satisfaction measure has been derived by using responses to the question "In general, how would you rate the overall quality of life in the Brisbane-South East Queensland region?" where 1 represents "very poor" and 5 represents "very good" on a 5-point scale.

Assessments of *regional attributes* are all *multi-item measures*. Many of the items come from the question "Thinking about things in general in Brisbane and the SEQ region, we want to know the degree to which you think something is good or bad."

A 5-point scale was used where 1 represents "very bad" and 5 represents "very good." The "regional pollution" measure is the average of air pollution, water pollution in rivers and the level of noise pollution; "regional costs" is the average of the cost of housing and the current cost of living, and "regional services" is the average of the provision of educational services and the provision of health services. "Regional population" is the average of three items from two questions. One item came from the question above and was on the level of population growth (reverse coded). The two other items came from the question "... consider some general issues about things in Brisbane and the SEQ region as a whole. How strongly do you agree or disagree with each of the following statements." A 5-point scale was used where 1 represents "strongly disagree" and 5 represents "strongly agree." The statements were "there are too many people in South East Queensland" and "urban sprawl is a problem in the South East Queensland region." The "regional transport" measure is the average of responses from three items to the previous question: "I think that traffic congestion is a problem"; "more roads and bridges are needed to enable people to travel around the Brisbane region in general"; and "we need better public transport in the South East Queensland region."

Assessments of *neighborhood attributes* are also multi-item measures. Three of the measures are based on a question asking respondents how much they agreed or disagreed with various statements about their neighborhood using a 5-point scale where 1 represents "strongly disagree" and 5 represents "strongly agree." The "neighborhood interaction" measure is the average of two items: "I have little to do with people in this neighborhood" (reverse coded) and "people in this neighborhood are willing to help each other out." "Neighborhood transport" is also the average of two items: "living in this area would be difficult without a car" (reverse coded) and "public transport in this area is adequate for my needs." "Neighborhood crime" is the average of three items "vandalism is a problem in this neighborhood"; "breaking and entering is a problem in this neighborhood"; and "I feel safe walking around this neighborhood after dark" (reverse coded). "Neighborhood access" is measured using the question "How satisfied or dissatisfied are you with your access to the following facilities" where 1 was "very dissatisfied," and 5 was "very satisfied." This measure is the average of 17 items such as shops, post offices, banks, health practitioners, hospitals, child care, schools, sporting facilities, and religious facilities. "Neighborhood services" was measured on the same scale with the question "How satisfied are you with the following services provided by your local council." The items used were public parks and gardens, swimming pools, libraries, community centers, street cleaning, street lighting, road maintenance, bus services, water and sewerage services, garbage collection, and recycling services.

Assessments of *housing attributes* use *single item measures*. "Housing size" was the number of bedrooms in the resident's dwelling. "Housing temperature" is based on whether the home was insulated or had air conditioning. If it had either, then it was coded as having temperature control. "Housing age" is the log of the estimated age of the resident's dwelling. The log was taken to normalize the distribution of sample scores. Finally, "housing ownership" has three levels of housing ownership: renting, purchasing, and fully owned.

## Results

### *Descriptive Statistics*

Most people in the SEQ region report being either "satisfied" or "very satisfied" in each of the three urban domains, and only a small percentage of residents are "dissatisfied" in any urban domain (see Table 15.1). This phenomena is common for measures of satisfaction with urban domains (see, for example, Marans 2002) and also with overall life satisfaction measures. The average assessments of urban attributes vary (see Table 15.2). The most favorably evaluated attribute of urban living is neighborhood access to facilities (like shops, schools, health services, and recreational facilities), followed by publicly provided neighborhood services (like parks and gardens, libraries, etc.). The least favorably evaluated attribute of urban living is regional transport, followed by neighborhood crime.

**Table 15.1** Percentage of residents satisfied with each urban domain and overall life

| | Housing | Neighborhood | Region | Overall life |
|---|---|---|---|---|
| Very dissatisfied | 1.0 | 1.3 | 1.2 | 0.4 |
| Dissatisfied | 6.5 | 6.5 | 4.5 | 2.3 |
| Neither satisfied nor dissatisfied | 4.2 | 3.4 | 9.0 | 4.7 |
| Satisfied | 63.7 | 57.3 | 51.3 | 60.1 |
| Very satisfied | 24.6 | 31.4 | 34.0 | 32.5 |
| Mean | 4.0 | 4.1 | 4.1 | 4.2 |

Source: The authors
*N*=1,347

**Table 15.2** Descriptive statistics for urban attributes

| Variable | Missing | Mean | SD | Skewness | Scale |
|---|---|---|---|---|---|
| Regional pollution | 4 | 2.9 | 0.8 | 0.1 | 5 point |
| Regional costs | 5 | 3.2 | 0.9 | –0.5 | 5 point |
| Regional services | 16 | 3.4 | 0.9 | –0.5 | 5 point |
| Regional population* | 7 | 2.9 | 0.8 | 0.3 | 5 point |
| Regional transport* | 12 | 3.6 | 0.7 | –0.4 | 5 point |
| Neighborhood interaction | 3 | 3.3 | 0.9 | –0.3 | 5 point |
| Neighborhood transport | 9 | 2.9 | 1.2 | –0.2 | 5 point |
| Neighborhood crime* | 7 | 2.7 | 1.0 | 0.5 | 5 point |
| Neighborhood access | 5 | 3.9 | 0.5 | –0.8 | 5 point |
| Neighborhood services | 5 | 3.6 | 0.5 | –0.5 | 5 point |
| Housing size | 14 | 3.2 | 0.9 | –0.1 | Interval |
| Housing temperature | 0 | 0.5 | 0.5 | 0.1 | Dichotomous |
| Housing age (prior to log) | 117 | 22.9 | 22.8 | 1.7 | Ratio |
| Housing ownership | 48 | 2.1 | 0.8 | –0.1 | 3 point |

Source: The authors
*Higher values indicate less favorable assessments for those measures with an asterisk. On other 5-point scales, higher values indicate more favorable assessments

There were a large number of missing values for housing age as many respondents (95) indicated they did not know the age of their home. Therefore, the model has been run including and excluding this variable to see the impact of its exclusion. It has been found to be a relatively important indicator of satisfaction with housing and so has been retained in the model. The only notable change in other parameter estimates when home age is excluded is that home temperature then becomes the most important predictor of housing satisfaction rather than home ownership.

## Testing the Model

The model tested is shown in Fig. 15.2. As mentioned previously, assessments of urban attributes predict satisfaction in relevant urban domains. No cross paths are initially tested; however, cross paths are later added to the model which significantly improves model fit, and this model is then moderated by demographic variables. The interrelationships between urban domains are reflected by bidirectional paths between the urban domains. Finally, satisfaction in different urban domains is used to predict overall life satisfaction, which has been done to establish the relative importance of the three urban domains rather than predicting overall life satisfaction per se.

Using the LISREL statistical package (version 8) (Joreskog and Sorbom 1996), the model has been tested by *path analysis* because of the ability of that analytical technique to test:

- Both *mediated* and *moderated* relationships
- *Two-way relationships*
- *Model fit*

Path analysis is a convenient way to test a complex model. After specifying the model, the actual covariance or correlation matrix is compared to that expected from the model. If they match closely, then the data are said to fit the model.

The data fitted the model in Fig. 15.2 very well (normed $\chi^2=2.32$, GFI=.99, AGFI=.98, RMSEA=.03, CFI=.99). Asymptotic covariance matrices have been used as recommended by Joreskog and Sorbom (1996) when variables are skewed and the sample size large (see Joreskog and Sorbom 1996). The total effective sample size for the model is 1,165 using list wise deletion. The $R^2$ statistics in Fig. 15.2 shows that the variations in satisfaction measures explained by the model are 23% for *housing satisfaction*, 24% for *neighborhood satisfaction*, 22% for *regional satisfaction* and 23% for *overall life satisfaction.*

## Path Coefficients

*Predictors of overall life satisfaction.* Housing satisfaction ($\beta=.36$) is found to be a better predictor of overall life satisfaction than regional satisfaction ($\beta=.23$), while

**Table 15.3** Descriptive statistics for urban characteristics

| Satisfaction measure (Predictor) | Beta coefficient |
|---|---|
| Housing | .15* |
| Neighborhood | –.003 |
| Region | .13* |
| Employment situation | .14* |
| Money available personally | .08* |
| Time to do things | –.03 |
| Relationship with partner | .17* |
| Relationship with children | .12* |
| Independence or freedom | .32* |

Source: The authors
Note: * denotes $p < .001$

neighborhood satisfaction ($\beta = .07$) is not a significant predictor of overall life satisfaction. However, other life domains are also important in predicting overall life satisfaction, as reflected by only 23% of the variation in overall life satisfaction being explained by satisfaction in the three urban domains. Therefore, an additional regression analysis was conducted to see whether there is a change in the relative importance (or path coefficients) of the three urban domains after adding in other life satisfaction domains (i.e., employment situation, money available personally, time to do things, relationship with partner, relationship with children, independence, or freedom). The variation in overall life satisfaction explained when including these other life domains was 38%. In this analysis, the housing satisfaction coefficient reduces to approximately the same importance as regional satisfaction (see Table 15.3). The reduction of the coefficient for housing satisfaction is due to housing satisfaction being moderately correlated with satisfaction with the amount of money available personally ($r = .30$, $p < .001$) and satisfaction with independence or freedom ($r = .28$, $p < .001$). Neighborhood satisfaction is still non-significant.

Although neighborhood satisfaction is not an important "direct" predictor of overall life satisfaction in the model, it does appear to have an "indirect" effect via housing satisfaction and regional satisfaction (see Fig. 15.2). The relationship between neighborhood satisfaction and overall life satisfaction is mediated by housing satisfaction and regional satisfaction. Note also that the paths from housing satisfaction and regional satisfaction to neighborhood satisfaction are both insignificant. This suggests that the relationships between satisfaction in different urban domains are essentially unidirectional, which turns out to be an unexpected finding.

*Predictors of satisfaction in urban domains.* With regard to predicting *regional satisfaction*, environmental considerations are found to be less important than provision of services and cost of living. Regional pollution, regional population, and regional transport are not significant predictors of regional satisfaction, while regional costs and regional services (that is, education and health services) are significant (see Fig. 15.2).

The most important predictor of *neighborhood satisfaction* is neighborhood interaction. The next most important is neighborhood crime, followed by neighborhood services provided by local government. The path coefficient between neighborhood transport and neighborhood satisfaction is negative. It may seem counter intuitive that better public transport is associated with lower neighborhood satisfaction; however, the small negative coefficient may have arisen if better public transport is associated with less desirable neighborhood characteristics not included in the model such as higher housing densities. Even if this is the case, any positive effects of good public transport on neighborhood satisfaction are likely to be small for most residents. Surprisingly, it has also been found that access to facilities such as shops, banks, and health practitioners is not a significant predictor of neighborhood satisfaction.

As already mentioned, *housing satisfaction* is the most important urban domain for predicting overall life satisfaction of people living in the SEQ region, although this importance decreases to about the same level as regional satisfaction when controlling for other life domains. In turn, housing satisfaction is predicted by housing ownership, housing age and housing temperature. Housing size, as measured by the number of bedrooms, is not a significant predictor of housing satisfaction, although later moderated analysis shows that this depends on lifecycle and lifestyle considerations.

The percentage of variation explained in each of the urban domains is 23% for *housing satisfaction*, 24% for *neighborhood satisfaction*, and 22% for *regional satisfaction* (see $R^2$ statistics in Fig. 15.2). The moderate levels of explained variance in the urban domains indicate that other urban attributes not included in the model also contribute to satisfaction with urban living. This of course is consistent with the conclusions that emerged in the literature review. However, the focus in this study is not on maximizing the prediction of satisfaction in different urban domains but rather on the possible moderating effects of demographic characteristics on particular urban attributes.

*Respecifying the model.* Modification indexes have been used to detect whether the model can be improved by specifying cross paths between urban attributes and different urban domains. Modification indexes suggested that the model can be improved by adding a cross path from regional services to neighborhood satisfaction ($\beta = .09$, $p < .01$) and another from neighborhood services to regional satisfaction ($\beta = .19$, $p < .001$). The addition of these two paths significantly improves the fit of the model ($\chi^2$ (2) = 2.32, $p < .001$).

To improve the stability of the model, two paths have been removed from the initial model. These are the insignificant paths from regional satisfaction and housing satisfaction to neighborhood satisfaction. It has been found in subsequent moderated analyses that the path coefficients for those paths have high standard errors and are fairly volatile.

## The Moderated Analysis

Three personal characteristics have been selected as potential moderators in the model. They are age, gender and family status. Rather than hypothesizing that particular paths are moderated by personal characteristics, this study takes an

**Table 15.4** Regression coefficients in path model by age group

| *Criterion*/predictors | Generation X | Baby boomers | Older people |
|---|---|---|---|
| *Regional satisfaction* | | | |
| Regional pollution | .13* a | .05 | .02 a |
| Regional transport | .09 a | −.07a | .01 |
| *Neighborhood satisfaction* | | | |
| Neighborhood interaction | .30* a | .30* b | .43* a b |
| Neighborhood crime | −.28* a | −.18* a | −.19* |
| *Housing satisfaction* | | | |
| Housing size | −.14* a b | .11* a | .09 b |

Source: The authors

*Significantly different from zero ($p<.05$)

Note: estimates with the same letter are significantly different from each other ($p<.05$). For example, the estimate for housing size for generation X is significantly different from baby boomers and older people, though the latter two are not significantly different from each other

exploratory approach by using an omnibus test for finding moderated paths in the model. As such, any findings will need to be confirmed with follow-up studies (one of which is being undertaken by the authors using data from the 2003 QOL survey in SEQ). The "multi-group moderated path analysis" technique is used which involves dividing the sample into groups and comparing the "free model" with the "constrained model." In the free model, the path coefficients are free to vary between the groups, while in the constrained model the paths are constrained to be equal for all groups. If the free model fits the data significantly better than the constrained model, then one or more of the paths in the model are moderated by the grouping variable. Using this approach to testing for moderating effects does require that variables be grouped into categories; thus, age was grouped into three categories rather than being considered as a continuous variable, while gender and family status are categorical variables by nature.

Age and family status are significant moderators with one or more paths in the model being significantly moderated by these variables. However, gender is not found to significantly moderate the model, ($\chi^2$ (128)=138.75, ns). This means—perhaps surprisingly—that none of the many paths in the model are moderated by gender. In other words, the importance of various urban attributes in predicting toward satisfaction with urban living does not vary significantly between males and females. It also means that the importance of different urban domains in predicting overall life satisfaction does not vary significantly between males and females. It would appear that both males and females place the same importance on various aspects of the urban environment in contributing to satisfaction with urban living and overall life.

*Age moderation.* The age of residents based on three broad generational age groups (*generation X* aged 18–32 years, *baby boomers* aged 33–51 years, and *older people* aged 52 years and over) does significantly moderate the model ($\chi^2$ (260)=323.85, $p<.01$). Some of the paths between urban attributes and urban domains are moderated by age (see Table 15.4); however none of the paths between the urban domains and overall life satisfaction are significantly moderated by age.

Positive assessments of pollution levels are a significant predictor of increased regional satisfaction for the younger generation X group only. It seems that regional pollution is less important for older people in predicting regional satisfaction. However, it might be expected that regional pollution could become an important factor for the wider population as the younger cohort ages and the older cohort dies. The path coefficients for transport systems are not significantly different from zero; however, the path coefficient for generation X is significantly different from and in the opposite direction to the path coefficient for baby boomers. This means that unfavorable assessments of the transportation system were more likely to lead to lower regional dissatisfaction for baby boomers than for generation Xers.

Table 15.4 also shows that neighborhood interaction is significantly more important for older people, while neighborhood crime is more important for generation X. The lower regression coefficient for neighborhood crime for older persons is despite a small positive correlation existing between age and neighborhood crime ($r = .07$, $p < .05$). The latter indicates that older people tend to rate perceive neighborhood crime at a higher level, though the lower regression coefficient for older people means that despite this fact, variations in crime perceptions are not as important in explaining neighborhood satisfaction among older people as it is for generation X. Neither neighborhood transport nor neighborhood access to facilities is a significant predictor of neighborhood satisfaction.

In predicting housing satisfaction, the path coefficient for home size is negative for generation X while it is positive for baby boomers and older people. It seems that younger people prefer smaller homes while baby boomers and older people prefer larger homes.

*Family status moderation.* Family status categories can be derived from determining whether or not a respondent is in a couple relationship, and whether or not they have children living at home. Couples have been classified as those married or in a *de facto* relationship (conversely, singles are unmarried, separated, divorced, or widowed), while parents have been classified as those residents with children living at home. The four family status categories are: singles, couples, single parents, and couple parents. The first two categories have no children living at home. Each of the categories has more than 400 survey respondents except single parents. The single parents category is considered to be too small for modeling purposes, so the moderated model has been run excluding this category as well as combining it with couple parents. The pattern of results is the same in both cases, and the analyses presented below are for the combined group which is called "parents."

The family status of residents significantly moderates the model ($\chi^2$ (260) = 421.37, $p < .001$). The model explains about double the variation in overall life satisfaction for singles compared to couples and parents (36%, 19%, and 18%, respectively). This may be due to the fact that couples and parents with children have an extra source of satisfaction in their lives (that is, their partners and children). As such, these relationships may displace urban living environments as sources of life satisfaction. Such a perspective is supported by ANOVA (analysis of variance) which shows that single people with no children at home have a significantly lower overall life satisfaction ($M = 4.07$) than is the case for the other two categories

**Table 15.5** Percentage in each age grouping by family status

| | Singles | Couples | Parents | Total |
|---|---|---|---|---|
| Generation X | 51 | 23 | 22 | 31 |
| Baby boomers | 23 | 36 | 68 | 44 |
| Older people | 26 | 41 | 11 | 25 |

Source: The authors
$N = 1,347$

**Table 15.6** Regression coefficients in path model by family status

| *Criterion*/predictors | Singles | Couples | Parents |
|---|---|---|---|
| *Overall life satisfaction* | | | |
| Regional satisfaction | .34* a b | .20* a | .17* b |
| Neighborhood satisfaction | .003 a | .06 | .11 * a |
| *Regional satisfaction* | | | |
| Regional pollution | .07 | –.01 a | .11* a |
| Regional population | –.03 a | –.16* a b | .01 b |
| *Neighborhood satisfaction* | | | |
| Neighborhood interaction | .28* a | .44* a b | .34* b |
| Neighborhood transportation | .02 a | –.05 | –.09* a |
| Neighborhood crime | –.29* a b | –.14* a | –.18* b |
| Neighborhood access | .07 | –.10 a | .9* a |
| *Housing satisfaction* | | | |
| Housing size | –.06 a | .06 b | .15* a b |
| Housing temperature | .07 a | .15* | .22* a |
| Housing age | –.14* a | –.18* b | –.10 a b |
| Housing ownership | .17* a | .35* a | .06 a |

Source: The authors
*Significantly different from zero ($p < .05$)
Note: estimates with the same letter are significantly different from each other ($p < .05$). For example, the estimates for housing ownership are all significantly different from each other

($M = 4.29$, $t(1344) = 5.62$, $p < .001$). In other words, close relationships generally add to QOL. This relationship also holds when controlling for age ($F(2, 1343) = 13.78$, $p < .001$).

Before analyzing the moderated path coefficients, it is useful to note the general relationship between family status and age groupings (see Table 15.5). Singles are much more likely to be younger, couples are likely to be somewhat older and parents are likely to be baby boomers. Understanding those general relationships can assist in interpreting the moderated analysis for family status.

In the model moderated by family status, the predictors of overall life satisfaction generally follow the same pattern as in previous analyses. The most important predictor of overall life satisfaction is housing satisfaction, followed by regional satisfaction and then neighborhood satisfaction. Unlike previous analyses, neighborhood satisfaction is a significant predictor of overall life satisfaction, but only for parents (see Table 15.6). In contrast, singles derive little overall life satisfaction from their neighborhood satisfaction and have a significantly higher path coefficient for regional satisfaction. Those single people are also likely to be relatively young (that is, aged between their late teens and early 1930s).

The importance of regional pollution in predicting regional satisfaction is generally low. However, pollution is a significant predictor of regional satisfaction for parents (with children living at home). Negative assessments of population levels, growth, and urban sprawl are significantly associated with reduced regional satisfaction for couples (with no children living at home). This result is not explained by age differences in the three groups (see the age moderation analysis). It seems that couples contain a substantial proportion of people who are concerned about population issues.

As with the unmoderated model, neighborhood interaction and neighborhood crime are the most important indicators of neighborhood satisfaction for family groups. Although neighborhood interaction is important for all family types, it is more important for couples. This may be due to the fact that couples are more likely to be older people (see Table 15.5), and as such, they may spend more time in their neighborhood. Perceived neighborhood crime is more important to single people, and this is significantly higher than for couples, and also parents. These results are consistent with those for age since about half of the single people without children living with them are relatively young (between 18 and 32 years of age). However, a quarter of that group are baby boomers, and another quarter are older people, so that the importance of perceived crime probably also relates to living alone.

Neighborhood access to facilities and the adequacy of neighborhood transport generally are of low importance in predicting neighborhood satisfaction for the family status groups. However, both these predictors are significant for parents with children living at home. Neighborhood access is significantly higher for parents than that for couples with no children living at home. This may be because parents use more local facilities associated with their children (e.g. schools, childcare, and recreation facilities). Adequacy of public transport is also significant for parents with children living at home, such that better public transport is associated with lower neighborhood satisfaction. This result is not intuitive and is perhaps due to good public transport being associated with other less desirable neighborhood characteristics, though we need to see if this result is replicated when analyzing the 2003 dataset.

Parents with children living with them prefer larger homes than couples and singles. It is interesting to note that the path coefficient for singles is not significantly different from zero; however, in the earlier age moderation analysis, generation X preferred smaller housing sizes. This means that the preference for smaller homes by generation X cannot be explained by the fact that they are more likely to be single (see Table 15.5) and therefore implies a generational shift toward smaller housing. Housing temperature is significant for couples with no children living at home and parents with children living at home, but not for single persons.

The pattern of path coefficients for home ownership is unexpected. The relative importance of home ownership for couples with no children living at home is large in comparison to the other family types, and home ownership is not a significant predictor of housing satisfaction for parents with children living at home. This pattern is the same even when single parents are excluded from the analysis and when ownership is simplified to two levels (renting and purchasing/fully owned housing). It may be that parents with children attach less importance to home ownership than the policy makers would have us believe.

## Discussion

### *Summary of Results*

This study of SEQ has tested a model of satisfaction with urban living with a focus on regional satisfaction and the moderating effects of selected demographic characteristics. Regional satisfaction is less commonly studied than satisfaction with other urban domains; however, regional satisfaction is an important urban domain in predicting overall life satisfaction. It has been found to be more important than neighborhood satisfaction, although not as important has housing satisfaction. However, regional satisfaction becomes as important as housing satisfaction when other non-urban life domains are controlled.

Generally speaking, *neighborhood satisfaction* is *not* an important indicator of *overall life satisfaction*. The exception is for parents where satisfaction with their neighborhood is significantly associated with overall life satisfaction. However, neighborhood satisfaction also contributes *indirectly* to overall life satisfaction through housing and regional satisfaction. The mediated effects of neighborhood satisfaction on overall life satisfaction by other urban domains are also supported by recent findings from Sirgy and Cornwell (2002).

The most important *predictors of regional satisfaction* in the SEQ region are assessments of regional services (education and health services) and the cost of living. This finding aligns with Rogerson's (1999) view that consumption experiences are important determinants of the QOL of people living in cities. Generally speaking, aspects of urban living relating to sustainability issues (for example, pollution, population levels, and transport systems) are found to be less important in predicting regional satisfaction. However, negative assessments of those aspects of urban living do impact on the level of regional satisfaction for the generation X group more than is the case for those from older generational groups. In fact, the model path coefficients suggest expanding transport systems is viewed positively by the baby boomers and negatively by generation X respondents. This suggests planners might be advised to focus on devising new ways of using existing infrastructure more effectively rather than expanding infrastructure when considering the transportation needs of future generations. Couples without children are the only group significantly concerned with problems relating to population growth to the extent that it is significantly associated with their level of regional satisfaction.

*Neighborhood satisfaction* is best predicted by assessments of neighborhood interactions and neighborhood crime. Other studies have also found perceptions of crime and neighborhood interaction to be important predictors of neighborhood satisfaction (see, for example, Bruin and Cook 1997; Campbell et al. 1976; Greenberg et al. 1994; Taylor 1995). In this study of SEQ, the moderated analysis also found that neighborhood interaction is more important for older people, while neighborhood crime is more important for single people and for younger people in predicting levels of neighborhood satisfaction.

It seems unlikely that the negative path coefficient between neighborhood transport and neighborhood satisfaction means that good public transport is viewed negatively. Notwithstanding this, any positive effects of good public transport on neighborhood satisfaction are likely to be small for most residents.

Like regional satisfaction, the provision of neighborhood services (for example, parks, pools, and libraries) is an important predictor of neighborhood satisfaction, and this was not moderated by any of the selected demographic characteristics. Neighborhood transport also does not seem to play an important role in neighborhood satisfaction, although in this study only public transport is evaluated for neighborhoods.

Somewhat surprisingly, it has been found that access to facilities (for example, shops, post offices, banks, and health practitioners) does not predict neighborhood satisfaction very well. This could be because the survey data showed most people to be very happy with their access to those services, although it does raise the question about whether satisfaction with urban areas is more associated with the provision of public services than it is with the provision of private services. Notwithstanding this generalization, access to facilities is a significant predictor of neighborhood satisfaction for parents with children.

*Housing satisfaction* is predicted by level of housing ownership, housing age and housing temperature. The size of the home is not an important indicator in the unmoderated model despite the fact that other studies show many people prefer to live in low-density environments (Audirac et al. 1990). However, this study shows that larger homes are highly valued by some, while smaller homes are highly valued by others living in the SEQ region. In the moderated analysis, the baby boomers, couples, parents, and older people valued larger homes more, while singles and those from generation X value smaller homes more. That finding emphasizes the importance for planners and developers of matching demographic information with housing stock characteristics to enhance overall life satisfaction of residents of cities.

## Study Limitations and Future Directions

There are a number of limitations to the study reported here:

(a) *The survey data are correlational*, and thus causation from the objective attributes of urban environments through to overall life satisfaction is only assumed (bottom-up causation). But some of the relationships may be accounted for by top-down causation. For example, there may be individual personality traits that result in positive QOL assessments and positive assessments of the urban environment generally.

(b) *The model did not include community satisfaction*, and including this may have reduced the relative importance of regional satisfaction. It is recommended that future studies examining satisfaction with urban living include four levels of urban living or domains (housing, neighborhood, community, and region) to confirm the relative importance of regional satisfaction.
(c) *The moderated analysis is exploratory rather than being based on a priori hypotheses*. Given that, the findings need to be replicated in later studies to ensure that some of the results are not due to chance. Despite this, the approach using multi-group moderated path analysis appears to be a useful tool to explore the ways in which relationships in complex models may vary with personal characteristics.
(d) *These results also need to be replicated with other populations*. Important features of the urban environment may vary between different regions and cultures, while some remain in common. Equally important, youths and children may have different perspectives on the urban environment, and those groups were not included in the sample for this study.

As mentioned earlier, further analysis is planned by the authors using the 2003 QOL in SEQ dataset to confirm the importance of regional satisfaction in contributing to overall life satisfaction and the moderating effects of selected demographic characteristics.

## Main Implications of Findings

The study reported here has found that regional satisfaction and housing satisfaction are important urban domains in predicting overall life satisfaction compared to neighborhood satisfaction. This suggests that planners and policy makers might avoid focusing on neighborhood projects at the expense of housing issues and broader urban issues.

As mentioned, the urban attributes relating to sustainability issues are more important to generation X, although couples are also concerned with population growth. As older cohorts die and more couples decide not to have children, there is likely to be increasing pressure to design environmentally friendly, sustainable and livable urban environments with less focus on growth. Continuing urban growth has even been associated with a reduction in perceived quality of life (Baldassare and Wilson 1995), and planners in the twenty-first century will need to shift to emphasizing sustainability issues as well as considering optimal sizes for urban centers and optimum centrality. Optimal centrality relates to the optimal size of a city such that the net benefits of centralization are maximized taking into account any loss in livability due to that centralization (Cicerchia 1999).

The study also reveals how people can differ in their values, experiences, perceptions, and assessments of urban living. In other words, something that is important to one demographic group may not be important to another in contributing to their satisfaction with the urban environment and overall QOL. Changing demographic profiles and population values need to be incorporated into long-term urban planning to match future urban environments with future population needs and values. It seems that over the foreseeable future, the fundamental aspects of urban living—such as service provision and the cost of living—will continue to hold primary importance in contributing to satisfaction with urban living and overall QOL. However, it also seems that the emphasis on providing environmentally friendly and sustainable urban environments will grow in importance and as well that there will be an increased demand for smaller homes.

**Acknowledgments** The survey data on which the analysis in this study is based was collected in 1997 with funding support from the Australian Research Council, the Queensland Treasury and the Queensland Department of Local Government and Planning through Australian Research Council Collaborative Project #C595301132. The analysis undertaken for this study has been made possible through funding for Australian Research Council Discovery Project #DP0209146.

## References

Audirac, I., Shermyen, A. H., & Smith, M. T. (1990). Ideal urban form and visions of the good life – Florida growth management dilemma. *Journal of the American Planning Association, 56*(4), 470–482.

Baldassare, M., & Wilson, G. (1995). More trouble in paradise – Urbanization and the decline in suburban quality-of-life ratings. *Urban Affairs Review, 30*, 690–708.

Boyer, R., & Savageau, D. (1981). *Places rated almanac*. Chicago: Rand McNally.

Brotchie, J., Newton, P., Hall, P., & Nukamp, P. (1985). *The future of urban form*. London: Coom, Helm, and Nichols.

Bruin, M. J., & Cook, C. C. (1997). Understanding constraints and residential satisfaction among low-income single-parent families. *Environment and Behavior, 29*, 532–553.

Campbell, A., Converse, P., & Rodgers, W. (1976). *The quality of American life: Perceptions, evaluations and satisfaction*. New York: Russell Sage.

Cicerchia, A. (1999). Measures of optimal centrality: indicators of city effect and urban overloading. *Social Indicators Research, 46*, 273–299.

Diener, E. (1984). Subjective well-being. *Psychological Bulletin, 95*, 542–575.

Glaeser, E., Kolko, J., & Saiz, A. (2000). *Consumer city* (Working Paper 7790). Cambridge: National Bureau of Economic Research.

Grayson, L., & Young, K. (1994). *Quality of life in cities*. London: British Library.

Greenberg, M., Schneider, D., & Choi, D. W. (1994). Neighborhood quality. *Geographical Review, 84*, 1–15.

Jeffres, L. W., & Dobos, J. (1995). Separating peoples satisfaction with life and public perceptions of the quality-of-life in the environment. *Social Indicators Research, 34*, 181–211.

Joreskog, K., & Sorbom, D. (1996). *LISREL8: User's reference guide*. Chicago: Scientific Software International.

Kahneman, D. (1999). Objective happiness. In D. Kahneman, E. Diener, & N. Schawrz (Eds.), *Well-being: The foundations of hedonic psychology*. New York: Russell Sage.

Keeble, D. (1990). Small firms, new firms, and uneven regional development in the United Kingdom. *Area, 22*, 234–245.

Kemp, D., Manicaros, M., Mullins, P., Simpson, R., Stimson, R., & Western, J. (1997). *Urban metabolism: A framework for evaluating the viability, livability and sustainability of South East Queensland*. Brisbane: The Australian Housing and Urban Research Institute.

Lance, C. E., Mallard, A. G., & Michalos, A. C. (1995). Tests of the causal directions of global life facet satisfaction relationships. *Social Indicators Research, 34*, 69–92.

Ley, D. (1996). *The new middle class and the remaking of the central city*. Oxford: Oxford University Press.

Liaw, K. L., Frey, W. H., & Lin, J. P. (2002). Location of adult children as an attraction for black and white elderly primary migrants in the United States. *Environment and Planning A, 34*, 191–216.

Lu, M. (1999). Determinants of residential satisfaction: Ordered logit vs. regression models. *Growth and Change, 30*, 264–287.

Marans, R. W. (2002). *Detroit area study: Presentations and selected findings*. http://www.tcaup.umich.edu/workfolio/DAS2001/findings/WebDistributionbyCty.html

Marans, R. W., & Rodgers, W. (1975). Toward an understanding of community satisfaction. In A. Hawley & V. Rock (Eds.), *Metropolitan America in contemporary perspective*. New York: Halsted.

Michalos, A. C. (1985). Multiple discrepancies theory (MDT). *Social Indicators Research, 16*, 347–413.

Parkes, A., Kearns, A., & Atkinson, R. (2002). What makes people dissatisfied with their neighborhoods? *Urban Studies, 39*, 2413–2438.

Pierce, R. M. (1985). Rating America's metropolitan areas. *American Demographics, 7*, 20–25.

Rogerson, R. (1999). Quality of life and city competitiveness. *Urban Studies, 36*(5), 319–328.

Schawrz, N., & Strack, F. (1999). Reports of subjective well-being: Judgmental processes and their methodological implications. In D. Kahneman, E. Diener, & N. Schawrz (Eds.), *Well-being: The foundations of hedonic psychology*. New York: Russell Sage.

Sirgy, M. J., & Cornwell, T. (2002). How neighborhood features affect quality of life. *Social Indicators Research, 59*, 79–114.

Sirgy, M. J., Rahtz, D., Cicic, M., & Underwood, R. (2000). A method for assessing residents' satisfaction with community-based services: A quality-of-life perspective. *Social Indicators Research, 49*, 279–316.

Taylor, R. B. (1995). The impact of crime on communities. *The Annals of the American Academy of Political and Social Science, 539*, 28–45.

Turksever, A. N. E., & Atalik, G. (2001). Possibilities and limitations for the measurement of the quality of life in urban areas. *Social Indicators Research, 53*, 163–198.

# Chapter 16
# Disaggregating the Measurement of Quality of Urban Life Dimensions Across a Complex Metro Region: The Case of Metro Detroit

**Byoung-Suk Kweon and Robert W. Marans**

## Introduction

As other chapters in this book have demonstrated, the quality of an individual's life is influenced by an array of socioeconomic and psychological factors. It may also be influenced by geographic factors. Geographic factors can be considered at the micro- and macrolevels. The microlevel often includes attributes of the built or the natural environment within which households are located. At a macrolevel, it includes the size of the individual's community, the type of political jurisdiction that defines that community, or the amount and extent of development in that community. Although there is a growing understanding of the association between socioeconomic and demographic factors such as income, education, gender, and race and quality of life (QOL), relationships between individual well-being and these geographic factors are not fully understood.

In this chapter, we investigate the geographic patterns of QOL, and we explore appropriate scales for examining spatial patterns showing QOL differences. Specifically, we identify QOL studies that have been conducted at different geographic scales (for example, nation, city, suburb, and neighborhood) and suggest a new approach for considering geographic scale in quality of life research. This approach is tested with the data from the quality of urban life (QOUL) study of the Detroit metropolitan area (see Chap. 7). The region contains nearly 4,600 square miles and

B.-S. Kweon (✉)
Department of Plant Science & Landscape Architecture, University of Maryland,
2102 Plant Sciences Building, College Park, MD 20742, USA
e-mail: kweonb@umd.edu

R.W. Marans
Institute for Social Research, University of Michigan, POB 1248, Ann Arbor,
MI 48106, USA
e-mail: marans@umich.edu

R.W. Marans and R. Stimson (eds.), *Investigating Quality of Urban Life: Theory, Methods, and Empirical Research*, Social Indicators Research Series 45,
DOI 10.1007/978-94-007-1742-8_16, 

has over 400 independent governmental jurisdictions (that is, cities, townships, villages, and counties) that vary in size, age, and period of development.

## Geographic Scale of Quality of Life Studies

Geographical scale refers to "the [spatial] dimensions of identifiable social or physical features of a landscape" (Reardon et al. 2008: p. 489). It differs from a cartographic scale that describes a unitless "ratio of distances on a map to distances in the real world" (Reardon et al. 2008: p. 489). Understandably, cartographic scales have not been used in QOL studies since it would be difficult to interpret research findings of such studies without referring to a "unit" or phenomenon. Cartographic scale also differs from methodological scale that is somewhat arbitrarily divided into units, such as census tracts. The US census has a framework that separates the country into divisions: states, metropolitan statistical areas (MSAs), counties, municipalities, census tracts, block groups, and blocks. The units of states, counties, and municipalities are political boundaries and could be considered geographic scales. However, the divisions of census tracts, block groups, and blocks are somewhat arbitrary spatial units. Even though census tract and block boundaries follow "permanent, visible feature, such as streets, roads, highways, rivers, canals, railroads and, high-tension power lines," the environmental characteristics of one track may be the same as an adjacent track. Therefore, census tracts may not be suitable measures to detect different patterns of QOL and of QOUL that occur at different geographic scales. Geographic scale, on the other hand, is a spatial scale that represents the geographical "extent" or scope of a study area (McMaster and Sheppard 2004). What are needed are geographic scale measures that are more appropriate for quality of life studies.

QOL studies can be scale-dependent. That is, certain patterns of QOL measures can only be found at specific geographic scales (Walsh et al. 2004). For example, the QOUL within different neighborhoods could be important in the allocation of resources, but the differences would not be discernable if QOUL were considered only at the city level of scale. Research findings in turn may have alternative policy implications depending on the size and type of geographic unit under study. Consequently, it is important to study QOUL at several scales simultaneously (for example, neighborhoods and city) in order to improve the understanding of its meaning according to different geographic patterns.

One limitation of many previous QOL studies is that they have relied largely on only one geographic scale, such as a nation, city, or county. Levin's (1992: p. 1944) Hierarchy Theory claims that "there is no single correct scale on which to describe populations." Rather than single-scale studies, Levin recommended "cross-scale" studies that quantify patterns of findings at broad scales as well as fine scales within the broader ones. The theory is used to understand how lower level and higher level systems interact between and within the system levels (McMaster and Sheppard 2004). Lower level systems act as wholes with respect to lower levels of hierarchy, but are parts of units at higher levels. The concept is explained by Reardon et al.

(2008) using macro- and microscale patterns. Those authors indicate that there are good reasons to think that both the causes and consequences of macroscale QOL may differ from those of microscale QOL. Hierarchy theory provides an important framework for QOUL research by establishing a rationale for studying multiple scales simultaneously.

## Examples of Using Geographic Scales in Quality of Life Research

QOL has been studied at discrete geographic scales, including nations, cities, suburbs neighborhoods, and communities. At the national level, Frijters et al. (2004) investigated life satisfaction of residents living in the former East Germany. This longitudinal study aimed at examining change of life satisfaction before and after reunification. In another national study, Flanagan (1978) looked at the QOL of residents throughout the USA. Although Flanagan sampled study participants from different states, the author only reported national, although data were available at the state level. The intention was to study Americans as a whole rather than a multiscale (that is, nation and state) phenomena. The US General Accounting Office (2003) has developed key national indicators assessing the nation's current situation as well as changes over time. Similar to the US General Accounting Office's approach, Johansson (2002) discussed the Swedish Level of Living Surveys to access the nation's current well-being using social indicators and social reporting. All of these QOL studies presented findings covering a single geographic unit, (the nation), rather than taking a multiscale perspective where findings cover states and cities in addition to the nation. The objectives of most national scale studies were to monitor change and to improve the well-being of their citizens, along with assessing current situations.

### *City*

Geographic scale is also used in numerous popular publications to compare different cities and to create a list of best places to live and retire inferring a high quality of living. Although there is criticism of using "proprietary" survey methods, *Money Magazine* has consistently compared quality of living in different cities in the USA (Guterbock 1997).

Elgin et al. (1974) conducted a city-scale study comparing how city size influenced citizens' QOL. Here they found that city size had negative impacts on residents' quality of life measured across a number of dimensions (for example, social, environmental, political, and systemic). In other words, larger cities had conditions of life that detract from what was ideal, such as congestion, pollution, ill-health, crime, loss of community, and governmental fragmentation. Those authors showed their concerns on rapidly growing big cities and called for greater needs of balanced urban growth and future urbanization.

Urban ecosystem services (for example, street trees, parks, forests, water bodies) have also been shown to have an impact on residents' QOL (Bolund and Hunhammar 1999).

In summary, the objective of city-scale studies seemed to be cross-sectional and included only one geographic scale (that is, city). The goal was to find the cities that provide high QOUL by comparing many factors, such as city size, crime rate, employment rate, cost of living, and so on. However, the results of the comparison were highly volatile depending on the evaluating factors. In addition, QOL may vary substantially among different areas within a city, or different cities within a state, etc.

New methods of measuring QOL at multiple scales have been developed in recent years (Epley and Menon 2008). This method allows comparison of cities as well as states with indicators that are available for both city and state scales and that contribute to the quality of life in the regions.

## *Suburb*

A limited number of studies have investigated how suburban environments influence residents' QOL, and the results of these studies send mixed messages.

Moller (1992) studied 1,200 youth living in nine suburbs in three major metropolitan areas in South Africa. The author concluded that geographical factors influence residents' QOL by finding that residents in Umlazi in the Durban suburb were more satisfied with their life than those in other suburbs. Moller also indicated that regional difference influence overall life satisfaction and youth outlook on the future. However, the author did not elaborate on why living in different suburban areas played a role in well-being.

Lloyd and Auld (2002) also examined place-centered variables (for example, clean air, water, noise, congestion, etc.) in the multiple suburbs. They found that the influence of place-centered variables on QOL was minimal.

In contrast, Baker and Palmer (2006) focused on one suburb of a major metropolitan area in the Southwest of the USA and found that physical conditions of residential environments (for example, housing upkeep, crowding and noise, access to facilities, etc.) significantly influenced residents' well-being.

## *Neighborhood/Dwelling*

Neighborhood open spaces play an important role in enhancing residents' QOL. Previous studies found that distance, pleasantness, and safety of neighborhood open spaces influence residents' quality of life (Sugiyama, Thompson & Alves 2009). Also, the amount of nature within census tracts can have a positive impact on QOUL

(Li and Weng 2007). Conversely, accessibility problems in home environments can have negative impacts on residents' well-being (Wahl et al. 2009).

## *Quality of Life Studies in Multiple Geographic Scales*

A number of previous studies considered multiple geographic scales to investigate QOL. Researchers found that residents living in urban areas were less satisfied with their communities than those living in nonurban areas (Dillman and Dobash 1972; Dillman and Tremblay 1977). Beck et al. (2009) compared health-related QOL among rural, semi-rural, and urban areas and found no differences in QOL between the three geographic areas.

Using data from a national survey of households in the USA, Marans et al. (1980) compared five different geographic areas: large urban, small urban, small town, large rural, and small rural. Although the sample size for some groups was small, a clear pattern of geographic influence was found. People in small rural areas reported higher levels of life satisfaction than those living in large urban areas.

## A New Approach to Investigating the Geographic Scale of Quality of Life

As discussed in Chap. 7, data from DAS2001 were reported for the entire region and for each of its seven counties. Because the city of Detroit is the largest political jurisdiction in the region and part of one county (Wayne), QOUL data covering the city were reported separately. We also wanted to report data at different geographic scales to reflect the different types of settlements or places that made up the region. Our approach therefore was to use US Census of Population data for each of the region's minor civil divisions (MCD) in conjunction with the MCD's distinctive geographic characteristics to create a hierarchical structure reflecting the *type of place* where people lived. Types of places identified were (see Table 16.1 and Fig. 16.1):

- The urban core (also referred to as the region's major city, namely Detroit)
- Detroit's older suburbs
- Detroit's newer suburbs
- Other large cities in the Metro Detroit area that are not contiguous to the urban core
- Midsize cities not contiguous to the urban core
- Small towns and rural settings throughout the metro area that were not contiguous to the urban core

**Table 16.1** Type of place classification

| Types of place | Size | Examples |
|---|---|---|
| Urban core | 950,000 | Detroit |
| Larger cities | 50,000–949,999 | Ann Arbor, Pontiac |
| Midsize cities | 15,000–49,999 | Mt. Clements, Port Huron, Monroe, Ypsilanti |
| Older suburbs | Varies | Warren, Southfield, Dearborn, Royal Oak, Melvindale, Ferndale |
| New suburbs | Varies | Canton, Brownstown, Clinton, Pittsfield, Bedford |
| Small towns | Less than 15,000 | Plymouth, Brighton, Howell, Chelsea |
| Rural | Varies | Green Oak, Scio, Springfields |

Source: The authors

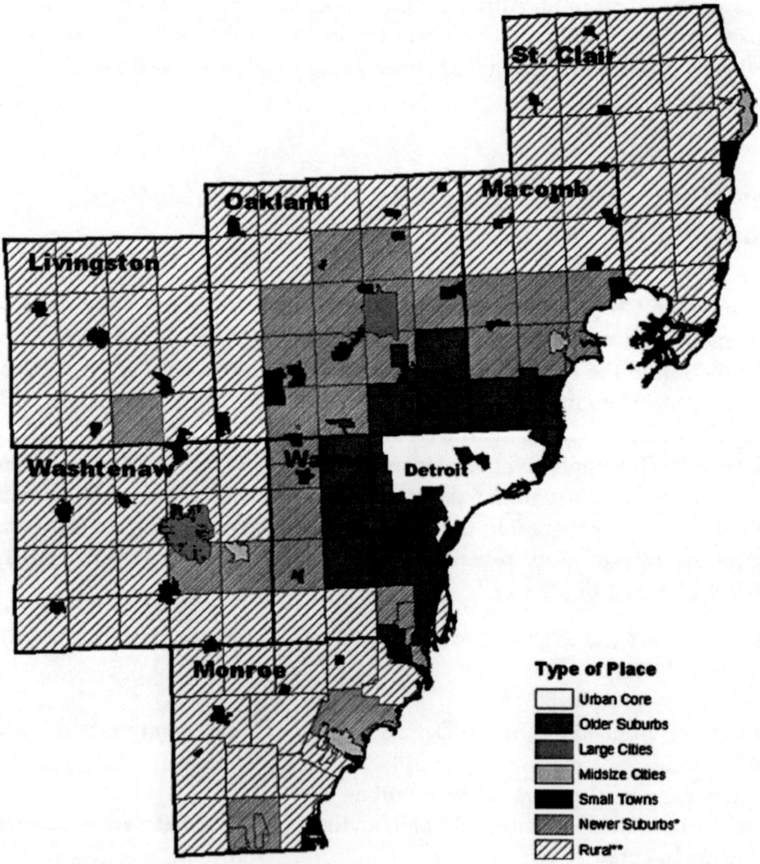


**Fig. 16.1** Type of place

Several steps were taken in creating the type of place classification:

(a) Initially, minor civil divisions (MCD) in Metro Detroit were sorted into two groups: Incorporated cities and villages, and unincorporated townships. Incorporated cities and villages were then grouped according to the size of the population, while unincorporated townships were divided into two groups according to the population size.
(b) Next, the incorporated MCD's were classified into places that were adjacent or contiguous to Detroit and those that were far from Detroit and relatively self-contained (that is, Ann Arbor, Howell, Monroe, and Port Huron). Self-contained places were those settlements whose spatial pattern was independent of surrounding settlements. Unincorporated townships where respondents lived were classified either as places where there was residential development (high density) or places that were largely rural or agricultural in character (low density). Low-density portions of township were places where the density was less than 0.5 housing units per acre.

The number of places where respondents live was then reduced to the seven descriptive groupings.

We examine several outcome measures from the Detroit Area Study at two geographic scales: type of place and county of residence. The outcome measures discussed in greater detail in

Chapter 7 included an overall QOL measure and measures capturing ratings of specific physical attributes of urban environments.

Overall, QOL was measured by asking Detroit area residents: *"How satisfied are you with your life as a whole?"* Participants responded by using 7-point response scale (where 1 = "completely satisfied" to 7 = "completely satisfied").

The questions *"How satisfied are you with your dwelling?"* and *"How satisfied are you with this neighborhood as a place to live?"* also used the same 7-point response scale.

Assessments of the quality of local parks, public schools, and nearby public transportation were determined by asking a series of questions enabling respondents to rate these services on a 5-scale ranging from "very good" (5) to "not good at all" (1).

## Findings

Findings are presented at two geographic scales:

- The types of places
- The counties where the sampled households were located.

As mentioned above, Metro Detroit contains seven different types of places which are distributed across the seven southeast Michigan counties. We also examine the QOL of residents in the seven places *within* each county. Resident ratings of

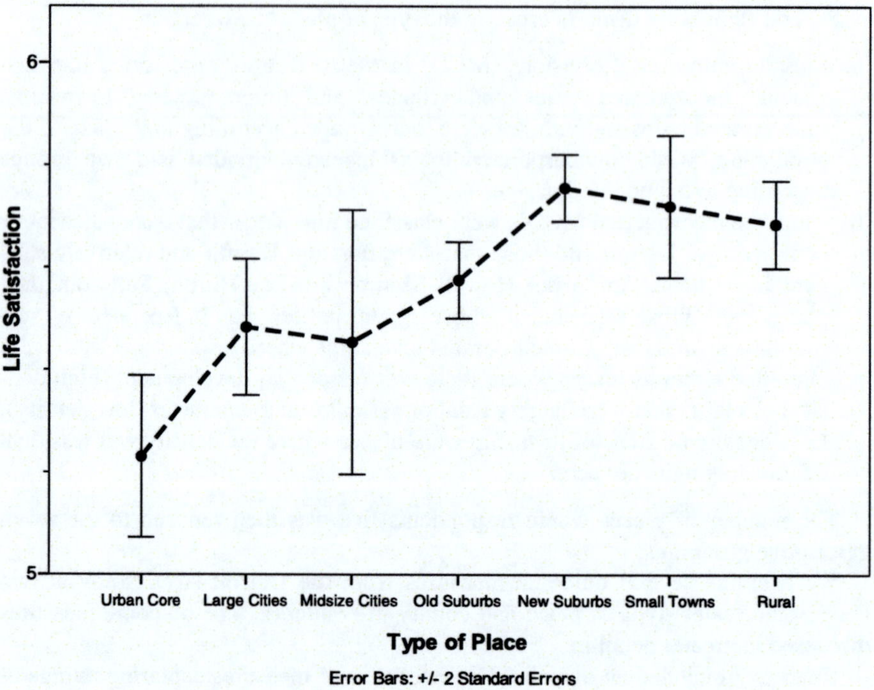


**Fig. 16.2** Ratings of life satisfaction by type of place (mean score) (Source: The authors)

their dwellings, neighborhoods, local parks, schools, and public transportation systems within each type of place and each county are also presented.

## *Life Satisfaction*

Figure 16.2 presents mean satisfaction scores (on a 7-point Likert scale) for each type of place in Metro Detroit. The figure shows that, overall, people tended to be satisfied with their lives (mean = 5.61). However, people living in the larger places were less likely to be satisfied, while those living in the suburbs (older and newer), small towns, and rural areas were more satisfied. The figure also shows that within each type of place, there were modest variations in responses to the life satisfaction question. Nonetheless, the satisfaction level in urban core (mean = 5.23) was significantly lower than satisfaction levels in the old and new suburbs, small towns, and rural areas. Similarly, the satisfaction levels in new suburbs (mean = 5.76) were significantly higher than those found in the urban core, large cities, and old suburbs.

Figure 16.3 shows that there were variations in satisfaction scores for people living in each of the seven counties. Residents of Wayne County, including the city

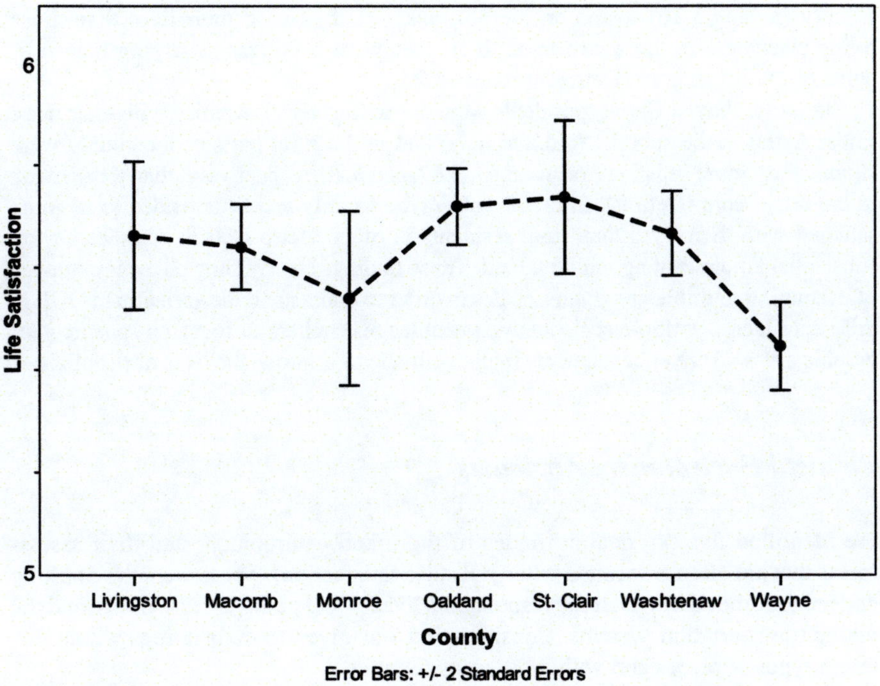


**Fig. 16.3** Ratings of life satisfaction by county (mean score) (Source: The authors)

**Table 16.2** Life satisfaction for type of place within each county

| | Livingston | Macomb | Monroe | Oakland | St. Clair | Washtenaw | Wayne |
|---|---|---|---|---|---|---|---|
| Urban core | | | | | | | 5.23 |
| Large cities | | | | 5.31 | | 5.49 | |
| Midsize cities | | 5.45 | 5.33 | | 5.70 | 5.29 | |
| Old suburbs | | 5.49 | | 5.70 | | | 5.57 |
| New suburbs | 5.83 | 5.80 | 5.83 | 5.79 | | 5.70 | 5.57 |
| Small towns | 5.64 | 5.80 | 5.16 | 5.70 | 5.77 | 5.95 | 5.38 |
| Rural | 5.64 | 5.57 | 5.37 | 5.53 | 5.73 | 5.89 | 5.70 |

Source: The authors

of Detroit, were the least satisfied (mean = 5.45), whereas those living in St. Clair were most satisfied (mean = 5.74). High levels of satisfaction were also expressed in Macomb, Oakland, Livingston, and Washtenaw Counties. With the exception of Livingston County, these expressions of satisfaction were significantly different than those expressed by Wayne County residents.

When we consider satisfaction scores for people living in different types of places *within* each county, a somewhat different pattern emerges (see Table 16.2). For example, Table 16.2 showed that residents living in small towns reported the highest mean satisfaction scores. However, small town residents living in Monroe

County (mean = 5.16) were considerably less satisfied with their lives than those living elsewhere in the metro area; their lives were even less satisfying than residents of the urban core (Detroit) (mean = 5.23).

Similarly, Wayne County residents were not uniformly dissatisfied with their lives. Those living in the suburbs (old and new) and in the rural parts of the county were significantly more satisfied (mean = 5.57, 5.57, and 5.70, respectively) than those living in the urban core (Detroit). In fact, rural Wayne County residents tended to be more satisfied with their lives than rural residents in other Metro Detroit counties having substantial amounts of agricultural land. These findings clearly showed the importance of examining multiple geographic scales in order to understand the variation in QOL in different places. Multiple-scale analysis enables researchers to focus on specific geographic areas where enhancements might be made to improve the lives of people.

## *Physical Environment Attributes*

We identified five physical attributes of the urban environment and their assessments that are often associated with QOL (life satisfaction). These were the dwelling and neighborhood in which the respondents lived, local parks, local schools, and the public transportation systems. Consideration was given to their ratings within different types of places and within different counties.

### Dwellings

Figure 16.4 shows a gradual increase in levels of housing satisfaction with dwelling as the geography changes from the urban core to rural areas (urban core mean = 4.90; rural mean = 5.68). The residents in urban core (Detroit) were significantly less satisfied with their dwelling than all other types of places. At the same time, levels of satisfaction in Wayne County (mean = 5.14) were significantly lower than those of all other counties (see Fig. 16.5). From a geographical perspective, people living in low-density settlements such as small towns and rural areas were more likely to view their housing favorably than people in the more densely populated areas, including the urban core.

However, the residents in the large cities in Oakland and the midsize cities in Macomb were less satisfied with their dwellings than were the residents in Wayne County (see Table 16.3). The thoughts of residents in large cities in Oakland and the midsize cities in Macomb should not be ignored by only looking at a type of place or county-level analysis. Their lower satisfaction levels could only be detected by looking at both place data and county level data simultaneously. Also, the residents in the newer suburbs and rural areas in Wayne County were significantly more satisfied with their dwellings than residents in Wayne's urban core. In other words, not everyone in Wayne County was dissatisfied with their dwellings.

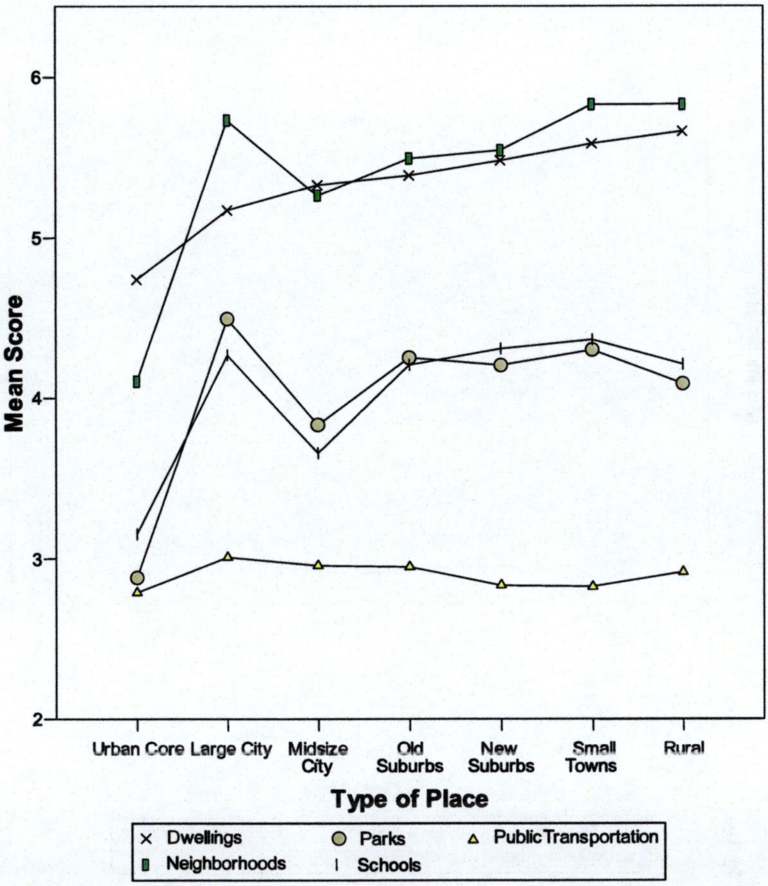


**Fig. 16.4** Ratings of physical attributes by type of place (mean score) (Source: The authors)

## Neighborhoods

As seen in Fig. 16.5, Wayne County residents rated their neighborhoods significantly lower than residents in the other counties. At the same time, Livingston residents had the highest ratings which were significantly different from the ratings of Wayne County residents (5.92 versus 5.07). Similarly, residents living in urban core rated their neighborhoods significantly lower than residents from other places (see Fig. 16.4). Neighborhoods in the urban core had the lowest ratings (mean = 4.31), while those in rural areas had the highest ratings (mean = 5.84). When considering either county of residence and type of place, the differences are statistically significant ($p < .0001$).

In examining neighborhood satisfaction by county and type of place simultaneously, variations in the satisfaction scores can be found within each county (see Table 16.3). For instance, Wayne County shows significant differences between the

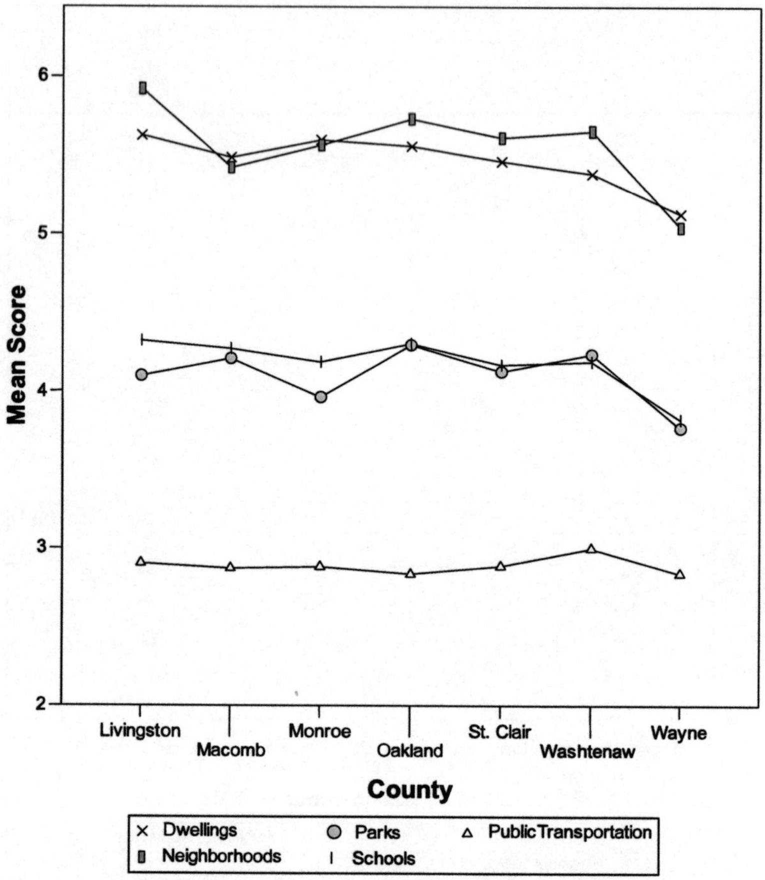


**Fig. 16.5** Ratings of physical attributes by county (mean score) (Source: The authors)

satisfaction score in the urban core and the satisfaction scores of residents living in the old and new suburbs, small towns, and rural areas. In fact, residents living in Wayne County's small towns had the highest neighborhood satisfaction level in the metro area (mean = 6.13). Similarly, residents in Oakland were also very satisfied with their neighborhoods (mean = 5.74); however, residents in the county's large city were much less satisfied with their neighborhoods (mean = 4.77). In fact, large city residents in Oakland were significantly less satisfied with their neighborhoods than residents living in all other places in the county except in the small towns ($p < .05$).

Table 16.3 also shows county differences in ratings of new suburban neighborhoods. Among those living in the new suburbs, Washtenaw residents were significantly less satisfied with their neighborhoods than residents from other counties. These variations only can be detected by considering both place and county geographic scales simultaneously.

**Table 16.3** Ratings of physical attributes by county and type of place (mean score)

| | | | | | | | | |
|---|---|---|---|---|---|---|---|---|
| Urban core | Dwellings | | | | | | | 4.90 |
| | Neighborhoods | | | | | | | 4.31 |
| | Parks | | | | | | | 2.91 |
| | Schools | | | | | | | 3.16 |
| | Transportation | | | | | | | 2.85 |
| Large cities | Dwellings | | | | 4.85 | | 5.19 | |
| | Neighborhoods | | | | 4.77 | | 5.75 | |
| | Parks | | | | 3.33 | | 4.54 | |
| | Schools | | | | 3.50 | | 4.31 | |
| | Transportation | | | | 2.64 | | 3.03 | |
| Mid-size cities | Dwellings | | 4.89 | 5.48 | | 5.42 | 5.55 | |
| | Neighborhoods | | 5.20 | 5.43 | | 5.36 | 5.38 | |
| | Parks | | 3.67 | 3.86 | | 4.13 | 3.66 | |
| | Schools | | 3.24 | 3.84 | | 4.03 | 3.38 | |
| | Transportation | | 2.21 | 2.68 | | 3.35 | 3.00 | |
| Old suburbs | Dwellings | | 5.41 | | 5.53 | | | 5.28 |
| | Neighborhoods | | 5.42 | | 5.77 | | | 5.48 |
| | Parks | | 4.17 | | 4.33 | | | 4.28 |
| | Schools | | 4.19 | | 4.31 | | | 4.11 |
| | Transportation | | 3.00 | | 2.87 | | | 2.88 |
| New suburbs | Dwellings | 5.90 | 5.52 | 5.75 | 5.59 | | 5.11 | 5.29 |
| | Neighborhoods | 6.06 | 5.48 | 5.72 | 5.76 | | 5.15 | 5.66 |
| | Parks | 4.20 | 4.28 | 4.02 | 4.26 | | 4.01 | 4.19 |
| | Schools | 4.32 | 4.39 | 4.35 | 4.38 | | 4.00 | 4.22 |
| | Transportation | 2.97 | 2.85 | 2.71 | 2.75 | | 3.05 | 2.72 |
| Small towns | Dwellings | 5.54 | 5.29 | 5.06 | 5.41 | 5.60 | 5.79 | 5.63 |
| | Neighborhoods | 5.92 | 5.58 | 5.72 | 5.61 | 5.88 | 5.88 | 6.13 |
| | Parks | 4.11 | 3.79 | 3.89 | 4.40 | 4.42 | 4.43 | 4.38 |
| | Schools | 4.42 | 4.11 | 4.11 | 4.02 | 4.53 | 4.61 | 4.50 |
| | Transportation | 3.03 | 2.68 | 3.06 | 2.66 | 2.58 | 3.02 | 2.85 |
| Rural | Dwellings | 5.66 | 6.02 | 5.66 | 5.64 | 5.54 | 5.73 | 5.52 |
| | Neighborhoods | 5.90 | 6.07 | 5.77 | 5.79 | 5.66 | 5.87 | 6.07 |
| | Parks | 4.08 | 4.16 | 3.88 | 4.31 | 4.02 | 4.09 | 4.58 |
| | Schools | 4.34 | 4.43 | 4.23 | 4.04 | 4.08 | 4.18 | 4.40 |
| | Transportation | 2.94 | 2.56 | 2.97 | 2.86 | 2.84 | 3.03 | 3.04 |

Source: The authors

## Local Parks

As seen in Fig. 16.4, parks in the urban core were rated significantly lower than were parks in other places ($p<.0001$). Similarly, the figure shows that parks in large cities were ranked significantly higher than parks in other places ($p<.001$) except small towns ($p=.06$). However, high ratings were not found in all large cities. The mean for the large city in Oakland was low (mean=3.33), while the mean for the large city in Washtenaw County was significantly higher (mean=4.54) (see

Table 16.3). This significant difference might be overlooked without considering park ratings at both county and place scales.

Figure 16.5 shows that Oakland and Washtenaw Counties had significantly higher park mean scores than Wayne County (4.29 and 4.24 versus 3.78). When examining the type of place scores within Wayne, however, low park ratings are shown for residents living in the urban core (mean=2.91), while the remaining places in the county had ratings comparable to Oakland and Washtenaw Counties. In fact, the highest ratings of local parks in the region were reported by rural area residents of Wayne County (mean=4.58) (see Table 16.3).

## Schools

Residents in Wayne County rated their schools significantly lower (mean=3.78) than did residents in the other counties (see Fig. 16.5). The differences in mean scores between Wayne and all other Counties were significant at $p<.0001$ level. At the other extreme, schools in Livingston had the highest ratings (mean=4.33). When considering the type of place, it shows that urban core had significantly lower school ratings than other places (see Fig. 16.4). The mean differences between urban core and all other types of places were significant at $p<.0001$ level.

While Oakland County and Washtenaw County residents appear to be comparable in their rating of schools, the pattern of responses in the two counties is similar to that of park ratings when considering the type of place where residents live. Washtenaw residents living in a large city and small towns had significantly higher mean scores than their counterparts in Oakland County. At the same time, Oakland County residents in the new suburbs gave higher ratings to schools than Washtenaw residents living in new suburbs. Table 16.3 also shows that Washtenaw County residents living in midsize cities gave their local schools relatively low marks. In fact, their ratings were significantly lower than the ratings of midsize city residents from St. Clair County.

## Public Transportation

Very few people were satisfied with their transportation system. Figures 16.4 and 16.5 indicate that public transportation had the lowest ratings, irrespective of the type of place or the county where people lived. The most positive ratings were given in the large cities (mean=2.88), while the lowest ratings were in midsize cities (mean=2.68). It also seems that the mean of Washtenaw County was slightly higher than those of the other counties. Washtenaw County was the only county in which the means of all different types of places are above 3.0. It might be worthwhile to investigate the public transportation system further as this could provide a precedent for the Detroit metropolitan areas.

## Conclusion

In this chapter we have examined the geographic patterns of QOL and have explored appropriate scales for examining spatial patterns showing QOL and QOUL differences. Many earlier studies focused on changes in quality of life at a single scale (for example, nation, city, suburb, and neighborhood). However, the variation in QOL can differ when considering multiple scales simultaneously. The chapter described how one geographic scale – type of place – was developed and used in the DAS2001 QOL survey in southeast Michigan. The type of place measure was used in conjunction with another geographic scale measure – county of residence – in analyzing life satisfaction of ratings of selected physical environmental attributes associated with where people live. The analyses revealed that a more accurate picture of QOL differences within a large metropolitan area can be portrayed when considering both geographic scales together. These differences may be important to policy makers when allocating limited resources aimed at enhancing the quality of peoples' lives.

## References

Baker, D. A., & Palmer, R. J. (2006). Examing the effects of perceptions of community and recreation participation on quality of life. *Social Indicators Research, 32*, 395–418.

Beck, S. L., Towsley, G. L., Caserta, M. S., Lindau, K., & Dudley, W. N. (2009). Symptom experiences and quality of life of rural and urban older adult cancer survivors. *Cancer Nursing, 32*(5), 359–369.

Bolund, P., & Hunhammar, S. (1999). Ecosystem services in urban areas. *Ecological Economics, 29*, 293–301.

Dillman, D. A., & Dobash, R. P. (1972). *Preferences for community living and their implication for population redistribution*. Pullman: Washington State University Agricultural Experimentation Station. Bulletin 764.

Dillman, D. A., & Tremblay, K. R. (1977). The quality of life in rural America. *The Annals of the American Academy of Political and Social Science, 429*, 115–129.

Elgin, D., Thomas, T., Logothetti, T., & Cox, S. (1974). *City size and the quality of life*. Washington, DC: The Stanford Research Institute.

Epley, D. R., & Menon, M. (2008). A method of assembling cross-sectional indicators into a community quality of life. *Social Indicators Research, 88*, 281–296.

Flanagan, J. (1978). A research approach to improving our quality of life, *American Psychologist*, February, pp. 138–147.

Frijters, P., Haisken-DeNew, J. P., & Shields, M. A. (2004). Money does matter! Evidence from increasing real income and life satisfaction in East Germany following reunification. *American Economic Review, 94*(3), 730–740.

Guterbock, T. M. (1997). The polls-review: Why Money Magazine's "Best Places" keep changing. *Public Opinion Quarterly, 61*(2), 339–355.

Johansson, S. (2002). Conceptualizing and measuring quality of life for national policy: From the Swedish level of living survey to an epistemology of the democratic process. *Social Indicators Research, 58*(1–3), 13–32.

Levin, S. A. (1992). The problem of pattern and scale in ecology. *Ecology, 73*(6), 1943–1967.

Li, G., & Weng, Q. (2007). Measuring the quality of life in city of Indianapolis by integration of remote sensing and census data. *International Journal of Remote Sensing, 28*(2), 249–267.

Lloyd, K., & Auld, C. (2002). The role of leisure in determining quality of life: Issues of content and measurement. *Social Indicators Research, 58*, 43–71.

Marans, R. W., Dillman, D., & Keller, J. (1980). *Perception of life quality in rural America: An analysis of survey data from Four studies*. Ann Arbor: The University of Michigan, Survey Research Center, Institute for Social Research.

McMaster, R. B., & Sheppard, E. S. (2004). Introduction: Scale and geographic inquiry. In E. S. Sheppard & R. B. McMaster (Eds.), *Scale and geographic inquiry* (pp. 1–22). Oxford: Blackwell Publishing.

Moller, V. (1992). Spare time use and perceived well-being among black South African youth. *Social Indicators Research, 26*, 309–351.

Reardon, S. F., Matthews, S. A., O'Sullivan, D., Lee, B. A., Firebaugh, G., Farrell, C. R., et al. (2008). The geographic scale of metropolitan racial segregation. *Demography, 45*(3), 489–514.

Sugiyama, T., Thompson, C. W., & Alves, S. (2009). Associations between neighborhood open space attributes and quality of life for older people in Britain. *Environment and Behavior, 41*(1), 3–21.

The United States General Accounting Office. (2003). *Forum on key national indicators: Assessing the nation's position and progress*. Washington, DC: GAO and The National Academies.

Wahl, H.-W., Schilling, O., Oswald, F., & Iwarsson, S. (2009). The home environment and quality of life-related outcomes in advanced old age: Findings of the ENABLE_AGE project. *European Journal of Ageing, 6*, 101–111.

Walsh, S. J., Crews-Meyer, K. A., Crawford, T. W., & Welsh, W. F. (2004). Population and environment interactions: Spatial considerations in landscape characterization and modeling. In E. S. Sheppard & R. B. McMaster (Eds.), *Scale and geographic inquiry* (pp. 41–65). Oxford: Blackwell Publishing.

# Chapter 17
# A Spatial Clustering Approach Analyzing Types of Objective Quality of Urban Life Using Spatial Data for Survey Respondents: South East Queensland, Australia

**Rod McCrea**

## Introduction

In this chapter, it is illustrated how an integrated approach using spatial objective data for residents who responded to a quality of life (QOL) survey can be employed to develop a statistical model to analyze types of objective quality of urban life (QOUL) for these respondents in a large-scale urban setting.

It was discussed in Chap. 2 how two approaches are commonly used to measure objective QOUL:

- The (territorial) social indicators approach
- The weighted index approach

With the social indicators approach, various objective indicators of QOUL are separately measured and monitored over time to detect improvement or decline in specific aspects of urban QOL (for example, Cicerchia 1996; D'Andrea 1998; Perz 2000; Archibugi 2001). With the weighted index approach, separate objective indicators of QOUL are weighted and combined to form an index which may then be used as an overall measure of QOUL for different places and for ranking them (see Liu 1975; Boyer and Savageau 1981, 1985, 1989; Cutter 1985; Savageau and D'Agostino 1999).

This chapter takes a different approach to examining objective QOUL. Rather than estimating *levels* of objective QOUL, the focus here is on using spatial clustering of objective indicators to identify different *types* of objective QOUL using data for the Brisbane-South East Queensland (SEQ) region, Australia. The

R. McCrea (✉)
Institute for Social Science Research, The University of Queensland, Level 4, GPN3 Building, Campbell Road, St Lucia, Brisbane, QLD 4072, Australia
e-mail: r.mccrea@uq.edu.au



R.W. Marans and R. Stimson (eds.), *Investigating Quality of Urban Life: Theory, Methods, and Empirical Research*, Social Indicators Research Series 45,
DOI 10.1007/978-94-007-1742-8_17, 

different types of objective QOUL relate to the residential locations of residents responding to the 2003 SEQ Quality of Life Survey (2003 SEQQOL). That study integrates this residential location data with spatial objective information derived from a number of sources. GIS tools are used to integrate the survey data with the spatial objective data, and using cluster analysis, typologies of objective QOUL are derived.

Objective indicators of urban environments may be measured at broad or at more specific levels. For example, distance from home to a post office is a specific characteristic of residential locations, while access to services and facilities more generally is a broad *dimension*. This chapter focuses on broad dimensions of the objective urban environment. This has two main advantages: First, it is easier to encapsulate QOUL using a limited number of broad dimensions rather than many specific objective characteristics of urban environments; second, broad dimensions of the urban environment are more related to urban form and therefore are more suitable for identifying broad types of QOUL in SEQ.

Ten broad objective dimensions of the urban environment are examined. Although other possible broad objective dimensions may also have been examined, these ten cover main dimensions in both the physical and social environments underlying the urban structure in SEQ, and they relate to readily available secondary data.

Four objective dimensions of the physical environment are examined which relate to the built and natural environments:

- *Objective access* which relates to distances between the survey residents' homes and services and facilities
- *Objective density* which relates to population, dwelling, and road densities
- The *objective rural environment* which relates to distances between the survey residents' homes and rural or semi-rural land
- The *objective coastal environment* which relates to the distance from the survey residents' homes to the coast

Six objective dimensions of the social environment are also examined and were derived from a study of the socio-spatial structure of SEQ (Western and Larnach 1998). They relate to percentages of neighborhoods with various social dimensions, namely:

- Younger non-nuclear household
- Nuclear family household
- Older non-nuclear household
- Socioeconomic status
- Disadvantage
- Ethnicity

These ten objective physical and social dimensions have been related to the QOUL literature.

In this chapter, the focus is on describing how these dimensions may cluster together to create objective types of QOUL across the SEQ study region.

## Quality of Urban Life and Objective Dimensions of the Urban Environment

Throughout the chapters in Part II of this book, there was detailed and specific discussion of the theoretical underpinnings of approaches to the study of QOUL. That included approaches that incorporated a focus on characteristics of the *physical environment* and their relationships to QOUL and on approaches that incorporated a focus on the *social environment* and their relationships to QOUL.

### *The Physical Environment and QOUL*

The physical environment includes both the *built environment* and the *natural environment*.

#### The Built Environment

Theories relating to a consideration of the built environment include *Optimal Centrality Theory* (Cicerchia 1999; Archibugi 2001), which relates urban density, access to services and facilities and overloading of urban structure to QOUL. It postulates that there is an optimum urban scale or urban size which maximizes trade-offs between the benefits of "city effect" and costs of "urban load." (That was illustrated in Fig. 4.3 in Chap. 4.) This city (or urbanization) effect might mean improved access to opportunities, services, and facilities that are available by virtue of a city's size. In contrast, the urban load effect might mean negative consequences as a result of urban growth (for example, congestion, overcrowding, cost of housing, and environmental degradation). The theory postulates that there will be net benefits to QOUL as small urban centers grow and additional services and facilities are provided, while at the same time, relatively low costs are incurred in terms of increased urban load. However, as urban growth continues past an optimum level, the rate of increase in city effect slows, and the rate of increase in urban load quickens, eventually leading to urban "overload" where additional growth decreases QOUL.

Optimal Centrality Theory can be extended to include the influence of urban density on QOUL. As with urban scale, urban density might be associated both with increasing access to services and facilities (see Rogerson et al. 1989; Rogerson et al. 1996; Glaeser et al. 2000) and with increasing urban problems such as pollution, traffic congestion, and cost of housing. Using that extended theoretical framework, objective density may also related to access and overloading (as demonstrated, for example, by Schwirian et al. (1995); Perz 2000; Brown et al. 1997; Senecal and Hamel 2001; Cramer et al. 2004; Schwanen and Mokhtarian 2004; Filion et al. 2006), although in contradiction, some research (see, for example, Baldassare and Wilson 1995) suggests that high density and rapid population growth may be a strong predictor of subjective QOUL. It has also been shown that satisfaction with access to services and facilities is an important consideration in making residential location decisions (see, for example, Dokmeci and Berkoz 2000; Chiang and Hsu 2005; Mitrany 2005; Ge and Hokao 2006), although Optimal Centrality Theory suggests an optimal trade-off between access and urban density in terms of QOUL. Overall, however, there is evidence that increasing urbanization may be stimulated by QOUL considerations.

### The Natural Environment

Close proximity to natural environments (such as rural and coastal environments) has been found to facilitate recovery from the stress (Ulrich et al. 1991; Kaplan 1995; Berto 2005) in contrast to higher levels of stress frequently found in more dense and crowded urban environments (for a review, see Walmsley 1988). For example, preferences for suburban and low-density living may in part be explained by an attraction to natural environments for stress relief from urban living (van den Berg et al. 2007). Thus, as an example, close proximity to rural and coastal environments might be associated with higher QOUL, all other things being equal.

However, "all other things being equal" is never the case in urban environments because they represent a range of trade-offs between access, density, and natural environments. Neither are social characteristics homogenous across urban environments since residents have different preferences and constraints (Fredland 1974; Speare et al. 1975; Desbarats 1983; Ge and Hokao 2006) which are then reflected in spatial patterning of social characteristics in urban environments.

## *The Social Environment and QOUL*

In considering the roles of the social environment as an influence on QOUL, *Social Disorganization Theory* predicts that neighborhood social ties would be stronger (that is, more organized) in neighborhoods that are:

- More stable (for example, lower residential mobility)
- More affluent (for example, more community facilities and resources)
- Less disadvantaged (for example, fewer social problems)
- More ethnically homogeneous (for example, fewer ethnic minorities) (Shaw and McKay 1942; Sampson and Groves 1989; Lowenkamp et al. 2003)

Neighborhood social ties underlie social capital and a sense of community which is a main component of overall QOL (Cummins 1996).

In testing Social Disorganization Theory, relationships have been found between various objective social dimensions and subjective evaluations of the social environment. For example:

- Less social capital and sense of community have been found in more disadvantaged neighborhoods (Kawachi et al. 1999; Cantillon et al. 2003)

- Less social cohesion among neighbors has been found in disadvantaged and less residentially stable neighborhoods (Sampson et al. 1997).
- Higher neighborhood attachment and involvement have been found in higher class and more residentially stable neighborhoods (Taylor 1996).

Socio-spatial patterns may also be associated with different subcultures, and *Subculture Theory* postulates that larger urban environments have sufficient population for the formation of subcultures to manifest spatially by allowing residents with similar social backgrounds and lifestyles to live in close proximity (Savage et al. 2003). The formation of subcultures might reflect consumption of similar services and facilities by similar residents, as discussed, for example, in the work of Lees (2000) and Clark (2005) on gentrification in cities whereby residents of higher socioeconomic status displace those of lower socioeconomic status so as to access services and facilities associated with high end consumption patterns.

However, more generally speaking, residents with similar life course and housing careers tend to choose similar areas in which to live (for example, Clark and Huang 2003; Clark et al. 2006). Subcultures may then become associated with local areas via the spatial concentration of residents with similar social backgrounds, lifestyles, values, and consumption preferences. Not surprisingly, where residents live becomes an important source of identity for individuals (Butler 2007).

Once areas become associated with particular subcultures, subcultures can become a factor in residential location decisions (Glavac and Waldorf 1998; Savage et al. 2003). This is driven by a preference of many residents to live in neighborhoods with similar others, which is a form of social homophily (see Lazarsfeld and Merton 1954; McPherson et al. 2001; Savage et al. 2005). This in turn might enhance socio-spatial variation (Fischer 1984).

## Types of Objective QOUL and a Spatial Clustering Approach

We have seen that objective dimensions of the urban environment related to QOUL are not independent. With the physical environment, more access to services and facilities might be associated with higher urban density and less access to the natural environment. Moreover, there are trade-offs between different dimensions of the physical environment such that is unlikely for neighborhoods to score well on all dimensions of the urban environment (for example, to have good access to services and facilities, low urban density and good access to the natural environment). Conversely, it is unlikely for neighborhoods to score poorly on all dimensions of the urban environment. Hence, particular combinations of the physical urban environment will be more common than others, giving rise to the idea of different types of objective QOUL which are spatially clustered.

This idea of different types of QOUL takes on additional momentum given that different combinations of the physical environment are associated with different social environments (for example, in some places, such as Australia's big cities, higher socioeconomic status is associated with inner city living, which is not necessarily the case, for example, in most US cities). Residents with different social characteristics have different life courses, preferences, and constraints which are reflected in their residential location decisions. So, different types of QOUL will have particular combinations of both physical and social dimensions.

These different *types* of objective QOUL do not identify different *levels* of QOUL, as with the more common social indicators and weighted average approaches to objective QOUL. If one were to experience a different type of QOUL, one would not necessarily be experiencing different levels of QOUL, but one would be having different QOUL experiences. This study aims to identify the main types of objective QOUL in the SEQ region.

## Method

### Data Sources

In the study reported here, data from various secondary data sets relating to SEQ, Australia have been used.[1] These data sets are: the 2003 SEQQOL survey; various GIS-based data sets; and Basic Community Profiles from the 2001 Census of Population and Housing (Australian Bureau of Statistics 2001a, b).

As discussed in Chap. 8, the 2003 SEQQOL survey collected mainly subjective data from 1,610 residents living in SEQ aged 18 years and over. However, the residential locations of respondents were geocoded and linked to objective GIS data sets and population census data. Thus, the respondents from this survey provided the sampling points for objective measures of QOUL. The 2003 SEQQOL survey used a geographically stratified random sampling methodology, and the spatial distribution of sampling points was shown in Fig. 8.1 in Chap. 8.

A GIS land use database was used to identify the land use for each land parcel in SEQ (Queensland Department of Local Government Planning Sport and Recreation 2002). This information was used to exclude rural environments from analyses as well as being used to calculate the distance of each resident to rural and semi-rural land use areas.

GIS layers from MapInfo Street Pro8 (MapInfo 2003) were also used:

- The "Features" layer was used to calculate distances from each resident's home to a range of services and facilities (for example, shopping centers, sporting facilities, hospitals, and schools).
- The "Streets" and "Highways and Main Roads" layers were used to calculate road density around each resident's home.
- Finally, the "Ocean" layer was used to calculate the distance of each resident's home to the coast.

[1]The study area comprised the Moreton and Brisbane Statistical Divisions in the Australian Standard Geographic Classification Australian Bureau of Statistics (2001b).

**Table 17.1** Eigenvalues for objective dimensions of the urban environment

| Objective dimension of the urban environment | Component 1 | Component 2 |
|---|---|---|
| Objective access | 3.73 | .84 |
| Objective density | 2.97 | .03 |
| Objective rural environment | 1.47 | .53 |
| Objective coastal environment | n.a. | n.a. |
| Objective younger non-nuclear environment | 2.31 | .40 |
| Objective nuclear family environment | 2.95 | .60 |
| Objective older non-nuclear household environment | 1.91 | .80 |
| Objective socioeconomic environment | 6.70 | .83 |
| Objective disadvantaged environment | 2.37 | .92 |
| Objective ethnic environment | 2.18 | .84 |

Source: The author

Notes: $N$=1,518; *n.a.* not applicable

Basic Community Profiles (BCPs) were extracted for each Census Collection District (CCD) from the population census (Australian Bureau of Statistics 2001a, b). The BCPs contained demographic and socioeconomic data for each CCD, as well as the CCD area for calculating population and dwelling densities. CCDs cover about 225 dwellings on average and can be thought of as neighborhood areas.

## *Deriving Measures*

The objective dimensions of the urban environment in SEQ were constructed from related items using *Principal Component Analysis* (PCA), except for the "objective coastal environment" which had only one item (see below). PCA was used to identify and measure broad dimensions underlying related items. Because PCA is sensitive to the size of correlations between items which are in turn sensitive to outliers (Tabachnick and Fidell 1996), the items associated with each dimension were initially transformed using logarithms and square roots where necessary so they were reasonably normally distributed. Separate PCAs were then conducted for each objective dimension using the items associated with that particular dimension.

Poorly loading items were discarded from each PCA, with the remaining items all loading well onto the first component of each objective dimension. Only the first component for each dimension was retained since the second components all had eigenvalues less than 1 (see Table 17.1).

Finally, PCA was used to weight the items associated with each dimension using the regression method in SPSS to construct a single standardized measure for each objective dimension of the urban environment. There were no missing values for these objective measures since they were constructed from items derived from GIS databases or from the population census. These objective dimensions are described below, grouped into objective dimensions of the physical environment (both built and natural environments) and the social environment.

## Objective Dimensions of the Physical Environment

Most of the objective physical environment measures used straight line distances as objective indicators of latent proximity between each resident's home and various aspects of the urban environment. However, objective density used the number of persons, dwellings, and road length in the immediate area surrounding a resident's home. Items for each objective dimension of the physical urban environment are detailed below, together with their PCA loadings in brackets.

(a) *Objective access* used straight line distances from a respondent's residence to their closest: neighborhood shopping center (.67), subregional shopping center (.69), regional shopping center (.63), commercial area (.62), sporting facility (.61), hospital (.78), primary school (.67), and high school (.76).
(b) *Objective density* used three items: person density (1.00), dwelling density (.99), and road density (1.00). Person density and dwelling density referred to the number of persons and dwellings per hectare within each resident's neighborhood (CCD), while road density referred to the total road length contained within 1 km of a resident's dwelling, calculated using GIS.
(c) *Objective rural environment* used two items: the straight line distance from each respondent's home to the closest land zoned for rural land use (.86), as well as to the closest zoned for rural-residential land use (.86).
(d) *Objective coastal environment* used one item: the shortest straight line distance between a resident's home and the coastline. Since only one item was used, it was standardized by subtracting each resident's score from the mean distance to the coast and dividing this by the standard deviation, rather than standardizing using PCA.

## Objective Dimensions of the Social Environment

As mentioned, objective dimensions of the social environment related to the socio-spatial structure of SEQ identified by Western and Larnach (1998). The first three dimensions related to the prevalence of different household structures in the social environment (that is: younger non-nuclear households, nuclear family households, and older non-nuclear households), and the last three measures related to socioeconomic, disadvantaged, and ethnic dimensions of social environments. These represent the six main objective dimensions by which neighborhoods vary in SEQ in terms of demographic and socioeconomic characteristics.

Each objective dimension of the urban environment was again constructed using PCA. The items used for objective dimensions of the social environment (shown below) refer to percentages of CCDs with various demographic and socioeconomic characteristics. The component loadings are again shown in brackets.

(a) *Objective younger non-nuclear households* used three items: persons never married (.86), dwellings rented (non-government) (.87), and group households (.90).

(b) *Objective nuclear family[2] households* used four items: two-parent family households (.86), persons aged 5–14 (.79), persons aged 0–4 (.52), and dwellings being purchased (78).
(c) *Objective older non-nuclear households* used three items: persons aged 65 or more (.78), lone-person households (.92), and persons divorced or separated (.68).
(d) *Objective socioeconomic environment* used ten items which captured the socioeconomic status of areas: labor force with graduate qualification (.94), managers and professionals (.92), females employed as professionals (.79), persons employed in finance, property, or business services (.80), households with annual income over $78,000 (.77), labor force with no qualifications (–.90), tradesperson (–.74), laborers (–.81), persons employed in manufacturing (–.69), and persons having left school under 15 years of age (–.79).
(e) *Objective disadvantaged environment* used five items: unemployed males (.81), unemployed females (.73), 15–19-year-old persons unemployed (.55), dwellings rented (government) (.63), and single-parent family households (.70).
(f) *Objective ethnic environment* used four items: persons of non-Christian religions (.86), born in South East Asia (.83), born in Southern and Eastern Europe (.70), and born in Central and South America (.52).

## Results

### *Spatial Distributions of Objective Dimensions*

Spatial distributions of the ten objective dimensions of the urban environment in SEQ were first examined. The following patterns were identified:

(a) Objective access and objective density were highest in the major urban centers (Brisbane, Ipswich, the Gold Coast, and the Sunshine Coast), lower in suburban areas and lowest in outer suburbs, rural-residential areas, and small towns. An inverse pattern was found for the objective rural environment, while the coastal environment was understandably near the coast.
(b) Younger non-nuclear households were concentrated in major urban centers; nuclear family households were concentrated in middle and outer suburbs, and older non-nuclear households were concentrated in older suburbs and near the coast.
(c) The higher socioeconomic areas were concentrated in Brisbane City (especially inner suburbs, western suburbs and along the Brisbane River) as well as being scattered along the coast amongst lower socioeconomic areas. Lower socioeconomic areas were concentrated in a strip of suburbs from Ipswich to Logan, and along the arterial growth corridors from Brisbane outskirts toward both the

[2]Nuclear families are couples with dependent children.

Sunshine Coast and the Gold Coast, as well as in rural-residential areas and small towns.

(d) The spatial concentration of disadvantaged areas was similar to that for lower socioeconomic areas except that not all low socioeconomic areas were disadvantaged. For example, the rural-residential areas and small towns south of Brisbane City and Ipswich (like Beaudesert and Mount Tamborine) showed relatively little disadvantage.

(e) Finally, areas with higher concentrations of ethnic populations were found primarily in the two major urban areas of Brisbane City and the Gold Coast.

Overall, there were clear spatial patterns in objective dimensions of the urban environment in SEQ which were interrelated with each other. Hence, it is likely that different types of objective QOUL characterized by unique combinations of these dimensions would also be spatially clustered.

## *Types of Objective QOUL in SEQ*

Using *cluster analysis*, these ten objective *dimensions* relating to QOUL were used to derive four objective *types* of QOUL in SEQ. There are three main cluster analytic methods used in the social sciences: hierarchical agglomerative methods, factor analytic methods, and iterative partitioning methods (Aldenderfer and Blashfield 1984). Since the types of urban environment were not conceptualized in a hierarchical way, a hierarchical method was not chosen. A factor analytic method may have been considered, but this option was not available in the SPSS statistical package being used. So, the *iterative partitioning method* was used with the *K-Means* option in SPSS.

Cluster analysis identifies relatively homogeneous *types* by identifying similar patterns of outcomes on a set of variables (that is, the objective dimensions of the urban environment).

(a) Firstly, initial locations for *K* cluster centers are based on an initial pass of the data which identifies *K* cases that are well separated in the space of the objective dimensions of the urban environment.

(b) Secondly, each case is assigned to a cluster based on Euclidean distances to the closest cluster center in the space of the objective dimensions. Then, each cluster center is recalculated using the mean value of the cases assigned to it. This second step is continually reiterated, with some cases being reassigned and cluster centers being recalculated, until the changes in cluster centers are negligible.

The number of clusters needs to be specified in *K*-mean cluster analysis. The number of clusters chosen in the final solution was based on each cluster having one or more distinctive characteristics. A cluster was deemed to have a distinctive characteristic on an objective dimension of the urban environment if the cluster mean on that dimension was approximately one standard deviation above or below the overall mean for that dimension. After testing a range of solutions from two to

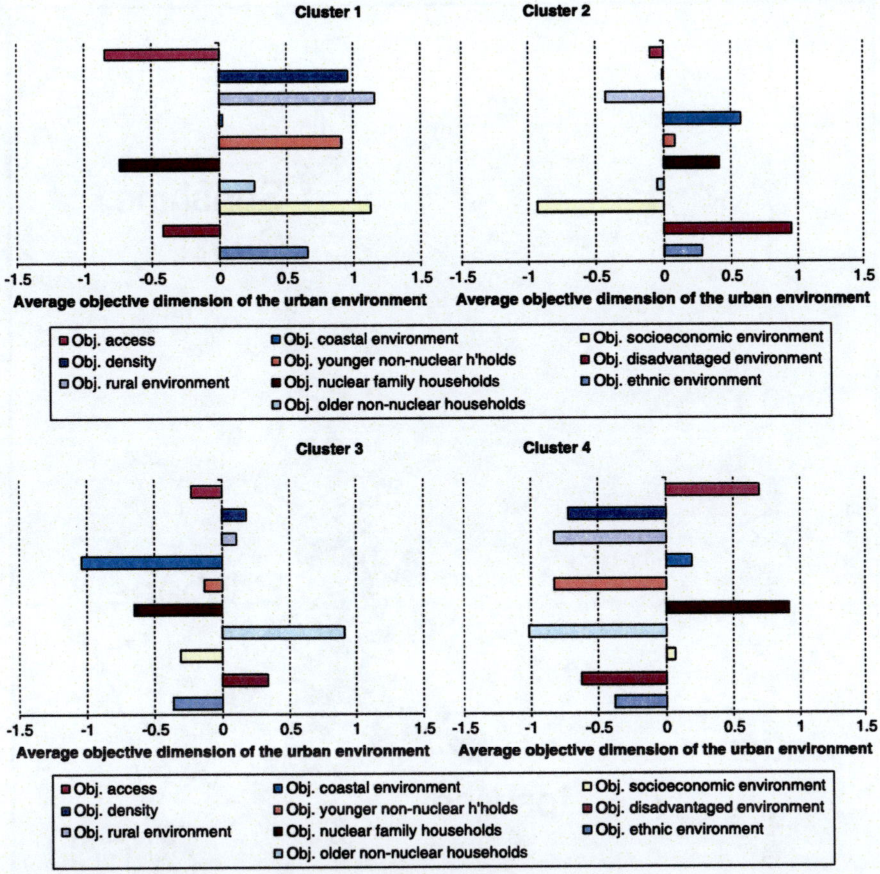


**Fig. 17.1** Mean cluster scores for types of objective urban environment in Brisbane-South East Queensland (Source: The author)

six clusters, the four-cluster solution was chosen because it was the most informative and produced the largest number of distinctive clusters.

The *four clusters* or *types of objective QOUL* in SEQ are described below with reference to the mean scores for each cluster on each of the objective dimensions (see Fig. 17.1[3]) together with the spatial distribution of the clusters (see Fig. 17.2). In Fig. 17.1 the first column in the legend refers to the top three bars in each bar chart; the second column refers to the middle four bars; and the last column refers to the bottom three bars in each bar chart.

(a) *Cluster 1* had good objective access and high objective density and was away from objective rural environments. In this cluster, there were relatively high

[3]Note that the bars in the graphs in Fig. 17.1 indicate the strength of the positive or negative association of an objective urban environment attribute with a cluster.

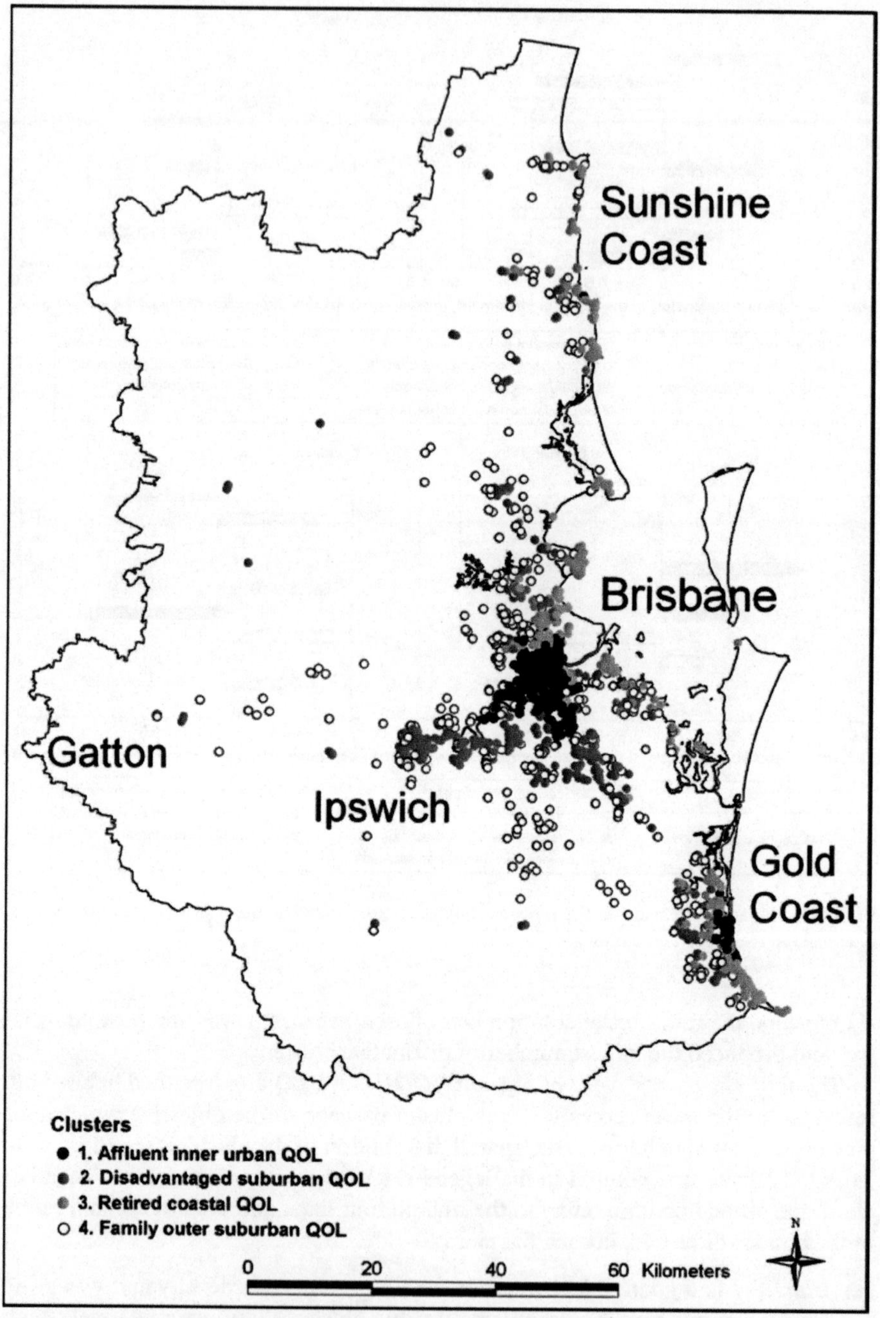


**Fig. 17.2** Spatial distributions of objective types of urban environment in Brisbane-South East Queensland (Source: The author)

proportions of younger non-nuclear households and relatively low proportions of nuclear family households. This cluster was also associated with relatively high socioeconomic environments. Cluster 1 was mainly found in the inner suburbs of Brisbane City and the Gold Coast, and this type of objective urban QOL was called *affluent inner QOUL*.

(b) *Cluster 2* had average levels of objective access and objective density. However, it was characterized by relatively low objective socioeconomic and relatively high disadvantaged scores. This cluster was spatially concentrated in a strip of suburbs running east from Ipswich and extending to the south-eastern growth corridor, though it was also scattered among outer suburbs and small towns. This cluster was called *disadvantaged suburban QOUL*.

(c) *Cluster 3* was found mainly along the coast, though also extended into some older suburbs north and east of Brisbane City. They were characterized by high proportions of older non-nuclear households and lower proportions of nuclear family households. Given the prevalence of older non-nuclear households near the coast, this cluster was called *retired coastal QOUL*.

(d) *Cluster 4* was characterized by high proportions of nuclear family households and low proportions of young non-nuclear and older non-nuclear households. This cluster was mainly found in the outer suburbs and rural-residential areas with lower density and lower access to services and facilities. This cluster was called *family outer suburban QOL*.

## Discussion

Most studies that have examined objective QOUL use a social indicators approach or weighted index for the purpose of monitoring, evaluating, and ranking different *levels* of objective QOUL.[4] However, the research reported in this chapter uses a spatial clustering approach to identify different *types* of objective QOUL. Using ten objective dimensions of the physical and social urban environments relating to objective QOUL, four main types of objective QOUL were identified for SEQ in Australia. They were:

- Affluent inner QOUL
- Disadvantaged suburban QOUL
- Retired coastal QOUL
- Family outer suburban QOUL

These are discussed in turn.

[4]That includes: Liu (1975), Boyer and Savageau (1981), Boyer and Savageau (1985), Cutter (1985), Boyer and Savageau (1989), Cicerchia (1996), D'Andrea (1998), Savageau and D'Agostino (1999), Perz (2000), and Archibugi (2001).

### *Affluent Inner QOUL*

This cluster was associated with relatively high socioeconomic areas concentrated near large urban centers. Consistent with Optimal Centrality Theory (Cicerchia 1999; Archibugi 2001), residents living in these areas would trade-off denser, congested, and overloaded urban environments for good access to services and facilities. These areas were commonly older inner urban areas, though there were not high proportions of older households living in these areas. In contrast, relatively high proportions of younger non-nuclear and high socioeconomic households lived in these areas, suggesting that they are undergoing or have undergone processes like gentrification. As such, these areas also provide good access to high end consumption opportunities like fine dining and entertainment as well as close proximity for professionals to inner city employment (see Lees 2000; Clark 2005).

### *Disadvantaged Suburban QOUL*

This cluster was spatially concentrated in a strip of suburbs south of Brisbane as well as scattered areas in outer suburbs and small towns. These areas were characterized by high proportions of unemployment, single-parent families, and public housing. Not surprisingly, the socioeconomic status of these areas was also low. According to Social Disorganization Theory, these disadvantages are also likely to have more social problems associated with less social capital, social cohesion, and sense of community (Sampson, et al. 1997; Kawachi, et al. 1999; Cantillon, et al. 2003).

While these areas had average access to services and facilities in terms of distance, there may still be accessibility problems for many disadvantaged residents living in these areas. For disadvantaged residents without cars, access would be difficult in these middle and outer suburbs as services and facilities are often beyond walking distance, and public transport services are less frequent. However, even for disadvantaged residents with cars, accessibility to services and facilities may still be limited in terms of affordability. So even though distance to services and facilities was average, consumption opportunities may be more limited in these areas, resulting in a more downmarket or even impoverished QOUL with concomitant social problems.

### *Retired Coastal QOUL*

This cluster was characterized as areas on or near the coast with high proportions of older non-nuclear households and relatively low proportions of nuclear family households (that is, couples with dependent children). They had reasonably good access to services and facilities and were not overly densely populated on average.[5] Many of these areas also have high natural amenity from beaches and the sea, so these areas represent a mild trade-off between access, overloading and natural amenity. Also, the natural coastal environment would facilitate a less stressful lifestyle (Ulrich et al. 1991; Kaplan 1995; Berto 2005). Many older non-nuclear households would have been able to choose these areas based on lifestyle considerations, being less bound by work and housing career considerations.

### *Family Outer Suburban QOUL*

This cluster was spatially concentrated in outer suburbs and small towns. Being on the urban fringes, they were generally close to rural-residential areas and had lower objective densities. However, they also had the lowest access to services and facilities of the four types of objective QOUL. Presumably, access to services and facilities was being traded-off against housing affordability since these areas were associated with both purchasing homes and having dependent children. The longer distances to access services and facilities and longer distances to employment opportunities mean that commuting would also be a feature of family outer suburban QOUL.

## Implications, Limitations, and Future Research

This study has identified four main *types of objective QOUL* in SEQ. One could hypothesize which types afford higher *levels* of objective QOUL. For example, based on assumptions about levels of constraints in choosing where to live, one may hypothesize that disadvantaged suburban QOUL affords the lowest level, followed by family outer suburban QOUL and then either affluent inner urban or retired coastal QOUL. While it may be possible to measure different QOL levels for each type of objective QOUL, the purpose of this study was simply to identify qualitatively different types of objective QOUL in SEQ. For example, affluent inner QOUL and retired coastal QOUL may have similar levels of objective QOUL; however, they are qualitatively different types of QOUL.

Complicating the issue of different levels of QOL for the different types of QOUL categories identified in this study is the notion of individual trade-offs in choosing where to live. Individual residents make different trade-offs in choosing where they live based on their own life circumstances, preferences, and constraints

[5] The coastal areas with very high densities were found primarily in inner areas of Gold Coast City and were categorized as affluent inner QOUL rather than retired coastal QOUL.

(for example, Fredland 1974; Speare, et al. 1975; Desbarats 1983; Ge and Hokao 2006). So, the level of urban QOL afforded by each type of objective QOUL will also differ between residents. This highlights limitations of measuring levels of QOUL within the context of different subjective preferences, life circumstances, and residential location choices. However, it does not subtract from the usefulness of different types of QOUL.

The four main types of objective QOUL found for SEQ are not definitive. Adding additional dimensions of the urban environment to the cluster analysis may help to identify other types of objective urban QOL. Subtypes of objective QOUL may also exist which may be investigated using hierarchical cluster analysis. Types of objective urban QOL may also change over time. However, the broad dimensions used in this analysis covering physical and social aspects of the urban environment are likely to have identified the main types of QOUL in SEQ.

Once types of objective QOUL have been identified, they can be further described in future research. As mentioned, subtypes may be identified by adding additional dimensions to the cluster analysis. Activities in urban environment may also be measured and associated with different types of objective QOUL (for example, different types of leisure activities or different commuting activities). Alternatively, qualitative research may be used to record narratives of lived experiences in each type. Such quantitative and qualitative research would help to more fully describe these different types of objective QOUL.

## Conclusion

While other studies have focused on measuring different levels of QOUL, this study used a spatial clustering approach to identify four main types of objective QOUL in SEQ, Australia. While other types or subtypes may exist, four main types were identified: an affluent inner urban QOUL, a disadvantaged suburban QOUL, a retired coastal QOUL, and a family outer suburban QOUL. These objective types of QOUL reflected unique combinations of the built, natural, demographic, and socioeconomic characteristics found in the urban environment.

Having qualitatively different types of objective QOUL suggests that residents have different lived experiences and lifestyles associated with living in different types of urban environments. Even so, levels of objective QOUL may vary between types. However, even if levels of objective QOUL were the same between any two types, residents would still have differing preferences for living in different types of objective QOUL. This highlights the importance of measuring different types of objective QOUL.

**Acknowledgment** This chapter is based on research funded by the Australian Research Council Discovery project # DP0209146.

## References


Aldenderfer, M. S., & Blashfield, R. K. (1984). *Cluster analysis*. London: Sage Publications.

Archibugi, F. (2001). City effect and urban overload as program indicators of the regional policy. *Social Indicators Research, 54*(2), 209–230.

Australian Bureau of Statistics. (2001a). *Census of population and housing: Basic community profiles*. Retrieved September 2004.

Australian Bureau of Statistics. (2001b). *Census of population and housing: Census geographic areas digital boundaries, Australia, 2001 (2923.0.30.001)*. Retrieved September 2004.

Baldassare, M., & Wilson, G. (1995). More trouble in paradise – Urbanization and the decline in suburban quality-of-life ratings. *Urban Affairs Review, 30*(5), 690–708.

Berto, R. (2005). Exposure to restorative environments helps restore attentional capacity. *Journal of Environmental Psychology, 25*(3), 249–259.

Boyer, R., & Savageau, D. (1981). *Places rated almanac*. Chicago: Rand McNally.

Boyer, R., & Savageau, D. (1985). *Places rated almanac*. Chicago: Rand McNally.

Boyer, R., & Savageau, D. (1989). *Places rated almanac*. New York: Prentice-Hall.

Brown, D. L., Fuguitt, G. V., Heaton, T. B., & Waseem, S. (1997). Continuities in size of place preferences in the United States, 1972–1992. *Rural Sociology, 62*(4), 408–428.

Butler, T. (2007). For gentrification? *Environment and Planning A, 39*(1), 162–181.

Cantillon, D., Davidson, W. S., & Schweitzer, J. H. (2003). Measuring community social organization: Sense of community as a mediator in social disorganization theory. *Journal of Criminal Justice, 31*(4), 321–339.

Chiang, L. N., & Hsu, J. R. (2005). Locational decisions and residential preferences of Taiwanese immigrants in Australia. *GeoJournal, 64*(1), 75–89.

Cicerchia, A. (1996). Indicators for the measurement of the quality of urban life: What is the appropriate territorial dimension? *Social Indicators Research, 39*(3), 321–358.

Cicerchia, A. (1999). Measures of optimal centrality: Indicators of city effect and urban overloading. *Social Indicators Research, 46*, 276–299.

Clark, E. (2005). The order and simplicity of gentrification – a political challenge. In R. Atkinson & G. Bridge (Eds.), *Gentrification in a global context: The new urban colonialism* (pp. 256–264). London: Routledge.

Clark, W. A. V., & Huang, Y. Q. (2003). The life course and residential mobility in British housing markets. *Environment and Planning A, 35*(2), 323–339.

Clark, W. A. V., Deurloo, M. C., & Dieleman, F. M. (2006). Residential mobility and neighbour hood outcomes. *Housing Studies, 21*(3), 323–342.

Cramer, V., Torgersen, S., & Kringlen, E. (2004). Quality of life in a city: The effect of population density. *Social Indicators Research, 69*(1), 103–116.

Cummins, R. (1996). The domains of life satisfaction: An attempt to order chaos. *Social Indicators Research, 38*(3), 303–328.

Cutter, S. (1985). *Rating places: A geographer's view on quality of life*. Washington D.C.: Association of American Geographers.

D'Andrea, S. S. (1998). Italian quality of life. *Social Indicators Research, 44*(1), 5–39.

Desbarats, J. (1983). Spatial choice and constraints on behaviour. *Annals of the Association of American Geographers, 73*(3), 340–357.

Dokmeci, V., & Berkoz, L. (2000). Residential-location preferences according to demographic characteristics in Istanbul. *Landscape and Urban Planning, 48*(1–2), 45–55.

Filion, P., McSpurren, K., & Appleby, B. (2006). Wasted density? The impact of Toronto's residential-density-distribution policies on public-transit use and walking. *Environment and Planning A, 38*(7), 1367–1392.

Fischer, C. S. (1984). *The urban experience*. New York: Harcourt, Brace Jovanovich.

Fredland, D. R. (1974). *Residential mobility and home purchase: A longitudinal perspective on the family life cycle and the housing market*. Lexington: Lexington Books.

Ge, J., & Hokao, K. (2006). Research on residential lifestyles in Japanese cities from the viewpoints of residential preference, residential choice and residential satisfaction. *Landscape and Urban Planning, 78*(3), 165–178.

Glaeser, E., Kolko, J., & Saiz, A. (2000) *Consumer city* (Working paper 7790). Cambridge: National Bureau of Economic Research.

Glavac, S. M., & Waldorf, B. (1998). Segregation and residential mobility of Vietnamese immigrants in Brisbane, Australia. *Professional Geographer, 50*(3), 344–357.

Kaplan, S. (1995). The restorative benefits of nature: Toward an integrated framework. *Journal of Environmental Psychology, 15*, 169–182.

Kawachi, I., Kennedy, B. P., & Wilkinson, R. G. (1999). Crime: Social disorganization and relative deprivation. *Social Science and Medicine, 48*(6), 719–731.

Lazarsfeld, P. F., & Merton, R. K. (1954). Friendship as a social process: As substantive and methodologicalaAnalysis. In M. Berger (Ed.), *Freedom and control in modern society* (pp. 18–66). New York: Van Nostrand.

Lees, L. (2000). A reappraisal of gentrification: Towards a 'geography of gentrification'. *Progress in Human Geography, 24*(3), 389–408.

Liu, B. (1975). *Quality of life indicators in U.S. metropolitan areas, 1970*. Washington: Washington Environmental Research Center.

Lowenkamp, C. T., Cullen, F. T., & Pratt, T. C. (2003). Replicating Sampson and Groves's test of social disorganization theory: Revisiting a criminological classic. *Journal of Research in Crime and Delinquency, 40*(4), 351–373.

MapInfo (2003) *User guide: MapInfo StreetPro 8*. Greenwich, New South Wales: MapInfo, Australia.

McPherson, M., Smith-Lovin, L., & Cook, J. M. (2001). Birds of a feather: Homophily in social networks. *Annual Review of Sociology, 27*, 415–444.

Mitrany, M. (2005). High density neighborhoods: Who enjoys them? *GeoJournal, 34*, 131–140.

Perz, S. G. (2000). The quality of urban environments in the Brazilian Amazon. *Social Indicators Research, 49*(2), 181–212.

Queensland Department of Local Government Planning Sport and Recreation (2002) *Land use zones for land parcels in South East Queensland* [unpublished data for Geographic Information Systems]. Brisbane: Queensland Department of Local Government, Planning, Sport and Recreation.

Rogerson, R. J., Findlay, A. M., Morris, A. S., & Coombes, M. G. (1989). Indicators of quality of life - some methodological issues. *Environment and Planning A, 21*(12), 1655–1666.

Rogerson, R. J., Findlay, A. M., Paddison, R., & Morris, A. S. (1996). Class, consumption and quality of life. *Progress in Planning, 45*, 1–66.

Sampson, R. J., & Groves, W. B. (1989). Community structure and srime – Testing Social-Disorganization Theory. *American Journal of Sociology, 94*(4), 774–802.

Sampson, R. J., Raudenbush, S. W., & Earls, F. (1997). Neighborhoods and violent crime: A multilevel study of collective efficacy. *Science, 277*(5328), 918–924.

Savage, M., Warde, A., & Ward, K. (2003). *Urban sociology capitalism and modernity*. New York: Palgrave MacMillan.

Savage, M., Bagnall, G., & Longhurst, B. (2005). *Globalisation and belonging*. London: Sage.

Savageau, D., & D'Agostino, R. (1999). *Places rated almanac*. New York: Wiley.

Schwanen, T., & Mokhtarian, P. L. (2004). The extent and determinants of dissonance between actual and preferred residential neighborhood type. *Environment and Planning B-Planning and Design, 31*(5), 759–784.

Schwirian, K. P., Nelson, A. L., & Schwirian, P. M. (1995). Modeling urbanism - economic, social and environmental-stress in cities. *Social Indicators Research, 35*(2), 201–223.

Senecal, G., & Hamel, P. J. (2001). Compact city and quality of life: Discussions of the Canadian approach to sustainability indicators. *Canadian Geographer, 45*(2), 306–318.

Shaw, C. R., & McKay, H. D. (1942). *Juvenile delinquency and urban areas*. Chicago: University of Chicago Press.

Speare, A., Goldstein, S., & Frey, W. H. (1975). *Residential mobility, migration, and metropolitan change*. Cambridge: Ballinger Pub. Co.

Tabachnick, B. G., & Fidell, L. S. (1996). *Using multivariate statistics*. New York: HarperCollins College Publishers.

Taylor, R. B. (1996). Neighborhood responses to disorder and local attachments: The systemic model of attachment, social disorganization, and neighborhood use value. *Sociological Forum, 11*(1), 41–74.

Ulrich, R. S., Simons, R. F., Losito, B. D., Fiorito, E., Miles, M. A., & Zelson, M. (1991). Stress recovery during exposure to natural and urban environments. *Journal of Environmental Psychology, 11*(3), 201–230.

van den Berg, A. E., Hartig, T., & Staats, H. (2007). Preference for nature in urbanized societies: Stress, restoration, and the pursuit of sustainability. *Journal of Social Issues, 63*(1), 79–96.

Walmsley, D. J. (1988). *Urban living: The individual in the city*. New York: Wiley.

Western, J., & Larnach, A. (1998). The social and spatial structure of South-East Queensland. *Australasian Journal of Regional Studies, 4*(2), 215–237.

# Chapter 18
# Using GIS to Derive Region-Wide Patterns of Quality of Urban Life Dimensions: Illustrated with Data from the Brisbane-SEQ Region

**Prem Chhetri, Robert Stimson, and John Western**

## Introduction

Variations in the characteristics of geographic space across a large-scale urban environment are likely to play a significant role in creating variation in people's *subjective assessment* of quality of urban life (QOUL) dimensions and their evaluations of phenomena, such as local "neighborhood attractiveness attributes," which may affect decisions such as choosing where to live within a city. If those differences are embedded within a geographic space (such as a metro-region) in terms of the environmental, economic and social characteristics of the living spaces across a large-scale urban region, then it may be useful to be able to translate the subjective evaluations of those dimensions impacting QOUL – which are typically derived from aggregating unit record survey-based data – into a generalized spatial pattern across the city region.

It has been quite commonplace for the scored subjective assessments of aspects of QOUL derived from sample survey data to be aggregated and to be treated as being indicative of the collective assessments of groups of people. For example, disaggregated data collected through surveys can be aggregated to measure specific quality of life (QOL) for particular groups, such as women, minorities, older people, and so on. In a similar way, aggregation can also be undertaken on a spatial basis for

P. Chhetri (✉)
School of Business IT and Logistics, RMIT University, GPO Box 2476V, Melbourne 3001, Australia
e-mail: prem.chhetri@rmit.edu.au

R. Stimson
Australian Urban Research Infrastructure Network (AURIN), Faculty of Architecture, Building and Planning, University of Melbourne, VIC 3010, Melbourne, Australia
e-mail: rstimson@unimelb.edu.au

J. Western (deceased)
School of Social Science, The University of Queensland, Brisbane, QLD 4072, Australia

R.W. Marans and R. Stimson (eds.), *Investigating Quality of Urban Life: Theory, Methods, and Empirical Research*, Social Indicators Research Series 45,
DOI 10.1007/978-94-007-1742-8_18, © Springer Science+Business Media B.V. 2011

areas or regions. Spatial statistical analysis of the aggregated data derived from the individual unit record survey data enables the spatial variability – those subjective assessments of QOUL dimensions – to be visualized through mapping. Potentially, that approach may have useful applications in urban and regional planning.

However, as was pointed out in Chap. 4, relatively few attempts have been made to aggregate individual-level survey data on the evaluation of QOUL into subregions of a city or across the continuous space of an urban region to identify spatial variability in those subjective perceptions of aspects. That is in spite of the availability of geographic information system (GIS) technology, which enables researchers to integrate spatial modeling with multivariate statistical tools of analysis.

In this chapter, we demonstrate the application of GIS technology combining statistical and spatial modeling techniques to derive region-wide spatial patterns of aspects of the subjective assessment of QOUL in the Brisbane-South East Queensland (SEQ) region, the third largest metro-region in Australia, using unit record data from the 2003 Brisbane-South East Queensland Quality of Life Survey (SEQQOL2003). In doing so, we draw on some of our already published works (Chhetri et al. 2005, 2006, 2007; Stimson et al. 2006) to which the reader is referred for full details of those applications.

Two applications are discussed using the subjective data derived from the 2003SEQQOL survey:

(a) In one, an "ordered weighted average" nonlinear aggregation technique has been used to derive generalized patterns of the subjective assessment of QOUL dimensions across subregions of the SEQ region.
(b) In the other, the objective was to identify and map the generalized spatial patterns of the underlying dimensions of the subjective assessment of "neighborhood attractiveness attributes" which may have affected the choices made by survey respondents in deciding where to live. Those patterns are simulated and mapped using the "neighborhood operation" function in GIS.

## The Case Study Region and the Survey

As discussed previously in Chap. 8, in SEQQOL2003 survey questionnaire was designed to collect data on the subjective assessments of the survey respondents on a range of QOUL domains. That included issues relating to QOUL dimensions at the scale of the SEQ region as a whole as well as QOUL issues relating to the local (including neighborhood) level of scale. Those subjective assessments were quantified using a 5-point Likert scale. The survey questionnaire also collected information on the normal range of demographic and socioeconomic variables along with locational information related to a range of the spatial behaviors of survey respondents.

Because of resource limitations, for some of the specialist topics covered in the QOL survey, the sample was split with only one half the survey respondents ($N=776$) being asked to complete specific sets of questions, including a section of survey questionnaire which collected information on residential location decision choices and the subjective assessment of a set of "neighborhood attractiveness attributes." In part of that survey, respondents were asked to indicate the relative importance of a list of items that were important in their decision to move into their current place of residence. "Importance" was defined as a state or quality of being significant, influential or worthy of note. Using a list of 13 *neighborhood attributes*, the perceived relative importance that respondents attach to those attributes was ascertained also using a 5-point Likert scale. The survey items used were developed by Marans (2003), and their dimensions have been tested by Vogt and Marans (2004).

An objective of the SEQQOL2003 study was to be able to interface the subjective and behavioral information collected from respondents to the survey with spatial objective data about the wider SEQ region as well as the local area (neighborhood) where respondents live. That necessitated the use of GIS technology facilitated through the "geocoding" of location of survey respondents (refer back to Fig. 8.1 in Chap. 8). Those locational data could then be displayed on a map of the SEQ region using MapInfo Professional GIS and MapInfo StreetPro database, while ESRI GIS software was used to join the subjective satisfaction assessment scores of survey respondents on QOL domains to each "geocoded" survey respondent to create maps and for visualization of derived generalized spatial patterns of those subjective phenomena across the study region.

With the "geocoding" of the location of survey respondents, it was possible to use those GIS-enabled tools to derive spatially generalized patterns of subjective assessment information derived through the statistical modeling of survey data. The two applications that are summarized in what follows are presented here by way of illustrating the potential of these methodologies to enhance our understanding of the nature of subjective assessment of QOUL in a broader geographic regional context.

## Application 1: Deriving Subregional Patterns of an Aggregated Measure of QOUL Dimensions

In the first application, Chhetri et al. (2007) employed a GIS methodology and statistical modeling techniques to derive subregional patterns of the subjective assessment of QOL dimensions using the 2003 SEQ QOL Survey data.

### *Context*

Chhetri et al. (2007) argued that the aggregated scored subjective assessments derived from the SEQQOL2003 survey data may be treated as a surrogate indicator of the collective "well-being" of people living in the subregional components of the SEQ region. For example, the individual-level disaggregated data collected through the survey may be aggregated to measure specific aspects of QOUL for particular groups, such as men, women, minorities, older people, and so on. In a similar way, aggregation can also be undertaken on a spatial basis for areas or regions. QOUL assessments, when aggregated for areas, can indirectly reflect the spatial variation in perceived QOUL. However, Marans (2003) had noted that QOUL in a place or a specific geographic setting is a subjective phenomenon, and therefore may vary according to an individual's gender, perception, cultural and ethnic background, socioeconomic status, educational level, family situation, health, disability, age and/or past experiences.

## *The Approach*

The approach developed by Chhetri et al. (2007) involved the following procedures:

- An analysis of individual ratings of QOUL dimensions provided by the survey data
- An aggregate of those ratings
- The subsequent derivation of average or collective scores for subregions within the SEQ region to represent spatial variations in levels of well-being reflected through those subjective assessments of QOUL dimensions

With subregions as the criterion for spatial aggregation, Chhetri et al. (2007) used a number of techniques to measure and map aggregated subjective assessments of QOUL across various life satisfaction domains.

## *Measurements*

In the 2003SEQQOL survey, life satisfaction was measured by asking survey respondents this question:

How satisfied are you with your overall quality of life?

It was measured on a 5-point Likert scale, where 1 represents "very dissatisfied" and 5, "very satisfied."

Twelve questions were asked to assess their level of satisfaction with aspects of the following QOL domains:

- Your employment situation
- The amount of money you have available to you personally
- Your housing: "the amount of time you have to do things you want to do"
- Your relationship with your partner
- Your relationship with your children
- Your independence or freedom

## *Aggregation Techniques*

As explained by Chhetri et al. (2007), the "weighted average operator" is a commonly used composite index that may be derived from a set of scaled responses to survey questions. The term "composite index" can be defined as an aggregation of the indicator values which collectively convey information about the quality of some complex aspects or components of a condition (Nijkamp et al. 1990). In the "weighted average operation," indicators are multiplied with their associated weights and then added together to construct the overall index. In quality-of-life (QOL) research and other research contexts, weights are often assessed by asking people to assign importance scores to each item assessed in the survey. The index can be expressed as

$$U = Y_1 X_1 + Y_2 X_2 + \cdots + Y_n X_n$$

where

$X_n$ is the *n*th indicator
$Y_n$ is its corresponding weight

The "ordered weighted average" (OWA) is often used for aggregating multiple indicators to form an overall score. This is a relatively new but flexible method of aggregation. It allows the users to decide upon the types of aggregation depending on the purpose of their decision-making (Filev and Yager 1998; Mendes and Motizuki 2001; Smith 2001; Yager 1988; 2004). The basic formula of OWA is given as

$$F = w_1 b_1 + w_2 b_2 + \cdots + w_j b_j$$

where

$b_1, b_2, \ldots, b_n$ are the positional values of the indicators
$w_1, w_2, \ldots, w_n$ are the weights of those positional values

OWA aggregation is a nonlinear aggregation because it uses an ordering process to aggregate the indicators' values. The value of the weights ranges between 0 and 1.

Yager (1988) had proposed "OWA operator," which can be used to evaluate the performance of urban areas on QOUL indicators (Mendes and Motizuki 2001). The overall performance rating depends upon the criteria used to assign weights to the indicators based on their importance in the subjective assessments of subjects. There can be a situation where all indicators hold the same importance, while in other situations, satisfaction with one of the indicators is all that is desired. For example, for older people, health status may be a far more important domain than employment status. Therefore, their QOL is heavily dependent on health-related indicators. On the other hand, weightings might not be the same for a young unemployed person. These combinations of situations can be adjusted through the use of "And" and "Or" operators, which combine the function that an indicator performs in the overall ranking (Yager 1988: p. 183).

According to Yager (1988), an "OWA operator" of dimension $n$ is a mapping function

$$f : R^n \rightarrow R$$

that has an associated weight vector $W$

$$W = [w_1, w_2, w_3, \cdots w_n]^T$$

such that

$$w_i \varepsilon [0,1],$$

$$\sum_{i=1} w_i = 1$$

The functional value $f(a_1 \ldots a_n)$ determines the aggregated value of arguments $a_1, a_2 \ldots a_n$ in such a manner that

$$f(a_1 \cdots a_n) = \sum_{j=1}^{n} w_j b_j \quad \text{(aggregation equation)},$$

where

$b_j$ is the $j$th largest element of the collection of the $n$ aggregated objects (here, indicators) $a_1, a_2 \ldots a_n$. (Filev and Yager 1998).

A fundamental aspect of the "OWA operator" involves ordering the indicator values. If we consider ranking the indicator scores from highest to lowest, the indicator with the highest score is given the first-order weight; the indicator with the next highest score is given the second-order weight, and so on. This has the effect of weighting an indicator based on its rank in a descending order from maximum to minimum. Thus, argument $a_i$ (a particular indicator value) is not associated with a particular weight $w_i$, but rather a weight $w_i$ is associated with a particular ordered position $i$ of the argument (Filev and Yager 1998: p. 158). A known property of the "OWA operators" is that they include the *Max*, *Min*, and arithmetic mean operators for the appropriate selection of the vector $W$:

$$\text{For } W = [1,0,0,\cdots 0], f(a_1 \cdots a_n) = \text{Max}_i\, a_i \left[\text{Optimistic OWA (OP} - \text{OWA)}\right]$$

$$\text{For } W = [0,0,0,\cdots 1], f(a_1 \cdots a_n) = \text{Min}_i\, a_i \left[\text{Pessimistic OWA (PE} - \text{OWA)}\right]$$

$$\text{For } W = [1/n, 1/n, \cdots 1/n], f(a_1 \cdots a_n) = \frac{1}{n}\sum_{i=1}^{n} a_i \left[\text{Average OWA (AV} - \text{OWA)}\right]$$

All "OWA operators" are bounded by the *Max* and *Min* operators

$$[\text{Min}_i a_i \leq f(a_1 \cdots a_n) \leq \text{Max}_i a_i]$$

(Filev and Yager 1995: p. 1998).

Recently, other types of OWA operators (that is, "maximum entropy OWA" and "exponential OWA") have been developed (Smith 2001; Filev and Yager 1998).

A somewhat simpler approach which does not require the solution of a nonlinear programming problem is the "exponential OWA operator" (EX-OWA) (Filev and Yager 1998). This method is an alternative solution to the constrained optimization problem. OWA weights can easily be generated according to either of the following equations:

$$w_1 = \theta; w_2 = \theta(1-\theta); w_3 = \theta(1-\theta)^2; \ldots w_{n-1} = \theta(1-\theta)^{n-2}; w_n = (1-\theta)^{n-1} [EX\text{-}OWA]$$

Here,

$\theta$ is a parameter of the indicators that belong to the unit interval $0 \leq \theta \leq 1$.

Yager (1988) used the dispersion or entropy (evenness) associated with a weighting vector. He used this measure to develop a procedure to generate the OWA weights that have a predefined degree of "orness" $\alpha$. That is, the weights will be as even as possible (maximizing entropy), subject to yielding a given level of "orness." These are called as "maximum entropy OWA" (ME-OWA) weights. This approach is based on the solution of the following constrained nonlinear optimization problem (Smith 2001):

$$\text{Maximized } E(W) = \sum_{i=1}^{n} w_i ln w_i \text{ [maximum entropy OWA]},$$

subject to:

$$\alpha = \frac{1}{n-1} \sum_{i=1}^{n} (n-i) w_i (\alpha = orness)$$

$$w_i \varepsilon [0,1], i = 1, \cdots, n$$

As discussed above, in the SEQQO2003 study, QOUL had been assessed on a range of life satisfaction dimensions. The aggregation techniques discussed above were used by Chhetri et al. (2007) to weight QOUL indicators described on various life satisfaction domains. In the following section, we will report on the results derived from those aggregation operators and analyze the inter-regional variability in QOUL.

An objective in the approach used by Chhetri et al. (2007) was to compare spatial patterns in the perceived QOUL of people at a collective level, as reflected in the application of these different weighting regimes.

## *Results*

Table 18.1 gives the means and standard deviations of the scores for the perceived life satisfaction dimensions referred to above. For satisfaction with social relationships, housing, independence/freedom, family life, friends, the means were greater

**Table 18.1** Descriptive statistics for quality-of-life dimensions

| Satisfaction with | Mean | S.D. |
|---|---|---|
| The amount of money you have available to you personally | **3.12** | 1.20 |
| The amount of time you have to do the things you want to | **3.29** | 1.22 |
| Your romantic relationships | **3.77** | 1.27 |
| Your employment situation | **3.78** | 1.15 |
| The way you spend your leisure time | **3.88** | 0.95 |
| Your health | **3.89** | 0.99 |
| With your social relationships | 4.01 | 0.88 |
| Your housing | 4.19 | 0.85 |
| Your independence or freedom | 4.20 | 0.91 |
| Your family life | 4.24 | 0.91 |
| Your friends | 4.26 | 0.75 |
| Living in the Brisbane-South East Queensland region | 4.36 | 0.71 |
| Your overall standard of living | 4.07 | 0.80 |
| **Your life as a whole** | **4.24** | **0.73** |

Source: Chhetri et al. (2007)
Scores on a 5-point scale

**Table 18.2** Transfer of location-specific QOL survey data into subregions: an example

| Region | Level of satisfaction | Employment (Count) | Employment (%) |
|---|---|---|---|
| R1 | Very dissatisfied | 9 | 16.67 |
| $n$=54 | Dissatisfied | 8 | 14.81 |
| | Neither satisfied nor dissatisfied | 14 | 25.93 |
| | Satisfied | 17 | 31.48 |
| | Very Satisfied | 5 | 9.26 |
| | | **54** | **100.00** |

Source: Chhetri et al. (2007)

than 4 on the 5-point Likert scale. With the overall standard of living in the SEQ region and life as a whole, the survey respondents gave high satisfaction scores. On the other side of the spectrum, for six life satisfaction domains – satisfactions with the amount of money, the amount of free time available, romantic relationships, employment situation, leisure time and health – the overall score reported by the respondents was slightly lower, being between 3.1 and 3.9. Note that items with relatively lower scores have high standard deviations, while for items with high satisfaction scores, lower standard deviations were found.

In order to aggregate data across 21 subregional areas in the SEQ region, subjective assessment scores on various QOUL dimensions need to be aggregated. As an example to illustrate the process involved, Table 18.2 shows that there were 54 respondents to the survey (that is, from the 50% split sample) in one of the 23 subregions, namely, subregion 1 (R1). In that subregion, the scores of survey respondents living on satisfaction with employment were aggregated and then converted into percentages. Those values for the different subregions were then transformed from 1 to 0 on the basis of their positional rankings. The procedure was replicated for all items in the survey. That produced a table with all the indicators sorted on the basis of their positional values.

In order to measure and compare the performance of 21 subregions across the SEQ region on the QOUL domains, OWA operators were used by Chhetri et al. (2007) to aggregate items derived from the survey data. Item aggregations by OWA operators were implemented using Visual Basic for Application (VBA) in MS Excel and MathCAD. The values of OWA operators varied on a scale between 1 (one) and 0 (zero). Here, 1 (one) means 100% satisfaction or the ideal condition of the indicators existed, and 0 (zero) means that the indicators revealed 100% dissatisfaction.

Under the OWA operators, while considering the positional values of the QOL items and nonlinear arithmetic and/or exponential aggregation, performance of subregions on the QOL measures varies between 30% and 97% of the ideal condition, depending on the model that was being considered:

(a) Under the MAX-OWA operator, the average score on the QOL item was about 90% of the ideal condition.
(b) In contrast, under the MIN-OWA operator, the average score on the QOL item was 39% of the ideal condition.
(c) AVERAGE-OWA reveals the trade-off among the QOL items, and the average score was about 70% of the ideal condition.
(d) Under the "Exponential-OWA operator," the average life satisfaction score calculated for this operator was 42% of the ideal condition (while considering the low level of "orness"), and this is 82% of the ideal condition (while considering the high level of "orness").
(e) The "maximum entropy-OWA operator" considers some constraints of the indicators, which means considering the desired degree of "orness" to maximize the evenness among the QOL indicators. Under this operator, the QOL score was about 60% of the ideal condition (while "orness" was low), and that situation was about 80% (while "orness" was high).

Chhetri et al. (2007) mapped the spatial patterns – across 21 subregions of the SEQ region – of the patterns derived from the application of the alternative "weighted average operators" discussed above and compared the resultant spatial. It was found that a much greater spatial differentiation started to emerge in the patterns for the "exponential operator" and the "maximum entropy operator." Some of the subregions – for example, Northern Outer, Eastern and Southern Inner in the Brisbane metropolitan area – performed better on those two regimes. But somewhat surprisingly, the tourism area of Noosa in the north of the SEQ region dropped its position, while Gold Coast Inner emerged to be a winner on both of these indices. The spatial patterns for the exponential and maximum entropy were somewhat similar (correlation coefficient $r^2$=0.72), although they still shared some similarity with Max, Min and Average operators in terms of high scores for the coastal and interior subregions within the SEQ region.

The generalized pattern of subregional variation in the level of subjective QOL across the SEQ region for the "maximum entropy operator" solution that is thus derived from the unit record survey data is shown in Fig. 18.1.

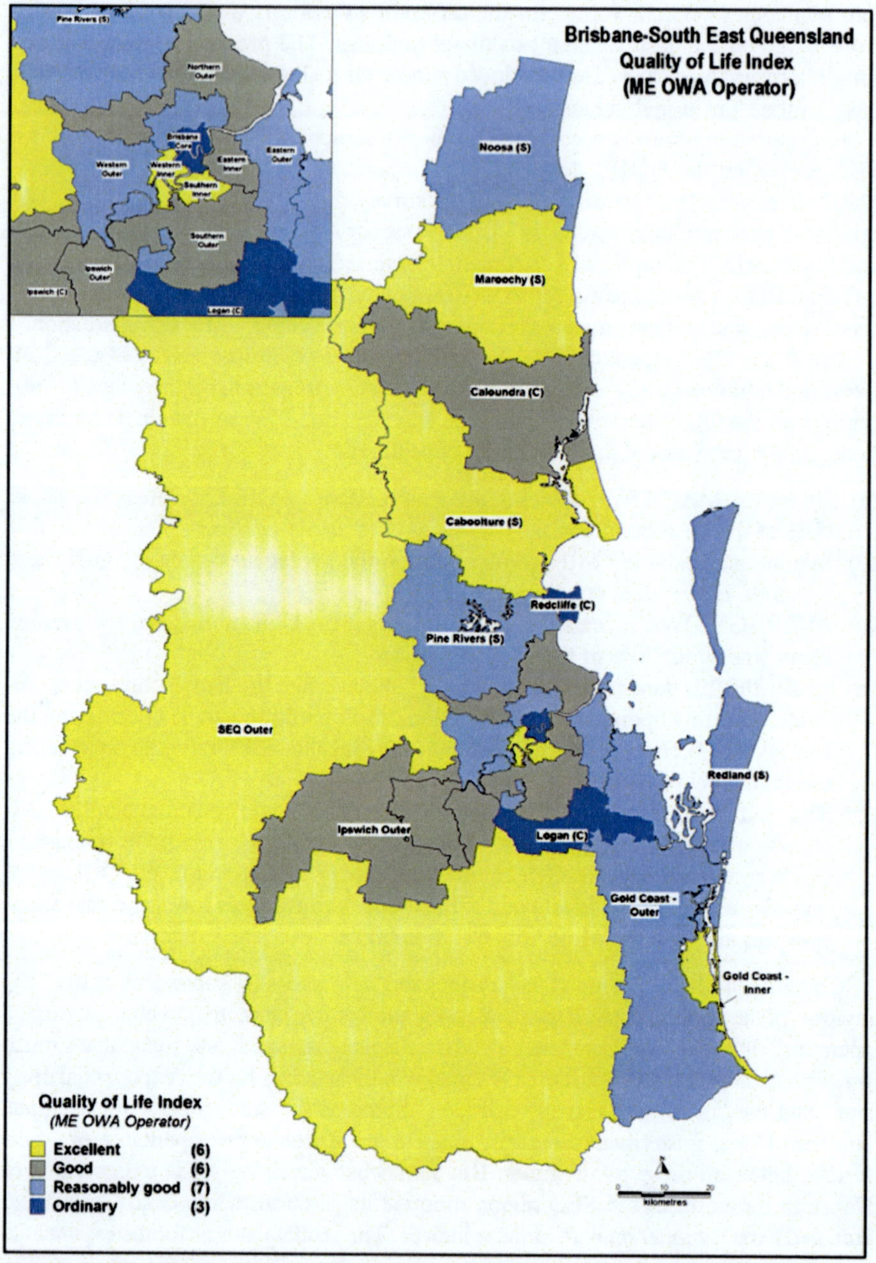


**Fig. 18.1** Subregional variability across the SEQ region in the subjective assessment of QOL dimensions using maximum entropy (Source: Chhetri et al. 2007)

## Application 2: Deriving Generalized Spatial Patterns of Subjective Assessments of "Neighborhood Attractiveness" Attributes and of Objective Measures of Neighborhood Characteristics

In the second application, Chherti et al. (2005, 2006) and Stimson et al. (2006) used a GIS methodology to model and visualize the generalized spatial patterns of the subjective assessment of "neighborhood attractiveness" attributes using data collected through the SEQ QOL2003 survey and of objective measures of neighborhood characteristics which might underpin residential location choice.

### *Context*

Residential location choices are typically conducted within the framework of decision-making by individuals and by households that incorporate their subjective preferences relating to attributes of residential environments in an urban area. The effects that a neighborhood and its characteristics have on residential location decision choices have been well documented in residential location decision and choice literature research over the last few decades (see, for example, Stimson 1978; Mingche and Brown 1980; Maher and Saunders 1994; Dökmeci and Berköz 2000; Vogt and Marans 2004). Both objective and subjective approaches have been adopted to investigate the importance of neighborhood characteristics in decision-making.

However, as was discussed in Chap. 4, relatively few studies investigating the subjective components of QOUL have explicitly tried to bridge this objective–subjective gap. In the late 1980s, a few studies did use a combined approach (Cutter 1985; Johnston 1988; Rogerson et al. 1989). But those studies attempting to capture those subjective preferences and the assessment of "neighborhood attractiveness" attributes using measures derived from perceptual data collected through surveys rarely sought to integrate that information within a spatial framework for analysis, other than to "geocode" the address of survey respondents, and to analyze spatial aspects of the residential location decision choice and process (such as the characteristics of the spatial search field during the residential location decision-making and choice process) and/or to relate self-rated preferences of residential areas and the reported levels of residential satisfaction (Golledge and Stimson 1997: pp. 464–476). Not often has an attempt been made in such research to examine the isomorphism of place identification.

It is, of course, difficult to arrive at a definitive list of neighborhood indicators that can reasonably predict residential location decision choices. Contextual or locational characteristics of neighborhoods – such as availability of amenities and access to services and facilities – are not given enough importance in hedonic price modeling studies, which many economists and regional scientists have used to model residential choices (Cheshire and Sheppard 1995). That is partly because the derivation of contextual information is often difficult to quantify, and it is conceptually more complex than data relating to structural attributes of dwellings and local housing market characteristics (Orford 1997).

Nevertheless, some additional dimensions have emerged in several studies, which perhaps can enhance the predictive power of traditional hedonic models, and that might include, for example, amenity, open space, jobs, family and friends and household composition have been repeatedly used in recent times.

However, even if we were to agree on a list of such objective indicators, the interactions between various components of neighborhoods and their effect on residential location decisions are still quite complex. Such interactions often develop "locational externalities" that have positive as well as negative influences on neighborhood quality (Pinch 1985). For example, a property close to a shopping center might facilitate ease of access but at the same time produce negative impacts, such as noise and traffic congestion.

Therefore, the integration of subjective and objective indicators is desirable because the effects of locational externalities on survey data points – the geocoded locations of the residences of survey respondents – and the data from QOUL surveys which give us information on the survey respondent's preferences and assessments "neighborhood attractiveness attributes" and of the reasons why they chose to live in their neighborhood.

It has been argued by Chhetri et al. (2005, 2006) and Stimson et al. (2006) that with the emergence of GIS technology, the mapping of neighborhood attractiveness reflected in both the "objective neighborhood characteristics" and in the subjective assessment of "neighborhood attractiveness attributes" derived from survey data may have the potential to reveal more about spatial processes that result in the social structuring of residential areas and the spatial patterning of the way people organize their activities, express their preferences and make their residential decision choices. The approach taken by Chhetri et al. (2005, 2006) and Stimson et al. (2006), which is discussed below, represents a methodological advance toward that end by using GIS to integrate subjective assessments of neighborhood attributes underpinning residential location decision choices collected via the SEQQOL2003 survey with objective socioeconomic and locational characteristics of neighborhoods derived.

## *Selecting Variables*

In the Chhetri et al. (2005) and (2006) study, the neighborhood was taken to be the suburb in which people live, and in the SEQ region, suburbs equate with Statistical Local Areas (SLAs) in the census. The subjective "neighborhood attractiveness attributes" and the objective neighborhood characteristics used are discussed below.

### Subjective Assessments of "Neighborhood Attractiveness Attributes"

The items included in the SEQQOL2003 survey which related to the subjective assessment of "neighborhood attractiveness" attributes in which survey respondents were asked to rate on a 5-point Likert scale with respect to how important the attribute was in affecting their residential location choice were as follows:

- Close to work
- Convenience to shopping centers and schools
- Good schools
- People similar to you
- Spaciousness of the area
- Close to natural areas (creeks, parks, beaches, etc.)
- Cost of housing
- Close to family and friends
- Familiarity with area
- Close to public transport
- Attractive appearance of neighborhood
- Lots of recreational opportunities
- Size of the community.

### Digital Data for Spatial Objective Neighborhood Characteristic

Digital data acquired to compile some spatial "objective neighborhood characteristics" were primarily obtained from three sources:

- MapInfo StreetPro (8.0.1) database
- Department of Local Government and Planning
- Australian Bureau of Statistics 2001 Census of Population and Housing.

StreetPro data contain numerous geographic layers with information on highways and main roads, railway lines and railway stations and on urban features, such as parks and reserves, (and drainage) with their attributes. More specific sub-layers, such as shopping centers, schools and recreation facilities (different types of sport facilities, entertainment venues and the like), may be extracted as separate layers from the "feature" layer in StreetPro database. The data from Queensland Department of Local Government and Planning were used to identify areas assigned as industrial and commercial zones. Census data related to Socio-Economic Indexes for Areas (SEIFA) and to the geography of Statistical Local Areas (SLA) – which are the equivalent of suburbs and which were taken to be a surrogate for neighborhood – such as dwellings, population and median monthly rent were also extracted.

## *The Modeling and Mapping*

In their study modeling the potential subjective neighborhood attributes and investigating the objective neighborhood characteristics potentially affecting the underlying residential location choices of the SEQQOL2003 survey respondents,

Chhetri et al. (2005) and (2006) used the *Principle Components Analysis* (PCA), which is commonly used as a data reduction tool.

Mapping the generalized patterns of the factors or dimensions thus derived was undertaken using the *neighborhood operation* function in GIS. That procedure allows spatial relations of point-based survey data to be quantified and visualized as a surface. Because the output generated by the "neighborhood operation" function stores data in a raster format, the manipulation, visualization and modeling of spatial relationships and patterns of interaction were easier and faster than those of more traditional methods. The data layer could be easily manipulated via standard statistical operations, such as mean or median, range and standard deviation, on a range of data formats, including nominal and ordinal.

That involved five sequential steps:

(a) Step 1 extracted the dimensions or factors underlying survey respondent assessments of "neighborhood attractiveness attributes" and their importance affecting respondent's choice of neighborhood to live
(b) Step 2 identified the objective indicators of neighborhood characteristics
(c) Step 3 standardized the objective indicators
(d) Step 4 developed a parameterized linear equation for each factor derived from Step 1
(e) Step 5 built a spatial model and produced maps showing the spatial patterns of the underlying dimensions across the entire SEQ region.

Those steps are discussed in what follows.

## Step 1: Extracting Underlying Dimensions of Neighborhood Characteristics

PCA facilitates the exploration of manifest data to identify latent components from a set of interrelated variables (see Hair et al. 1998). A reduced number of new variables, known as components, are obtained from highly correlated variables. Statements about variables (that is, the neighborhood attractiveness attributes) and observations (that is, the SEQQOL2003 survey respondents) were listed in columns and rows, respectively, using *R*-mode analysis (Nie et al. 1975). The calculated Kaiser-Meyer-Olkin (KMO) of 0.811 exceeded the $p \leq 0.5$ level of confidence considered to be acceptable for the use of PCA.

In order to maximize the specificity of the neighborhood attractiveness attributes, attributes were retained if their loadings on one of the factors were greater than 0.40. Items that did not fulfill this criterion were deleted. Two items "good school" and "street layout" were eliminated from the modeling. The eigenvalues derived from the PCA were plotted against the order of extraction and visual inspection using the rules of the scree test indicating the number of principal components for extraction. The varimax rotation procedure – which uses an orthogonal structure – was used to extract the appropriate number of factors. That technique generates a solution in which high factor loadings are maximized and low factor loadings are minimized.

**Table 18.3** Component loadings and reliability statistics of "neighborhood attractiveness attribute" importance scales derived from the PCA of the subjective assessments of neighborhood attractiveness attributes

| Factors (Bold) and Items (Italic) | Aesthetics | Amenities | Social interaction | Eigen-Value | Cronbach's Alpha |
|---|---|---|---|---|---|
| **Aesthetics factor** | | | | **3.4** | **.75** |
| *Openness/spaciousness of area* | ***.795*** | −.027 | .01 | | |
| *Close to natural areas* | ***.789*** | −.078 | .133 | | |
| *Attractive appearance of neighborhood* | ***.679*** | .199 | .098 | | |
| *Lots of recreational opportunities* | ***.548*** | .023 | .424 | | |
| *Community's size* | ***.480*** | .382 | .255 | | |
| **Amenities factor** | | | | **1.6** | **.68** |
| *Convenient to shopping centers, schools* | .063 | ***.793*** | .175 | | |
| *Close to public transport* | −.018 | ***.709*** | .155 | | |
| *Close to work* | −.006 | ***.582*** | .122 | | |
| *Housing costs* | .389 | ***.487*** | −.181 | | |
| **Social interaction factor** | | | | **1.1** | **.62** |
| *Familiarity with area* | .105 | .137 | ***.723*** | | |
| *Close to family and friends* | −.021 | .076 | ***.658*** | | |
| *People similar to you* | .292 | .229 | ***.626*** | | |

Source: Chhetri et al. (2006)

In this way, a total of three significant components with eigenvalues greater than 1 were extracted. These components were also tested for internal reliability using Cronbach's alpha. The results are summarized in Table 18.3.

The three significant components derived from the PCA were described as follows:

(a) The first component accounted for the greatest proportion of variance (eigenvalue = 3.4) and accounted for 21.37% of the total variance. The five items defining the component have high loadings from 0.48 to 0.795. This component has been named the *aesthetic factor*.
(b) The second component had an eigenvalue = 1.6. The four items defining the component had loadings ranging from 0.487 to 0.793. This component has been named the *amenity and accessibility factor.*
(c) The third component had an eigenvalue = 1.1. The three items defining the component had loadings ranging from 0.62 to 0.723. This component has been named the *social interaction factor*.

These components are descriptively, not normatively, labeled, and together they explained 53.83% of the variability in observations.

### Step 2: Identifying Objective Indicators for Survey Items

Objective indicators of neighborhood characteristics that might relate to the subjective neighborhood attractiveness attributes used in the SEQQOL2003 survey were quantified or approximated using digital data with some being directly derived from spatial and biophysical characteristics, such as distance, accessibility and proportion of open space to total area. Deriving others was more complicated (for example, occupational diversity and elevation diversity) or was reliant on surrogate-based indicators. Items in the survey for which surrogate indicators were used included: spaciousness, familiarity with the area; and close to family and friends.

Two layers were used to measure *attractive appearance of landscape*. That included an "open space index" and an "elevation diversity index." The "open space index" was calculated as the proportion of open space to the total area for each suburb. Earlier studies suggested that topography is one of the important dimensions of neighborhood attractiveness as it determined patterns and forms for many other landforms and land cover features (Linton 1968; Bishop and Hulse 1994). Using a digital terrain model, elevation diversity was calculated as the standard deviation of height within a 2.5 km radius of the spatial filter. The higher value indicates the greater elevation diversity. It was calculated as:

$$Elevation\ diversity = FOCALSTD\left(Height, KERNEL, KERNAL, 2.5km\right)$$

The "neighborhood operation" procedure in GIS was used to calculate two other objective indicators relating to the questions "*lots of recreational opportunity*" and "*access to shopping center and to school.*" The access measure based on the "neighborhood operation" procedure in GIS was preferred over other methods, such as the "container approach," which calculates the count of facilities within a suburb or SLA, or the "distance approach," which measures the shortest distance to the nearest facility. The container approach did not incorporate a spatial dimension (for example, if clustered, a large area would be unserved), and it did exclude areas, despite juxtaposition to a facility, from the measure if they fell into other suburbs (Talen and Anselin 1998). However, the access measure used included both the number and size of the opportunities accessible within a commutable distance.

The survey item "*lots of recreational opportunity*" was measured as the number of recreation facilities available for contact or interaction from a given point or location within its surrounding neighborhood. The radius for the neighborhood operation was defined as a limit beyond which people are less likely to travel to visit a facility. As people often drive to those facilities, a radius of 2.5 km was assumed as a commutable distance, which roughly corresponded to the average size of a neighborhood, except in rural parts of the SEQ region. Using the StreetPro database, neighborhood operation was conducted on a point data layer of sporting facilities, and a count for each cell in the output grid was calculated as:

$$Recreational\ opportunity = FOCALVARIETY\left(Sporting\ facilities, KERNAL, 2.5\ km\right)$$

The other item "*access to shopping center and school*" was calculated in a similar way, except that the element of size was also incorporated. Two layers of shopping center and school were used separately and were then overlaid onto a single layer. Shopping centers were ranked according to their sizes from 1 to 3 (Regional SC 3, Neighborhood SC 2 and Local SC 1). The attribute field was used instead of point feature to calculate the "sum" of the total value of an assigned field on shopping center layer within the defined radius of 2.5 km. It was calculated as:

$$\textit{Access to shopping center} = \textit{FOCALSUM}\left(\textit{Shopping Center Ranking, KERNAL, 2.5 km}\right)$$

The remaining three survey items – "close to work," "close to natural areas," and "proximity to transport" – were estimated using a measure based on Euclidean distance. Distance to the Brisbane CBD and the distance to industrial and commercial areas were calculated and merged to derive the item "close to work." For "proximity to transport" distance to the Brisbane City Cat Ferry route, distance to railway station and distance to main roads and motorway were calculated. Similarly, the survey item "close to natural areas" was also generated by amalgamating three layers: distance to creeks, distance to parks and reserves and distance from the coast.

A number of *surrogate measures* were generated using the census data available at the SLA level. *Dwelling density* was used to represent "*spaciousness*," which was calculated as the number of dwellings per square kilometer. Similarly, the survey item "*community size*" was calculated as the *number of people per square kilometer*, while the *median weekly rent* obtained from census data was a surrogate for the item "cost of housing." Monthly housing loan repayment was also considered as a potential candidate, but it was dropped due to its inappropriateness to represent more established areas where the monthly loan repayment for houses bought several years ago might not be a good indicator of the cost of housing at the time when the SEQQOL2003 survey was conducted.

The longer people are resident at one place, the greater will be their familiarity with the area, and therefore, the greater will be the likelihood of buying a home or staying at that locality. If this assumption is valid, the survey item was estimated by calculating the *proportion of people living in the same SLA for at least 5 years*. The survey item "close to family and friends" was also measured through a "friendliness index," which is defined as the proportion of people visiting others from the same suburb, a variable collected in the survey.

Two variables were combined to represent the item "*people similar to you.*" These include occupational diversity and lifecycle diversity. Using the *Simpson Diversity Index* (Simpson 1949), occupational categories were converted into the occupational diversity index. Simpson's Index (SDI) is the first of a set of nonparametric approaches to calculating heterogeneity. The index ranges from 0 to 1, where the closer it is to 1, the less diverse the community. It was calculated as:

$$\text{SDI} = \sum p_i^2$$

where

$$p_i = n_i / N$$

with

$N$ = the total number of individuals in the groups
$n_i$ = the number of groups
$P_i$ = the relative proportion of individuals in group$_i$

The SDI defines the probability of two individuals from a random sample to belong to the same category or type. In other words, if the index value is 0, then everyone in the population has a different occupation, and the probability of finding someone with the same occupation is zero. The nine major occupation groups from the one-digit ASCO code were reclassified into five groups based on their levels of skill. These included: managers, administrators and professionals; associate professionals; tradespersons and advanced clericals; intermediate; and elementary and laborers.

Using the same method, a *Lifecycle Diversity Index* (LDI) was calculated for: "couple with children," "couple without children," "one parent family" and "other family." Both indices were combined to represent a surrogate for the survey item "people similar to you." However, if the SEQ region has been culturally and linguistically more diverse, then the inclusion of these would have had higher priority.

## Step 3: Standardizing Objective Indicators for Modeling Factors of Importance

The GIS layers created in Step 2 were in different metrics. For example, distance cannot be readily compared with the number of shopping centers or with dwelling density. Therefore, the data were standardized in order to remove the effect of measurement units. The data were converted to standard scores by subtracting the mean and by dividing the standard deviation for each variable (Hair et al. 1998). This converts raw data into a standardized value with a mean of 0 and a standard deviation of 1.

Furthermore, few data layers, like distance from natural areas, proximity to transport network, dwelling density and closeness to work, were negatively related to the survey items. In other words, lower values indicate better access, greater proximity and spaciousness. Consequently, these variables were reversed on scoring. Mapping was conducted at the level of Statistical Local Areas (SLA) for two reasons:

(a) First, it roughly corresponds to suburbs, which the results of the 2003 SEQ QOL survey showed to be an areal unit that was commonly perceived as being the respondents' neighborhood.
(b) Second, census data are readily available. For those variables, which were in the grid format (for example, closeness to transport and natural areas and access to shopping centers etc.), data for the relevant layers were tabulated by "summarizing" them against the SLA geography of the region.

The *Summarized Function* in GIS calculates Poly-on-Grid Statistics, including the average of all cell values that lie within each SLA, and adds it as an attribute to the SLA geography in a tabular format.

### Step 4: Developing Linear Parameterized Equations for Mapping the Underlying Factors

To map the patterns generalized and derived from the three factors described in Step 1, the factor-score coefficients matrix was used by Chhetri et al. (2005) and (2006) to generate parameterized linear equations That matrix is an array of coefficients that are similar to "regression weights," and they were used to generate factor scores from the variables (Nie et al. 1975: p. 488; Tabachnick and Fidell 1996: p. 591). The use of factor scores for mapping is commonplace in the geographical literature (see, for example, Burley and Brown 1995; Western and Larnach 1998; Kliskey 2000), and using the factor score coefficients matrix as a set of linear equations is also commonplace (see, for example, Nie et al. 1975; Tabachnick and Fidell 1996; Arrowsmith and Inbakaran 2002; Chhetri and Arrowsmith 2002). The objective indicator data for each factor were simply multiplied by their respective factor coefficient scores and were then added. Three linear equations representing the factors were used, and three maps were produced.

The following three linear models were used:

(a) "Aesthetic factor" map

$$= -.068v_1 + .175v_2 + -.061v_3 + -.089v_4 + .150v_5 + .285v_6 + .176v_7 + .014v_8 + -.083v_9 + .352v_{10} + .372v_{11} + -.126v_{12}$$

(b) "Amenity/accessibility factor" map

$$= .317v_1 + .287v_2 + .425v_3 + .386v_4 + .150v_5 + .046v_6 + -.113v_7 + -.003v_8 + -.060v_9 + -.130v_{10} + -.079v_{11} + -.074v_{12}$$

(c) "Social interaction factor" map

$$= -.019v_1 + -.282v_2 + -.031v_3 + -.016v_4 + .027v_5 + -.081v_6 + 2.32v_7 + 3.51v_8 + .468v_9 + -.025v_{10} + -.121v_{11} + .454v_{12}$$

where

$v_1$ = standardized value of close to work
$v_2$ = standardized value of cost of housing
$v_3$ = standardized value of convenience to shopping and school
$v_4$ = standardized value of close to transport
$v_5$ = standardized value of lots of recreation opportunities
$v_6$ = standardized value of attractive appearance
$v_7$ = standardized value of size of community
$v_8$ = standardized value of similarity of people

$v_9$ = standardized value of familiarity with area
$v_{10}$ = standardized value of close to natural areas
$v_{11}$ = standardized value of openness/spaciousness
$v_{12}$ = standardized value of close to family and friends.

### Step 5: Mapping the Patterns and Associations of the Importance of Neighborhood Characteristics

Using the parameterized linear equations, three maps representing the factors were generated. In doing so, it was necessary to first determine the absence or presence of spatial patterns on the mapped data and second to establish the associations that spatial patterns held with socio-demographic and economic characteristics. If the patterns and characteristics were found to be associated, Chhetri et al. (2005) and (2006) used the term *socio-spatial structures*, and those could be mapped for each of the three factors, as shown later in Figs. 18.2–18.4.

Spatial patterns – in terms of clustering, randomness or dispersal – embedded in these maps were quantified by employing two common global indices of spatial autocorrelation: Moran's *I* statistic and Geary' *C* statistic. The Moran's *I* can range from +1 to −1, with values close to +1 indicating clustering and values near −1 indicating dispersion. Geary's *C* ranges from 0 to 2, and when *C* is less than 1, events which are close together are more similar than those which are further apart, while values greater than 1 indicate that events close together are dissimilar. Statistical significance of the observed clustering or dispersion can be determined by the use of a *z*-score. These indices were calculated using ArcView scripts developed by Lee and Wong (2001).

The binary connectivity was used to create a distance matrix, wherein adjacent polygons were given a score of 1 or otherwise 0. Table 18.4 shows that both the Moran *I* and Geary *C* values evidenced the presence of positive spatial autocorrelation for the aesthetic, amenity and social interaction factors. The suburbs that are close together have similar values than those that are further apart. It should also be noted that the amenity dimension, when compared to the others, exhibits a greater tendency of spatial clustering. The *z*-scores indicate that there is less than 1% likelihood that these clustering could be the result of random chance.

From the maps showing three spatial patterns thus produced by Chhetri et al. (2006), clear spatial patterning could be detected in the aesthetic factor, the amenity/accessibility factor and the social interaction factor maps. It is possible to investigate whether this patterning is a reflection of the presence of socio-spatial structures. To incorporate socio-demographic and economic characteristics in the analysis, the "Index of Relative Socio-Economic Disadvantage" from the Australian Bureau of Statistics "Socio-economic Indexes for Areas (SEIFA) Index" was used. In addition, a range of data sets, such as generated geographic layers and the "social structures" identified by Western and Larnach (1998) in the SEQ region, were also used to interpret socio-spatial structures based on the maps of aesthetic, amenity/accessibility and social interaction.

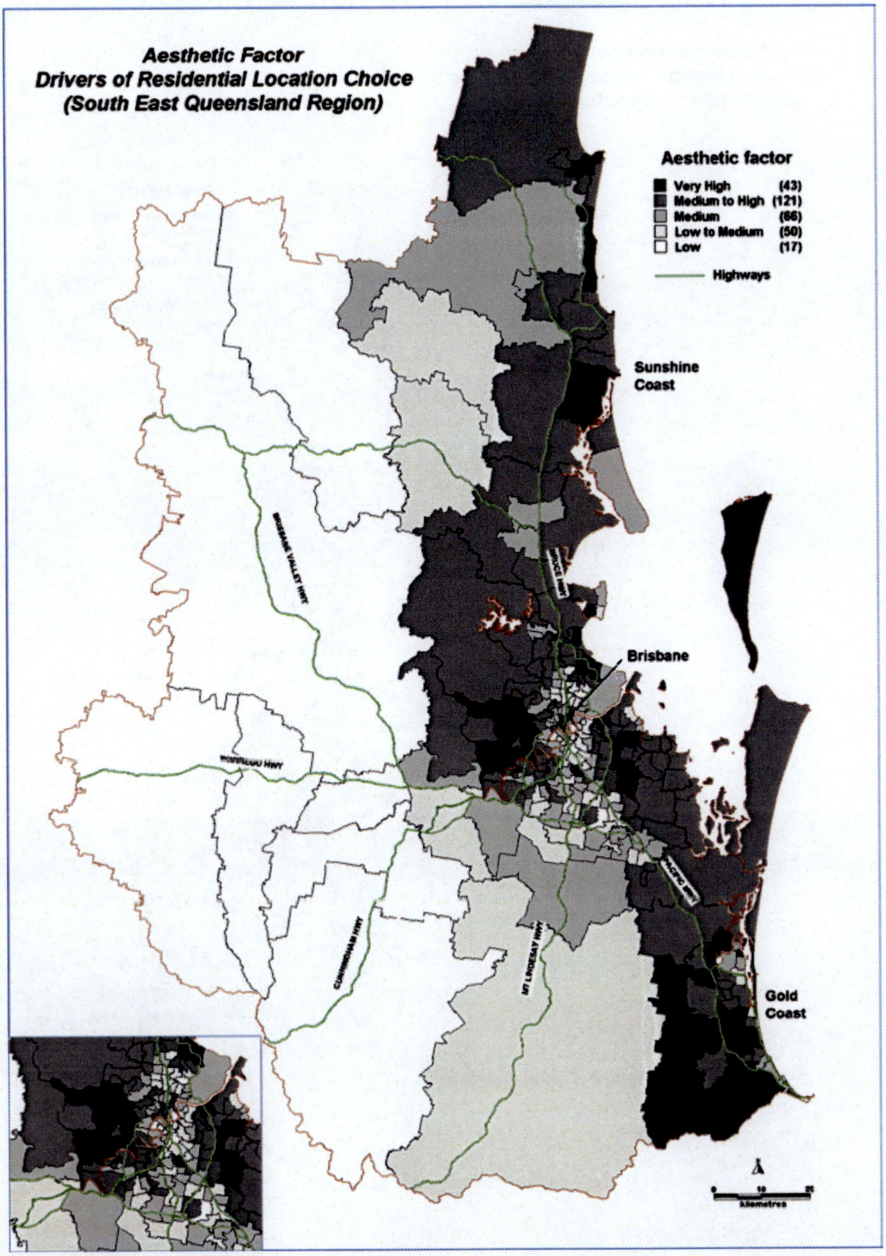


**Fig. 18.2** Geographic variation in the "aesthetic factor" across the SQ region's neighborhoods (Source: Chhetri et al. 2006)

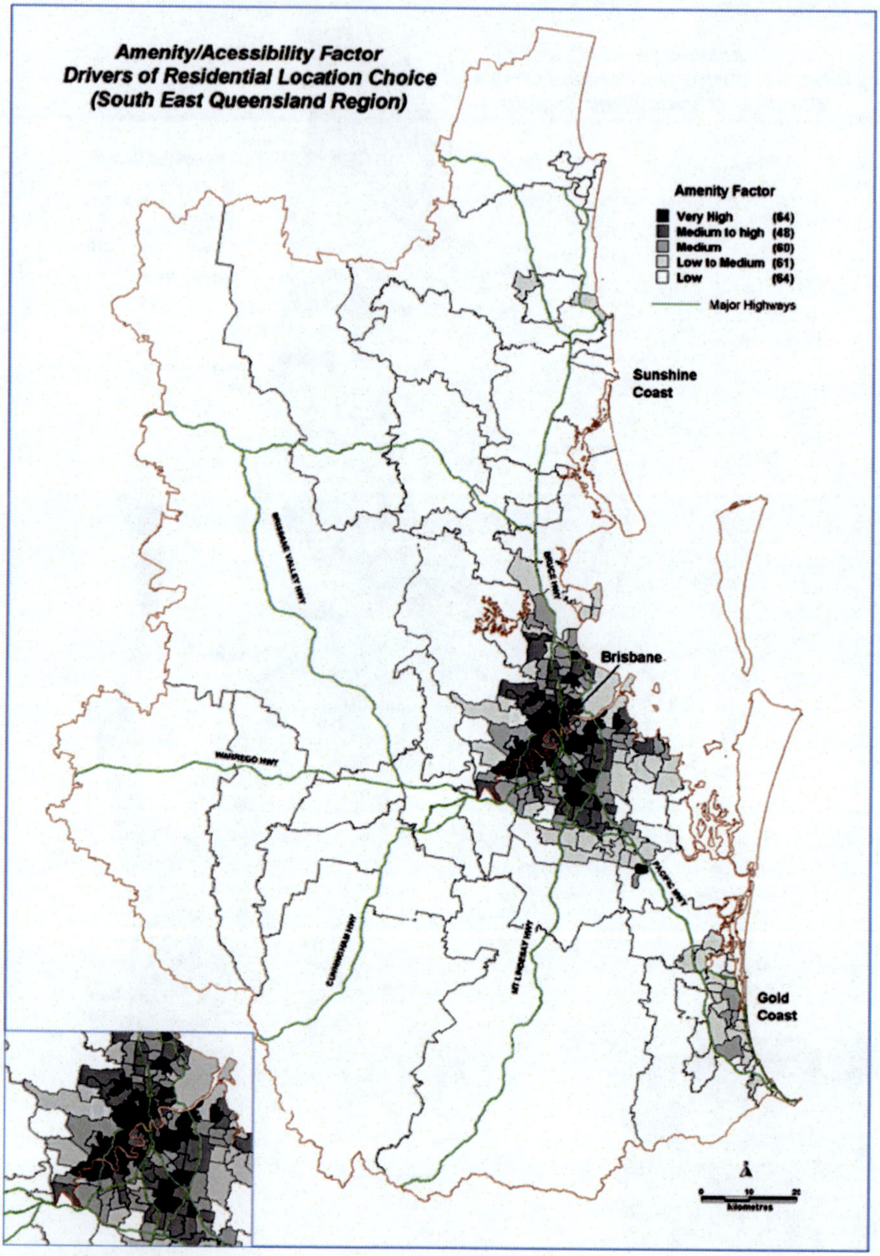


Fig. 18.3 Geographic variation in the "amenity and accessibility factor" across the SEQ region's neighborhoods (Source: Chhetri et al. 2006)

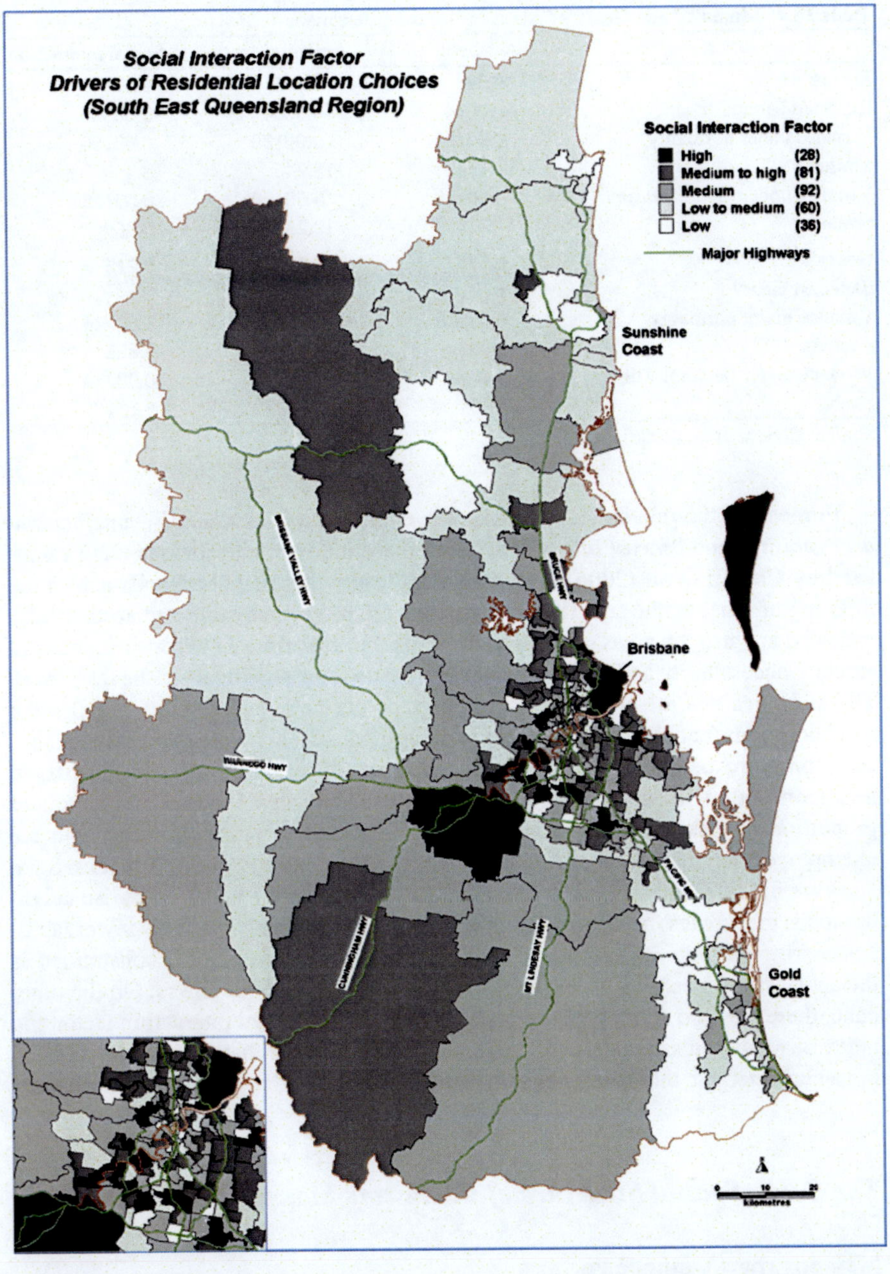


**Fig. 18.4** Geographic variation in the social interaction factor in the Brisbane-South East Queensland (Source: Chhetri et al. 2006)

**Table 18.4** Moran's *I* and Geary's *C* statistics for the factor maps

| | Aesthetic | Amenity | Social interaction |
|---|---|---|---|
| **Moran's I** | **0.398** | **0.655** | **0.298** |
| Expected Moran | –0.0034 | –0.0034 | –0.0034 |
| Variance under normality | 0.00130 | 0.00130 | 0.00130 |
| z-value | 11.134 | 18.246 | 8.359 |
| Variance under randomization | 0.00128 | 0.00129 | 0.00129 |
| *z*-value | 11.224 | 18.282 | 8.369 |
| **Geary's C** | **0.558** | **0.238** | **0.715** |
| Expected Geary | 1 | 1 | 1 |
| Variance under normality | 0.00236 | 0.00236 | 0.00236 |
| *z*-value | –9.093 | –15.675 | –5.862 |
| Variance under randomization | 0.00482 | 0.00295 | 0.00272 |
| *z*-value | –6.357 | –14.004 | –5.451 |

Source: Chhetri et al. (2006)

To measure the relationships of the derived factors of "*aesthetic*," "*amenity and accessibility*" and "*social interaction*" with these indexes, correlations were calculated by Chhetri et al. (2006). The results indicated that neighborhoods across the SEQ region (that is, the suburbs or SLAs) that are more accessible and aesthetically endowed are also the areas of relative advantage. Correlations between aesthetic and amenity/accessibility and SEIFA index were found to be significant at the 0.01 level. This indicates that the areas that scored high on aesthetic and amenity are also the neighborhoods that have high proportion of high household income or a skilled labor force or more people with higher education qualifications. More disadvantaged neighborhoods with attributes, such as families with low income and a greater proportion of less educated people working in unskilled occupations or who are unemployed, are more likely to have low scores on accessibility and aesthetics. Such areas are not only socially and economically disadvantaged but also comprise neighborhoods that are less aesthetically appealing and that are with relatively fewer facilities nearby. Willingness to move to a neighborhood of one's choice is constrained as the actual move needs to go through a range of socioeconomic filters. On the other hand, there were no significant correlations between the social interaction factor and index of relative disadvantage. It can therefore be inferred that social interaction is not a characteristic that is restricted to advantaged or disadvantaged neighborhoods.

## *The Socio-Spatial Structure of the Factor Maps*

### Mapping the "Aesthetic Factor"

Figure 18.2 shows the pattern of factor scores for the aesthetic factor. It indicates that neighborhoods with highest scores are neatly aligned along the coast and other waterways, indicating the effects of proximity to natural areas, such as creeks, beaches and parks. Some interior parts of the SEQ region, despite their closeness to natural areas, have low to moderate scores perhaps due to the presence of few recreational opportunities. Overall, it is the coastal areas – particularly around the Gold Coast and the Sunshine Coast – that seem to have the comparative advantage of considerable investment on recreation infrastructure. The growth corridors and the inland areas of the SEQ region – such as Ipswich in the western corridor and parts of Logan and Beaudesert in the southern corridor and the Caloundra (Hinterland) and Caboolture (Part B) in the northern corridor – had low scores on this factor.

### Mapping the "Amenity and Accessibility Factor"

Figure 18.3 shows the pattern for the amenity and accessibility factor. The map shows that the neighborhoods with the highest scores included the suburbs in inner Brisbane City that generally provide greater accessibility to job opportunities and amenities, while neighborhoods in outer suburban tend to have low scores and are relatively disadvantaged in this regard. The suburbs in the inner parts of Brisbane City had high concentrations of nontraditional households (for example, couples without children, single and group households) with a lot of rental and medium-density housing. The suburbs within Brisbane City's inner suburbs also had high land values, greater ethnic diversity and high socioeconomic status. It appears that the greater accessibility to jobs and amenities had attracted a particular socio-demographic profile. The Gold Coast and the surrounding suburbs also have moderately high scores. They had a higher proportion of rental and high-density housing along the coastal strip.

The outer suburban neighborhoods – and particularly those in the northern parts of Brisbane City and into the northern corridor – as well as neighborhoods in the interior parts of the SEQ region had some of the lowest amenity scores, indicating the need to better strategically plan infrastructure provision.

The overall pattern in this map was rather CBD-centric and exhibited a somewhat concentric pattern with distortions along the transport arteries producing a wedge-like structure.

### Mapping the "Social Interaction Factor"

The pattern for the social interaction factor is shown in Fig. 18.4. It exhibited a mixed spatial patterning. This factor represents survey respondents who considered family relations, friendliness and similarity with others as the most important reasons for moving to their current place of residence. Increasing social exclusion, individualism and independence often thrive in the inner city, which tends to have low scores. On the other hand, strong family ties, greater social capital, residential permanency and a stronger sense of belongingness were strong drivers in this factor, and the neighborhoods with high scores are found more in the outer suburban areas and, in particular, in rural–urban fringe areas of the SEQ region.

There were, however, low scores in many of the neighborhoods on the Sunshine Coast and on the Gold Coast possibly partly because of the presence of larger number of floating residents with stronger urban ties and cosmopolitan lifestyle.

### *Overall Socio-Spatial Structures*

Chhetri et al. (2006) have shown how the maps compiled to represent the three neighborhood attractiveness attribute factors discussed above could be merged to form a composite map of neighborhood attractiveness for SLAs across the SEQ region. This is shown in Fig. 18.5.

Chhetri et al. (2006) did that as follows. By modifying the "Max" function in Excel, a new categorical variable was created, which retained the name of the factor with the highest score of the three. Because those factors were orthogonal and uncorrelated, it was assumed that the likelihood of neighborhoods displaying high scores on all factors is low. The map reproduced in Fig. 18.5 reveals a more definitive spatial contiguity on these factors. Most neighborhoods along the coast were classified as "aesthetic," while a large majority of suburbs in inner Brisbane City and along the main road corridors are described as "amenity-oriented." On the other hand, "social interaction" characterizes those neighborhoods that were either predominately rural or were outer suburban places with cultural diversity and low scores on the index of disadvantage, particularly in parts of the western corridor.

Analysis of variance (ANOVA) was used by Chhetri et al. Chhetri et al. (2005) and (2006) to determine the degree to which those of "aesthetic," "amenity" and "social interaction" categories of neighborhoods were associated with advantage or disadvantage according to their scores on the SEIFA index of disadvantage. From the results reproduced in Table 18.5, it was evident that "aesthetic," "amenity" and "social interaction" did show significant differences at the 95% confidence level. It was also apparent that mean scores for neighborhoods with greater accessibility had scored well above the national average of 1,000 on the SEIFA index. But the neighborhoods with high "social interaction" scores had low mean scores on the SEIFA index. However, further analysis using a Bonferroni test indicated that the differences in the means of the "aesthetic" and "amenity" factors against the SEIFA index of relative disadvantage were not significant ($p \leq 0.05$). Overall, those results supported the argument that there were notable socioeconomic differences across suburbs.

## Conclusion

In this chapter, we have demonstrated the application of GIS methodologies that combine statistical and spatial modeling techniques to derive generalized or region-wide spatial patterns of aspects of the subjective assessment aspects of QOUL

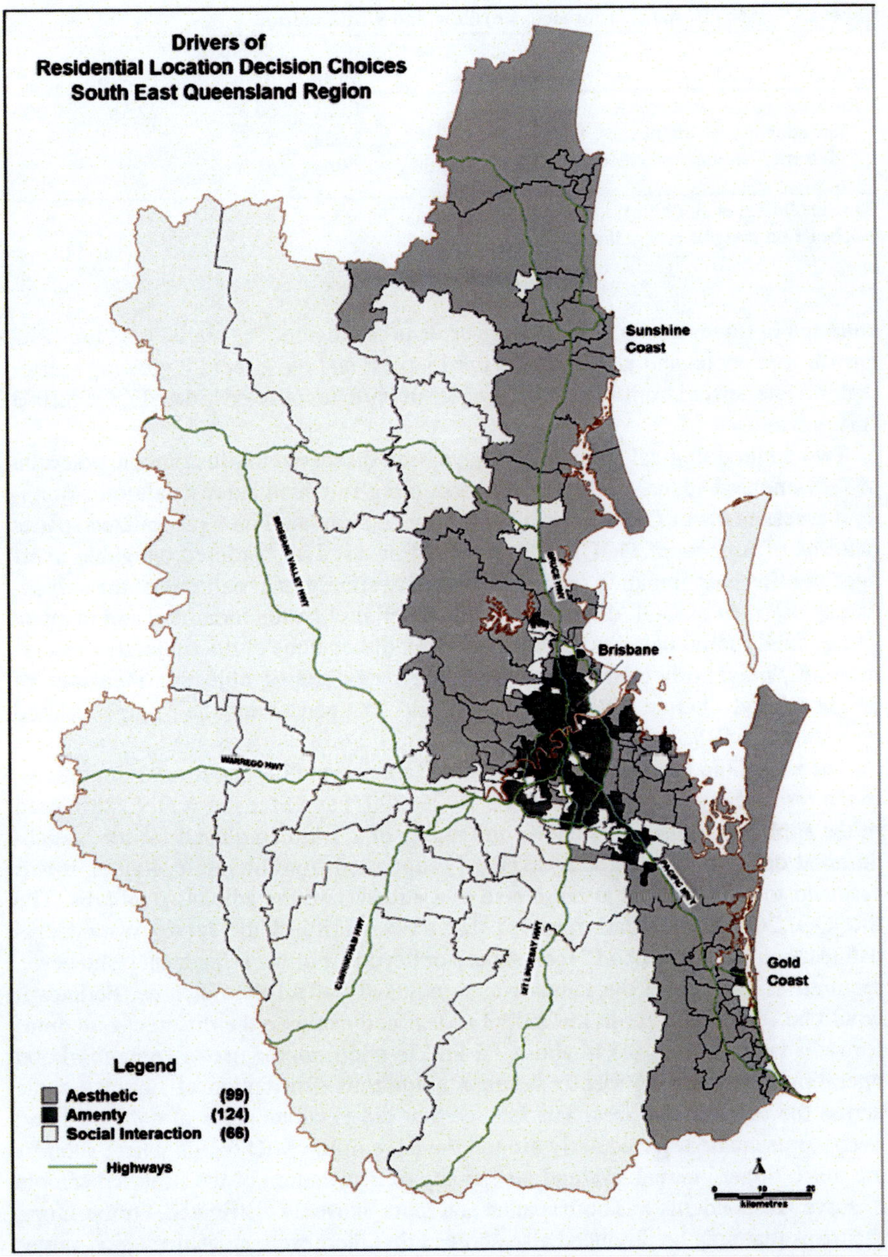


Fig. 18.5 Overall geographic variation in the factor distribution in the Brisbane-South East Queensland (Source: Chhetri et al. 2006)

**Table 18.5** ANOVA statistics for factors to move and SIEFA indexes

| Indexes | Factors | Mean | Standard deviation | F-ratio | Sig.* |
|---|---|---|---|---|---|
| *Disadvantage* (Low income, low educational attainment, high employment, unskilled occupation) | Aesthetic | 1,021 | 56.5 | 22.56 | 0.000 |
| | Amenity | 1,032 | 57.5 | | |
| | Social interaction | 968 | 81.3 | | |

Source: Chhetri et al. (2006)
*Correlation is significant at the 0.001 level (2-tailed)

collected in survey conducted in a large-scale urban region. The region is a "sun-belt" growth metropolis and is Australia's third largest and most rapidly growing metro-region. Data derived from unit record sample survey data collected in the SEQQOL2003 survey was used.

Two methodological applications have been discussed to illustrate the potential of GIS-enabled spatial and statistical modeling in which survey data on respondent assessments of QOUL dimensions may be simulated into generalized spatial patterns of aspects of QOUL. One application used an "ordered weighted average" nonlinear aggregation technique to derive subregional patterns of the subjective assessment of QOL dimensions. The other application identified and mapped generalized spatial patterns of the underlying dimensions of the subjective assessment of "neighborhood attractiveness" attributes and of objective measures of neighborhood characteristics simulated and mapped using the "neighborhood operation" in GIS.

The use of the "neighborhood operation" routine in GIS has several limitations. In the research by Chhetri et al. (2005, 2006, 2007) and Stimson et al. (2006), used in the application discussed above, the radius of 2.5 km employed for the "neighborhood operation" was not a "standard" supported from literature. Rather, it was assumed to represent the average size of a suburb and the adjoining suburbs. The SEQQOL2003 survey data revealed that the majority of the survey respondents defined their "neighborhood," their suburb or beyond. Further refinement is, however, required as the size of the radius can significantly affect the outcome. Perhaps it would be sensible to rerun the model to test and evaluate the different outcomes for radii varying from 0.5 to about 3.5 km. In addition, the use of "neighborhood operation" for mapping ideally requires ubiquitous distribution of survey sample across the entire study area. The lack of that did generate areas of data gaps and over-representation, particularly around the edge of the SEQ region where suburbs are much larger in areal size and the density of distribution of the stratified sample of survey respondents was both sparser and more unevenly distributed. Furthermore, the maps that may be produced also showed that there were areas for which values were calculated on the neighborhood attractiveness factors even if there were no respondents in the suburb.

Nonetheless, the exploratory applications of the methodologies discussed in this chapter demonstrating how the interface of sample survey data with GIS technology to derive spatial patterns of perceptual or attitudinal phenomena seem to have considerable potential. They also demonstrate the ways in which survey research and spatial analysis may be integrated to enhance the way assessments and evaluations collected via a sample survey may be generalized across the space of the region.

Those approaches may enable us to provide new perspectives to inform planning and development processes in cities. But these approaches require further experimentation, development and testing using similar survey data relating to the assessment of QOUL dimensions in a variety of spatial settings.

**Acknowledgement** This chapter is based on research funded by the Australian Research Council Discovery, project # DP0209146.

## References

Arrowsmith, C., & Inbakaran, R. (2002). Estimating environmental resiliency for the Grampians National Park. *Tourism Management, 23*(3), 295–309.

Bishop, I. D., & Hulse, D. W. (1994). Prediction of scenic beauty using mapped data and geographic information systems. *Landscape and Urban Planning, 30*, 59–70.

Burley, J. B., & Brown, T. J. (1995). Constructing interpretable environments from multidimensional data: GIS suitability overlays and principal component analysis. *Journal of Environmental Planning and Management, 38*(4), 537–550.

Cheshire, P., & Sheppard, S. (1995). Evaluating the impact of neighbourhood effects on house prices and land rents: results from an extended model. *Economica, 62*, 247–267.

Chhetri, P., & Arrowsmith, C. (2002). Developing a spatial model of hiking experiences in natural landscapes. *Cartography, 13*(2), 87–102.

Chhetri, P., Stimson, R. J., Western, J., & Shyy, T-K. (2005) What matters in a neighbourhood?. In *Proceedings of spatial sciences conference 2005: Spatial intelligence, innovation and praxis*. Melbourne: Spatial Sciences Institute.

Chhetri, P., Stimson, R. J., & Western, J. (2006). Modelling the factors of neighbourhood attractiveness reflected in residential location decision choices. *Studies in Regional Science, 36*(2), 35–45.

Chhetri, P., Stimson, R. J., Akbar, D., & Western, J. (2007). Developing perceived quality of life indices: an application of ordered weighted average operators. *Studies in Regional Science, 37*, 553–572.

Cutter, S. L. (1985). *Rating places: A geographer's view on quality of life*. Washington, DC: Association of American Geographers, Resource Publication in Geography.

Dökmeci, V., & Berköz, L. (2000). Residential-location preferences according to demographic characteristics in Istanbul. *Landscape and Urban Planning, 48*, 45–55.

Filev, D., & Yager, R. R. (1995). Analytical properties of maximum entropy OWA operators. *Information Sciences, 85*, 11–27.

Filev, D., & Yager, R. R. (1998). On the issue of obtaining OWA operator weights. *Fuzzy Sets and Systems, 94*, 157–169.

Golledge, R. G., & Stimson, R. J. (1997). *Spatial behaviour a geographer perspective*. New York: The Guilford Press.

Hair, J., Anderson, R., Tatham, R., & Black, W. (1998). *Multivariate data analysis: With readings*. Englewood Cliffs: Prentice-Hall.

Johnston, D. F. (1988). Toward a comprehensive 'quality of life' index. *Social Indicator Research, 20*, 190–214.

Kliskey, A. D. (2000). Recreation terrain suitability mapping: a spatially explicit methodology for determining recreation potential for resource use assessment. *Landscape and Urban Planning, 52*, 30–43.

Lee, J., & Wong, D. W. S. (2001). *Statistical analysis with ArcView GIS*. New York: Wiley.

Levinsohn, A., Langford, G., Rayner, M., Rintoul, J. & Eccles, R. (1987). A micro-computer based GIS for assessing recreation suitability. In *Proceedings of GIS 87* (pp. 739–747). Falls Church: American Society for Photogrammetry and Remote Sensing.

Linton, D. L. (1968). The assessment of scenery as a natural resource. *Scottish Geographical Magazine, 84*, 219–238.

Maher, C. A., & Saunders, E. (1994). *Changing patterns of relocation within a Metropolitan Area: Intraurban mobility in Melbourne 1971–9*. Paper presented to Workshop on Mobility, Melbourne.

Marans, R. W. (2003). Understanding environmental quality through quality of life studies: The 2001 DAS and its use of subjective and objective indicators. *Landscape and Urban Planning, 65*(1), 73–83.

Mendes, J. F. G., & Motizuki, W. S. (2001). Urban quality of life evaluation scenarios: The case of Sao Carlos in Brazil. *CTBUH Review, 1*(2), 1–11.

Mingche, M. L., & Brown, J. H. (1980). Micro-neighbourhood externalities and hedonic housing price. *Land Economics, 56*, 125–141.

Nie, N. H., Hull, C. G., Jenkins, J. G., Steinbrenner, K., & Bent, D. H. (1975). *Statistical package for the social sciences* (2nd ed.). New York: McGraw-Hill.

Nijkamp, P., Rietveld, P., & Voogd, H. (1990). *Multicriteria evaluation in physical planning*. Amsterdam: Elsevier Science Publishers.

Orford, S. (1997). *Valuing the built environment: A GIS approach to the Hedonic Modelling of housing markets*. Unpublished Ph.D. thesis, University of Bristol, Bristol.

Pinch, S. (1985). *Cities and services: The geography of collective consumption*. London: Routledge and Kegan Paul.

Rogerson, R. J., Findlay, A. M., & Morris, A. S. (1989). Indicators of quality of life: Some methodological issues. *Environment and Planning A, 21*(12), 1655–1666.

Simpson, E. H. (1949). Measurement of diversity. *Nature, 163*, 688.

Smith, P. N. (2001). Numeric ordered weighted averaging operators: Possibilities for environmental project evaluation. *Journal of Environmental System*; 2000–2001, *28*(3), 175–191.

Stimson, R.J, (1978). *Social space, preference space and residential location behaviour: A social geography of Adelaide*. Unpublished Doctoral thesis, Department of Geography, Flinders University, Adelaide.

Stimson, R. J., Chhetri, P., & Western, J. (2006, February). *Deriving metro-wide subjective assessments of neighbourhood attraction from sample survey data: An application in the Brisbane-South East Queensland region, Australia*. In Western regional science association, 45th Annual Conference, Santa Fe, February.

Tabachnick, B., & Fidell, L. (1996). *Using multivariate statistics*. New York: Harper Collins.

Talen, E., & Anselin, L. (1998). Assessing spatial equity: An evaluation of measures of accessibility to public playgrounds. *Environment and Planning A, 30*, 595–613.

Vogt, A. C., & Marans, R. W. (2004). Natural resources and open space in the residential decision process: A study of recent movers to fringe counties in southeast Michigan. *Landscape and Urban Planning, 69*, 255–269.

Western, J., & Larnach, A. (1998). The social and spatial structure of South-East Queensland. *Australasian Journal of Regional Studies, 4*(2), 215–237.

Yager, R. R. (1988). On ordered weighted averaging aggregation operators in multicriteria decision making. *IEEE Transactions on Systems, Man and Cybernetics, 18*(1), 183–190.

Yager, R. R. (2004). OWA aggregation over a continuous interval argument with applications to decision making. *IEEE Transactions on Systems, Man and Cybernetics-Part B: Cybernetic, 34*(5), 1951–1963.

# Part V
# The Way Forward

# Chapter 19
# Challenges for Quality of Urban Life Research

**Robert W. Marans and Robert Stimson**

In this chapter, we summarize the goals for this book and our efforts to achieve them. We then discuss challenges that lie ahead for its contributors and others interested in quality of urban life (QOUL) and related areas of research.

The book began with a detailed review of approaches to research on quality of life (QOL) and (QOUL) over the past 50 years. Much of that research was based on the collection of objective measures covering attributes of different types of places or on survey research involving the collection of subjective measures from individuals living in those places. Research that collected both objective and subjective measures has also been summarized. Case studies drawn from different parts of the world are then presented illustrating similarities, as well as differences, in situational context, methodological approaches, and in several instances, research findings. Finally, the book demonstrated a number of new methodological approaches to analyzing and modeling QOUL data, several of which integrate the objective and subjective approaches reviewed in the earlier chapters.

As we noted elsewhere, the book has not examined the extensive literature that has emerged in recent years on happiness, a concept closely linked to quality of life (for example, Bok 2010; Ferris 2010; Frey 2011; Gilbert 2006; Lyubomirsky 2008). Nor has it considered in any depth the related topics of community indicators (for example, Gahin and Paterson 2003; Sirgy et al. 2009; Phillips 2005) and urban sustainability indicators (Schaffer and Vollmer 2010). Similarly, we have not presented QOUL case

R.W. Marans (✉)
Institute for Social Research, University of Michigan, P.O. Box 1248, Ann Arbor, MI 48106, USA
e-mail: marans@umich.edu

R. Stimson
Australian Urban Research Infrastructure Network (AURIN), Faculty of Architecture, Building and Planning, University of Melbourne, VIC 3010, Melbourne, Australia
e-mail: rstimson@unimelb.edu.au

R.W. Marans and R. Stimson (eds.), *Investigating Quality of Urban Life: Theory, Methods, and Empirical Research*, Social Indicators Research Series 45,
DOI 10.1007/978-94-007-1742-8_19, 

studies from Latin America which has been examined in two recent publications (Lora et al. 2010; Paramo and Garcia-Gil 2010). Lora and his colleagues offer a compendium of work on QOL in several Latin America cities, most of which have used objective indicators, such as the market price of housing, as well as individual life satisfaction measures, to assess quality. Clearly, readers will find other limitations in what we have presented including gaps in the chapters of each of our contributors. In part, they are the result of chapter length guidelines imposed on our authors which limited discussion of their case studies and presentation of additional findings. But with several case studies, it also reflects work that their authors' have yet to perform. Accordingly, we offer some challenges to our collaborators as they continue working with existing data covering their respective places or as they pursue new QOUL research in other settings.

## Challenges for Our Contributors

Among the many opportunities for our collaborators are further analyses of the rich array of data collected during the course of their respective studies. This could include additional cross-sectional analyses showing how subsets of their samples respond to different environmental conditions. These subsets might include groups that differ by gender, age, educational level, income, and other sociodemographic and economic characteristics. The analyses may also include responses to QOUL questions for those living in different sections of the city, such as neighborhoods, school districts, or units, at other geographic scales. And in cases where objective conditions have been measured in addition to survey responses to those conditions, further analyses examining the degree of association between the two measures are warranted. Having both types of data, as well as computer mapping capabilities, create opportunities for both modeling (see discussion below) and expanding knowledge in the field of environment-behavior research.

A related challenge for our collaborators is examining cross-cultural differences or similarities in findings from surveys conducted in different parts of the world. Even when survey questions used in the different settings are not identical, it would be possible to determine the extent to which respondents in say, Bangladesh, view attributes of their residential environment differently than those in Istanbul or Detroit neighborhoods. In fact, the initial conception for the DAS 2001 was not only to launch a series of QOUL studies for metro Detroit that would be conducted every 5–6 years throughout the twenty-first century but also to seed similar QOUL studies in other urban areas allowing for cross-cultural research. DAS 2001 questions were indeed asked as part of an ongoing survey in the Brabant region in the Netherlands in 2002, and comparisons between the Detroit area and comparable Dutch urban settlements were reported to government officials from that region (Ester et al. 2002).

Still another challenge for our collaborators is finding additional outlets for their research. Besides international scholarly publications, efforts should be made to present study results in more popular forums, such as local media or on the Web. At the same time, efforts should be made to ensure that local government officials are made aware of study findings, including new analysis opportunities using the existing data, so as to inform the multitude of planning and policy decisions that need to be made in their respective jurisdictions.

## Challenges for QOUL Researchers Including Our Contributors

Based on experiences with our respective studies in Metro Detroit and Southeast Queensland and in compiling the book, we have identified a number of challenges in dealing with future quality of urban life research. These cover the collection of data, modeling of urban systems and the behaviors of urban residents, and the application of research in the realms of policy, planning, and environmental design.

### *Data Collection*

A significant issue in collecting data covering QOUL is cost. If objective measures reflecting urban life are readily available such as crime statistics, school achievement records, tax rates, housing prices, or census data, including demographic characteristics of the population living in a particular geographical area, the costs of compiling them are relatively inexpensive. However, if it is deemed desirable to collect other objective measures covering attributes of the built environment such as vegetation coverage, land use characteristics, or vacant or abandon buildings where such information is not available from other sources, data collection could be costly in terms of personnel and time. In the past, such information was obtained through field work when teams would visit a place and record the data accordingly (for example, land use, number of abandon buildings in a block, etc.) Some of that work could be expedited by using aerial photography, but only in places where such photography existed. More recently, the uses of satellite imagery and GIS technology have made the collection of such information easier and less costly. If the research were being conducted through a university and students were participating and receiving academic credit for their work, costs would also be reduced. But a constant challenge for researchers is finding the right balance between determining what is the optimal data necessary for their research, identifying the sources of those data, ensuring the data collection process is thorough and accurate, and compiling the data with minimal time and personnel costs.

A related challenge deals with the collection of data from people using social surveys. It is generally acknowledged that surveys of populations are expensive, particularly if they involve interviewers who must contact by telephone or in person a sample of the population and then conduct interviews with them. Self-administered questionnaires mailed to sample households are less expensive, but suffer other shortcomings, such as lower response rates and limitations in the number and length of questions to be asked.

A related issue with surveys is time. That is, it takes time to make repeated calls on sample participants, deal with non-responses, build data files, conduct analyses, and report findings. In recent years, these traditional survey problems have been addressed in part by new data management systems and the advent of surveys conducted over the Internet (Groves et al. 2009; Couper 2008). The major costs associated with Web surveys are the design and programming of the questionnaire, selecting and contacting the sample, using incentives to encourage participation, and tracking responses during the data collection process. Compared to mail questionnaires however, costs of Web surveys are significantly less expensive. Similarly, the turnaround time to administer the questionnaire and compile and report the data is considerably shorter than the time needed with other survey approaches. On the downside, Web surveys covering households have relatively low response rates and may not be appropriate for populations having high illiteracy rates and/or that lack access to computers or the Internet. In 2010, 80% of the U.S. population used the Internet whereas the Internet was only used by 45% of population in Turkey and less than 1% of the Bangladesh population (Internet World 2011, internetworldstats.com). These figures are likely to be higher in urban areas of these countries and should increase over time. As they do and when incentives are offered to increase response rates, Web surveys will become a more viable means of measuring QOUL.

A further issue to address when using surveys to investigate QOUL is the design of the sample. When researchers are interested in examining relationships between subjective information on QOL/QOUL and spatial objective information, a sampling design is often required that includes an appropriate spatial stratification and potentially differential sampling fractions within each spatial unit. This will ensure both the required geographic spread of the sample while providing a sufficient sample size within each spatial stratum that would be suitable for analysis. This issue is particularly important when researchers want to investigate variations in subjective assessments of QOL domains across different parts of a city, such as neighborhoods or administrative/local government units.

## *Modeling*

It should be apparent from some of the chapters in this book that there is increasing interest among researchers in developing more sophisticated models of QOUL outcomes. Such modeling efforts are needed to investigate the nature and complexities in the links between explanatory variables in predicting levels of satisfaction with QOL/QOUL domains and to investigate the links between the subjective assessment of QOL/QOUL and spatial and other objective measures. Those challenges were outlined in Chapter 4 and illustrations of how such modeling might be approached have been provided in chapters in Part V of this book. For example, it has been demonstrated in some of the research reported in those chapters how structural equation modeling can provide powerful insights into the nature and strength of relationships that might exist between an outcome measure and variables for which it is hypothesized might be explanatory factors.

It is evident that the increasing level of sophistication of geographic information systems (GIS) is opening new frontiers for modeling in QOUL research, especially through their enabling powers to integrate different forms of data and at different levels of spatial scale. In this regard, it is now essential that survey-based approaches to investigate subjective QOL/QOUL incorporate a capability to geocode the residential locations of the survey respondents as well as the locations of places in an urban environment with which they interact, (for example, place of employment, where they shop, the nearest transit stop, and so forth). This is necessary if an objective of the research is to integrate subjective and objective QOUL data.

There is also much benefit to be gained in QOUL research through the adoption of other modeling approaches that have not been addressed in any detail in the chapters to this book but which are becoming more widespread in social science research. Two of those which have an obvious utility for adoption in QOUL research are microsimulation (see, for example, Clarke 1996; Ballas et al. 2006) and hierarchical analysis.

Microsimulation methods are currently being used to merge survey data with spatial objective information such as that available through the census to generate "synthetic variables" which are estimates of the distribution of measures of behavioral phenomenon across small areas for which census data are available. Microsimulation modeling employs tools to relate the known occurrence of a behavioral measure available in survey data – differentiated among various demographic and socioeconomic groups – to the potential incidence of that behavior for those groups at a local level. In that way it would, for example, be possible to use data collected in a QOUL survey derived from a probability sample design to generate through microsimulation an estimate of subjective evaluations of specified QOUL domains for small areas across the total geographic space of a city or region.

Similarly, hierarchical analysis can be used to integrate data collected in different modes and at different levels of spatial scale to produce estimates of specified phenomena for which specific information has been collected at one level but is not explicitly available at another level.

Finally, there is considerable potential to incorporate agent-based modeling into QOUL research designs as suggested in Chapter 3 with the work of Fernandez et al. (2005).

## *Applications*

Another challenge for researchers wanting to measure the QOUL is ensuring that findings are recognized and used in a timely manner by those formulating urban policy, including elected officials and urban planners. For many engaged in such research, a motivating factor is the desire to generate data that would inform policy and environmental design decisions so as to optimize the QOL experience of urban residents. That is, it is believed that decision makers such as elected or appointed policy makers and public planners would have a better understanding of their constituents' thoughts and actions derived from scientific research and would be in a better position to respond to their interests and concerns. Similarly, data emanating from quality of life research could also be used in the nonpublic sector where organizational and business decisions regarding location, markets, and consumer behavior might take into account people's perceptions, satisfactions, and preferences as well as other indicators of urban quality. Indeed, our initial conception for this book was to use as a subtitle, *Implications for Planning and Policy*. Our own experiences, together with those reported by our collaborators, suggest that policy makers and planners in both the public and private sectors, while interested in the research at the outset, typically are not consummate users of research findings when deliberating or establishing policy. Or if policy makers and others have been exposed to study results, they are unable to articulate how those results have informed their decisions. Consequently, we altered the book to focus on theory, methods, and empirical work and subtitled it accordingly.

Lack of interest or use of research findings by policy makers is often attributable in part to the lapse between the time the study was conceived and launched and the time the researchers release study findings. In some instances, the issues that were deemed important to policy makers during the inception of the study may have been resolved by the time findings become available. In other instances, more pressing issues may emerge and become a higher priority for policy makers, drawing attention away from the issues that initially prompted the study.

Researchers also believe that findings from QOUL research, irrespective of the geographical or political context, should be disseminated through local media or via the Web so as to inform the general public. (See for example, Marans 2005). Although this periodically occurs with press releases announcing the completion of the research along with key findings, continuous media exposure is needed in order to sustain the interest of public officials and their constituents. Efforts should be made by researchers, working in collaboration with marketing and media personnel, to present findings for QOUL studies in repeated but small doses and in an interesting, understandable, and attractive manner.

Still another challenge for researchers interested in QOUL is to find ways to replicate studies over time. In their work in the Southeast Queensland region, Stimson and colleagues used surveys at two points in time and produced measures showing change in perceptions of several dimensions of urban life (See Chapter 8). A similar plan was proposed in connection with the quality of urban life study in Metro Detroit, but was not implemented (See Chapter 7). In fact, with the exception of the Southeast Queensland study, we are unaware of survey-based quality of urban life studies that have measured change, in large part, because of costs associated with conducting follow-up surveys. Yet, numerous studies involving social surveys have been designed and executed that follow populations over time. For example, the Panel Study for Income Dynamics (2011), conducted by the University of Michigan's Institute for Social Research, has collected data annually on the economic health and social behavior of U.S. families since 1968 (University of Michigan. Panel Study for Income Dynamics (2011) http://psidonline.isr.umich.

edu/). This longitudinal study has been replicated in several other countries creating opportunities for comparative analyses and informing policy makers in many nations. Other examples of ongoing major surveys including descriptions of sample sizes, modes of data collection, and target populations are reported in Groves et al. 2009.

Finally, we believe there are opportunities for QOUL researchers to pursue their interests within the context of the community indicator movement and especially with the measurement of sustainability indicators. The topic of community indicators has drawn the attention of scholars and policy makers since the early 1980s and is an outgrowth of social indicators activities that were prevalent in the 1970s. Numerous cities throughout the U.S., Canada, and in several European countries have developed community indicator programs designed to offer policy makers in the public and private sectors sets of metrics with which to gage progress in efforts to improve the lives of community residents. For the most part, metrics have been associated with economic development and healthy community efforts, but also cover other topics important to urban planners, including QOL (Gahin and Paterson 2003; Phillips 2003; Phillips 2005; Sirgy et al. 2009). More recently, sustainability has emerged as an important theme within community indicator projects and have focused on environmental indicators dealing with biodiversity, climate change, water and wastewater, and solid waste in addition to numerous social indicators (see Portney 2003; Sustainable Seattle, http://www.sustainableseattle.org; http://www.iisd.org/pdf/wpg.qoli.pdf; Jacksonville Community Council, http://www.jcci.org/jcciwebsite/pages/indicators.html). Although most indicator projects rely on information available from secondary data sources available from local, state, and federal governmental agencies, some have also incorporated data from social surveys. A recent effort to include survey data as indicators within a sustainability program is Seattle's Happiness Initiative (http://www.sustainableseattle.org/sahi). The survey has been patterned after work developed in Bhutan and adopted in other countries (Weiner 2008; Centre for Bhutan Studies 2008, http://www.grossnationalhappiness.com/Default.aspx) and is intended to measure various domains of life, including those dealing with the community and the natural environment. While the survey is intended to be representative of the population in the greater Seattle area, residents of the region and indeed anyone with access to the website can respond to survey questions.

As interest in urban sustainability programs grows and as efforts to assess the value of community indicator initiatives increases, there will be opportunities for QOUL researchers to participate bringing to bear their experiences and expertise in collecting and analyzing indicator data, and in working with decision makers in using the information to inform policy.

In this concluding chapter, we have offered on overview of the main themes of the book and presented a number of challenges for those wishing to pursue QOUL research. Those challenges deal with collecting objective and subjective data, using the data in statistical modeling, and efforts to use findings from QOUL research in formulating decisions in the public and private sectors. We believe these challenges present important opportunities for scholars with interests in urban policy, planning, and environmental design that impact the way people live, work, and play.

## References


Ballas, D., Clarke, G., Dorling, D., Rigby, J., & Wheeler, B. (2006). Using geographical information systems and spatial microsimulation for the analysis of health inequalities. *Health Informatics Journal, 12*, 65–79.

Bok, D. C. (2010). *The politics of happiness: What government can learn from the new research on well-being*. Princeton: Princeton University Press.

Centre for Bhutan Studies; Gross National Happiness (2008). http://www.grossnationalhappiness.com/Default.aspx

City of Winnipeg Quality of Life Indicators (n.d.). http://www.iisd.org/pdf/wpg.qoli.pdf

Clarke, G. P. (Ed.). (1996). *Microsimulation for urban and regional policy analysis*. London: Pion.

Couper, M. P. (2008). *Designing effective web surveys*. New York: Cambridge University Press.

Ester, P., Vinken, H., van Dun, L., Beckers, T., & Koning, J. (2002). *The contented citizen*. How *Brabant citizens evaluate their quality of life*. Tilburg: Globus (220 pp.)

Fernandez, L. E., Brown, D. G., Marans, R. W., & Nassauer, J. I. (2005). Characterizing locational preference in an exurban population: Implications for agent-based modeling. *Environment and Planning B, Planning and Design, 32*, 799–820.

Ferris, A. L. (2010). *Approaches to improving the quality of life: How to enhance the quality of life*. New York: Springer.

Frey, B. S. (2011). Happy people live longer. *Science, 331*(6017), 542–543.

Gahin, R., & Paterson, C. (2003). Community indicators: Past, present, and future. *National Civic Review, 90*, 347–361.

Gilbert, D. T. (2006). *Stumbling on happiness*. New York: Random House.

Groves, R. M., Fowler, F. J., Couper, M. P., Lepkowski, J. M., Singer, E., & Tourangeau, R. (2009). *Survey methodology* (2nd ed.). Hoboken: Wiley.

Internet World Stats: Usage and population statistics (2011). http://www.internetworldstats.com/

Jacksonville Community Council, Inc. Citizens building a better community (n.d.). http://www.jcci.org/jcciwebsite/pages/indicators.html

Lora, E., Powell, A., van Praag, B. M. S., & Sanguinetti, P. (2010). *The quality of life in Latin American cities: Markets and perception*. Washington, DC: The Inter-American Development Bank.

Lyubomirsky, S. (2008). *The how of happiness: A scientific approach to getting the life you want*. New York: The Penguin Press.

Marans, R. W. (2005). Modeling residential quality using subjective and objective measures. In D. U. Vestbro, Y. Hourol, & N. Wilkinson (Eds.), *Methodologies in housing research*. Gateshead, Tyne & Wear: The International Press.

Paramo, P., & Garcia-Gil, M. E. (Eds.). (2010). *La Dimension Social del Spacio Publico: Aportes para la Dalidad de Vida Urbana. to Human Dimensions of Public Space*. Bogota: Universidad Pedagógica Nacional and Universidad Santo Tomás.

Phillips, R. (2003). *Community Indicators: Planning Advisory Service Report 517* Washington, DC: American Planning Association.

Phillips, R. (Ed.). (2005). *Community indicators measuring systems*. Burlington: Ashgate Publishing Company.

Portney, K. E. (2003). *Taking sustainable cities seriously: Economic development, the environment, and quality of life in American cities*. Cambridge: The MIT Press.

Schaffer, D., & Vollmer (Rapporteurs), D. (2010). *Pathways to urban sustainability: Research and development on urban systems*. Washington DC: National Academy Press.

Sirgy, J. M., Phillips, R., & Rahtz, D. (Eds.). (2009). *Community quality-of-life indicators: Best cases IV*. New York: Springer.

Sustainable Seattle (n.d.). http://www.sustainableseattle.org

Sustainable Seattle's Happiness Initiative (n.d.). http://www.sustainableseattle.org/sahi

University of Michigan, Panel Study for Income Dynamics (2011). http://psidonline.isr.umich.edu/

Weiner, E. (2008). *The geography of bliss*. New York: Hachette book Group USA.

# About the Contributors

**Perver Korça Baran** is a research associate professor in both the College of Design and the College of Natural Resources at North Carolina State University. She has a Bachelor of Architecture degree from the University of Kiril i Methodij-Skopje and master's and Ph.D. in urban planning from Istanbul Technical University. Her current research interests center on measuring the quality of urban life, understanding urban and neighborhood spatial structure as it relates to health and socio-psychological processes, and GIS, in particular its potential for urban design and active living research.

**Fulin Bölen** is an architect and planner and professor emeritus in the Department of Urban and Regional Planning, at the Istanbul Technical University. She served as head of the Department from 2000 until 2003. Her principal research interests and publications deal with urban land values, land-use intensity, and the quality of residential environments. Between 2005 and 2007 she served as team leader of a research project on "The Objective Measures of Quality of Life in Residential Areas" in İstanbul that was used in preparing an environmental plan for the Greater Istanbul Municipal area.

**Ying Cao** is completing his Ph.D. in the College of Architecture and Environment, Sichuan University in Chengdu, China. From September 2008 to June 2010, he was a visiting scholar at Arizona State University where he collaborated on the Digital Phoenix project in the areas of future land use and water sustainability.

**John I. Carruthers** is the director of the Sustainable Urban Planning Program at The George Washington University in Washington, DC. His current research focuses on land use and urban form, estimating the value of environmental quality; urban quality of life and sustainability, and immigrant assimilation. His recent research has been published in the *Journal of Regional Science*, Papers in Regional Science, Regional Science and Urban Economics, and Urban Studies. Dr. Carruthers is a member of the RSAI Council, the governing board of the Regional Science Association International. He serves on the editorial boards of several journals,

R.W. Marans and R. Stimson (eds.), *Investigating Quality of Urban Life: Theory, Methods, and Empirical Research*, Social Indicators Research Series 45, DOI 10.1007/978-94-007-1742-8, © Springer Science+Business Media B.V. 2011 including the *Annals of Regional Science*, the *Journal of the American Planning Association*, and Papers in Regional Science.

**Prem Chhetri** is an associate professor of logistics and geographic systems and the deputy head of the School for Industry Engagement at the Royal Melbourne Institute of Technology (RMIT) in Melbourne Australia. He is also the manager for the Urban and Regional Sustainable Future Program within RMIT's Global Cities Institute. His research interests include urban quality of life, emergency logistics and policy response, and application of GIS and GPS technologies in urban modeling and logistics systems analysis.

**Donald Dillman** currently is Regents Professor in the Department of Sociology and serves as deputy director for Research in the Social and Economic Sciences Research Center at Washington State University, where he has been a faculty member since 1969. From 1991 to 1995 he was senior survey methodologist at the US Bureau of the Census, providing leadership for the development of the 2000 Decennial Census Form and implementation procedures. A past president of the American Association for Public Opinion Research, he is also the recipient of the 2002 Roger Herriot Award for Innovation in Federal Statistics. He has authored more than 250 publications. His 1978 book, *Mail and Telephone Surveys*, was the first to provide step-by-step procedures for conducting such surveys, and is now in its third edition as *Internet, Mail and Mixed-Mode Surveys: The Tailored Design Method* (Wiley 2009).

**Subhrajit Guhathakurta** is professor and associate director of the School of Geographical Sciences and Urban Planning at Arizona State University (ASU). He is also among the founding faculty members of ASU's School of Sustainability and holds a faculty appointment in that school. Prof. Guhathakurta serves as the lead member of several research centers at ASU including the GeoDa Center for Geospatial Analysis and Computation, Decision Center for a Desert City, and the project on Sustainable Urban Systems, Technologies, and Infrastructure (SUSTaIn). He is author of 5 books and monographs and over 70 scientific papers. He has held visiting appointments at the Center for Urban Spatial Analysis at University College London, the Indian Institute of Information Technology, Bangalore, and at the Center for Sustainable Urban and Regional Futures at the University of Queensland in Brisbane. More recently, he held the German National Science Foundation (DFG) Mercator Guest professorship at Technische Universitat Kaiserslautern, Germany.

**Alexander G. Keul** is trained as a meteorologist and environmental psychologist. He is currently an associate professor of psychology at Salzburg University in Austria and honorary professor at the Vienna University of Technology. He has been a strong advocate for housing evaluation studies throughout Austria and served as co-organizer of the 2004 International Association of People-Environment Studies (IAPS) 2004 conference in Vienna. Dr. Kuel has headed research projects for the Austrian Safety Research project (KIRAS) and the Austrian Ministry of Agriculture. More recently, he has conducted research on passive solar housing and user satisfaction with housing.

**Byoung-Suk Kweon** is an assistant professor in the Department of Plant Science & Landscape Architecture, College of Agriculture and Natural Resources at the University of Maryland. Previously, she was a research investigator at the University of Michigan's Institute for Social Research and a faculty member at Texas A and M University. She is a recipient of a national recognition award from the Council of Educators in Landscape Architecture for her research on improving environments for school children. Dr. Kweon is a board member for the Environmental Design Research Association.

**Robert W. Marans** is a research professor at the Institute for Social Research and a professor emeritus of architecture and urban planning in the Taubman College of Architecture and Urban Planning at the University of Michigan. Marans is a registered architect and is active in recreation policy and planning in southeastern Michigan. He is president of the Washtenaw County Parks and Recreation Commission and commissioner of the Huron-Clinton Metropolitan Authority (HCMA), the governing body responsible for the planning, development, and operations of the metroparks throughout Southeastern Michigan. Dr. Marans is the author or coauthor of 7 books and more than 100 articles and technical reports. He currently serves on the editorial boards of several professional journals and has lectured extensively throughout the USA, and in Europe, Asia, South Africa, South American, Australia, and the Middle East.

**Rod McCrea** is an adjunct research fellow with the Institute for Social Science Research at The University of Queensland. His Ph.D. dissertation modeled links between objective characteristics and subjective evaluations of urban quality of life, and it won the 2008 Best Dissertation Prize from the Australian and New Zealand Regional Science Association International. He has also published nine other refereed papers in the field of urban quality of life, including winning the 2007 Best Paper Prize in the journal *Applied Research in Quality of Life.* Rod also has other research interests in quality of life issues more generally including subjective well-being, work-life balance, community social processes, and community wellbeing.

**Benjamin L. Messer** is a Ph.D. candidate in the Department of Sociology and a research assistant in the Social and Economic Sciences Research Center at Washington State University. During his stay at Washington State, he has written and published papers covering aspects of survey methodology. His dissertation, research examines public attitudes toward energy resources and technologies used for electricity generation and consumption in urban and rural regions of the USA.

**Gary T. Moore** is an emeritus professor of Environment-Behavior Studies at the University of Sydney, Australia and former dean of the university's Faculty of Architecture. He has been a guest professor at the College of Architecture and Urban Planning, Tongji University, China. Prior to his tenure at the University of Sydney, he was professor of architecture and director of the NASA Wisconsin Space Grant Consortium at the University of Wisconsin-Milwaukee. Dr. Moore was an early pioneer in the environment-behavior field, publishing important books on environmental cognition, designing environments for handicapped children, emerging environmental design research directions, and a three-volume series, *Advances in Environment, Behavior and Design*. He is a fellow of the American Psychological Association, and currently serves as editor-in-chief of the *Architectural Science Review*.

**Abul Mukim Mozammel Haque Mridha** is an architect and educator who has spent most of his academic life as a faculty member at Khulna University, Bangladesh. He received his Ph.D. from the Faculty of Architecture, Design and Planning at the University of Sydney in 2010. He has been an active member of Environment, Behaviour and Society (EBS) Research Group and worked as a research associate for *Architectural Science Review* (*ASR*) while pursuing his doctoral studies. Dr. Mridha is currently a research assistant in the Urban Research Centre at the University of Western Sydney. His research interest and activities focus on architecture and the nexus between built environments and human behavior. He also has a number of publications in his credit.

**Gordon F. Mulligan** is professor emeritus in the School of Geography and Development at the University of Arizona. He is past editor of the *Journal of Regional Science*, recipient of the Boyce prize for service to regional science, and currently sits on numerous editorial boards in geography, regional science, and urban studies. His extensive research interests include location theory, spatial competition, shopping behavior, regional multiplier analysis, migration, urbanization, crime, and housing.

**Derya Oktay** is currently a professor of architecture and urban design at the Department of Architecture and director of the Urban Design Programme, Eastern Mediterranean University, North Cyprus, and the director of the Urban Research & Development Center at the same institution. Her teaching and research focus on sustainable urbanism, socio-spatial aspects of housing environments, architectural design in urban context, urban identity, public urban spaces, and quality of urban life. She was a visiting scholar in such universities as University of California – Berkeley, Politecnico di Milano, University of Michigan, and Oxford Brookes University. Dr. Oktay has authored nearly 100 publications including books, book chapters, journal articles, conference papers, and popular press articles. She currently serves on the editorial boards of *Urban Design International, Journal of Urban Design & Planning, and Asian Journal of Environment-Behaviour Studies.*

**Thomas Prinz** has served as primary researcher at the Research Studios Austria – iSPACE since 2002. He studied geography and applied geo-informatics at the University of Salzburg completing his Ph.D. in 2007. His doctoral thesis examined spatial indicators as planning criteria for settlement areas in Salzburg. Dr. Prinz is an analytical scientist with expertise in spatial analysis and modeling, cartography, and GIS-aided planning instruments and methods. His on-going research focuses on the use of spatial indicators and indicator systems to support sustainable planning and development.

**Ahmet Rustemli** retired from Eastern Mediterranean University (Northern Cyprus) in the summer of 2008 where he taught social psychology, research methodology, and social statistics. Prior to moving to Cyprus, Dr Rustemli was affiliated with Middle East Technical University in Ankara, Turkey. During his long career, Dr. Rustemli conducted several large social research projects in Cyprus and Turkey dealing with proxemics, attribution theory, and intergroup relations. He authored and coauthored numerous papers in Turkish and English covering these topics.

**Tung-Kai Shyy** is a research fellow in the Queensland Centre for Population Research (QCPR) in the School of Geography, Planning and Environmental Management at the University of Queensland. He was a member of ARCRNSISS and developed Web-based GIS applications as a shared research web resource. His research publications have focused on GIS-based spatial modeling, spatial optimization, crime mapping, Web-based GIS applications and urban quality of life. One of his co-authored paper received the "*Applied Research in Quality of Life (ARQOLS)* – Best Paper Award 2006" presented by the International Society for Quality of Life Studies (ISQOLS) and Springer Publications, December 2007.

**Robert Stimson** is the director of the Australian Urban Research Infrastructure Network (AURIN) at The University of Melbourne. He is also emeritus professor in the School of Geography, Planning and Environmental management at the University of Queensland where he was professor of Geographical Sciences and Planning from 1997 to January 2011. He is an analytical human geographer and regional scientist. He is a fellow of the Academy of the Social Sciences in Australia and a fellow of the Regional Science Association International. He is author of 48 books and monographs and more than 300 scientific papers. He was the convener of the Australian Research Council Research Network in Spatially Integrated Social Science from 2004 to 2009.

**Fatih Terzi** is a research assistant in the department of Urban and Regional Planning at Istanbul Technical University. He received his Ph.D. degree in Urban and Regional Planning (2010) at Istanbul Technical University. He participated in teaching and related academic activities in Department of Landscape Architecture at Clemson University, South Carolina, USA as a visiting research scholar. He also was affiliated with the Centre for Advanced Spatial Analysis (CASA) at University College London, as an Academic Affiliate. His areas of interest include urban spatial development and GIS implementation in urban planning and urban modeling.

**Handan Turkoğlu** is the director of the Urban and Regional Planning Department at Istanbul Technical University, Faculty of Architecture. She is currently teaching at the same department at Istanbul Technical University. She is author and coauthor of 3 books, 7 book chapters, and 78 scientific papers published internationally and nationally. Her areas of interest include GIS and urban planning, disaster mitigation, housing, and quality of life. She is a member of European Housing Research Network (ENHR) and International Association for People-Environment Studies Network.